Band Structure Spectroscopy of Metals and Alloys

Band Structure Spectroscopy of Metals and Alloys

edited by

D. J. FABIAN
L. M. WATSON

Department of Metallurgy
University of Strathclyde
Glasgow, Scotland

1973

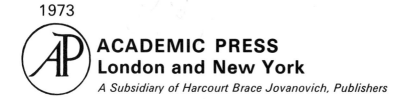

ACADEMIC PRESS
London and New York
A Subsidiary of Harcourt Brace Jovanovich, Publishers

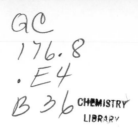

ACADEMIC PRESS INC. (LONDON) LTD.
24/28 Oval Road,
London NW1

United States Edition published by
ACADEMIC PRESS INC.
111 Fifth Avenue
New York, New York 10003

Library of Congress Catalog Card Number: 72-12268
ISBN: 0-12-247440-6

Printed in Great Britain by
J. W. Arrowsmith Ltd., Bristol

Contributors

ALESHIN, V.G., *Institute of Metals, Academy of Sciences of the Ukrainian SSR, 36 Vernadskogo Prospect, Kiev 142, USSR* (p.107)

ALLOTEY, F.K., *University of Science and Technology, Kumasi, Ghana* (p.361)

AZÁROFF, L.V., *Department of Physics, Institute for Materials, Science and Engineering, University of Connecticut, Connecticut 06268, USA* (p.491)

BABANOV, Y.A., *Institute of Metal Physics, Academy of Sciences of the USSR, Sverdlovsk, USSR* (p.431)

BARRIE, A., *Vacuum Generators Limited, Charlwoods Road, East Grinstead, Sussex, UK* (p.91)

BELASH, V.P., *Institute of Metal Physics, Academy of Sciences of the USSR, Sverdlovsk, USSR* (p.237)

BLOKHIN, M.A., *Department of Solid State Physics, University of Rostov, Rostov-on-the-Don, USSR* (p.321)

BONNELLE, C., *Laboratoire de Chimie Physique, De la Faculté des Sciences de Paris, 11, Rue Pierre et Marie Curie, Paris, France* (p.251)

BROUERS, F., *Laboratoire de Physique des Solides, Orsay 91, France* (p.407)

BROWN, R.S., *Department of Mathematics and Physical Sciences, Embrey Riddle Aeronautical University, Florida, USA* (p.491)

BUSCH, G., *Eidgenössische Technische Hochschule Zürich, Laboratorium für Festkörperphysik, CH-8049 Zürich, Hönggerberg, Switzerland* (p.579)

CRISP, R.S., *Department of Physics, University of Western Australia, Nedlands, Western Australia 6009* (pp. 229, 453)

CUTHILL, J.R., *Alloy Physics Section, US Department of Commerce, National Bureau of Standards, Washington DC 20234, USA* (p.191)

DEMANGEAT, C., *Université Louis Pasteur, Laboratoire de Structure Electronique des Solides, Strasbourg, France* (p.405)

DIMOND, R.K., *Department of Physics, University of Western Australia, Nedlands, Western Australia 6009* (p.229)

DOBBYN, R.C., *Alloy Physics Section, US Department of Commerce, National Bureau of Standards, Washington DC 20234, USA* (p.191)

DONAHUE, R.J., *Department of Metallurgy and Institute of Materials Science, University of Connecticut, Connecticut 06268, USA* (p.491)

ENDERBY, J.E., *Department of Physics, University of Leicester, Leicester LE1 7RH, UK* (p.609)

FABIAN, D.J., *Department of Metallurgy, Colville Building, 48, Portland Street, Glasgow, UK* (pp.91, 215)

FISCHER, D.W., *Analytical Branch, Materials Physics Division, Department of the Air Force, Air Force Materials Laboratory, (AFSC), Wright-Patterson Air Force Base, Ohio 45433, USA* (p.669)

FOMICHEV, V.A., *Leningrad State University, Institute of Physics, Leningrad B164, USSR* (p.259)

FUGGLE, J., *Department of Metallurgy, Colville Building, 48, Portland Street, Glasgow, UK* (p.91)

GAUTIER, F., *Université Louis Pasteur, Laboratoire de Structure Electronique des Solides, Strasbourg, France* (p.405)

GEGOOSIN, I.I., *Department of Solid State Physics, University of Rostov, Rostov-on-the-Don, USSR* (p.321)

GÜNTHERODT, H.J., *Eidgenössische Technische Hochschule Zürich, Laboratorium für Festkörperphysik, CH-8049 Zürich, Hönggerberg, Switzerland* (p.579)

GYORFFY, B.L., *University of Bristol, H.H. Wills Physics Laboratory, Royal Fort, Tyndall Avenue, Bristol BS8 1TL, UK* (pp.385, 641)

HAGSTRÖM, S.B.M., *Department of Physics, Linköping Institute of Technology, Linköping, Sweden* (p.73)

HAGUE, C.F., *Laboratoire de Chimie Physique, De la Faculté des Sciences de Paris, 11, Rue Pierre et Marie Curie, Paris, France* (p.251)

HEDIN, L., *Department of Theoretical Physics, Sölvegatan 24A, S-223 62 Lund, Sweden* (p.331)

HOLLIDAY, J.E., *McDonnell Douglas Research Laboratories, McDonnell Douglas Corporation, Department 224, Building 33, Room 171, PO Box 516, St. Louis, Missouri 63166, USA* (p.713)

ISHMUKHAMETOV, B.KH., *Institute of Metal Physics, Academy of Sciences of the USSR, Sverdlovsk, USSR* (p.237)

KÄLLNE, E., *Uppsala Universitet, Fysika Inst., Box 530, S75121 Uppsala 1, Sweden* (p.205)

KAPOOR, Q.S., *Department of Physics, Texas Tech University, PO Box 4180, Lubbock, Texas 79409, USA* (p.215)

KIESER, J., *Physikalisches Institut, Universität Karlsruhe (TH), 75 Karlsruhe 1, West Germany* (p.557)

KUNZ, C., *Deutsches Elektronen-Synchrotron, 2 Hamburg 52, (Gr. Flottbek), Notkesteig, 1, West Germany* (p.503)

KURMAEV, E.Z., *Institute of Metal Physics, Academy of Sciences of the USSR, Sverdlovsk, USSR* (p.237)

LATHAM, D., *Vacuum Generators Limited, Charlwoods Road, East Grinstead, Sussex, UK* (p.91)

LINDAU, I., *Chalmers University of Technology, Physics Department, S-40220 Göteborg 5, Sweden* (p.55)

LONGE, P., *Université de Liège, Faculté des Sciences, Physique Théorique, Sart Tilman, Liège 1, Belgium* (p.341)

MCALISTER, A.J., *Alloy Physics Section, US Department of Commerce, National Bureau of Standards, Washington DC 20234, USA* p.191)

MARCH, N.H., *Department of Physics, The University, Sheffield S10 2TN, UK* (p.297)

MERZ, H., *Physikalisches Institut, Universität Karlsruhe (TH), 75 Karlsruhe 1, West Germany* (p.543)

NAGEL, D.J., *Code 7685, X-Ray Optics Branch, Naval Research Laboratories, Washington DC 20390, USA* (pp.457, 567)

NEDDERMEYER, H., *Sektion Physik der Universität München, Lehrstuhl Professor Faessler, 8 München 22, Geschwister-Scholl-Platz 1, West Germany* (p.153)

NEMNONOV, S.A., *Institute of Metal Physics, Academy of Sciences of the USSR, Sverdlovsk, USSR* (p.237)

NEMOSHKALENKO, V.V., *Institute of Metals, Academy of Sciences of the Ukrainian SSR, 36 Vernadskogo Prospect, Kiev 142, USSR* (p.107)

NIKIFOROV, I.Y., *Department of Solid State Physics, University of Rostov, Rostov-on-the-Don, USSR* (p.321)

NILSSON, P.O., *Chalmers University of Technology, Physics Department, S-40220 Göteborg 5, Sweden* (p.55)

NORRIS, P.R., *Department of Metallurgy, Colville Building, 48, Portland Street, Glasgow, UK* (p.229)

RICHTER, J., *Department of Solid State Physics, University of Rostov, Rostov-on-the-Don, USSR* (p.375)

RIEDINGER, R., *Université Louis Pasteur, Laboratoire de Structure Electronique des Solides, Strasbourg, France* (p.405)

RUDNEV, A.V., *Leningrad State University, Institute of Physics, Leningrad B164, USSR* (p.259)

SACHENCO, V.P., *Department of Solid State Physics, University of Rostov, Rostov-on-the-Don, USSR* (p.375)

SAWADA, M., *Osaka University, Department of Solid State Electronics, 525, Ikejiri, Sayama, Minamikawachi, Osaka, Japan* (p.725)

SENKEVICH, A., *Institute of Metals, Academy of Sciences of the Ukrainian SSR, 36 Vernadskogo Prospect, Kiev 142, USSR* (p.107)

SLATER, R.A., *Department of Chemistry, Queen Mary College, Mile End, London, E.1., UK* (p.655)

SOKOLOV, O.B., *Institute of Metal Physics, Academy of Sciences of the USSR, Sverdlovsk, USSR* (p.431)

SOMMER, H., *Department of Solid State Physics, University of Rostov, Rostov-on-the-Don, USSR* (p.321)

SPICER, W.E., *Stanford University, Stanford Electronics Laboratories, Stanford, California 93405, USA* (p.7)

STOTT, M.J., *Queen's University, Department of Physics, Stirling Hall, Kingston, Ontario, Canada* (p.385)

ULMER, K., *Physikalisches Institut, Universität Karlsruhe (TH), 75 Karlsruhe 1, West Germany* (p.521)

URCH, D.S., *Department of Chemistry, Queen Mary College, Mile End, London, E.1., UK* (p.655)

VEDRINSKI, R.V., *Department of Solid State Physics, University of Rostov, Rostov-on-the-Don, USSR* (p.375)

VEDYAYEV, A.V., *Department of Physics, State University of Moscow, USSR* (407)

WATSON, L.M., *Department of Metallurgy, Colville Building, 48, Portland Street, Glasgow, UK* (pp.91, 125, 215)

WIECH, G., *Sektion Physik der Universität München, 8 München 22, Geschwister-Scholl-Platz 1, West Germany* (pp.173, 629)

WILLIAMS, M.L., *Alloy Physics Section, US Department of Commerce, National Bureau of Standards, Washington DC 20234, USA* (p.191)

ZIMKINA, T.M., *Leningrad State University, Institute of Physics, Leningrad B164, USSR* (p.259)

ZÖPF, E., *Sektion Physik der Universität München, 8 München 22, Geschwister-Scholl-Platz 1, West Germany* (pp.173, 629)

Preface

The present Volume is based on the proceedings of the international meeting held at the University of Strathclyde, 26-30 September 1971, on electronic structure of metals and alloys studied by the methods of "band structure spectroscopy". This is the term we have adopted to describe those spectroscopies of the solid state that probe the energy distribution of electron states in the valence or conduction band of a material. Both experimental and theoretical aspects were covered, continuing the theme of the first such conference at Strathclyde in 1967, at which the underlying purpose was to promote an informal yet critical exchange of ideas and concepts among experimentalists and theoreticians.

A vital function of these meetings is for experimentalists on the one hand to become more aware of the theoretical limitations embodied in the interpretation of measured spectra, and to familiarize theoreticians on the other with the wealth of experimental data now becoming available, particularly on alloys. It is abundantly clear that band structure spectroscopy has an important rôle to play in the development of a theory of alloys; a development which we note commenced with the work of Mott and Jones in the early 1930's. It is additionally interesting to recall that – following the first observation of band spectra chiefly by Skinner and coworkers – cooperation between experimentalists and theoreticians was there at the beginning with the interpretation by Mott, Jones and Skinner in 1936 of the K and L emission spectra of simple metals. It was therefore a particular pleasure to all of us at this conference to have in our midst Sir Nevill Mott himself actively participating during the entire proceedings, which in fact encompassed his penultimate days as Director of the Cavendish Laboratories. Unanimously the present contributors wish to express their recognition of this participation, from the 'beginning' to the present, by dedicating this book to Sir Nevill.

In shaping the book we have laid emphasis on review of experimental and theoretical band-structure techniques applied to alloys. Historically, soft x-ray emission and absorption, and one-electron calculations of the valence and conduction bands, gave a start. Currently the methods of UV and x-ray photoemission spectroscopy are becoming equally if not more important; while on the theoretical side many-electron effects, and scattering theory such as that embodied in the coherent potential approximation, probably form the platform from which a theory of alloys will be built. With this in mind the field has been extended to include liquid and amorphous metals and alloys, and an examination of where band spectroscopy might help in understanding the electronic structure of these condensed phases.

Discussion at the meeting was extensive and informal. We make no attempt to reproduce here all the material covered in discussion; authors have been invited to incorporate important aspects into their manuscripts, which then reflect the mood of the conference. In a few instances therefore a nonrigorous 'state of the art' viewpoint has been adopted; we believe that this enhances the value of the text. We express our thanks to the contributors who have cooperated in this manner.

It is a pleasure to thank The Royal Society European Programme, The British Council, The Science Research Council, and the University Court for financial support of the symposium, and the University Principal Sir Samuel Curran for sponsoring the meeting. We record also our debt to colleagues Q.S. Kapoor, D.G. Hart, C.A.W. Marshall and the late Professor E.C. Ellwood for their enthusiasm, help and encouragement in organizing the conference. Finally, it is a pleasure also to express our immense gratitude to Mrs. S. Bhalla for her invaluable and painstaking assistance in the typing and preparation of this text.

October 1972
Strathclyde

Derek Fabian
Lewis Watson

Contents

Part 3

THEORY AND MANY-BODY EFFECTS OF SOFT X-RAY AND PHOTOELECTRON EMISSION FROM METALS AND FROM DISORDERED ALLOYS

To Professor Sir Nevill Mott

INTRODUCTORY COMMENT

N.F. Mott
Cavendish Laboratory, University of Cambridge.

I should like to say something about my recollections of the beginning of this subject, as I saw it in Bristol before the last World War. Of the two band-structure spectroscopies covered in this Volume, photoelectron emission is a fairly new development, but soft x-ray emission and absorption go back a long way, almost as far as the band theory of solids itself. Perhaps photoemission has turned out to be the simpler technique for obtaining results that can be interpreted in terms of band theory; one is not worried by the positive hole in the x-ray level, or by the problem of whether it disturbs the density of states - a problem that has proved extremely instructive as the papers presented here show. However, historically soft x-ray spectroscopy was of the greatest importance, particularly for our understanding of electrons in metals; and I shall start by explaining how this arose.

When quantum mechanics and the Schrödinger equation were first formulated in 1924-26, it was of course a splendid time for the small group of theorists who were then active in atomic physics, and they had something of a race to see who could first use it to explain the long-standing puzzles of physics. The spectrum of the helium atom, the homopolar bond in the hydrogen molecule, and the alpha decay of radioactive nuclei were all fitted into the new theory within two or three years. The first application to solids was made in 1928 by Sommerfeld, who showed why the electrons in metals do not each contribute 3/2k to the heat capacity, and formulated the concept of a boundary between occupied and unoccupied states. Very soon afterwards Bloch, Brillouin, Peierls, and Wilson described electrons moving in a crystalline field, introduced the concept of 'bands' and 'Brillouin zones', and explained in principle why there is a sharp difference between metals and insulators. In 1933 there appeared two astonishingly comprehensive papers by Sommerfeld and Bethe in the Handbuch der Physik (Vol 24-II); the first by Bethe on one-electron and two-electron systems, and the second on electrons in metals. The second article seemed, with one limitation, to answer almost every question that could be asked about electrons in solids. It even gave pictures of what we now call a Fermi surface in k-space, and we find these to look very much like the shapes they have since turned out to be. But the one major thing not answered in the Sommerfeld and Bethe paper was the effect of interaction between electrons; everything was worked out for non-interacting electrons, and the interaction e^2/r_{12} between a pair of electrons was neglected except in so far as it could be averaged in the Hartree-Fock sense. So nobody knew if these surfaces in k-space were just mathematical fiction, the result of a simple approximation, or whether they were something that could really be measured.

Perhaps the first experimental clue to solving this problem was provided by the work of the late Herbert Skinner. I remember, when I went from Cambridge to a Chair of theoretical physics at Bristol in 1933, that there was Skinner in the newly built H.H. Wills Physics Laboratory, amid a cloud of cigarette ash and – as it seemed to us with three pairs of hands – producing x-ray emission bands from aluminium and magnesium; what stood out a mile was that these emission bands had sharp upper limits, just as the Sommerfeld-Bethe theory predicted. Perhaps we ought not to have been surprised; after all, the linear specific heat showed just the same thing. But in 1934 Harry Jones, Skinner and I put our heads together and showed, by considering Auger transitions, that broadening of the sharp upper limit was not to be expected. Our argument was hardly different from that used later by Landau to demonstrate the physical reality of a Fermi surface; however, although the Fermi-surface concept was frequently used in pre-war days, for instance in the explanation in 1934 by Jones of the Hume-Rothery rule for alloys and in the 'two-band' theories of transition metals, I do not remember any confident belief that we were talking about something with physical reality.

Soft x-ray spectroscopy, together with photoemission techniques, have come a long way since then, and fortunately for us the problems are by no means easy to answer. It seems to me that the subject has never been so lively nor so promising, with new problems continually coming to the forefront, particularly those presented by alloys and amorphous materials. The recent work of de Dominicis and Nozières (1969), and by Friedel (1969), in helping us to understand the effect of the positive hole on x-ray emission, has been particularly exciting. And this takes us back to Skinner and the 1930's; his work showed that the L_3-emission spectrum of magnesium has a peak just below the Fermi level. We thought that this was due to overlapping s and p bands; overlap being necessary in a divalent material if it is to be metallic. But now the Nozières theory says that we can have a peak because of the positive hole. I came to the Strathclyde conference expecting to learn which kind of peak this was. But apparently we do not yet know. There is still plenty to be done.

REFERENCES

Friedel, J. (1969); Comments on Solid State Physics 2, 21.

Jones, H., Mott, N.F. and Skinner, H.W.B. (1934); Phys. Rev. 45, 370.

Jones, H. (1934); Proc. Roy. Soc. A 144, 225.

Nozières, P. and de Dominicis, C.T. (1969); Phys. RiO. 178, 1097.

Sommerfeld, A. and Bethe, H. (1933); in "Handbuch der Physik", Vol. 24-II.

Part 1

UV AND X-RAY PHOTOELECTRON SPECTROSCOPY OF METALS AND ALLOYS

ULTRAVIOLET PHOTOEMISSION STUDIES OF ALLOYS AND DISORDERED SYSTEMS

W.E. Spicer

Stanford University, Stanford, California, U.S.A.

1. INTRODUCTION

In less than a decade, photoemission techniques have been developed sufficiently to become a major tool for the experimental studies of solids. Once the usefulness of photoemission had been established for the simpler crystalline materials, the exciting possibility arose of using it to investigate the more complex materials such as alloys. It also became attractive for the study of the effects of changes in, or destruction of, crystalline order. For example, the changes on melting or transformation from crystalline to amorphous structure could be examined, as could the more subtle changes that occur in the ferromagnetic to paramagnetic transformation in Ni.

In this review, emphasis will be focused on the results obtained from alloys, liquids, and amorphous solids. However, when necessary pure crystalline materials will be discussed in order to place the results in a better perspective.

2. METALLIC ALLOYS

The first comprehensive study of d-band metals was that reported by Berglund and Spicer (1964). From this study, it became clear that one could easily distinguish between the filled d-bands and the predominantly s and p derived conduction bands. For this case, figure 1 presents photoemission energy-distribution curves (EDCs) for several copper samples prepared in different ways (Seib and Spicer 1970); experimental details are described elsewhere (Eden 1970, Derbenwick *et al* 1972). Such curves illustrate the energy distribution of the photoemitted electrons; the number of electrons within a given energy range is plotted vertically and their energy is plotted

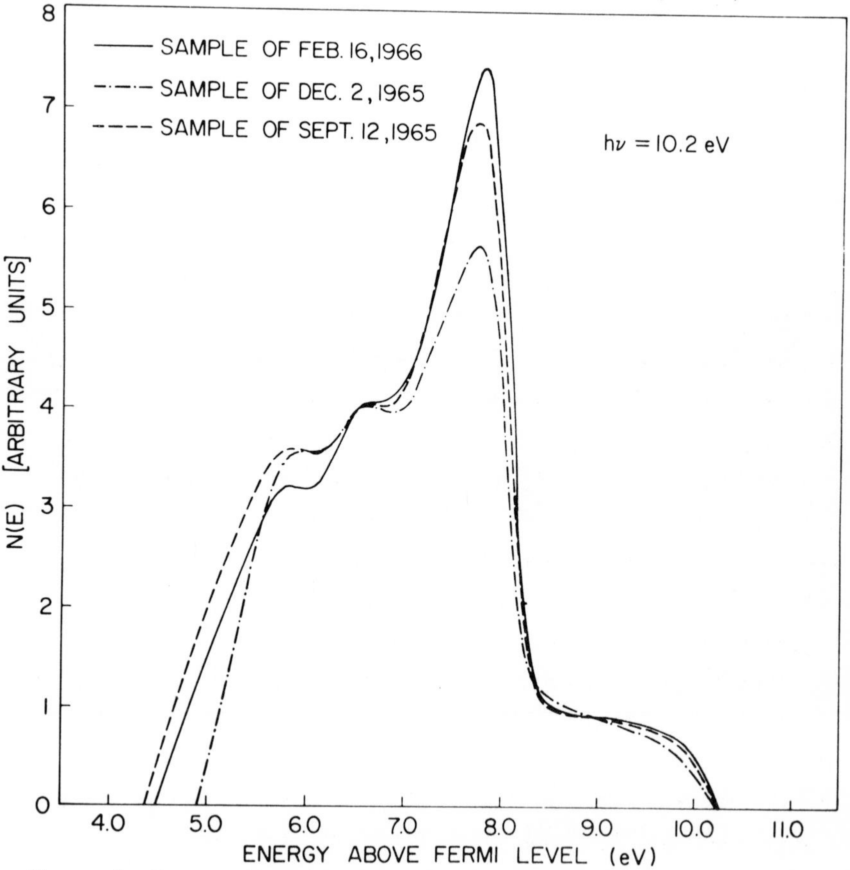

Figure 1 Energy distribution (EDCs) of the photoemitted electrons from three different heat-cleaned samples of copper, using photons of 10.2eV. The sharp rise at approximately 8eV is caused by excitation from the valence d-band.

The energy of the escaping electrons can be specified in various ways. It may be described in terms of the kinetic energy of the escaped electron, with the zero taken at the Fermi level (as in figure 1), while frequently the zero is considered to be the vacuum level (as for example in figure 4). When the structure in the EDCs is the result of valence-band structure, it is often useful to employ an energy scale in terms of the energy of the states from which the electrons are excited (figure 1). This is accomplished by subtracting the exciting photon energy from the energy of the excited electron.

Taking the zero of energy at the Fermi level gives (in the one-electron approximation) the energy of the state from which the electron was excited; that is, the energy of the hole produced by the excitation. An example of data plotted in this manner is those for copper and Cu-Ni alloys in figure 3.

The structure in the EDCs for copper shown in figure 1 corresponds well with the gross features of the copper density of states (Spicer 1972). For example, the sharp rise at approximately 8.4eV is caused by the onset of emission from the copper d-bands. The sharp peak at 7.7eV corresponds to a sharp peak in the density of states at 2.8eV below the Fermi level (the relationship between EDCs and the density of states are detailed by Spicer 1972, by Smith 1971, and Doniach 1970).

The region between the top of the d-band and the Fermi level is of particular interest in the study of the effect of alloying transition metals with noble metals, where predominantly s and p derived conduction states are to be seen (Spicer 1972). These states can be observed equally well in the noble metals silver (Walldén 1970) and gold (Krolikowski and Spicer 1970, Nilsson, Norris and Walldén 1970). By focusing attention on these states when a transition metal is alloyed with a noble metal, we can determine whether the rigid-band or virtual-bound-state model holds for the given alloy system. This is discussed in detail in the next section.

A. Alloys of Noble Metals with Transition Metals

Noble-metal-rich alloys When photoemission investigations of transition-metal and noble-metal alloys began, two principal theoretical models existed. One, the rigid-band model (Mott 1935, 1936), assumed that the common d and s-p bands were the two components in the alloy. Thus, for example, as nickel was added to copper, the Fermi level would fall towards the d-band, as shown schematically in figure 2. The Fermi energy will drop by approximately 1eV for a 25% Ni sample provided the bottom for each of the Ni and Cu s-p bands is unchanged by alloying.

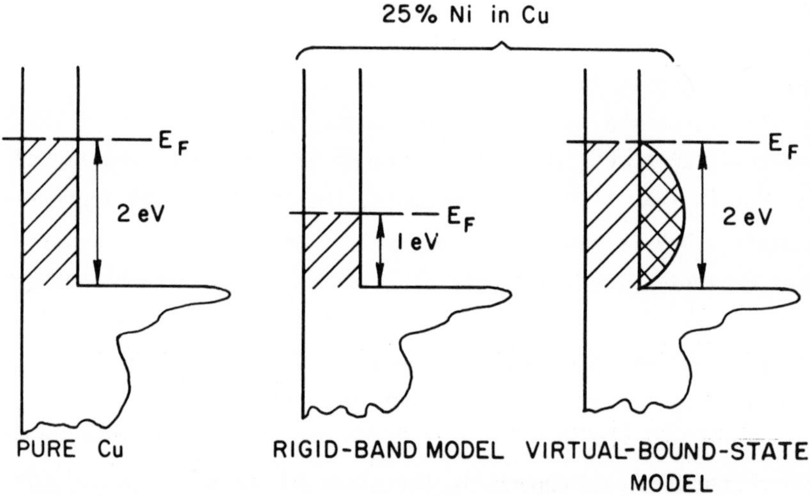

Figure 2 Schematics showing changes in the density of states of copper due to alloys with 25% Ni according to two alloy models. The rigid-band model was a simple one in which the s-p band is held rigid with respect to the d-band.

The second model is the Friedel and Anderson virtual-bound-state model (Friedel 1956 and 1958, Anderson 1961) in which no common copper and nickel bands exist, but instead the nickel atoms could cause a virtual-bound (resonant) state to form in the s-p bands of copper. To a first approximation, no shift of electrons would occur from the Ni to Cu sites; therefore, the energy difference between the top of the copper d-bands and the Fermi level would not be changed appreciably by the alloying (see figure 2).

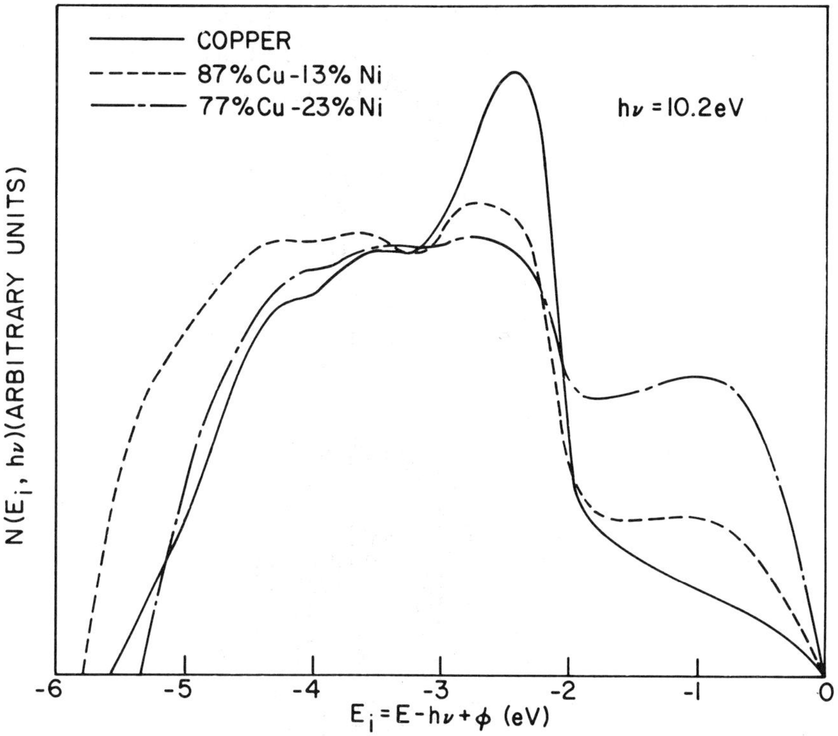

Figure 3 Energy distribution curves (EDCs) from pure copper and copper containing nickel (Seib and Spicer). The behaviour is to be expected from the virtual-bound-state model.

Figure 3 illustrates the experimental results for 13% and 23% nickel in copper (Seib and Spicer 1970); they can be seen to agree well with the virtual-bound-state model. On alloying, the Fermi level does not move; instead, increased state density is built up between the top of the copper d-bands and the Fermi level. The Gothenburg group (Norris and Nilsson 1968, Norris and Walldén 1969, Walldén 1970) has made extensive studies of other alloys. Figures 4, 5 and 6 present photoemission data for three alloy systems of transition metals with noble metals; in each the virtual bound state caused by the transition metal can be observed clearly. If these results are compared with those obtained for the Cu-Ni system, it can be seen that the details of the virtual bound state depend on the transition-metal atom involved in the alloying. This is in agreement with theory (Anderson 1961).

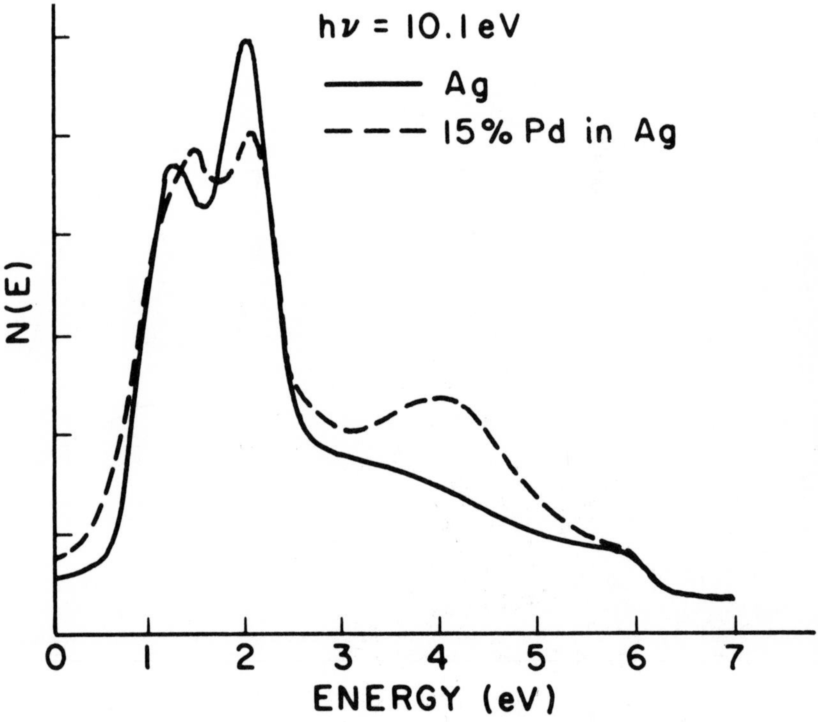

Figure 4 shows the results obtained by Norris and Nilsson (1968) for palladium in silver. The contrast of Ni-Cu to the Pd-Ag is interesting. The virtual bound state for nickel in copper lies close to the Fermi level; in fact, the experimental data suggest that it may extend across the Fermi level. The

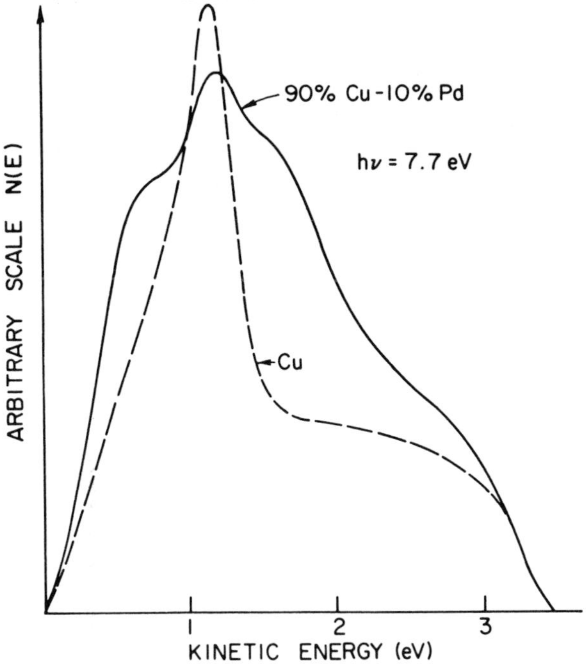

Figure 5 EDCs for copper and a Cu-Pd alloy (Walldén).

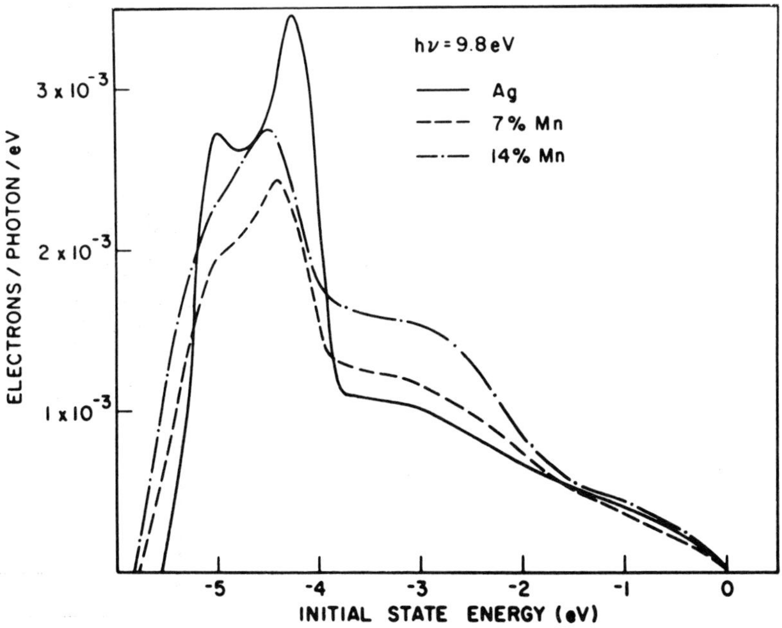

Figure 6 EDCs for manganese in silver (Walldén).

virtual bound state for palladium in silver lies much deeper and appears to have little or no strength near the Fermi level. Therefore, a simple inspection of the data reveals that palladium produces a well-defined filled resonant state; whereas, nickel may produce a state that is not completely filled. This correlates well with the magnetic behaviour of these alloys. It has been found also that Ni-Cu is paramagnetic while Pd-Ag is diamagnetic.

In discussing the occurrence of ferromagnetism in nickel (following Mott 1964) it is worth noting that, whereas it is difficult to calculate the exchange and correlation effects in metals, some insight is obtained by examining the atomic configuration of these metals. For example palladium, which is not ferromagnetic, has a $4d^{10}$ configuration, but atomic nickel has a $4s^2 3d^8$ configuration and *is* ferromagnetic. As a result, palladium prefers a filled d-shell, but nickel does not. We can see from figures 3 and 4 that this certainly seems to hold for the alloys in which palladium appears to have a filled d-orbital. A rather different view of these materials is presented in a discussion of the optical properties of palladium in silver by Kjöllerstrom (1969), and of nickel in copper by Meyers *et al* (1969). To show that these results are caused primarily by the transition metal and not by the host noble metal, data obtained from Pd-Cu (Norris and Walldén 1969) are shown in figure 5. Similar to Pd-Ag the maximum strength resulting from the palladium virtual bound state lies at least 2eV below the Fermi level and there is little or no added strength at the Fermi level.

Simple views such as those used in discussing palladium in silver and in copper can also be applied for manganese in these noble metals. Because manganese is known to contribute a magnetic moment of five electron spins in the alloys, it clearly has five filled and five empty d-states. Figure 5 shows the

EDCs obtained by Wallden (1970) for various amounts of manganese in silver. Here, the virtual bound state caused by the Mn is located several eV below the Fermi level and, within the accuracy of the measurement, has no strength near the Fermi level. As a result the manganese d-states form a broad virtual bound state centred approximately 3eV below the Fermi level, and the empty d-states must lie above this level because the exchange and correlation forces have split them off from the filled states by several eV. However, it is dangerous to attempt to describe systems in which many-electron forces are important (e.g. correlation and exchange) in terms of one-electron models.

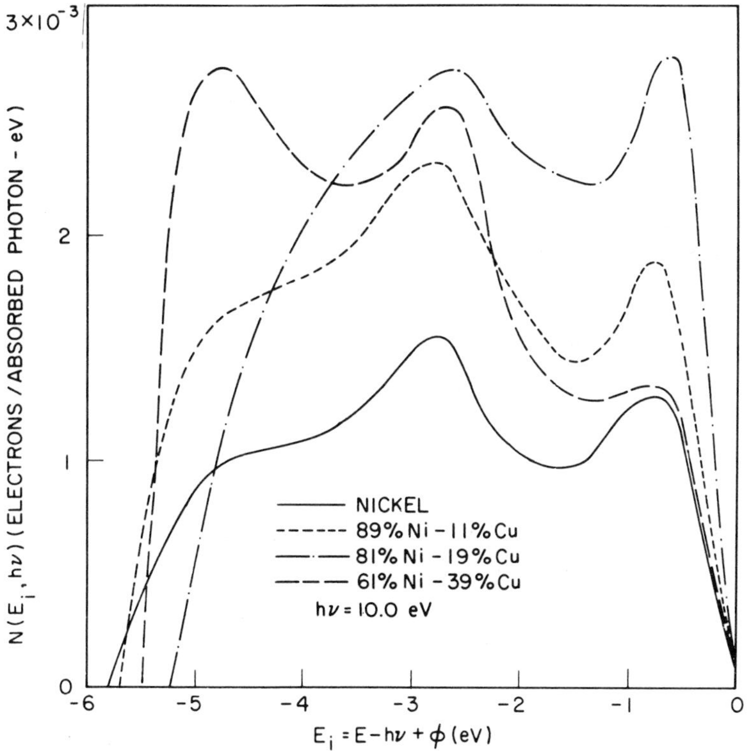

Figure 7 EDCs for nickel and a series of nickel rich Cu-Ni alloys (Seib and Spicer 1970).The vertical scale is absolute.

Nickel-rich Cu-Ni alloys Figure 7 shows EDCs for a series of Ni-rich alloys (Seib and Spicer 1970). These results are striking because of the small changes produced by adding relatively large amounts of copper to nickel. If the effects of changes in the workfunction caused by alloying are discounted, the EDCs will differ significantly from those of pure nickel only when copper contents of 20-40% are reached.

Fortunately, at the same time as the experimental work, theoretical study was under way; the Harvard group (Ehrenreich and co-workers) applied the coherent potential approximation to Cu-Ni alloys. Results obtained for a series of these alloys by Kirkpatrick *et al* (1970) are displayed in figure 8, where the curves indicate the t_{2g} component of the nickel density of states. It can be seen that the principal features in the nickel density of states (the strong peaks at 0.4 and 2.0 eV, respectively, corresponding to structure in the EDCs) are reduced in strength but not destroyed by the alloying. Furthermore, no new structure appears; again, in agreement with the photoemission results.

More recent calculations have been made by Stocks *et al* (1971), in which one parameter in the CPA theory was fitted to yield optimum agreement with the experimental EDCs. Again, the principal features of the EDCs were well-reproduced. Figure 9 is a typical set of curves, in which the calculated density-of-states contributions from both the copper and nickel in the alloy are shown. It can be seen that the density of states of copper provides a smeared-out featureless background; by contrast, the nickel structure is essentially unaffected on alloying.

It is interesting to compare this behaviour to that of nickel in the copper-rich Ni-Cu alloys. When nickel is placed in copper, the Ni atoms cause a virtual-bound state to form in the 2eV region between the top of the copper d-states and the Fermi

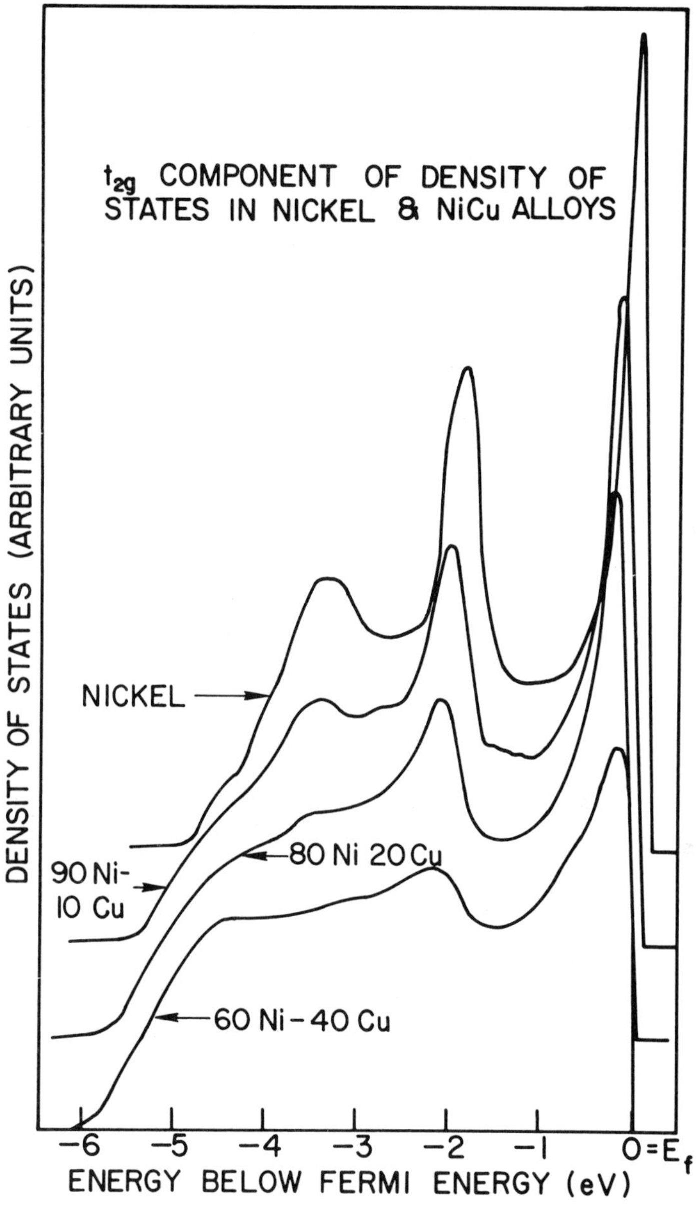

Figure 8 Density of states of nickel and Ni-Cu alloys calculated by Kirkpatrick, Velicky, and Ehrenreich using the CPA method.

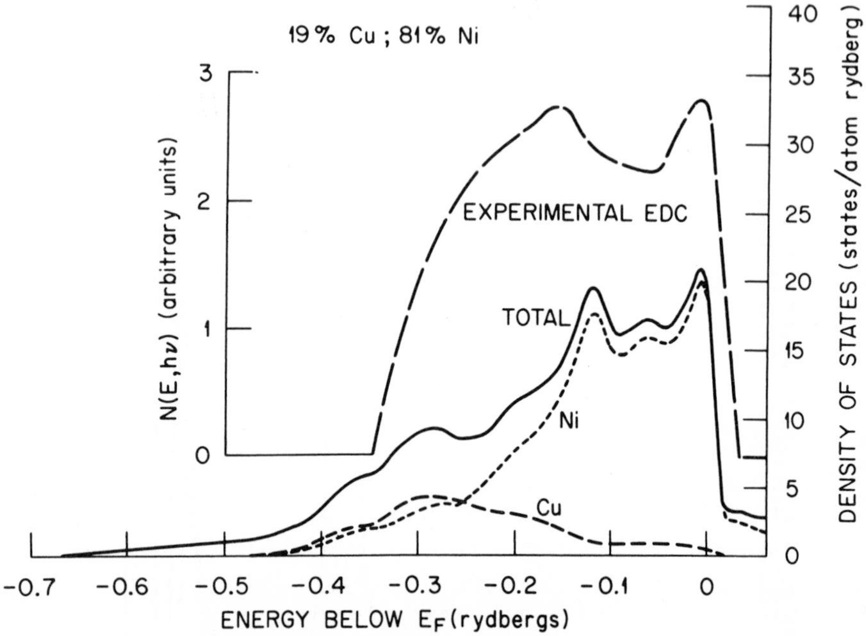

Figure 9 Experimental EDC compared with total density of states calculated by Stocks *et al* (1971) for a nickel-rich Ni-Cu alloy. The dashed curves indicate the contribution of the density of states from the copper and nickel.

level (figures 1 to 3). Because the density of states is low in this region, the virtual bound states formed by the nickel are easily observed; however, no such region exists for nickel and therefore the copper d-states must overlap with the nickel d-states. This adds to the width of the copper states and makes them difficult to observe because of the strong background of nickel d-states.

Figure 10 compares the calculations by Stocks *et al* (1972) with the experimental EDCs for 23% nickel in copper. From these calculations the nickel states can be seen to be almost completely concentrated between the top of the copper d-band and the Fermi level. As before, relatively good agreement is obtained between the measured EDCs and the calculated curves;

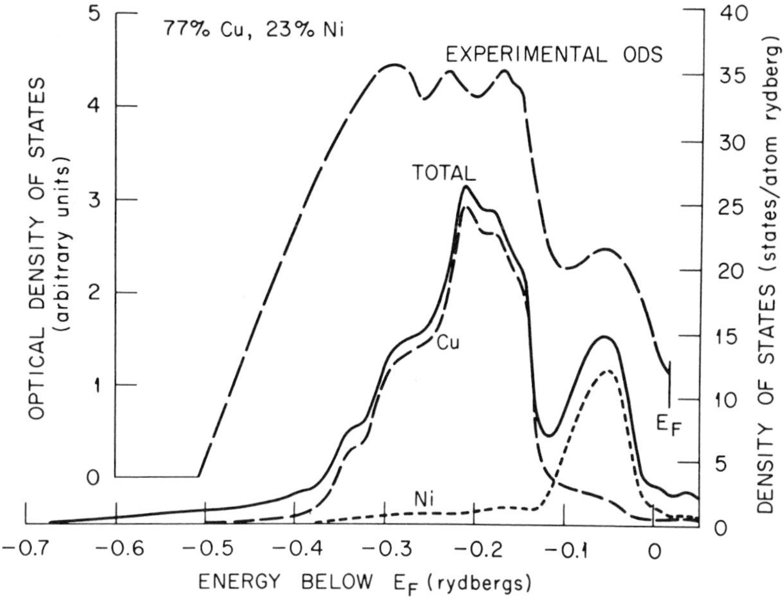

Figure 10 Measured optical density of states (ODS) obtained by Seib and Spicer (1970) compared with the calculated density of states obtained by Stocks *et al* (1972) for a copper-rich Cu-Ni alloy.

the poorest agreement occurs near 40% copper in nickel. In figure 11, the calculations by Stocks *et al* (1972) are compared with the experimental EDCs. Here it is interesting to note that no strong structure is predicted, but at least one strong feature appears in the experimental EDCs. This may be caused by the fact that the photoemission experiment averages over a small region of the crystal in which quasi-bands of nickel and copper may be set up. By contrast, the calculations average over the entire crystal and therefore do not emphasize the effects that may occur since local regions are rich in copper or nickel. Such regions will appear statistically, even in an ideally miscible alloy.

It is interesting to compare the results of ultraviolet

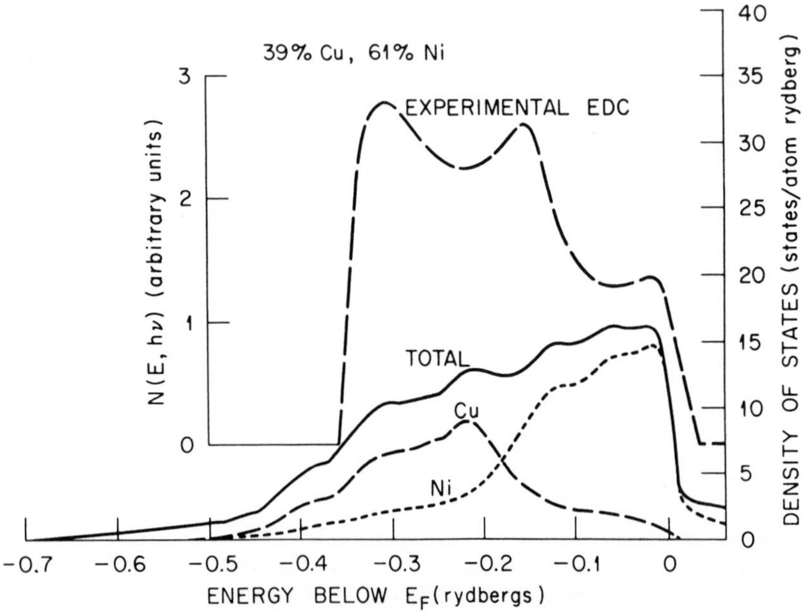

Figure 11 Experimental EDC for a 39%Cu-61% Ni alloy compared with the calculations by Stocks *et al* (1971). The separate contributions of copper and nickel are indicated by the dashed curves.

photoemission (UPS) and the CPA calculations with the results by Clift *et al* (1963) using soft x-ray emission (SXS). Figures 12 and 13 are emission spectra for nickel, copper and various Cu-Ni alloys. Clift *et al* found (figure 14) that they could reproduce the alloy spectra to a high degree of accuracy simply by adding, with proper weighting, the spectra of the pure materials. These results largely contradict the changes in density of states predicted by CPA theory and found by UPS experiment The nickel density of states, as predicted by CPA calculations (Stocks *et al* 197 ; figure 15), varies greatly and by an amount that should lie within the resolution of the SXS experiment (figure 14). Although it would be useful to repeat this experiment using more recent technique and higher

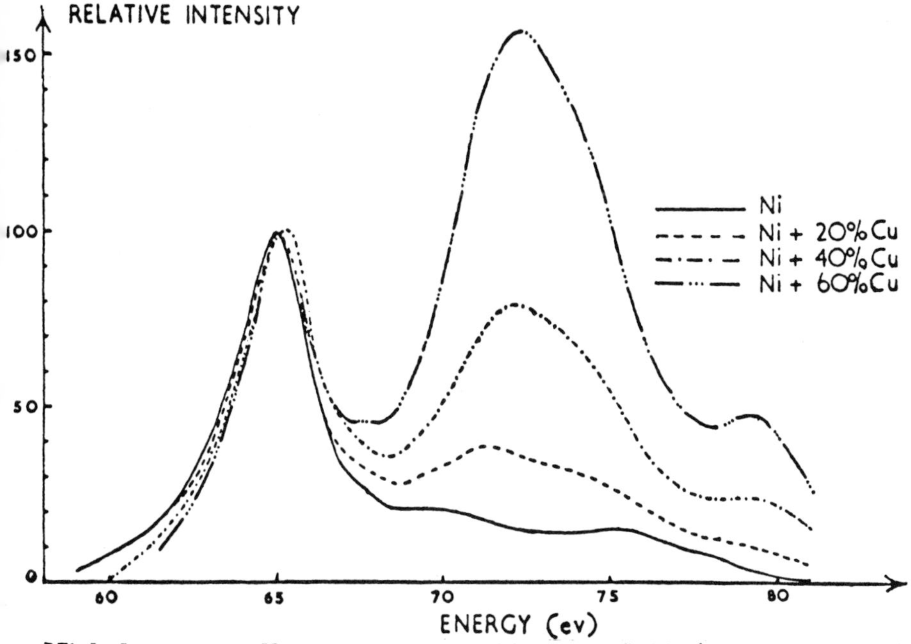

Nickel–copper alloy spectra : 0, 20, 40 and 60% copper in nickel.

Figure 12 X-ray emission spectra from nickel and nickel-rich Ni-Cu alloys (Clift *et al* 1963).

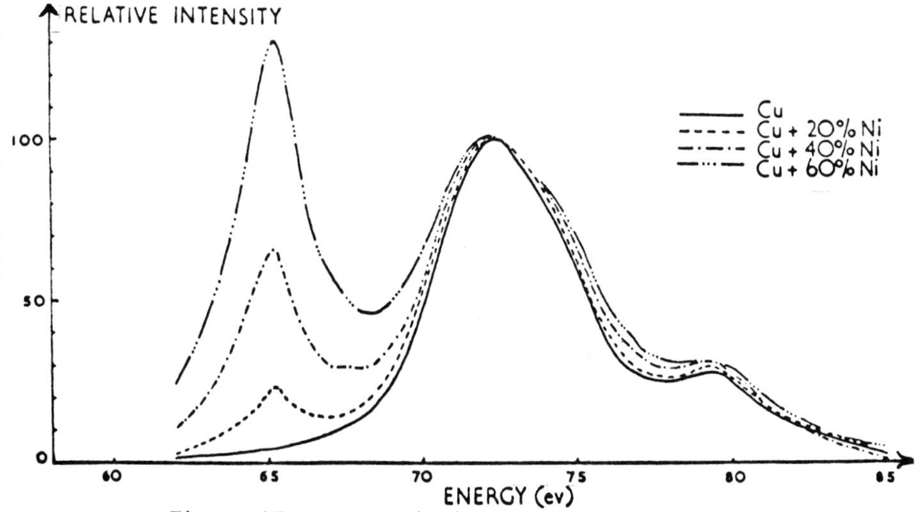

Figure 13 X-ray emission spectra from copper and copper-rich Cu-Ni alloys (Clift *et al* (1963).

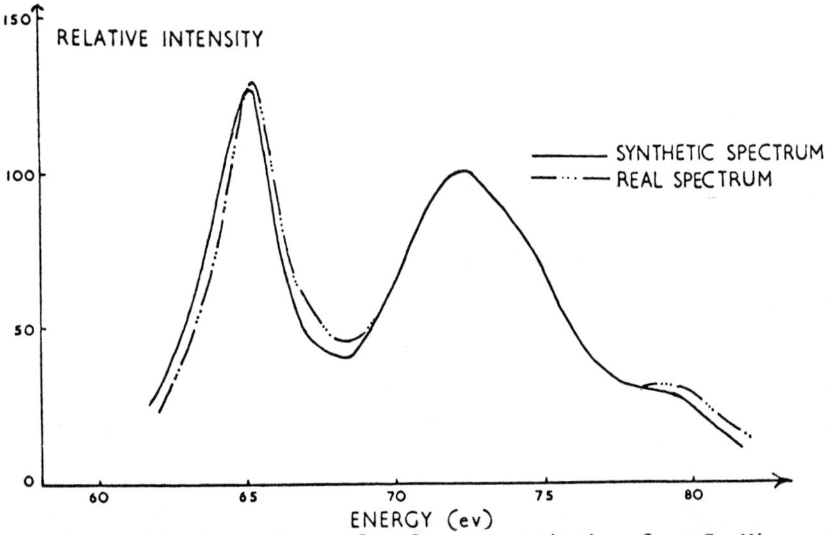

Figure 14 Comparison of soft x-ray emission from Cu-Ni (Clift *et al* 1963) with predicted spectrum obtained by adding spectra of the pure materials.

resolution, the results indicate that the effect of exchange and other many-body interactions may be much more significant in determining the x-ray emission spectrum than is the density of valence states. This is consistent with the x-ray absorption spectra for nickel and other 3d transition metals reported by the DESY group (Deutsches Elektronen-Syncnrotron; Sonntag, Haensel and Kunz 1967). If this is the case, it places in doubt the use of SXS for determining the density of states of transition metals.

B. Photoemission Studies of Other Alloy Systems

Nilsson (1970) has investigated the noble-metal alloy Ag-Au and alloys of noble metals with simple metals. Figure 16 shows the EDCs for the Ag-Au system. Assuming that these curves can be related to the density of states for the alloys, Nilsson (1970) found the experiments in agreement with the CPA theory, but generally not with other theories.

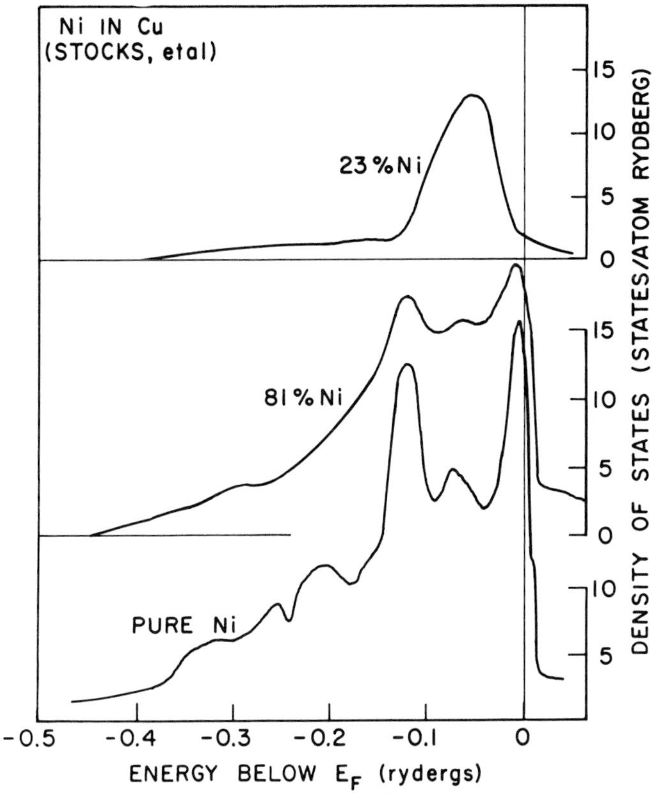

Figure 15 Density of states for pure nickel and for nickel in various Cu-Ni alloys, according to the calculation of Stocks *et al* (1971).

In figure 17 Nilsson compares EDCs obtained for silver and for a 15% In-Ag alloy. The principal effect is a downward shift of the silver d-states by approximately 0.5eV, which Nilsson attributes partly to a decrease of the silver bandwidth and partly to a shift upwards of the Fermi level, resulting from an increase in electron concentration.

Nilsson and Lindau (1972) also studied β-brass. Their results (figure 18) indicate that copper d-states are located approximately 2.3eV below the Fermi level; previous investigators (Amar and Johnson 1966, Arlinghaus 1969) have placed

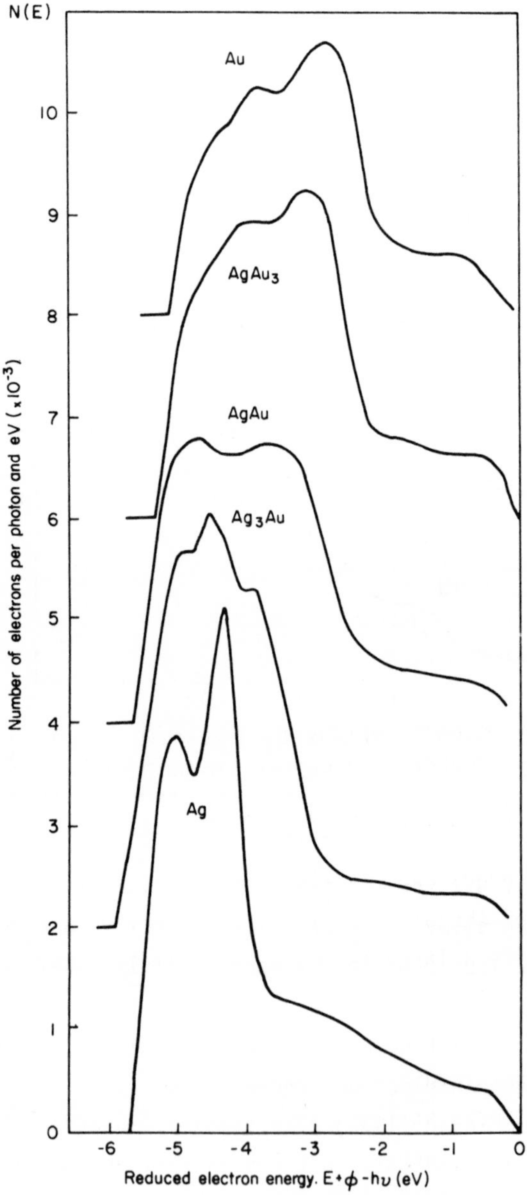

Figure 16 EDCs for silver, gold and their alloys (Nilsson *et al*, 1970).

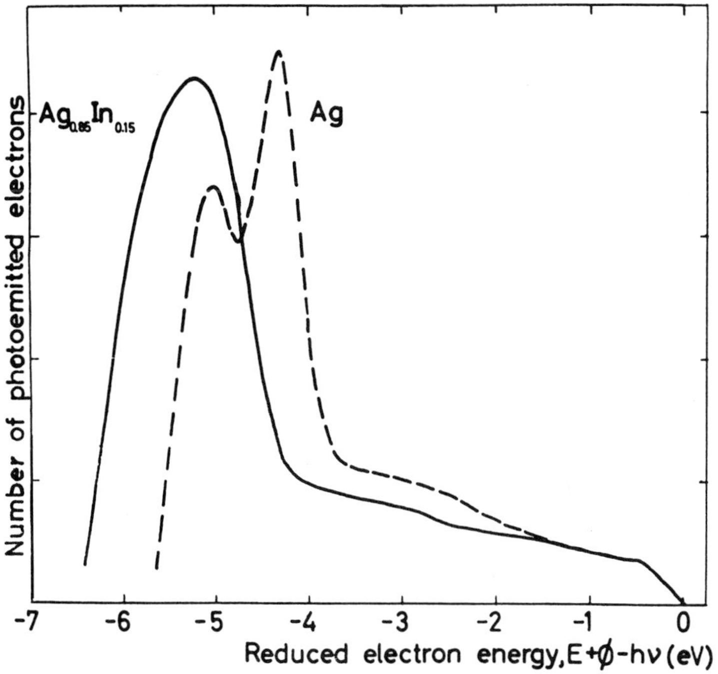

Figure 17 EDCs for Ag and an Ag-In alloy (Nilsson 1970).

these states considerably lower in the band. However, the results are in reasonable agreement with the band described by Arlinhaus (1967).

The significance of photoemission data is that they provide the most unambiguous experimental information available on certain features of the electronic structure of solids. Consider β-brass: despite a large amount of optical data, band calculations and additional information, it was previously impossible to locate the copper d-states correctly; however, photoemission data obtained by Nilsson and Lindau (1971) unambiguously located these states. Thus photoemission studies are expected to play an increasingly significant rôle in the understanding of alloys.

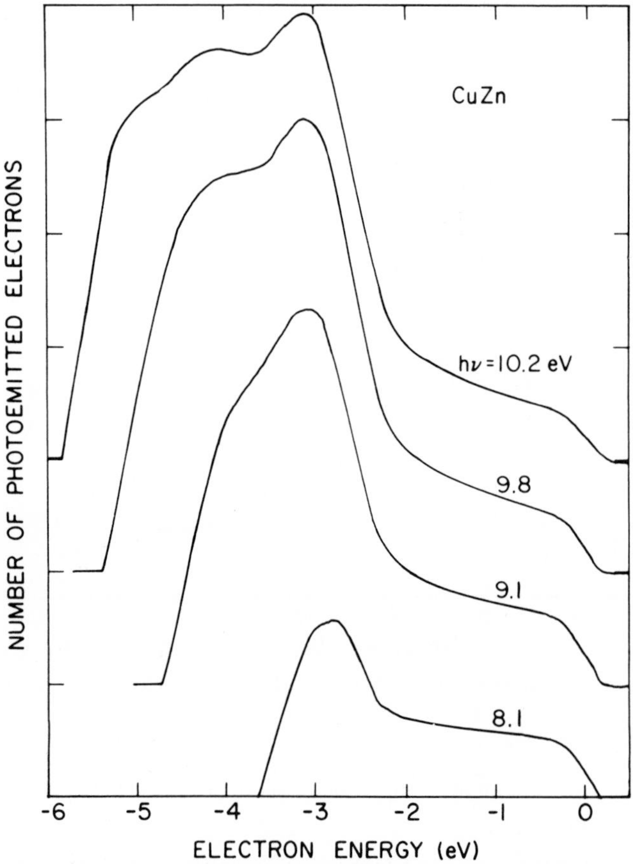

Figure 18 EDCs for β-brass (Nilsson and Lindau 1971).

3. CRYSTALLINE AND AMORPHOUS GERMANIUM

Because the electronic structure of solids is normally associated with the crystalline order, it is interesting to destroy this order and to investigate the changes produced in electronic structure. We should then be able to identify those features of the electronic structure that are associated with long-range crystalline order, and also features that depend only on local order and chemical bonding.

Germanium and silicon are valuable materials for such study because their band structures are well known. In the amorphous form, which can be produced in thin films, the covalent bond is retained (Polk 1971, Turnbull and Polk 1972); therefore, the chemical bonding and local order of the crystalline material is maintained in the amorphous material, but the long-range order is eliminated.

A. Crystalline Germanium

One of our objectives is to compare the results obtained for amorphous germanium and silicon with those obtained for the crystalline materials. We shall first examine typical photoemission results from crystalline germanium and then note how these results depend on details of the band structure, and thus on long-range order.

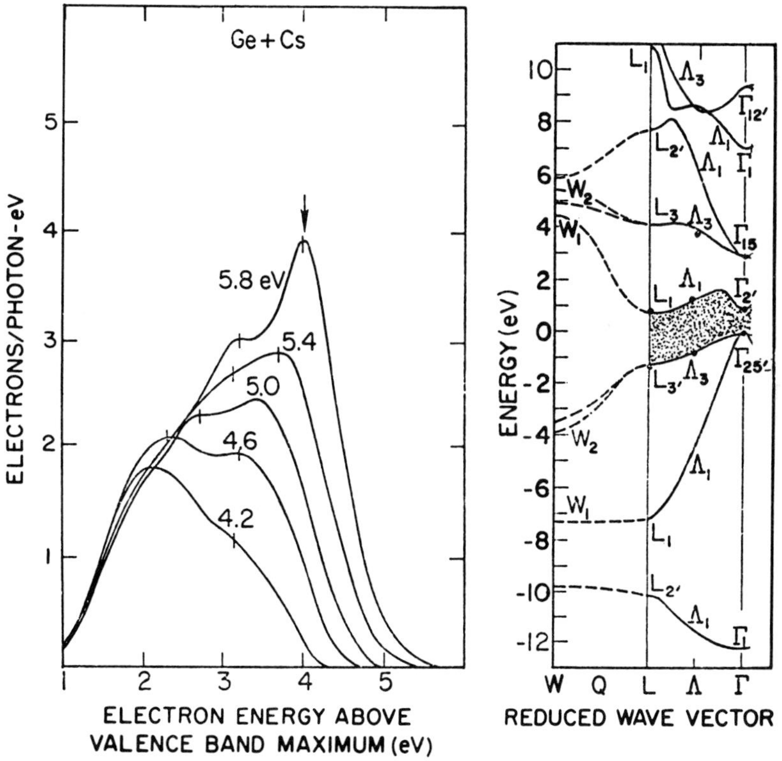

Figure 19 EDCs for germanium with workfunction reduced by a monolayer of cesium on the surface (Donovan, Matzusaki and Spicer, to be published). The energy of the exciting photons is indicated on each curve. A portion of the band structure of germanium is included.

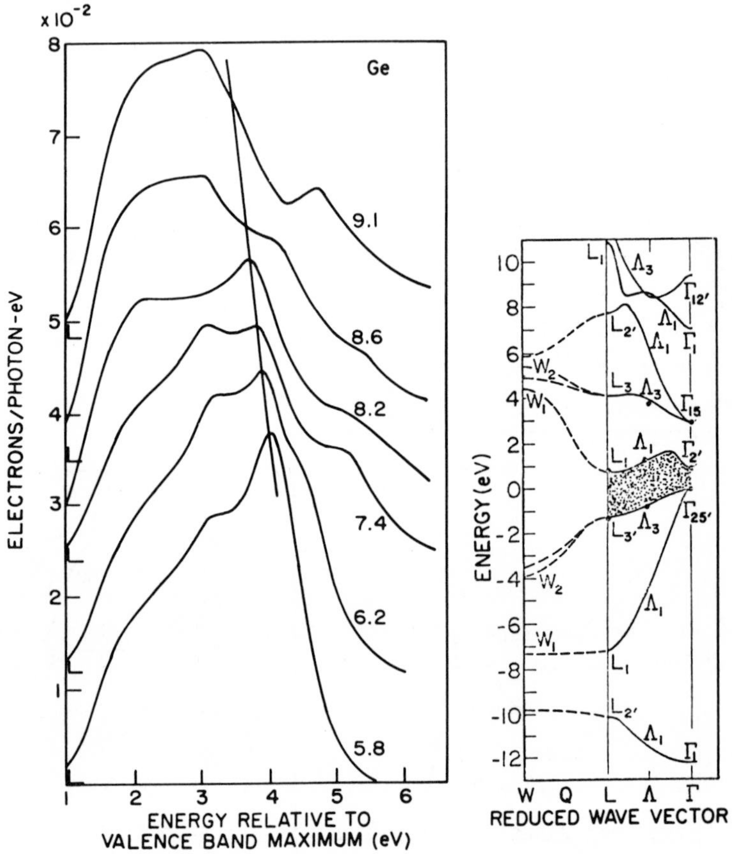

Figure 20 EDCs for germanium with workfunction reduced by a monolayer of cesium on the surface. The energy of the exciting photons is indicated on each curve. A portion of the band structure of germanium is included.

Figures 19 and 20 show the EDCs obtained by Donovan *et al* (to be published) crystalline germanium on which a monolayer of cesium was deposited to reduce the workfunction. We examine one feature in the EDCs in detail to illustrate how band-structure information can be obtained from photoemission data. One portion of the band structure has been placed in each figure so that the optical transitions can be followed easily. Because **k**-conservation provides an important selection rule in crystalline germanium, we need only consider vertical transitions in the E(**k**) diagram. Of particular interest are the transitions between the highest valence bands and the second set of conduction bands that run from Γ_{15} to L_3 and W_1. The onset for these transitions has been located accurately at 2.92±0.05eV (Donovan *et al* 1971). In the lowest-energy EDC (i.e. $h\nu$=4.2eV) shown in figure 19, the number of states that are coupled between the two bands is still reasonably weak; therefore, the intensity in the EDC above 3.0eV is fairly small. As $h\nu$ is increased, the states involved in the transition move out toward the zone faces where the number of available states increases, (the number in a given range of k increases as k^2, starting from k=0 at Γ). As a result, the intensity increases above 3.0eV and a peak begins to appear. With $h\nu$=5.8eV, a sharp peak appears at approximately 4.1eV due to the coupling of states near the zone face (the transitions from the vicinity of $L_{3'}$ to the vicinity of L_3). It should be remembered that the E(**k**) diagrams are only one-dimensional projections of the four-dimensional E-**k** space. Therefore, the peak we are following must involve transitions off the symmetry axis used to plot **k** in the E(**k**) diagrams. These diagrams, however, are useful guidelines. Because the bands become flat near the zone face, a peak is produced in the optical curves as well as in the corresponding EDCs, Only from photoemission measurements can we obtain the absolute energy of the initial and final states involved in the optical transition.

In figure 20, the photon energy is increased further. We can see from the band diagram that the initial state must move to a lower energy with increasing $h\nu$, because the valence band moves downward from $L_{3'}$ to W_2. The peak actually moves to a slightly lower energy as $h\nu$ is increased. However, the significant feature for the peak in the EDCs of figure 20 is that it loses strength as $h\nu$ is increased from 5.8 to 7.4 eV. This is to be expected from the E(**k**) curve because the slope of the initial state increases on moving from $L_{3'}$ down towards W_2. When $h\nu$= 8.2eV, the peak is again pronounced and then disappears completely, which is to be expected when the bottom of the band at W_2 is reached. Again, the bands become flat so that the number of states and thus the transition strength increases. The peak abruptly disappears for $h\nu$>8.2eV because the bottom of the band occurs at the W-point and therefore no states are available below that energy.

The fact that the peak moves to slightly lower energy with increasing $h\nu$ suggests (and we have supporting evidence) that the final transition accounting for the peak is to the vicinity of W_1 and not W_2; therefore W_1 is located at 3.9eV and the band minimum near W_2 is placed at 4.3eV. A detailed listing of some of the critical points in the band structure located by photoemission is given by Spicer and Edén (1969).

Several observations should be noted with respect to the photoemission studies of crystalline materials, where **k** provides an important selection rule. First, as $h\nu$ is increased the peaks in the EDCs frequently shift with energy in an irregular manner; the amplitude of the peaks can also be expected to change in an irregular fashion. These effects are caused primarily by the **k**-conservation condition. For example, in figures 19 and 20 the initial state for the transition moves to lower energy as $h\nu$ is increased, due to the constraint that **k** must be conserved in the optical transition.

Because of long-range order **k** is a good quantum number for a crystalline material, but is not if long-range order is destroyed. Similarly, we can use the reduced-zone approach and speak in terms of vertical transitions because of long-range order and the Bragg-reflection condition (Spicer and Donovan 1971). Two of the most interesting observations to be made concerning an amorphous material are in fact the questions of whether there is evidence of **k**-conservation as a selection rule, and whether the Bragg-reflection condition is significant.

B. Amorphous Germanium

In the study of amorphous germanium, sample preparation has been found to be critical. Because the first objective of such an investigation is to determine the intrinsic properties of the material, a standard is necessary by which the perfection of a sample can be judged (Polk 1971, Polk and Turnbull 1972). The Polk-Turnbull model provides such a standard. In this model, the four covalent bonds of each germanium atom are satisfied; there are no unsatisfied bonds and no empty voids. It appears that a large number of such bonds may occur if insufficient care is taken in sample preparation. It has been found that these effects can be minimized by: (1) maintaining a sufficiently low pressure during evaporation (less than $\sim 5 \times 10^{-6}$ torr), (2) sustaining a sufficiently low rate of deposition (usually $\sim$5 Å/sec), and (3) preserving a sufficiently large evaporation-to-substrate distance (normally 40cm). If the substrate is kept at approximately 50°C below the minimum temperature for crystallization, the microvoids can be essentially eliminated (Donovan *et al* 1970; Donovan and Heinemann 1972; Bauer *et al* 1972). If these precautions are not followed, the samples might possibly contain such a large density of microvoids that extrinsic rather than intrinsic properties will be measured (Théye 1970).

<u>Photoemission studies</u> Figure 21 shows the EDCs obtained from single-crystal and amorphous germanium (Donovan 1970). The curves are strikingly different. For crystalline germanium, the peaks appear and disappear irregularly as the photon energy is increased, which - as illustrated above - is a consequence of **k**-conservation. In contrast, no such behaviour appears in the EDCs for amorphous germanium; instead, these

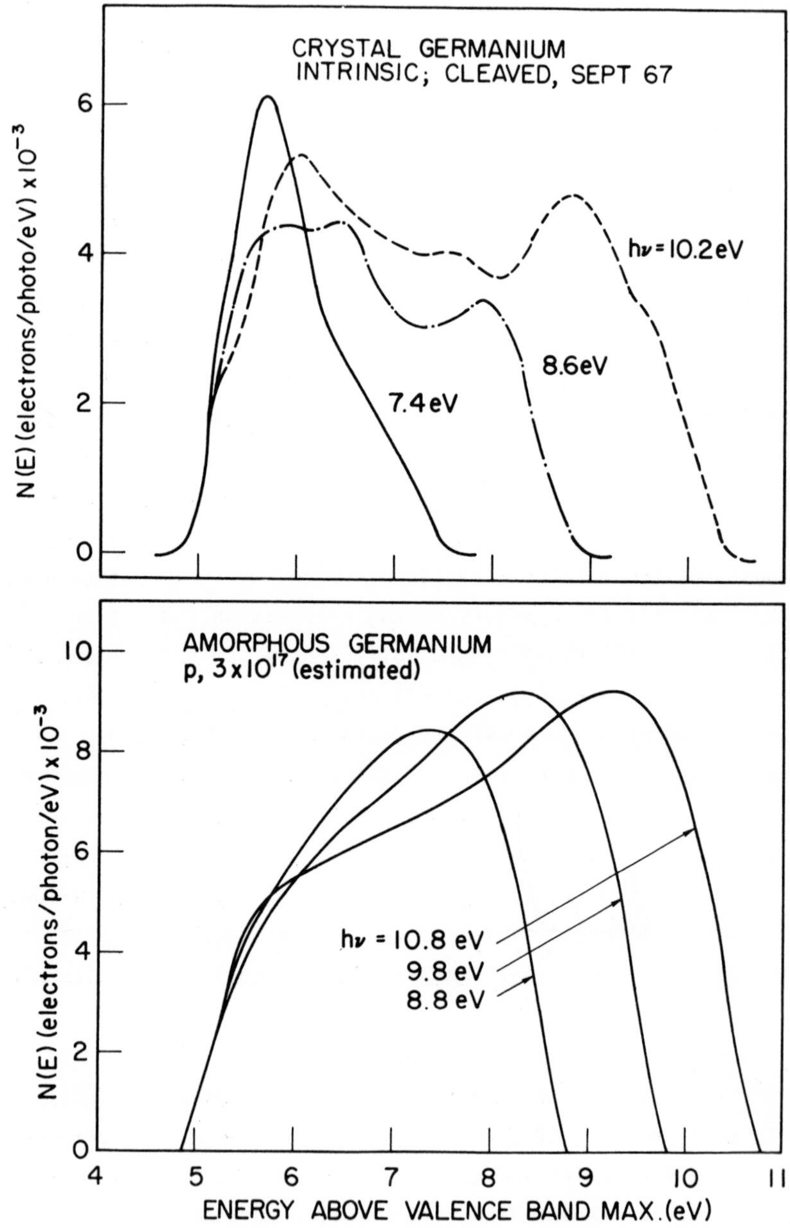

Figure 21 Comparison of EDCs for crystalline and amorphous germanium.

curves exhibit a broad feature that moves monotonically to higher energy as the photon energy is increased. We suggest that this feature is produced by structure in the valence band of the solid, and that k-conservation provides no constraint for the optical transitions, thus leaving conservation of energy as the significant optical selection rule.

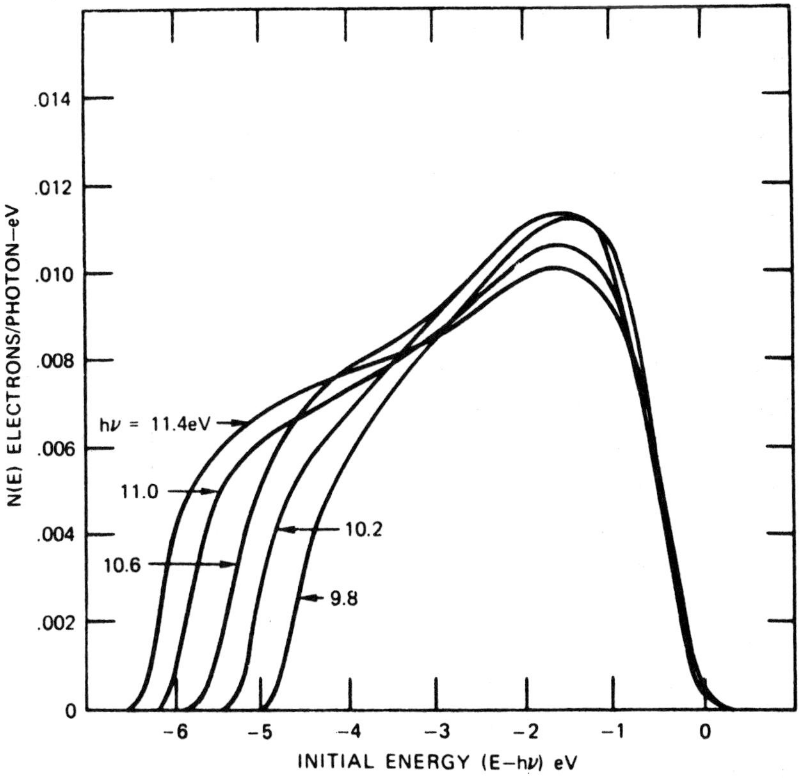

Figure 22 EDCs for amorphous germanium, vs the energy of the state from which the electron has been excited (initial state). The ordinate is on an absolute scale.

To confirm this, EDCs were plotted vs the initial energy of the states from which the electrons were excited (figure 22). In this figure the vertical scale is absolute in terms of the number of electrons emitted in unit energy range per absorbed photon. The variation in intensity is caused by the variation in escape probability with final-state energy (Krolikowski and Spicer 1969). We can see that the curves superimpose relatively well, which confirms that the structure in the EDCs is the result of structure in the valence band.

If k-conservation does not provide a major optical selection rule, then the nondirect model gives the simplest approximation to the optical excitation process. In this model, we assume that conservation of energy provides the only rule for optical selection. All matrix elements are taken to be equal and independent of photon energy. As a result, the probability of excitation into a final state as energy E, by a photon energy $h\nu$, is

$$P(E,h\nu) \propto N_f(E)\, N_i(E_i = E - h\nu) \qquad (1)$$

where $N_f(E)$ and $N_i(E_i)$ are respectively the final and initial densities of states (Krolikowski and Spicer 1969, Spicer 1957). Clearly this model is over-simplified. For example, the matrix elements cannot be constant over a large range of energy and at the same time satisfy the sum rules (Spicer and Donovan 1970); however, as long as the spectral range over which the model is applied is small compared to the energy range over which the sum rule is satisfied, it should yield a useful first approximation.

In applying equation (1) to a solid, we must find initial and final densities of states that, when substituted into equation (1), will reproduce the optical and photoemission data as closely as possible.

Figure 23 shows the density of states for amorphous germanium (Donovan 1970, Spicer and Donovan 1970, Donovan and Spicer 1968 and 1971) obtained by applying the nondirect model. A typical fit between the measured EDCs and those calculated by using the nondirect model is illustrated in figure 24. In these calculations, the escape probabilities of the electrons

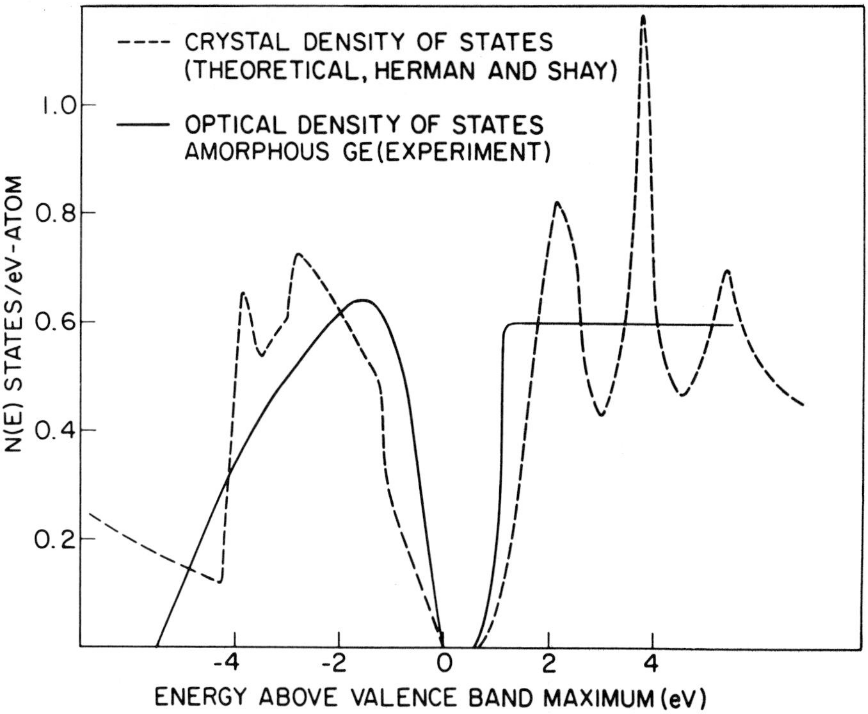

Figure 23 Density of states obtained from amorphous germanium by photoemission and optical experiments, compared with the calculated density of states for crystalline germanium.

excited in the solid are taken into account (Krolikowski and Spicer,1969).

Because the density of states in figure 23 is only a first approximation, the reliability of various features should be known. This is particularly important when comparing densities of states for the amorphous and crystalline materials (see also figure 23). For the amorphous material the conduction-band density of states rises rather abruptly and then levels out to a constant value. These features are linked by a fairly sharp 'corner', which could possibly be a peak rather than a corner. In fact, either a peak or enhanced matrix elements would produce better agreement between the calculated

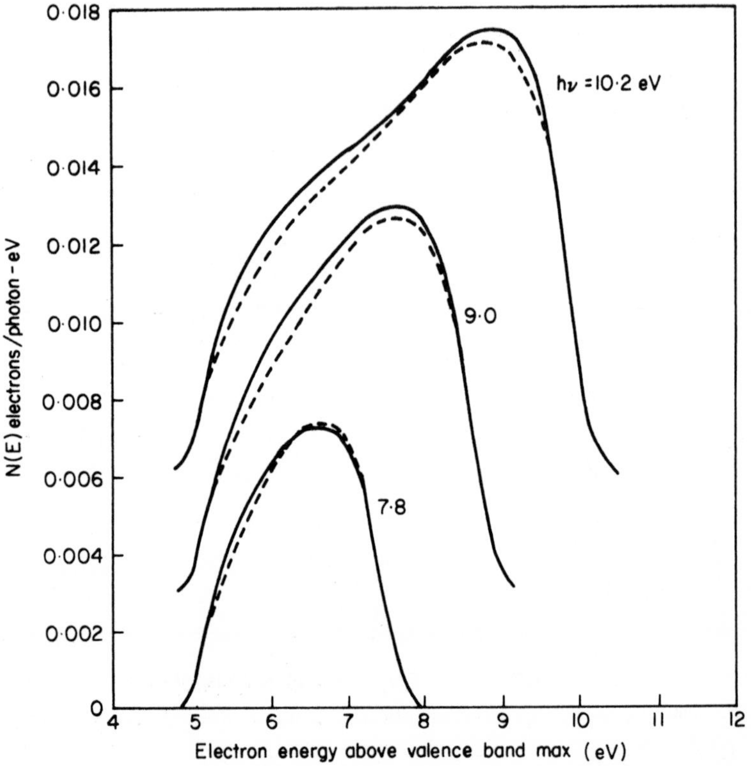

Figure 24 Comparison of measured EDCs with ones calculated from the density of states of figure 23, using the nondirect model.

and measured optical properties. However, above the 'corner' no significant additional structure appears in the conduction-band optical density of states. Such additional structure would produce a modulation in the EDC at fixed final-state energy (c.f. equation 1) and no such modulation is evident. By placing cesium on the surface the electron affinity can be reduced, but no structure was found down to 2eV. It can therefore be concluded that no strong structure appears in the conduction band above 2eV. The conduction-band density of states is observed to be flat; however, it may increase or decrease somewhat with increasing energy.

The structure in the EDCs is the result of the valence band. The position of the single valence-band peak and its general configuration, can therefore be deduced directly, and there is little uncertainty in this feature. It is more difficult to determine the detailed shape of the low-energy portion of the valence-band density of states (the region below -4eV, figure 23) because of the presence of inelastically scattered electrons that appear at higher values of $h\nu$ (Donovan 1970).

Character of the band edges Optical absorption of the band edge and an examination of the leading edge of the EDCs indicate a sharp well-defined band edge for amorphous germanium, when carefully prepared (Donovan *et al* 1969, Théye 1970, Spicer and Donovan 1970, Donovan *et al* 1970, Chopra and Bahl 1970, Spicer *et al* 1972). In figure 24a, the absorption from a single crystal of germanium is compared with the absorption from an amorphous sample that approaches the perfection of the Polk-Turnbull standard. We can see that the sharpness of the absorption edges for the two samples is comparable.

The amorphous sample, from which the data for figure 25 were taken, was formed by following the procedures already outlined; in addition, the substrate was kept at 50-100^{0}C below the crystallization temperature during evaporation. Studies on similar samples indicated a material density within a few percent of the crystalline density, and an absence of microvoids (Donovan and Heinemann 1972, Bauer *et al* 1972). If sample preparation is unchanged but the substrate temperature is lowered, appreciable numbers of microvoids (Théye 1970) occur in the film; however, the absorption edge remains sharp and is shifted to lower energy. If the necessary precautions are not followed, sharp edges do not form during evaporation, presumably because the defect (microvoid) density is too high. A sharp edge can be produced by annealing such a film; however, the edge occurs at relatively large energies, approximately 1.0eV

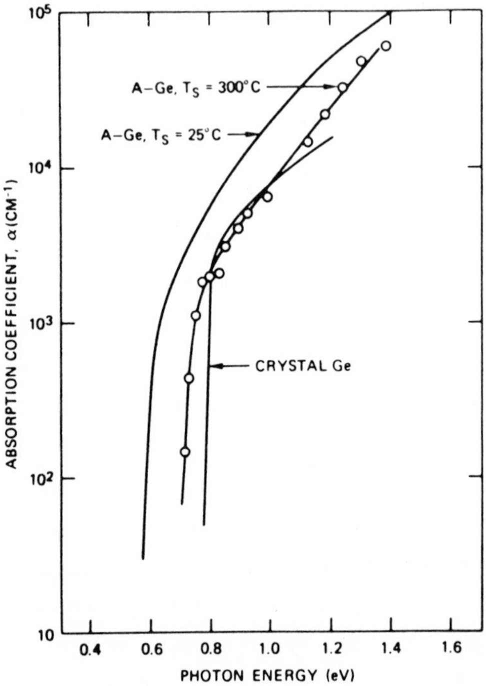

Figure 25 Comparison of the optical absorption edges obtained for crystalline germanium and for an amorphous sample with very few defects (A-Ge, T_S=300°C). The curve marked T_S=25°C is for an amorphous sample containing a larger density of defects.

(Théye 1971). Apparently, large densities of defects are difficult to remove completely by annealing; instead, annealing may cause small microvoids to collect together to form larger microvoids.

Figure 26 illustrates schematically the meaning of sharp band edges such as those found in properly prepared amorphous germanium. Recently, such edges have been predicted theoretically by Weaire (1971). The right-hand side indicates the type of band edge that has been postulated frequently (Cohen 1971) for amorphous materials, including germanium. Mobilities as a function of energy for the two models are also illustrated.

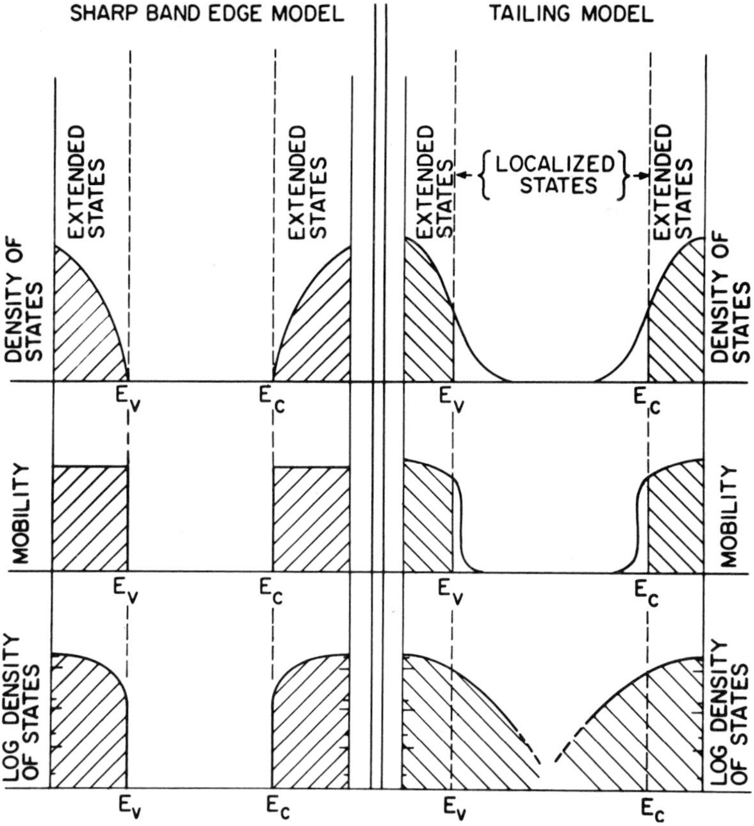

Figure 26 Comparison of the sharp band edges found for properly prepared samples of amorphous germanium (left-hand curves) with the band edges (right-hand curves), showing the tailing often discussed theoretically.

Photoconductivity measurements (Fischer and Donovan 1971) on amorphous germanium samples, sufficiently close to ideal, indicate that the density of states and 'mobility edges' coincide quite closely.

C. Densities of States for Amorphous and Crystalline Germanium

Where the amorphous and crystalline densities of states (figure 23) are compared, it can be seen that the sharp structure observed for the crystalline material disappears in the amorphous density of states. At first it is curious to note that the sharp structure in the density of states is lost but that sharp band edges remain. It has been suggested (Spicer and Donovan 1971) that this behaviour can be understood if a distinction is made between Bragg band-gaps caused by the Bragg-reflection condition and chemical band-gaps caused by chemical effects (for example, the covalent bond in germanium). It has been further suggested that the sharp structure in the crystalline density of states is the result primarily of Bragg band-gaps, whereas the gap between the valence and conduction bands is essentially a chemical band-gap. The structure remains sharp in the amorphous material because the covalent bond is well defined; however, it is affected by the Bragg-reflection condition, and thus its energy may change when long-range order is destroyed.

D. Summary

When the long-range order of the crystal is destroyed by transforming germanium or silicon (Pierce *et al* 1971) into their amorphous forms, parameters such as **k**, and structure that depends on long-range order disappear; however, quantities such as chemical band-gaps and valence bandwidth (caused by overlap or wavefunctions from adjacent sites) do not disappear. The actual value of the band-gaps in the crystalline material may depend on both the covalent bond and the Bragg-reflection condition. Therefore, the numerical value of the band-gap may change when the material is transformed from crystalline to amorphous.

4. OTHER PHOTOEMISSION STUDIES OF PHASE CHANGES

A. Nickel above and below the Curie Point

In the simple band model for ferromagnetic nickel, the bands associated with 'spin-up' and 'spin-down' electrons are assumed to be split by the exchange interaction (see figure 27). In this model, as applied to nickel, ferromagnetism results from

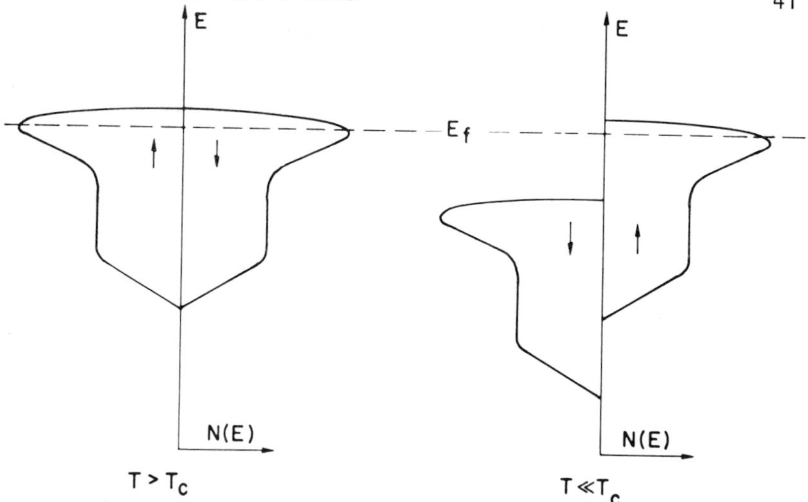

Figure 27 Simple band model for ferromagnetic nickel. T_c is the Curie temperature.

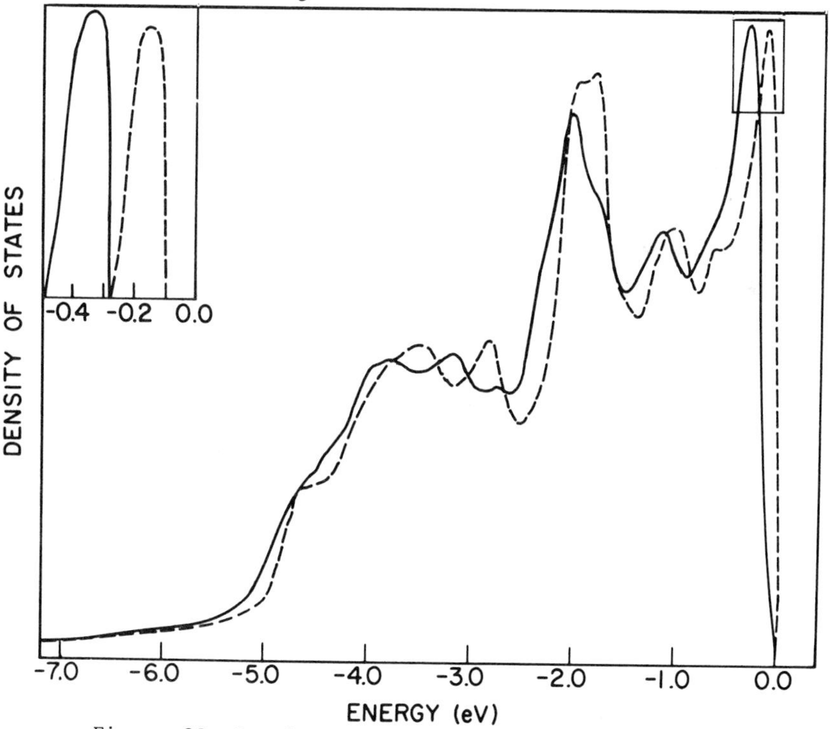

Figure 28 Density of states for nickel calculated using the simple band model of ferromagnetism in the ferromagnetic state ($T \ll T_c$, solid curve) and paramagnetic state ($T > T_c$, dashed curve).

the fact that one band (spin-up) is filled, whereas the spin-down band is only partially filled; therefore, the net density of spin-up electrons produces ferromagnetism. According to this model, the magnitude of the splitting decreases as the temperature is increased, and it disappears at the Curie temperature, $T_c = 631^0K$ for nickel. As a result, if photoemission measurements are made at room temperature (where we should have complete exchange splitting) and above T_c, then structure in the EDCs should change in accord with any change in the band structure. Figure 28 illustrates the results of calculations of the densities of states with and without an average exchange splitting of 0.29eV (Pierce 1970, Pierce and Spicer 1970). These calculations were based on those of Hodges *et al* 1966); similar calculations were made by Pierce and Spicer on the changes to be expected in the EDCs on the basis of either direct or nondirect optical transitions. In both cases the predicted changes were approximately 0.2eV, well within the experimental resolution of 0.05eV.

Figure 29 shows typical EDCs above and below the Curie temperature (Pierce and Spicer 1970). Particular attention should be paid to the sharp peak near the leading edge. This peak, in agreement with band calculations, should move by approximately 0.2eV (see figure 28); however, no movement was detected despite the fact that the resolution (0.05eV) was more than sufficient to detect such a change. We can see from figure 29 that the leading edge in the upper-temperature EDC is slightly higher in energy than the room-temperature EDC, due to the increase in the thermal energy of electrons near the Fermi surface. Recently, Rowe and Tracy (1971) have used a very sensitive technique, which they developed, to study the motion of the first peak in the nickel EDCs as a function of temperature. Their results are in agreement with those of Pierce and Spicer.

It should not be surprising that the changes predicted by

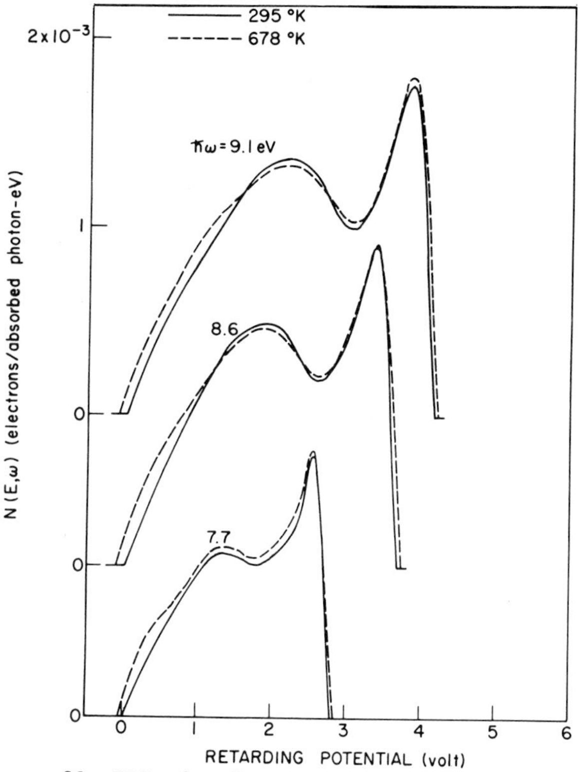

Figure 29 EDCs for ferromagnetic (solid line) and paramagnetic (dashed line) nickel. The temperature of the sample is indicated on each curve.

the simple band theory do not take place. If this theory is correct, nickel would not follow a Curie-Weiss law above T_C, because there would be no magnetic moments for temperatures above T_C and such moments are necessary for the Curie-Weiss law to apply. Recent experiments (Bänninger *et al* 1970), in which the magnetization of photoemitted electrons also measured, were found to disagree with the simple band model of magnetism.

One can obtain a consistent picture of the temperature dependence of the magnetic and photoemission results, if it is assumed that: (1) only a small region of the nickel is sampled by photoemission, and (2) within this small region approximately the *same* net magnetization and electronic structure appear in both ferromagnetic and paramagnetic nickel. In this model, the ferromagnetic order disappears above the Curie temperature because long-range order is destroyed. A Curie-Weiss law exists above T_c because of the moments associated with the small regions. These regions apparently fluctuate with time; however, the fluctuations are probably on a time scale which is long compared with the optical excitation time (10^{-14}-10^{-15}sec).

Doniach (1972) has suggested that the absence of change in the EDCs, and the electron-polarization effects, can be explained in terms of many-body effects involved in electron excitation. This is closely related to the nondirect model of optical excitation and may explain the large differences between EDCs calculated on the basis of direct transitions and those actually observed (Pierce 1970).

B. Crystalline and Liquid Metals

We have examined the effects of the loss of long-range order on covalent germanium. In this section, the relatively small amount of available data form the basis from which to discuss the changes that take place in the electronic structure of metals when long-range order is destroyed by melting.

When germanium is melted, it changes from covalent to metallic bonding; however, in contrast, metals remain metallic after melting and only long-range order is lost. Therefore with metals, as in the case of amorphous and crystalline germanium, long-range order is destroyed but local order is not greatly changed on melting. Two cases will be discussed: (1) noble metals, where the electronic structure is dominated by the rather tightly bound d-electrons, and (2) simple metals, where there are no such electrons.

Noble metals The most striking feature in the optical properties of the noble metals is the threshold for transitions from the d-bands to the Fermi level. These transitions are responsible for the characteristic colours of the noble metals.

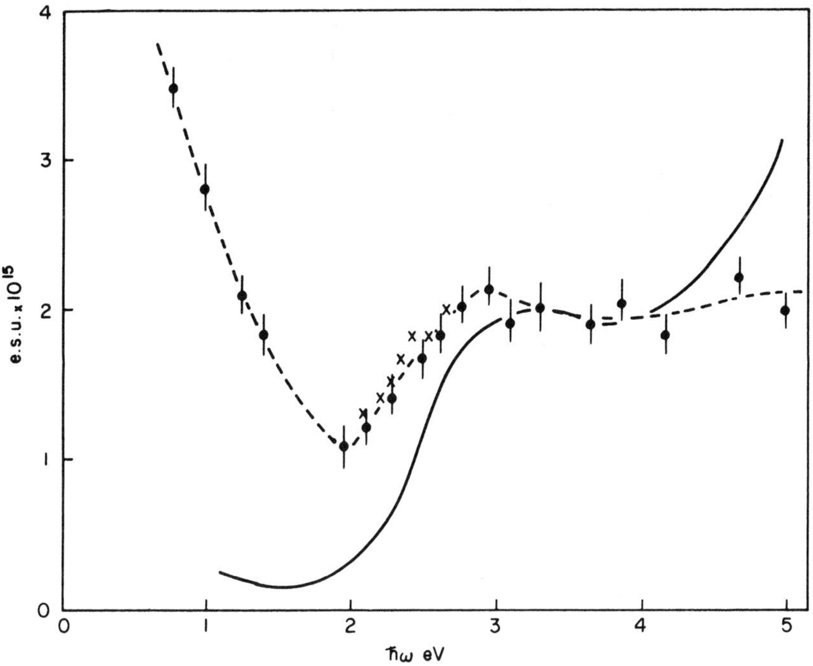

Figure 30 The optical conductivity for liquid and crystalline copper from the work of Miller. The heavy curve represents the crystalline copper and the dashed curve, with experimental points, represents the liquid. Note that the d-band absorption starting at approximately 2.0eV is not greatly changed. The increase for $h\nu < 2.0$eV in the liquid is due to increased electron scattering.

Figure 30 presents data obtained by Miller (1969) for liquid and crystalline copper. The threshold for transitions from the d-bands occurs at approximately 2.0eV. These transitions are not greatly disturbed by the destruction of long-range order; however, the free-electron or Drude absorption for $h\nu < 2$eV is strongly affected. These data suggest, to first order, that the transitions from the d-band are not greatly perturbed by the loss of long-range order, but that the free-electron absorption is affected because of reduced electron scattering length in the disordered material. Again, this observation is rather surprising; however, it would appear more reasonable if distinctions such as those between Bragg and chemical band-gaps (c.f. §3C) are considered because the energy difference between the top of the d-states and the Fermi level is, to the first approximation, a chemical band-gap.

Figure 31 shows EDCs obtained by Eastman (1971) for liquid and crystalline gold, with four values of photon energy from 16.8 to 40.8eV. The fine structure in the curves for the crystalline samples is easily explained in terms of direct (**k**-conserving) transitions. The distinct changes between the crystalline and liquid samples can be seen clearly; however, the changes are surprisingly small. This is even more striking when it is remembered that the curves of figure 31 are compared on an absolute basis (there is little change in the probability of excitation or escape on melting the material). This suggests that matrix elements may be insensitive to the loss of long-range order, and the experimental results are surprising if the conventional band theory is taken too seriously. Until recently, to calculate band structure and densities of states one had to use band theory, which depends on long-range crystalline order. Thus, we might expect structure in the density of states to disappear on melting; this however, is not always the case.

Figure 32 illustrates the comparison made by Eastman (1971) of a broadened density of states obtained by the band calculations of Connolly (unpublished) with the optical density of states obtained by a nondirect analysis (equation 1) of the EDCs for liquid gold. Again, there is a striking similarity between the two densities of states; which is a positive indication that the density of states of gold is not overwhelmingly dependent on long-range order. Instead, it is strongly

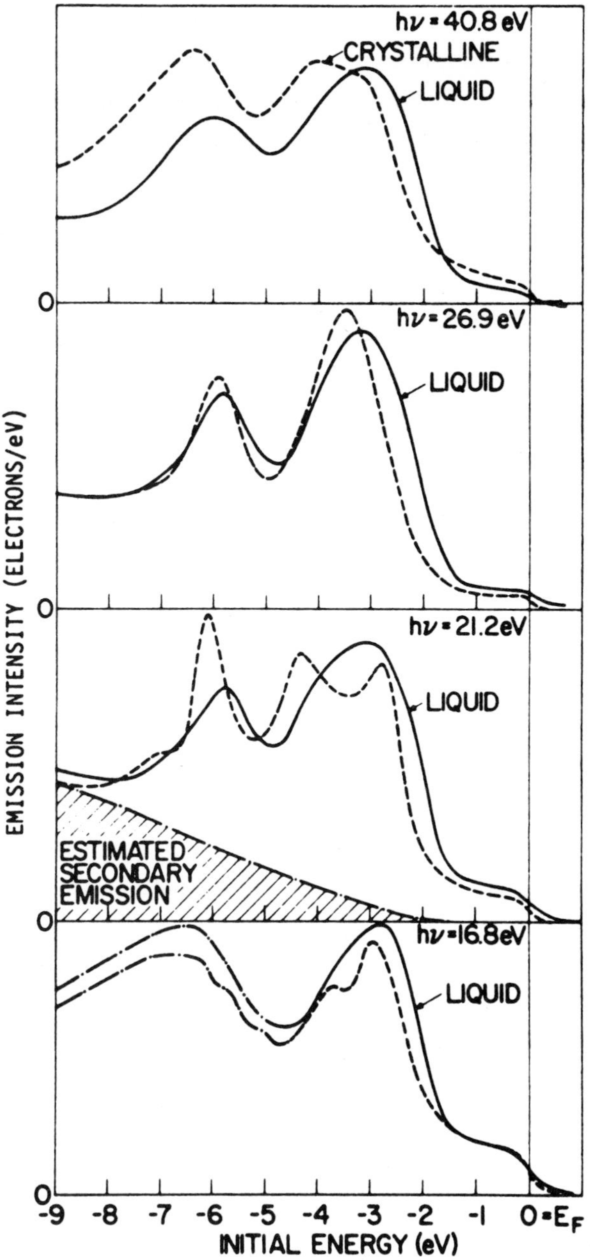

Figure 31 EDCs for crystalline and liquid gold.

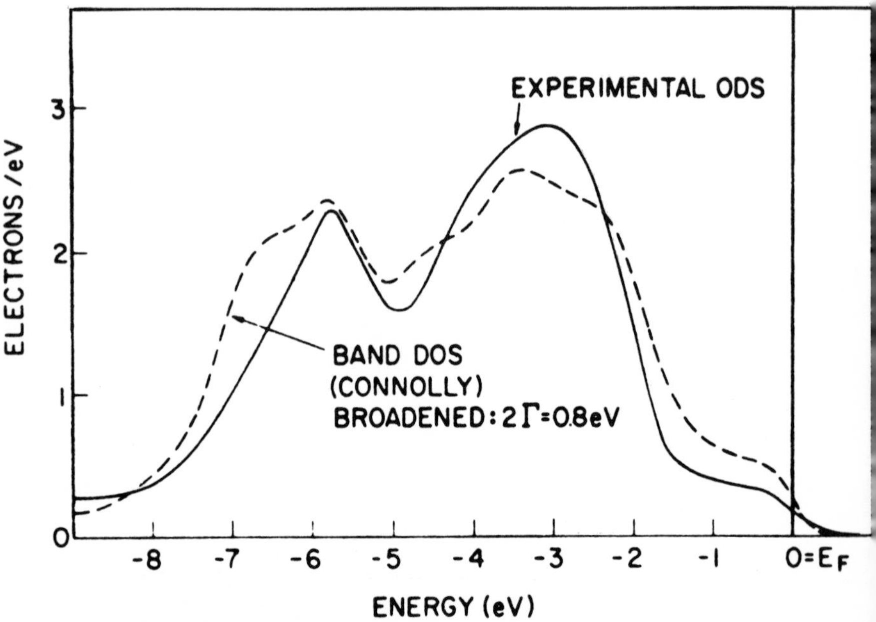

Figure 32 Optical density of states deduced from EDCs for liquid gold (Eastman 1971) compared with the density of states of gold calculated by Connolly (to be published).

affected by such factors as the local overlap between d-states on adjacent gold atoms. Long-range order provides a somewhat smaller effect. If a distinction is once again made between Bragg and chemical band-gaps, it could be concluded that chemical gaps are dominant in determining the density of states for the d-bands of gold.

Simple Metals Let us now examine some of the data available from simple metals (those that do not have d-states within a few electrovolts at the Fermi level). Figure 33 presents the data obtained by Choyke *et al* (1971) on crystalline and liquid mercury. A marked effect is apparent in the optical properties of this material when melted. The peak in ε_2 near 1.6eV is characteristic of many of the Group II and Group III simple metals, due to direct transitions across the Fermi surface.

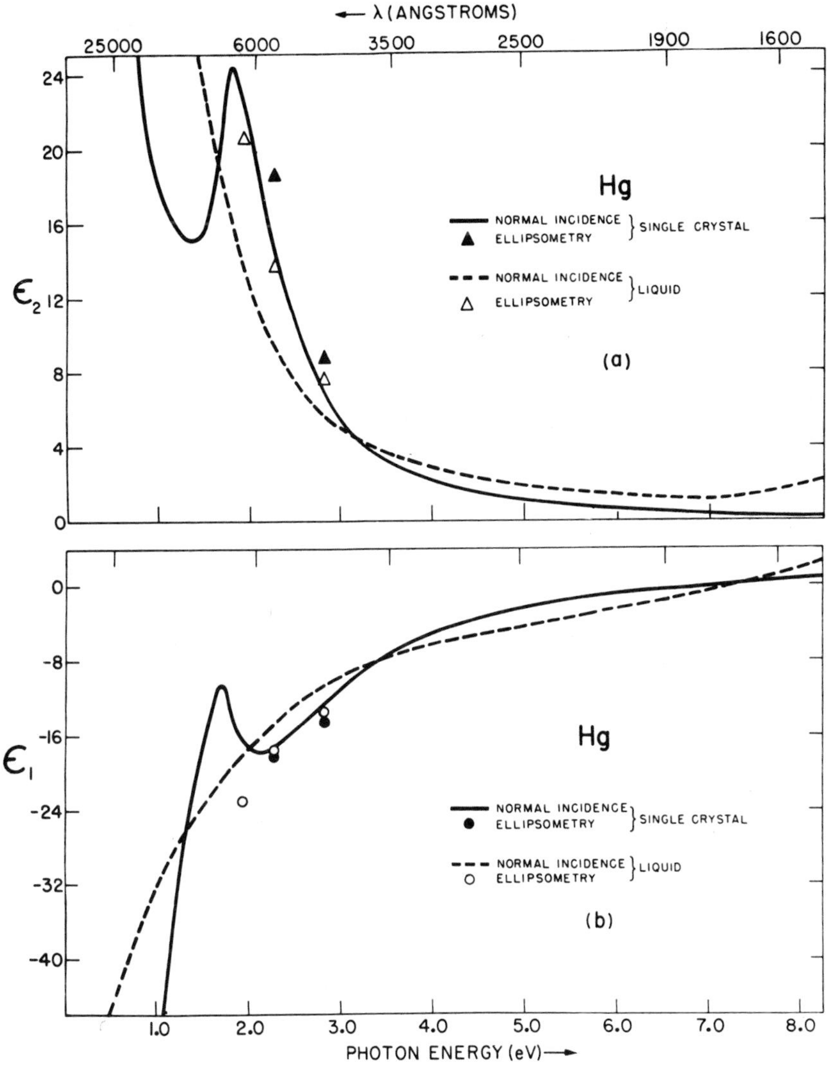

Figure 33 Real (ε_1) and imaginary (ε_2) parts of the complex dielectric constant for mercury in the crystalline (solid curve) and liquid (dashed curve) states. Note that the pronounced structure at approximately 1.6eV in the curves for the crystalline form disappears entirely in the liquid.

For mercury and other metals this peak disappears on melting, which is to be expected because the structure is the result of a Bragg band-gap, with the existence of a reduced-zone scheme (Ehrenreich *et al* 1963).

Indium is one of the metals similar to mercury, for which the optical peak near 1.6eV disappears on melting (Hodgson 1962). Koyana and Spicer (1971) have studied photoemission from this metal in the crystalline and liquid forms. The structure in the EDC for crystalline indium is the result of structure at the zone face caused by the so-called Bragg band-gaps. Surprisingly, some of this structure appears to persist in the liquid; however, because of electron scattering and other peculiarities, this is not as clearcut a case as that for gold.

With simple metals, therefore, apparently the reduced-zone scheme is destroyed on melting the material; however, certain structure in the densities of states associated with Bragg band-gaps may remain. Presumably this can be explained in terms of the existence of sufficient short-range order to support some Bragg-like scattering.

4. SUMMARY

It has been shown that the electron structure of solids and liquids can be effectively probed using photoemission. Striking characteristics of this technique are its versatility and the ease with which the changes produced by alloying or structural disorder can be investigated.

Acknowledgment

The support of the Advanced Research Projects Agency through the Office of Army Research-Durham, and of the National Science Foundation is gratefully acknowledged.

REFERENCES

Amar, H. and Johnson, K. (1966); in "Optical Properties and Electronic Structure of Metals and Alloys" ed. F. Abèles; North Holland, Amsterdam, 586.

Anderson, P.W. (1961); Phys. Rev. 124, 41.

Arlinghaus, F.J. (1967); Phys. Rev. 157, 491.

Arlinghaus, F.J. (1969); Phys. Rev. 186, 609.

Bänninger, U., Busch, O., Campagna, M. and Siegman, H.C. (1970); Phys. Rev. 25, 585.

Bauer, R.S., Galeener, F.L. and Spicer, W.E. (1972); Proc. of Int. Conf. on Amorphous and Liquid Semiconductors, Ann Arbor, Michigan 1971, *in press*.

Berglund, C.N. and Spicer, W.E. (1964); Phys. Rev. 136, A1030, A1044.

Chopra, K.L. and Bahl, S.K. (1970); Phys. Rev. B1, 2545.

Choyke, W.J., Vosko, S.H. and O'Keeffe, T.W. (1971); Sol. St. Comm. 9, 361.

Clife, J., Curry, C. and Thompson, B.J. (1963); Phil. Mag. 8, 593.

Cohen, M.H. (1971); Phys. Today, 24, 26.

Derbenwick, G., Pierce, D. and Spicer, W.E. (1972); "Methods of Experimental Physics", ed. Marton; Academic Press, *in press*.

Doniach, S. (1970); Phys. Rev., B2, 3898.

Doniach, S. (1972); "Proc. 17th Conf. on Magnetism and Magnetic Materials", Amer. Inst. Phys., *in press*.

Donovan, T.M. (1970); Dissertation, Stanford University.

Donovan, T.M., Ashley, E.J. and Spicer, W.E. (1970); Phys. Lett. 32A, 85.

Donovan, T.M. and Heinemann, K. (1972); Phys. Rev. Lett. *in press*.

Donovan, T.M., Matzusaki, J. and Spicer, W.E. to be published.

Donovan, T.M. and Spicer, W.E. (1968); Phys. Rev. Lett. 21, 1572.

Donovan, T.M., Spicer, W.E., Bennett, J.M. and Ashley, E.J. (1970); Phys. Rev. B2, 397.

Donovan, T.M., Spicer, W.E. and Bennett, J.M. (1969); Phys. Rev. Lett. 22, 1058.

Donovan, T.M., Fischer, J.E., Matzusaki, J. and Spicer, W.E. (1971); Phys. Rev. B3, 4292.

Eastman, D.E. (1971); Phys. Rev. Lett. 26, 1108.

Edén, R. (1970); Rev. Sci. Instr. 41, 252.

Ehrenreich, H. and Hodges, L. "Methods in Computation Phys.", ed. B. Adler; Academic Press, N.Y., 8, 179.

Ehrenreich, H., Philipp, H.R. and Segall, B. (1963); Phys. Rev. 132, 1918.

Fischer, J.E. and Donovan, T.M. (1972); Optics Comm. 3, 116; Proc. Int. Conf. on Amorphous and Liquid Semiconductors, Ann Arbor, Michigan, 1971, *in press*.

Friedel, J. (1956); Cond. J. Phys. 34, 1190.

Friedel, J. (1958); J. Phys. Radium, 19, 573.

Herman, F., Kortum, R.L. and Shay, J.L. (1967); Proc. of Int. Conf. of II-VI Semiconducting Compounds, Providence; Benjamin, N.Y.

Hodges, L., Ehrenreich, H. and Lond, N.D. (1966); Phys. Rev. 152, 505.

Hodgson, J.N. (1962); Phil. Mag. 1, 229.

Kirkpatrick, S., Velicky, B. and Ehrenreich, H. (1970); Phys. Rev. B1, 3250.

Kjöllerstvam, B. (1969); Phil. Mag. 19, 1207.

Koyama, R.Y. and Spicer, W.E. (1971); Phys. Rev. 4, 4318.

Krolikowski, W. and Spicer, W.E. (1970); Phys. Rev. B1, 478.

Krolikowski, W.F. and Spicer, W.E. (1969); Phys. Rev. 185, 882.

Meyers, H.P., Norris, C. and Walldén, L. (1969); Sol. St. Comm. 7, 1539.

Miller, J.C. (1969); Phil. Mag. 20, 1115.

Mott, N.F. (1935); Proc. Phys. Soc. (London), 47, 571.

Mott, N.F. (1936); Phil. Mag. 22, 287.

Mott, N.F. (1964); Adv. in Phys. 13, 325.

Norris, C. and Nilsson, P.O. (1968); Solid State Comm. 6, 649.

Norris, C. and Walldén, L. (1969); Solid State Comm. 7, 99.

Nilsson, P.O., Norris, C. and Walldén, L. (1970); Phys. Kondens. Materie, 11, 220.

Nilsson, P.O. (1970); Physica Scripta, 1, 189.

Nilsson, P.O. and Lindau, I. (1971); J. Phys. F. 1, 854.

Pierce, D.T. (1970); Dissertation, Stanford University.

Pierce, D.T., Ribbing, C.G. and Spicer, W.E. (1972); Proc.Int. Conf. on Amorphous and Liquid Semiconductors, Ann Arbor, Michigan, (1971), *in press*.

Pierce, D.T. and Spicer, W.E. (1970); Phys. Rev. Lett. 25, 581.

Pierce, D.T. and Spicer, W.E. (1972); Phys. Rev. *in press*.

Polk, D.E. (1971); J. Noncryst. Solid 5, 365.

Rowe, J.E. and Tracy, J.C. (1973); Proc. Int. Conf. on Electron Spectroscopy (Pacific Grove, Calif., 1971), ed. D.A. Shirley, p.55; North Holland.

Seib, D.H. and Spicer, W.E. (1970a); Phys. Rev. B2, 1676.

Seib, D.H. and Spicer, W.E. (1970b); Phys. Rev. B2, 1694.

Smith, N.V. (1971); Critical Reviews in Sol. Stat. Phys. 2, 45.

Sonntag, B., Haensel, R. and Kunz, C. (1967); Sol. Stat. Comm. 7, 597.

Soven, P. (1967); Phys. Rev. 156, 809.

Soven, P. (1969); Phys. Rev. 178, 1136.

Spicer, W.E. (1967); Phys. Rev. 154, 385.

Spicer, W.E. (1972); in Proc. 3rd IMR Symposium Electronic Density of States, Nat. Bur. Stand. (U.S.), Spec. Publ. 323, ed. L. Bennett, p.139.

Spicer, W.E. and Edén, R.C. (1969); Proc. of the 9th Int. Conf. on Phys. of Semiconductors (Nauka, Leningrad), p.65.

Spicer, W.E. and Donovan, T.M. (1970); Jour. Non-Crystalline Solids 2, 66.

Spicer, W.E. and Donovan, T.M. (1971); Phys. Rev. Lett. 36A, 459.

Spicer, W.E., Donovan, T.M. and Fischer, J.E. (1972); Int. Conf. on Amorphous and Liquid Semiconductors, Ann Arbor, Michigan, (1971) *in press*.

Stocks, G.M., Williams, R.W. and Faulkner, J.S. (1971); Phys. Rev. B4, 4390.

Théye, M.L. (1970); Optics Comm. 2, 329.

Théye, M.L. (1971); Mat. Res. Bull. 6, 103.

Turnbull, D. and Polk, D.E. (1972); Int. Conf. on Amorphous and Liquid Semiconductors, Ann Arbor, Michigan, (1971) *in press*.

Walldén, L. (C970); Phil. Mag. 21, 571.

Weaire, D. (1971); Phys. Rev. Lett. 26, 1541.

OBSERVATION AND CALCULATION OF INELASTIC SCATTERING IN UV-PHOTOEMISSION SPECTRA

P.O. Nilsson and I. Lindau

Department of Physics, Chalmers University of Technology, Gothenburg, Sweden.

1. INTRODUCTION

During the last decade UV-photoemission has been used to obtain information about the band structure of solids. The photoemission process has usually been viewed as a multistep process (see, for example, Berglund and Spicer 1964): optical excitation of an electron, transport through the material, and finally possible escape from the solid. Although this formalism today is questioned in principle (Schaich and Ashcroft 1970, 1971; Mahan 1970; Sutton 1970) it is in many cases a good first approximation for analysis of experimental data. One complication that might contribute is the effect of the surface, which distorts the bulk wavefunction.

During the process of transport through the bulk solid the electron may encounter scattering events of various kinds, giving rise to a distribution of secondary electrons. The scattering mechanisms, for electrons a few eV above the Fermi energy, are dominated by two processes: creation of electron-hole pairs and creation of plasmons. We first assume the interaction with plasmons to be weak, which is true provided there is no well-defined plasmon or if the photon energy is lower than approximately the plasmon energy. Further, electron-hole scattering (the Auger process) is not important at the photon energies usually employed. The scattering is then thus mostly due to inelastic electron-electron interactions. The scattering rate, for production of electron-hole pairs by an excited electron at energy E, can be obtained using first-order perturbation theory (Kane 1967). The result is

$$W(E) = \frac{V^2}{4\pi^2 h} \frac{\int dk_c dk_{c'} dk_{v'} \, \delta(E_c - E)}{\int dk_c \, \delta(E_c - E)} \delta(E_c + E_v - E_{c'} - E_{v'}) \, |M(c,c',v,v')|^2 \qquad (1)$$

V is the volume of the specimen. The scattering process involves interaction of the primary electron in state c with an electron in state v. The resulting electron pair is in state c', v'. The energy conservation is illustrated in figure 1.

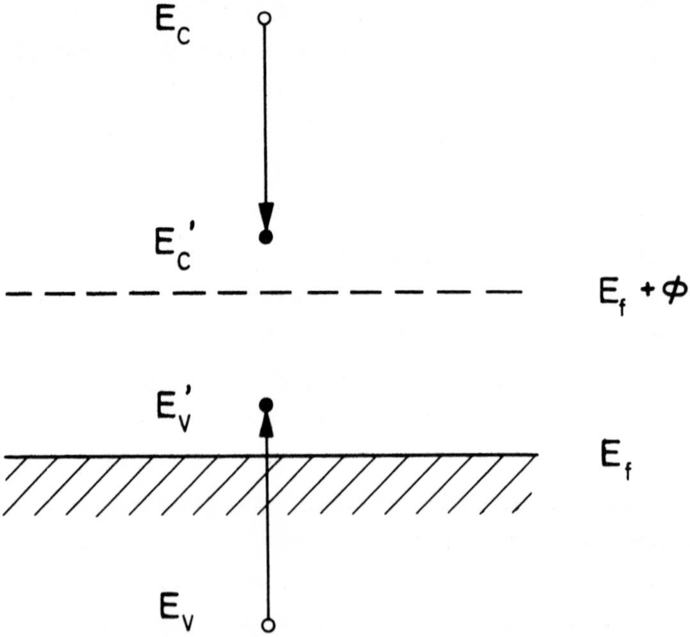

Figure 1 Schematic energy diagram of the electron-hole pair creation process. The primary electron is in state E_c.

Momentum is conserved according to

$$k_c + k_v = k_{c'} + k_{v'} + \mathbf{g} \qquad (2)$$

where **g** is a reciprocal lattice vector. The matrix element M is taken with respect to the screened Coulomb interaction, and consists of two terms if exchange is included. Kane (1967) calculated the scattering rate in silicon using this formalism, but also using a simpler so-called 'random-**k**' approximation where momentum conservation is ignored. The two results were identical to within the accuracy of the Monte Carlo method used.

The 'random-k' approximation follows if k_v in equation (1) is regarded as a random variable so that the probability of having $\mathbf{k}_v$ at energy E is proportional to the density of states $N(E_v)$. One then obtains

$$W(E) = A \int dE_{c'} N(E_{c'}) \int dE_v N(E_v) N(E_{v'}) = \int dE_{c'} P(E_c, E_{c'}) \qquad (3)$$

A includes an averaged value of $|M|^2$. The mean free path is now given by

$$l(E) = \langle v_g(E) \rangle / W(E) \qquad (4)$$

where $\langle v_g(E) \rangle$ is the average group velocity at energy E. It is now possible to derive the energy distribution curve (EDC) of photoemitted once-scattered secondaries. This was done initially by Berglund and Spicer (1964), who obtained

$$N_s(E_{c'}, \omega) = T(E_{c'}) \int dE_c \; 2 \frac{P(E_c, E_{c'})}{W(E_c)} N_o(E_o, \omega) \; C(\alpha, \ell, \ell') \qquad (5)$$

$N_o(E,\omega)$ is the excited internal energy distribution of primary electrons for photon frequency ω, and α is the optical absorption coefficient. T is a transmission function which varies smoothly from zero at the vacuum level to $\alpha\ell'/4$ at large energies:

$$T(E) = T_o(E)[1 + ln(1 - 2\alpha\ell T_o / (1 + \alpha\ell)]/2T_o\alpha\ell \quad (6)$$

$$T_o(E) = \tfrac{1}{2}\{1 - \sqrt{[(E_F + \phi)/E]}\} \quad (7)$$

C in equation (5) arises from geometrical considerations; it is a slowly varying function, approximately equal to 1 for $\alpha\ell' << 1$ and $\ell'/\ell << 1$:

$$C(\alpha,\ell,\ell') = \tfrac{1}{2}[ln(1 + \alpha\ell')/\alpha\ell' + (\ell/\ell')\, ln(1 + \ell'/\ell)] \quad (8)$$

We observe that $N_s(E_{c'},\omega)$ does not contain the matrix element but only the density of states and other known quantities.

We have used equation (5) on a simple model of the density of states to illustrate the typical shape of the secondary distribution. The result is shown in figure 2. We observe

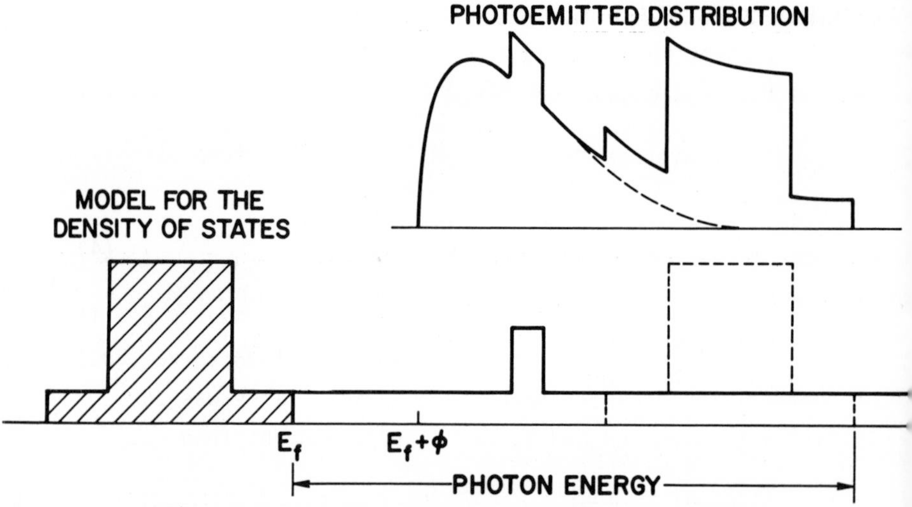

Figure 2 A simple density of states, and the corresponding photoemission energy distribution.

a broad asymmetric hump with its maximum a few eV above the vacuum level. It is also clear that a feature in the conduction-band density of states is reproduced in the spectrum. This fact is also directly evident from equation (5), because $N(E_{c'})$ in $P(E_c,E_{c'})$ can be taken outside the integral.

One aim of the present work was to investigate whether density-of-states effects can be detected in an experimental secondary spectrum. Mainly copper was used for this study because of its well-established band structure, and because it is relatively easy to prepare. The results indicate that more than one scattering event per electron must generally be considered. Because analytical expressions then become very complicated a Monte Carlo method was developed, which was shown also to account easily for surface scattering. In analysis of the data other possible surface effects were also considered.

2. EXPERIMENTAL TECHNIQUE

To obtain a spectrum with well separated primary and secondary distributions relatively high photon energy must be used. In the present investigation this was achieved with a gas discharge lamp using neon and helium, at respectively 16.8 and 21.2 eV. Although the lamp was directly attached to the experimental chamber it was possible with differential pumping to keep the chamber pressure in the region of 10^{-10} torr during the measurements. Energy dispersion of the photoemitted electrons was achieved with either a cylindrical retarding field or a spherically deflecting analyzer. The copper specimens were mainly evaporated films. During evaporation the pressure increased but was always below $5x10^{-10}$ torr. Some measurements were also performed on 100 , 110 and 111 orientations of bulk single crystals, which were cleaned by heating and by argon bombardment.

Photoemission from oxidized copper films was also studied. For this the chamber was first flushed several times with oxygen to prevent emission of other gases previously adsorbed on the walls. A clean copper surface was then produced by evaporation. Finally clean oxygen was introduced to the chamber to a known pressure and pumped off after a certain time.

3. RESULTS

In figure 3 two EDCs for copper are shown for photon energy 21.2eV. On the vertical axis the number of photoemitted electrons are indicated per energy unit. The electron energy on

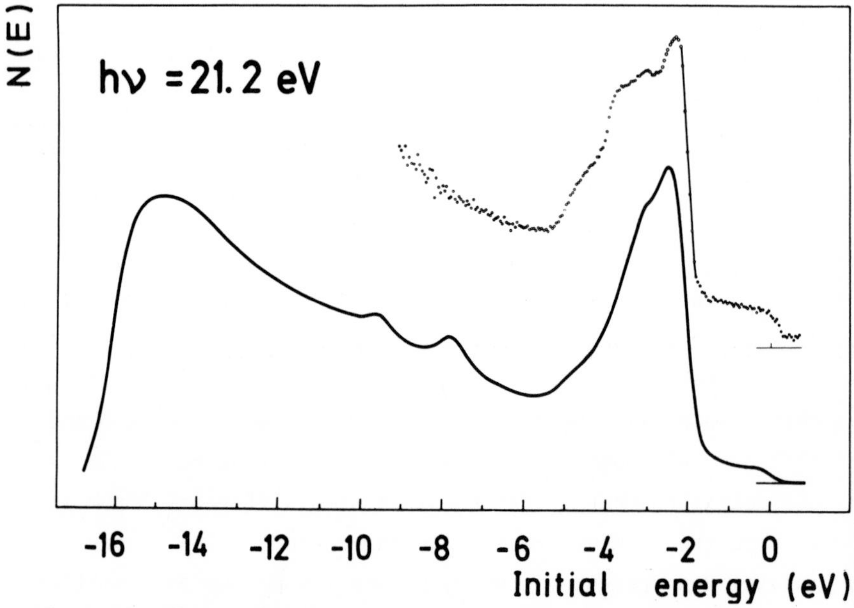

Figure 3 Photoemission energy distribution curves (EDCs) for copper, with 21.2eV photon energy. The upper curve was obtained with a deflecting electrostatic analyzer and the lower one with a retarding field analyzer.

the horizontal axis is shown on a reduced energy scale, with the Fermi energy chosen as zero. The upper curve was obtained with a deflecting field analyzer and the lower one with a retarding field analyzer. It is apparent that the deflecting type analyzer gives better resolution at high kinetic energies; more detail is resolved in the d-band. However, the position and width of the d-band is found to be the same for both analyzers. It starts at -2.0eV and extends to -5.0eV;

i.e. approximately 3eV wide in agreement with the results of Eastman and Cashion (1970). The background observed in the lower curve at the bottom of the d-band is less than that reported in previous measurements with hν=21.2eV; this should be noted. The broad peak at -15eV is due to the modulation of the distribution of inelastically scattered electrons with the escape function, as discussed above. Small corrections (∿5%) have been made for contributions from lower-energy lines in the lamp spectrum, contributing to the hump at -15eV. For this purpose, aluminium and lithiumfluoride filters were used. We also checked that the neon line at 16.8eV did not contribute. Two peaks below the d-band, at respectively -7.8 and -9.6eV, are also observed. These disappeared quickly with slight contamination of the copper.

Curve A in figure 4 is again an EDC for clean copper. Curve

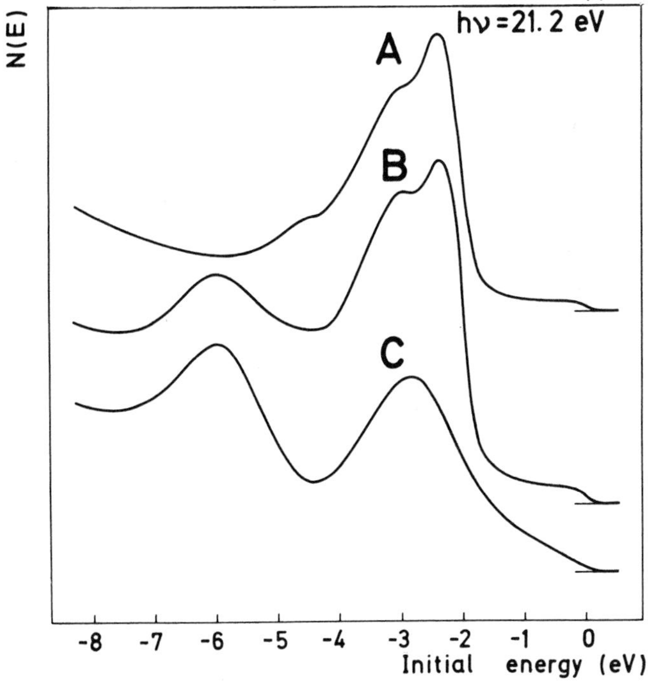

Figure 4 EDCs at 21.2eV for (A) clean, (B) partially oxidized, and (C) fully oxidized copper.

B was obtained after exposure to oxygen at a partial pressure of $5x10^{-9}$torr for 30 seconds. A hump is now found 6eV below the Fermi level, and the structure in the d-band is less defined. Curve C shows a film which has been exposed to a partial oxygen pressure of $1x10^{-6}$torr for 30 seconds. The peak at -6eV is now as dominant as the smeared-out d band, and no sharp Fermi level is observable. The spectrum is almost completely representative of Cu_2O.

4. CALCULATIONS AND DISCUSSIONS

Photoemission from copper has been studied previously by several investigators (see, for example, Berglund and Spicer 1964, Krolikowski and Spicer 1969, Smith and Traum 1970). Most measurements were with photon energies of <11.6eV. For higher photon energies there are data available from the work of Vehse and Arakawa (1968), Krolikowski and Spicer (1969) and Eastman and Cashion (1970). However, these investigations were not performed with ultra-high vacuum but at pressures in the region 10^{-8}torr. Eastman and Cashion (1970) attempted to avoid contamination effects by taking measurements during continuous evaporation of the copper; their data may be regarded as representative of clean copper. However, no results have been reported for the low-energy part of a high-photon EDC for clean copper.

We shall not discuss the unscattered part of the spectrum in detail. We only remark that this distribution qualitatively fits calculations of the photoemission joint density of states (Williams 1971).

To calculate the scattered distribution, the formalism of equations 3-5 was used. The density of states was taken from Williams *et al* (1971); figure 5. The primary distribution was calculated from non-direct formalism, i.e. as a convolution of initial and final states. As noted above this is not

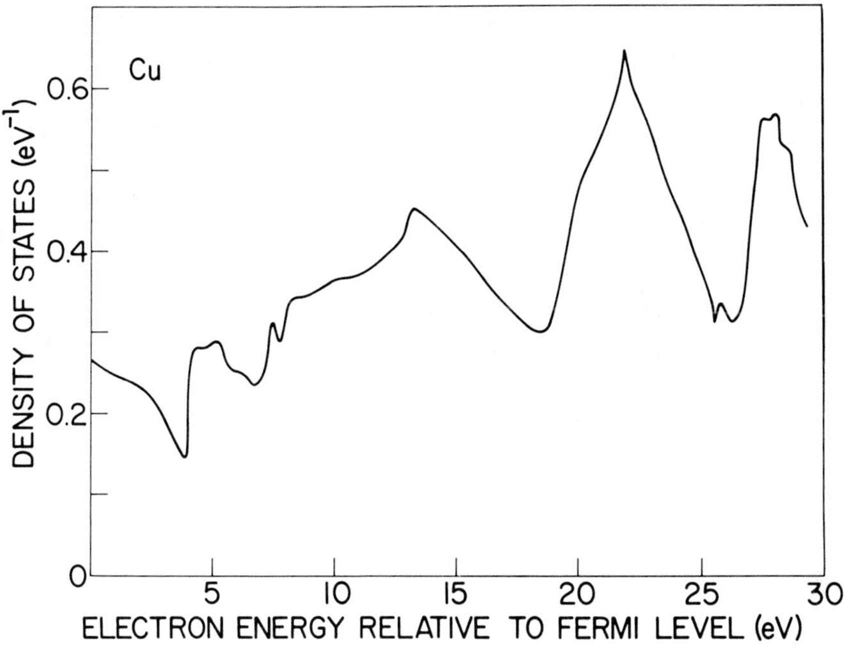

Figure 5 The density of states for copper above the vacuum level (Williams, Janak and Moruzzi, 1972).

strictly correct; however, the scattered part is not sensitive to small details in the primary part. The scattering length was normalized to 22Å at 8.6eV above the Fermi level (Krolikowski and Spicer 1969), and α was set to $12 \times 10^{-5} cm^{-1}$ for $h\nu = 21.2$ eV (Ehrenreich and Philipp 1962). The ratio of scattered to unscattered electrons was almost insensitive to this choice. The measured and calculated primary distributions were normalized to the same area.

The calculated yield of secondaries was found to be only 40% of the measured yield. A large additional contribution could come from electrons multiply scattered. For calculations of this effect a Monte Carlo programme was designed. The main principles are as follows.

A random depth is chosen for the optical excitation relative to the surface, $z = \ln u/\alpha$, with u arbitrary between zero and one. All possible excited states E_c are then traversed, with a given energy mesh dE, and for a given state E_c the optical transition strength is calculated. A random transport length is generated according to $x=-\ell(E)\ln u$, where $\ell(E)$ is calculated using equations (3) and (4) and the free-electron value $<v_g>$.

A random angle for the direction of the velocity of the electron is taken as $\Theta=2\arcsin\sqrt{u}$. If $z+x\cos\Theta$ falls inside the crystal, the electron scatters to a random energy $E_{c'}=E_F+(E_c-E_F)u$, with a normalized probability $P_N(E_c,E_{c'})$ calculated from:

$$P_N(E_c,E_{c'}) = \frac{2\ (E_c - E_F)}{dE} \frac{P(E_c,E_{c'})}{W(E_c)} \tag{9}$$

E_F is the Fermi energy. However, if $z+x\cos\Theta$ falls outside the crystal the electron escapes, provided it comes within the classical escape cone given by $\Theta_c = \arccos\sqrt{[(E_F+\phi+E_B)/(E_{c'}-E_B)]}$. ϕ is the workfunction and E_B an 'inner potential'. In the case $\Theta>\Theta_c$, the electron is assumed to reflect diffusely at the surface.

The results of the Monte Carlo calculation are shown in figure 6, which shows the total distribution obtained and also the contributions from electrons scattered respectively once, twice and three times. We obtain the result that 73% of the emitted electrons are scattered. Experimentally the result is approximately 77%. Out of the scattered electrons 61% are scattered once, 32% twice and 6% three times. We thus conclude that the analytical expression in equation (5) derived for once-scattered electrons cannot be applied at such high photon energies as are used here.

It should be noted that no essential parameters have been chosen in the calculations. If for example the scattering length is taken to be half the value used, then the same results for the relative amount of scattered electrons are obtained, although this changes the absolute yield to approximately half its value. Further, we should note that the matrix element M in equation (1) is not any longer present in equation (6) and thus does not influence the results obtained.

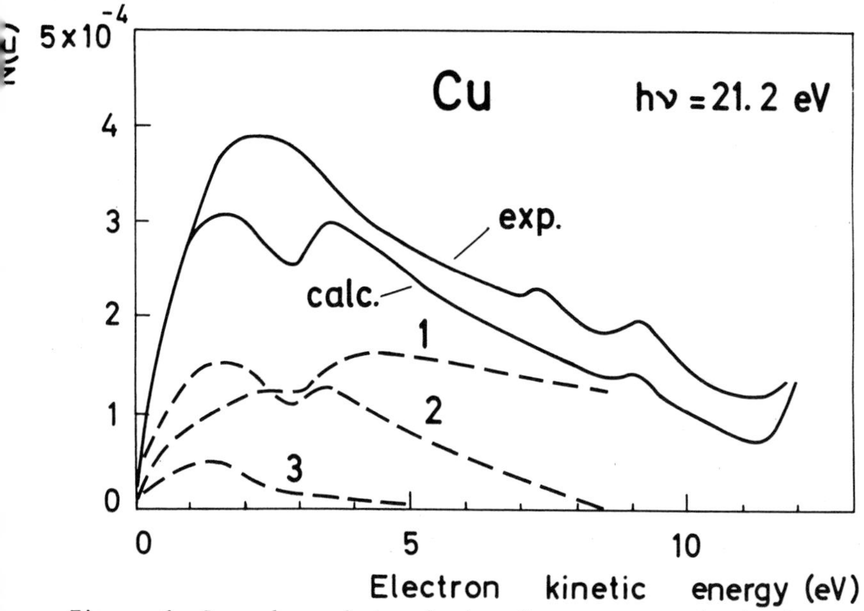

Figure 6 Secondary photoemission from copper; obtained experimentally and by Monte Carlo calculation. The calculated once-scattered, twice-scattered, and thrice-scattered electron distributions are indicated by curves 1, 2 and 3 respectively.

As a test of the Monte Carlo programme the scattering at the surface was neglected, in one run, whereby the result for once-scattered electrons should agree with the analytical result. The portion of scattered electrons was 61.2% in both cases. If the surface scattering was included in the Monte Carlo programme we obtained 62.1%. The effect of the surface is thus negligible, in contradiction to the findings by Krolikowski (1967).

Although the presented results may be changed by more refined calculations the present work shows that: (1) twice and even three-times scattered electrons contribute considerably to the EDC of copper at 21.2eV photon energy and (2) that electrons scattered at the surface contribute negligibly to the escape distribution.

The calculated dip at low kinetic energies, reflecting the conduction-band density of states, is not observed experimentally. A definite reason for this cannot be given at present, but some suggestions can be made. Firstly, the density-of-states calculation may not in principle be accurate enough. Secondly, density-of-states effects in the final state of the photoemission process will tend not to show up strongly in observed spectra. This occurs because a peak in the density of states is often associated with a small group velocity of the electron and thus a small photocurrent. Thirdly, if the expected feature is weak it may be difficult to resolve it due to the subtraction made in this energy region from the contribution of low photon energies. Finally, as many as three scattering events may not be well described in detail by step processes.

There is one calculated peak at about 9eV kinetic energy. This is very close to an observed peak, and there it is possible that we are observing the calculated peak, which originates from a peak in the conduction band density of states. The reason why this should be observed, but not the low-energy one discussed above, may be due to the fact that it mainly originates from once-scattered electrons.

For the 16.8eV distribution curve the main part of the spectrum was again reproduced by Monte Carlo calculations. However, the two small peaks were not observed. The reasons for this may be several. Firstly, the effect will be diminished because the 16.8eV resonance emission consists of two lines 0.2eV apart. Secondly, slight contamination could wash out any trace of the feature, as we found at 21.2eV. Further experiments are now in progress.

This last effect also suggests an explanation of the -7.8eV peak as a surface phenomenon of some kind. Surface plasmons have shown up recently in photoemission experiments in terms of characteristic energy losses (Smith and Spicer 1969). The so-called loss function can be obtained from optical data (Ehrenreich and Philipp 1962). A convolution of the observed EDC with the surface loss function gives in fact a peak around -10eV, as shown in figure 7. The calculated width is much

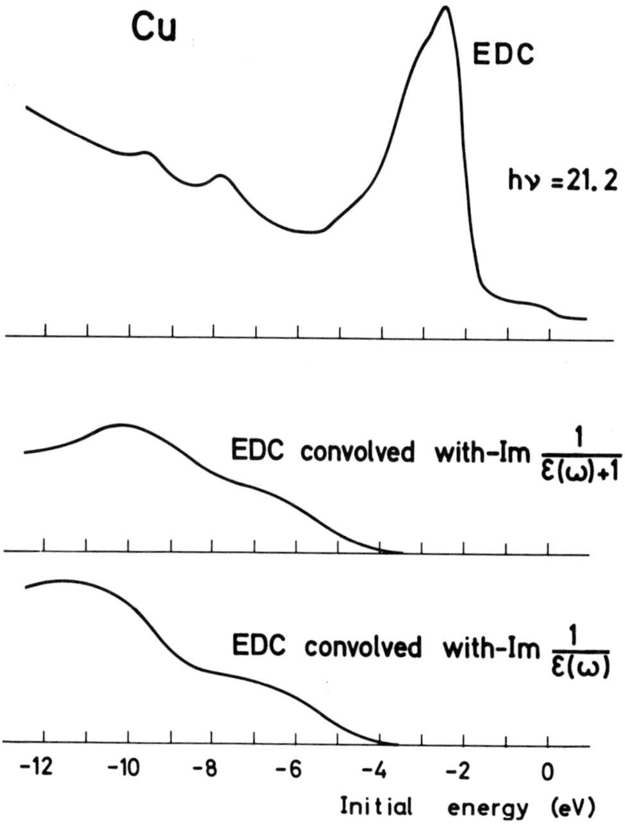

Figure 7 The unscattered part of the distribution curve convoluted with the loss functions $-Im\,\frac{1}{\varepsilon(\omega)}$ and $-Im\,\frac{1}{\varepsilon(\omega)+1}$ compared with the measured distribution.

larger than observed. However the simplified model described cannot be expected to give detailed information about the shape.

Forstmann and Pendry (1970) and Forstmann and Heine (1970) have discussed the possibility of localized electronic surface states in metals. These states would give rise to anomalous structure in the photoemission spectra from a (100) face of copper 5-6eV below the Fermi energy. No structure is observed in our experiment at this energy. The position of the peak at -7.8eV was the same for both polycrystalline films and for single crystals of different orientations. However, earlier photoemission data (Berglund and Spicer 1964) contained a 2-3eV broad peak centred about 6eV below the Fermi level. This additional structure could not be explained on the basis of band structure calculations. Suggestions were made pointing to many-body resonances (Philips 1965), two-electron processes (Cuthill *et al* 1966), d-band resonances (Berglund and Spicer 1964), and - as discussed above - surface states (Forstmann and Pendry 1970, Forstmann and Heine 1970). In later photoemission experiments this hump was not observed (Krowlikowski and Spicer 1969; Eastman and Cashion 1970; Eastman 1971). In our experiment the energy region around -6eV is well separated from the peak of scattered electrons and can thus be studied in detail. In figure 4 we show EDCs for clean and for oxidized polycrystalline copper surfaces. The experimental conditions were discussed above. Our experiments show that the hump at 6eV originates from the oxygen p-levels. As the oxygen partial pressure was relatively low during the oxidation we believe that curve C arises from Cu_2O (Seyboldt 1963). The results are in good agreement with predictions from the band structure of Cu_2O calculated by Dahl and Switendick (1966); they find a separation of the oxygen and copper levels of approximately 4eV compared with 3.2eV obtained in our experiment.

This observation of energy levels of adsorbed oxygen suggests that the peak at -7.8eV can originate from other adsorbed gases. However, this is unlikely in view of the experimental conditions described above. Further, the most probable gas would be hydrogen. We exposed the copper sample to 1 torr of hydrogen gas for 30 minutes. The scattered part of the spectrum was increased slightly but no hydrogen level appeared.

Discussion about electron scattering in the present report has mainly been confined to copper. The behaviour observed is however characteristic also of noble and transition metals. On changing the photon energy the scattered distribution generally changes smoothly according to the formalism described.

A different behaviour can be observed in other materials. By way of example we show in figure 8 EDCs for evaporated zinc

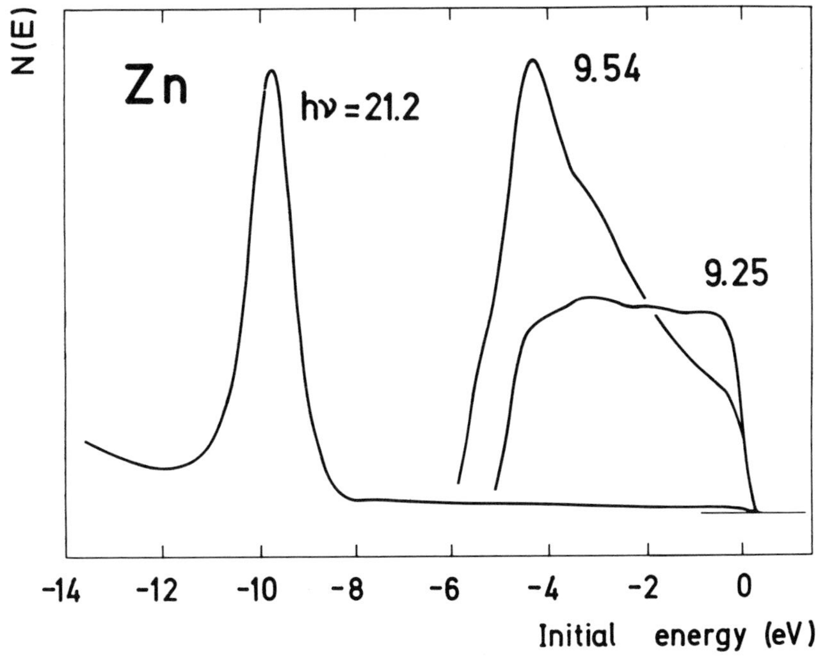

Figure 8 Photoemission energy distributions (not normalized) for zinc at different photon energies.

films (Nilsson and Lindau 1971, Nilsson and Eastman unpublished). We observe a very rapid increase in the fraction of scattered electrons when the photon energy is changed from 9.25 to 9.54 eV (i.e. only 0.3eV). A similar effect was observed by Mosteller *et al* (1969). Measurements of the yield showed a local increase in this energy region. The effect can be associated with excitations of surface plasmons (Endriz 1971). From optical data (Mosteller and Wooten 1968) we have deduced the surface plasmon energy to be approximately 9.5eV. At energies below this extra optical absorption is attributed to creation of surface plasmons by photons; while above the surface plasmon energy this process is not possible, which leads to a lower effective absorption. The electrons are then excited from deeper within the material and the relative amount of scattered electrons is higher. For a perfectly smooth film no surface plasmons can be excited and the effect should not show up. A similar effect would however occur at the bulk plasma frequency. There is no excitation of plasmons, but the optical absorption has a general decrease at that energy.

With 21.2eV photons a strong peak is observed in the EDC at -9.5eV initial energy. This originates from the zinc 3d-band. No effect of scattering is observed above this peak. This is because there are few primary electrons and the secondaries are distributed over as much as 16eV. However, the d-band peak is seen to contribute secondary electrons.

Very strong characteristic energy losses have recently been observed in EDCs for lithium and aluminium (Lindau *et al* 1971). No such contribution is observed in the 21.2eV spectrum of zinc However, a small peak may be hidden inside the d-band peak. The difference in behaviour can be attributed to non-free-electron behaviour of zinc (Juras *et al* 1971).

5. CONCLUSIONS

We have measured the photoemission distribution of clean copper using 21.2eV photon energy. The overall shape and relative magnitude of the scattered part of the spectrum can be reproduced by Monte Carlo calculations within the 'random-k' approximation. 60% of the scattered electrons were found to be scattered once. The scattering at the surface had negligible influence. A peak at 7.8eV below the Fermi energy was attributed to a feature in the conduction band density of states. A peak at 9.6eV was associated with a surface plasmon loss. No structure was found to be associated with predicted surface states. A previously observed hump at 6eV was shown to appear with oxidation of the copper surface. Rapid changes in the amount of inelastic scattering in zinc with photon energy is attributed to excitation of surface plasmons. Very low scattering was observed in zinc at high kinetic energies in the 21.2eV spectrum.

Acknowledgments

We wish to thank Professor G. Brogren for his support during this work. We are indebted to G. Brodén for many valuable discussions; and to the Swedish Natural Science Research Council for financial support.

REFERENCES

Berglund, C.N. and Spicer, W.E. (1964); Phys. Rev. 136, A1030 and A1044.

Cuthill, J.R., McAlister, A.J. and Williams, M.L. (1966); Phys. Rev. Letters 16, 230.

Dahl, J.P. and Switendick, A.C. (1966); J. Phys. Chem. Solids 27, 931.

Eastman, D.E. and Cashion, J.K. (1970); Phys. Rev. Letters 24, 310.

Eastman, D.E. (1971); Phys. Rev. B3, 1769.

Ehrenreich, H. and Philipp, H.R. (1962); Phys. Rev. 128, 1622.

Endriz, J.G. and Spicer, W.E. (1971); Phys, Rev. Letters 27, 570.

Ertl, G. (1967); Surface Sci. 6, 208.

Forstmann, F. and Heine, V. (1970); Phys. Rev. Letters 24, 1419.

Forstmann, F. and Pendry, J.B. (1970); Z. Physik 235, 75.

Juras, G.E., Segall, B. and Sommers, C.B. (1971); Solid State Commun. 10, 727.

Kane, E.O. (1967); Phys. Rev. 159, 624.

Krolikowski, W.F. (1967); Ph.D. Thesis, Stanford Electronic Laboratories, USA.

Krolikowski, W.F. and Spicer, W.E. (1969); Phys. Rev. 185, 882.

Lindau, I., Löfgren, H. and Walldén, L. (1971); Phys. Letters, 36A, 293.

Mahan, G.D. (1970); Phys. Rev. B2, 4334.

Marklund, I., Andersson, S. and Martinsson, J. (1968); Arkiv för Fysik 37, 127.

Mosteller, L.P. and Wooten, F. (1968); Phys. Rev. 171, 743.

Mosteller, L.P., Huen, T. and Wooten, F. (1969); Phys. Rev. 184, 364.

Nilsson, P.O. and Lindau, I. (1971); J. Phys. F: Metal Phys. 1, 854.

Philips, J.C. (1965); Phys. Rev. 146, A1254.

Schaich, W.L. and Ashcroft, N.W. (1970); Solid State Commun. 8, 1959.

Schaich, W.L. and Ashcroft, N.W. (1971); Phys. Rev. B3, 2452.

Seyboldt, A.U. (1963); Adv. in Physics 12, 1.

Simmons, G.W., Mitchell, D.F. and Lawtess, K.R. (1967); Surface Sci. 8, 130.

Smith, N.V. and Spicer, W.E. (1969); Phys. Rev. 188, 593.

Smith, N.V. and Traum, M.M. (1970); Phys. Rev. Letters 25, 1017.

Vehse, R.E. and Arakawa, E.T. (1968); Thesis, Oak Ridge National Laboratory, ORNL-TM-2240.

Williams, A.R. (1971); Bull. Am. Phys. Soc. 16, 414.

Williams, A.R., Janak, J.F. and Moruzzi, V.L. (1972); Phys. Rev. Letters 28, 671; and private communication.

X-RAY PHOTOELECTRON SPECTROSCOPY IN THE STUDY OF THE BAND STRUCTURE OF METALS

S.B.M. Hagström

Department of Physics, Linköping University, Sweden.

1. INTRODUCTION

The advance of photoelectron spectroscopy during the last decade, in the study of electronic structure and properties of matter is due to the progress in both experimental techniques and in the relevant theory. Although the method is used in a number of applications, there are still many unsolved problems connected with understanding the basic mechanism of the photoelectric process. It appears therefore that photoelectron spectroscopy will continue as an active research field, both in the finding of new applications and in the development of more fundamental knowledge.

For mostly experimental reasons the field has developed around two ranges of photon energies, namely the UV and x-ray ranges. Chiefly, studies using x-ray photoelectron spectroscopy (XPS) have involved the core electron levels, while the direct study of valence electrons has been made using ultraviolet photoelectron spectroscopy (UPS). However, XPS data contain also information on the structure of the valence electrons (Hagström 1972). In many cases it has been found that XPS and UPS results provide valuable complementary information. The present review will emphasize the kind of information to be obtained from XPS on the electron energy-band structure of solids, and in particular of metals. In §2 some of the principles and experimental procedures are reviewed, particularly pertinent to XPS studies of metals. In §3 we cover studies of some d-transition metals; and §4 describes one area in which the XPS method has provided unique information on band structure, the rare-earth metals with their 4f electron levels. A more general discussion of the method is given in §5.

2. PRINCIPLES AND EXPERIMENTAL PROCEDURES

The electrons emitted from a sample by a flux of photons of known energy can be studied with respect to the following three parameters: the *kinetic energy distribution*, the *angular distribution*, and the *spin distribution*. Most XPS studies have been made by measuring the kinetic energy distribution only, and in this review we shall emphasize this parameter.

The fundamental equation for the excitation of one electron with a photon of known energy $h\nu$, can be written

$$h\nu = E^f - E^i + E_{kin} \tag{1}$$

where E^f and E^i are the energies of respectively the final and initial states, and E_{kin} is the kinetic energy of the unscattered electron.

The excitation in XPS is usually by radiation obtained from a conventional x-ray tube, and an x-ray line in the radiation spectrum is used as a 'monochromatic' photon source. The most commonly used lines are the Mg $K\alpha_{1,2}$ at 1.25KeV and Al $K\alpha_{1,2}$ at 1.49KeV. The natural widths of these lines are between 0.7 and 0.8 eV, which contributes considerably to the linewidth of the photoelectron spectrum. Furthermore the photoelectron spectrum contains contributions other than the additional lines such as the $K\alpha_{3,4}$ satellites, and from bremsstrahlung.

Using a quartz crystal x-ray monochromator it has been possible to obtain photoelectron line widths of about 0.3-0.4eV, and spectra that are essentially free from the influence of satellites and bremsstrahlung. This can also be achieved, if the incident x-ray spectrum is well known, by applying a Fourier deconvolution.

The final-state energy E^f is defined as the total energy of the photoionized system minus the kinetic energy of the photoelectron. The simplest case to consider is the one in which the system consists of the unperturbed state with just one electron missing. However, experiments indicate that the final state is usually more complicated. Relaxation effects tend to decrease E^f and to increase the electron kinetic energy E_{kin}. An accurate estimation of the size of this effect has not yet been made. The hole state may also couple with localized valence electrons, giving rise to complicated multiplet effects in the band spectra - as we discuss in the results on the rare earth metals. On the other hand, the initial state energy E^i is usually a well defined ground state energy of the system.

In the simplest case for the final state, cited above, the binding energy of an electron is a well defined quantity. If the binding energy is referred to the Fermi level as zero, $(BE)_F$, the energy conservation equation for a metallic sample can be written

$$h\nu = (BE)_F + E'_{kin} + \phi_{sample} \qquad (2)$$

in which ϕ_{sample} is the workfunction of the sample. However, the kinetic energy E_{kin} is usually referred to the vacuum level of the spectrometer, and it is therefore the workfunction of the spectrometer material ϕ_{sp} that enters

$$h\nu = (BE)_F + E_{kin} + \phi_{sp} \tag{3}$$

Here E_{kin} is the kinetic energy of the elastically scattered electrons as measured in the electron spectrometer, and usually this is operated in a manner such that the instrumental contribution to the resolution is $\sim$0.5-1.0eV. Most spectra for solids are therefore recorded with an inherent resolution of $\sim$1.5eV, which is considerably larger than the case with UPS where the resolution is typically 0.2-0.3eV.

The kinetic energy distribution is modified by inelastic scattering of the photoelectrons before they emerge from the sample. This gives rise to structure in the electron spectrum to the low-energy side of a peak, and is an effect of particular importance when studying band spectra containing genuine features that arise from the electronic structure of the sample. In the case of XPS the two effects can be distinguished by studying also a sharp core level, for which any structure can be due only to the inelastic scattering mechanism. This provides a correction procedure which has been applied in the study of band structure for a series of transition metals (Fadley and Shirley 1970). Figure 1 shows the correction applied to the measured valence band of copper.

Inelastic scattering can occur both in the bulk material and at surface impurities. The surface atoms may present an additional chemical composition, due for example, to surface oxidation. It is therefore essential that the sample surface be kept clean, and with metals it is usually necessary to prepare the sample surface *in situ*. The importance of cleaning, with even a noble-metal such as gold, is demonstrated in figure 2 (Wilson, unpublished) which shows the band structure obtained for a gold sample cleaned by sputtering. The photoelectron spectrum shown here was recorded using direct Mg Kα

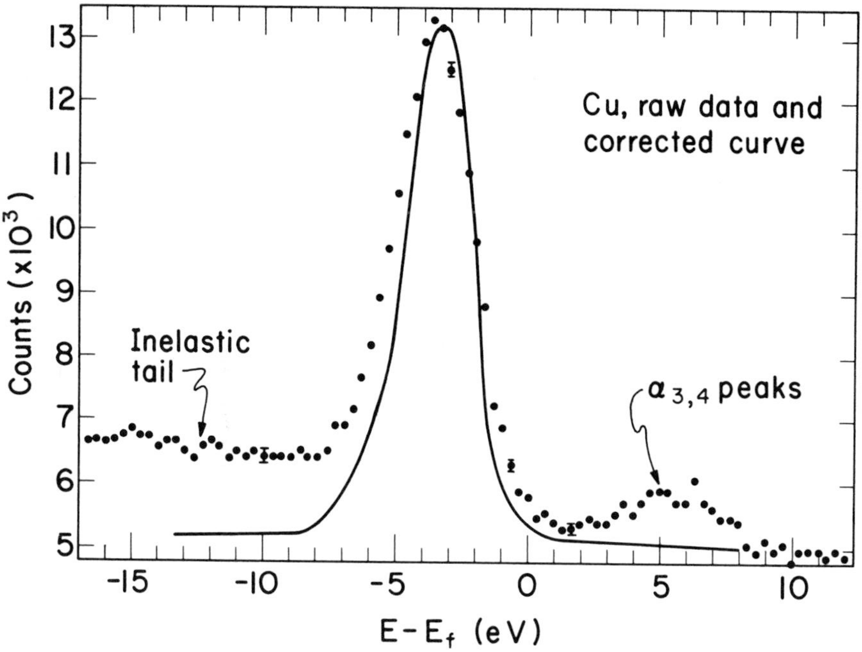

Figure 1 The valence band of copper in an XPS spectrum obtained with Mg Kα excitation. The full line represents the spectrum after a correction has been applied to the primary data (dotted line) for the effects of inelastic scattering and Mg $K\alpha_{3,4}$ satellite x-radiation. (Fadley and Shirley 1970).

excitation from an x-ray tube; for comparison the corresponding spectrum obtained from a monochromatic x-ray source is shown in figure 3 (Melera, unpublished).

The escape depth from the sample of elastically scattered electrons varies with energy and with material. There are relatively few experimental data available, and further work in this field is to be encouraged. For gold, experimental evidence (Baer *et al* 1970) indicates an escape depth of ∿20Å for XPS measurements, while with carbon the corresponding figure is reported to 70-80Å (Steinhardt *et al* 1972).

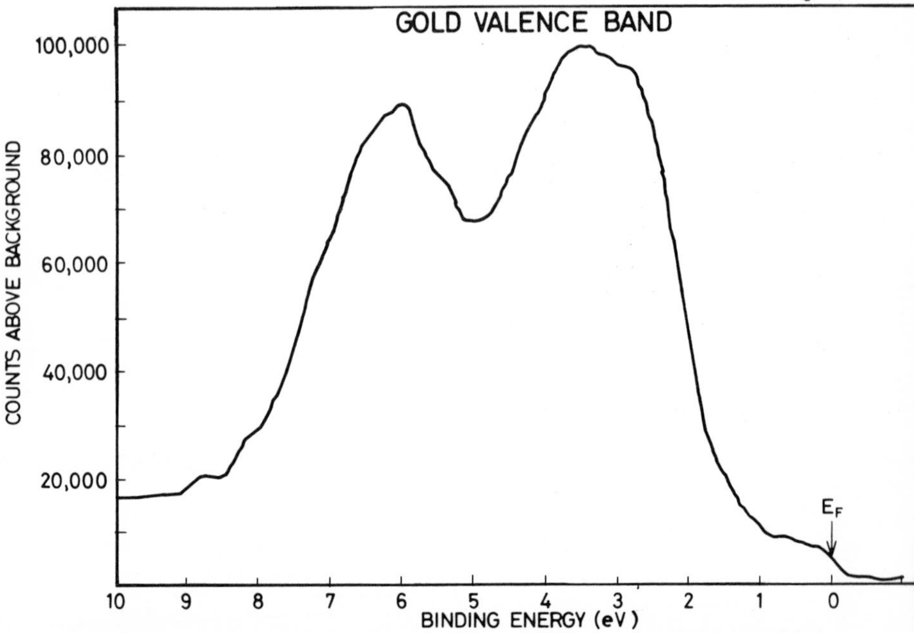

Figure 2 The effect of sample surface preparation demonstrated for a gold sample. The XPS spectrum of the gold valence band shows considerably more detail after cleaning by sputtering *in situ* (Wilson unpublished); c.f. figure 5 which shows the same spectrum from a sample exposed to air.

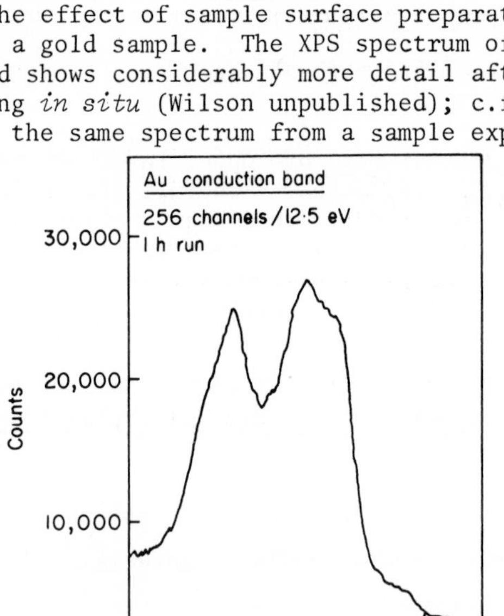

Figure 3 The XPS spectrum of the valence band of gold recorded using monochromatic x-radiation obtained from a quartz crystal monochromator (Malera, unpublished).

Sample charging, which for conducting samples such as metals and alloys presents no difficulty, can be a serious problem for non-conductors when one is interested for example in locating core levels accurately. With such samples the question of the reference level is also still unsolved.

3. TRANSITION METALS

The first metals to be studied for band structure using XPS were the transition metals. This was done in two independent investigations, both using unfiltered radiation and only moderate vacuum conditions. The sample surface were kept clean by employing elevated sample temperatures.

The metals studied (Baer *et al* 1970a) were the 3d-elements iron, cobalt, nickel and copper. In the 4d series the elements rutherium, rhodium, palladium and silver were investigated; and the metals osmium, iridium, platinum and gold from the 5d series. In the investigation reported by Fadley and Shirley (1970), the elements zinc, cadmium and mercury were also studied in their chemical compounds ZnS, $CdCl_2$, and HgO.

Figure 4 provides a summary of the results for the elements studied by Fadley and Shirley. The pronounced feature in the spectra arises from the d-bands. All the spectra have been corrected for inelastic-scattering by recording a sharp core-level, as already described.

The inelastic scattering is particularly pronounced for the elements with unfilled d-states, while the noble metal configurations give rise to small electron-electron scattering possibilities. This is illustrated in figure 5 (Baer *et al* 1970a) which shows the uncorrected XPS spectrum from the gold 5d-band. The inset figure is the spectrum of the Au core levels $4f_{7/2}$ and $4f_{5/2}$, and indicates the small contribution of inelastically scattered electrons to the spectrum.

Compared with UPS spectra, the XPS results appear to be

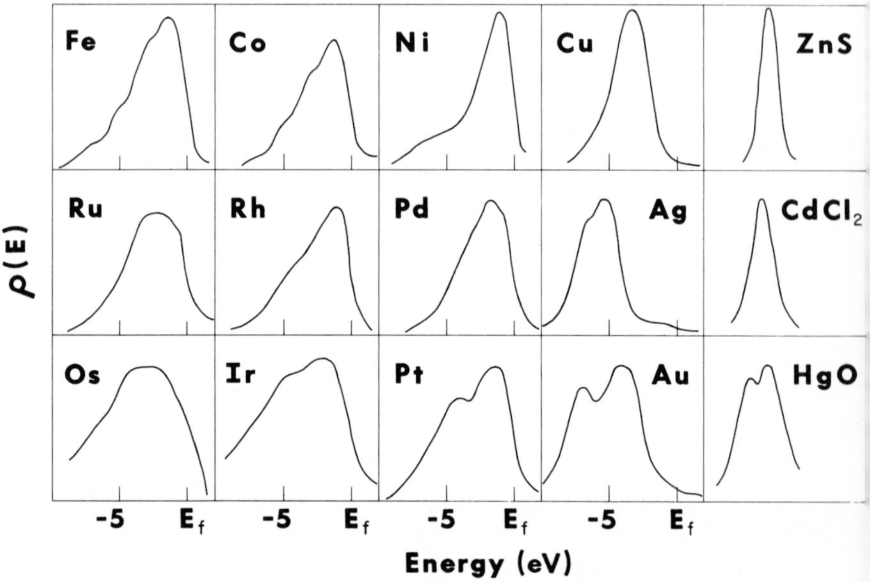

Figure 4 Summary of the XPS spectra on the d-bands of the transition elements studied by Fadley and Shirley (1970). Because of the low resolution the spectra appear quite structureless, compared with UPS spectra. The overall bandwidths obtained using the two methods are in reasonable agreement.

relatively structureless. The spectra in these investigations were recorded with instruments giving an inherent broadening of ∿1.5eV. Minor structure in the bands will therefore be smeared-out. However, some of the features observed in the UPS spectra of these metals are due to the strong influence of the final-state and crystal-momentum conservation restrictions on low-energy photoelectron spectra. This fact is illustrated by figure 6 which shows UPS spectra of the valence band of gold (Eastman 1972) recorded for different photon energies.

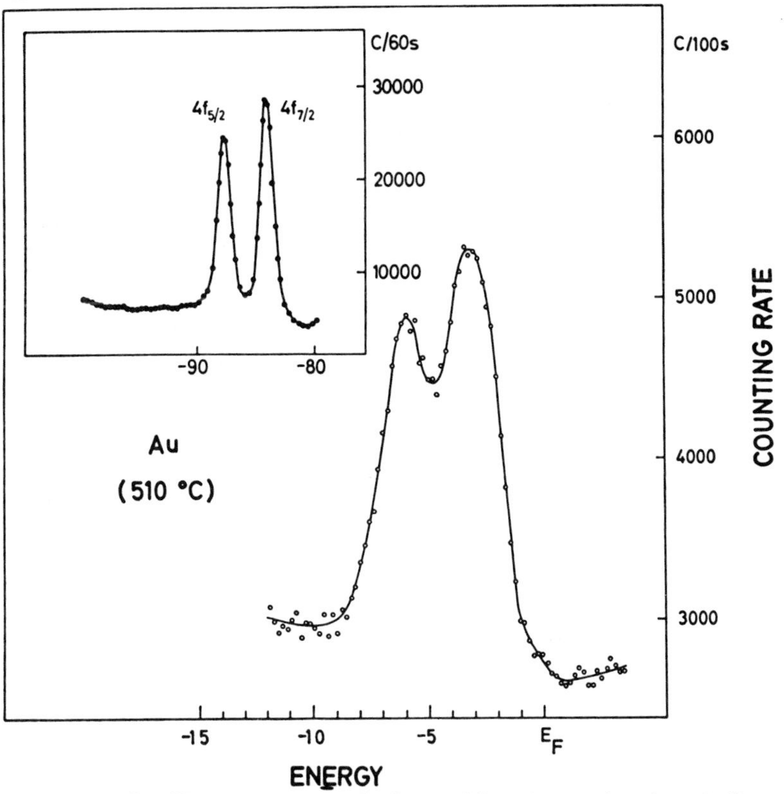

Figure 5 The XPS spectrum of the gold valence band and (inset) the core-level doublet $4f_{7/2}$ and $4f_{5/2}$ (Baer *et al* 1970a).

The point is further verified by comparison of the UPS spectra of crystalline and liquid samples of gold (figure 7; Eastman 1971).

When comparing UPS and XPS spectra the best agreement is obtained between high-energy UPS spectra and Mg Kα or Al Kα. For obtaining structural details in XPS spectra it is desirable to work with monochromatic x-rays but the difference between spectra obtained with filtered and un-filtered x-radiation is not dramatic, as seen in figure 2; the cleanliness of the sample surface rather seems to be the important factor.

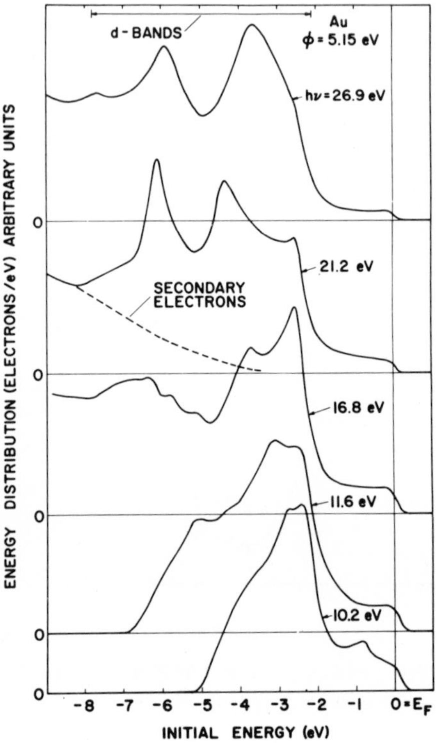

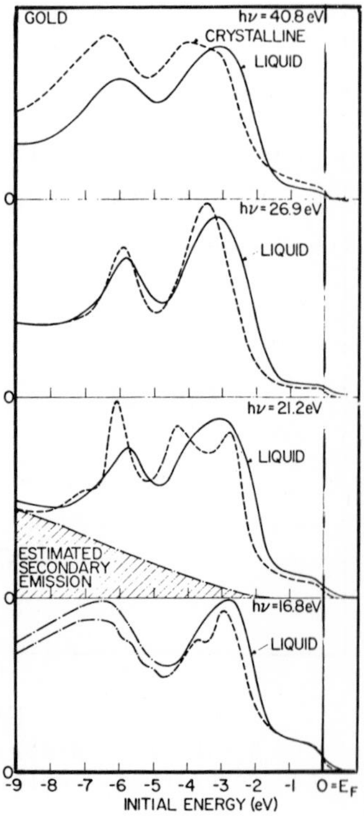

Figure 6 UPS spectra of the valence band of Au for different photon energies. The strong dependence of the observed structure on photon energy is attributed to the influence of the final state on the UPS spectra (Eastman 1971).

Figure 7 Comparison of UPS spectra of the valence band of Au for the crystalline state and the liquid state (Eastman 1971).

Good agreement is usually obtained on comparing the widths of the d-bands deduced from UPS and XPS measurements. Some ambiguities arise from the difficulty of separating-out the tail of inelastically scattered electrons.

4. RARE-EARTH METALS

The rare-earth metals form an interesting series of elements with many physical and chemical properties related to the existence of the partly filled 4f shell. The 4f electrons are energetically located in the valence band region, but due to their high principal quantum number they retain many atomic properties. The XPS technique has been applied to the study of the outer levels in the rare earths both in the form of metals (see Hagström 1972) and as chemical compounds (Wertheim *et al* 1972). The main purpose of these investigations was to study the 4f electrons. Previous attempts to study these with optical and UPS methods gave little evidence of their presence due to the small oscillator strengths of the corresponding transtions.

In the following the studies of the rare earths as metals will be discussed. Their normal electron configurations are $4f^n5d^16s^2$; but the tendency to retain filled or half-filled electron shells give rise to departures from this configuration in cerium, which has a tendency to lose its single 4f electron, and in europium and ytterbium, which form the configurations $4f^76s^2$ and $4f^{14}6s^2$ respectively in the solid state and thereby give rise to the same 4f-electron configurations as gadolinium and lutetium.

The XPS spectrometer employed in these investigations was of conventional design with an operating pressure in the range of 10^{-7} torr. Samples were prepared by evaporation of the metals *in situ*, which is necessary due to the high reactivity of these rare-earth metals. The XPS technique was first used as an analytical tool to monitor the presence of oxygen on the sample surface, by recording the oxygen photoelectron spectrum. The presence of an oxygen signal could usually be observed ∿5-10 minutes after evaporation; and a repeat evaporation was then made. Even with these precautions slight contamination of the sample surface with oxide cannot be excluded; this may be particularly true for the metals with high evaporation temperatures.

The XPS spectra of the metals with a half-filled (Eu, Gd) or a filled (Yb, Lu) 4f-shell have a simple and symmetrical structure; see figures 8 and 9 (Hedén, Löfgren and Hagström 1972). The dominant structure is due to the 4f electrons. The rest of the elements have a much more complex structure; see figure 10 (Hagström 1972).

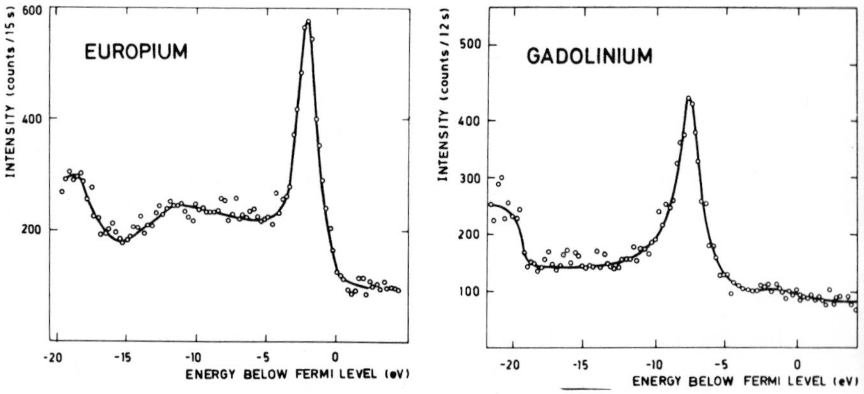

Figure 8 XPS spectra of the valence band region of Eu and Gd, both of which have a half-filled 4f electron shell (Hedén, Löfgren and Hagström 1972).

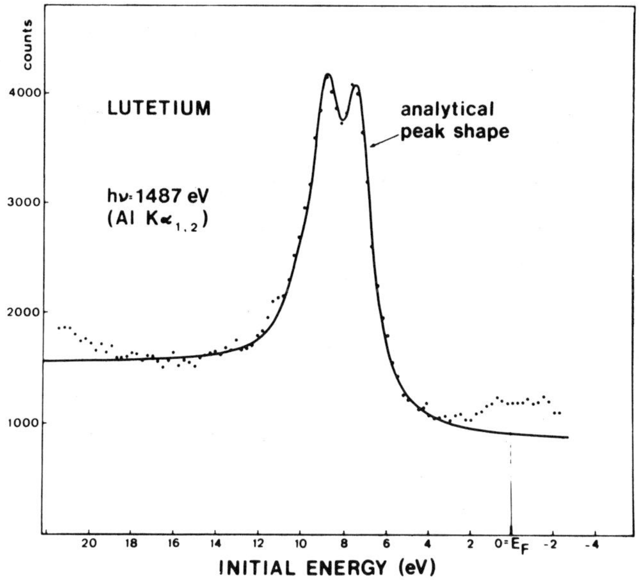

Figure 9 XPS spectrum of Lu, which has a filled 4f shell. The same form of spectrum is exhibited by Yb, but with the 4f electrons located closer to the Fermi level.

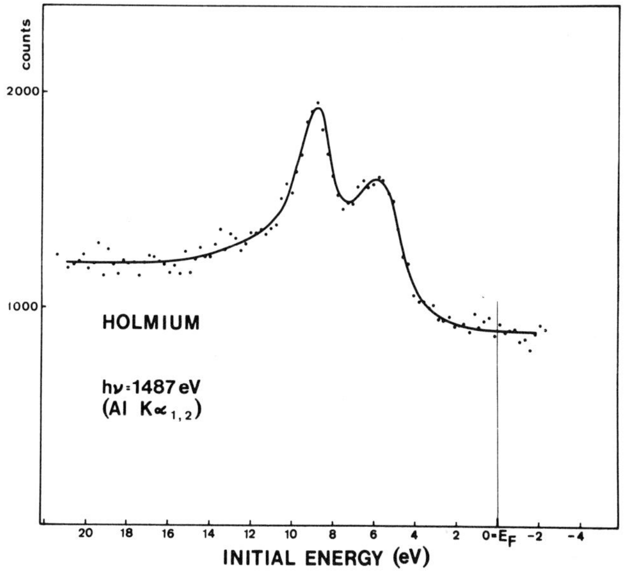

Figure 10 XPS spectrum of Ho. This form of spectrum is typical for those rare-earth metals that do not have a half-filled or a filled 4f shell (Hagström 1972).

The first feature to be noted is the intensity with which the 4f electrons show-up in the spectra. This is quite different from low-energy photon UPS measurements, in which the 4f electrons are barely detected. This fact is explained on the basis of the high principal quantum number of the 4f electrons, which causes the cross-section maximum to appear several hundred electron-volts above the threshold energy.

The spectra of europium and gadolinium are very similar, with a fairly narrow single line arising from the 4f electrons. Their energy is located quite close to the Fermi level in europium, due to the already noted fact that europium acquires a half-filled 4f shell through transfer of its 5d electron into the 4f shell. This location of the 4f electrons for europium is 2.1±0.4eV below the Fermi level, while the corresponding figure for Gd is 7.7±0.4eV.

Both ytterbium and lutetium exhibit similar spectra but with a fairly well resolved double peak, located at respectively 1.4eV and 2.7eV for ytterbium and at 7.2eV and 8.8eV for lutetium. These double peaks are due to spin-orbit splitting of the 4f electron levels.

The spectra of the metals with non-symmetrically filled 4f shells exhibit a much more complex structure originating from the 4f electrons and extending over an energy range of 10-15eV. As already noted some of this structure may be due to effects caused by a slight oxidation of the sample surface. However, it is obvious that the spectra from the two groups of elements have principally different spectra.

As indicated by equation (1) the photoelectron spectrum will reflect the energy of the final state of the photoionization process. This is quite obvious in the case of molecular photoelectron spectroscopy with UV excitation; however, in the case of XPS studies on solids it has often been assumed that the final state differs from the initial state by an inner shell hole only. This simplified model is in fact quite rarely accurate enough, and in order to explain details in XPS spectra one must consider the influence of the final state.

In the case of the rare earths it is known from spectroscopic data that the final state has a complex spectrum of multiplet configurations (Dieke 1968). In the case of the rare earth europium these energy levels are closely spaced within $\sim$0.6eV, which explains the narrow 4f peak observed in the XPS spectrum for this element. In the case of the rare earths with an incompletely filled or half-filled 4f shell these multiplet configurations extend over a wide energy range, which explains the complex structure in the XPS spectra of these elements. However, further experimental data obtained with better sample surface control are necessary to establish the origin of the details in the XPS spectra of the rare-earth metals.

5. CONCLUSION

The XPS technique is a powerful tool for determination of the density of states $\rho(E)$ of occupied valence electron levels in solids. The information obtained to-date has been limited in structural details due to insufficiencies in the experimental conditions, mainly in resolution. However, with the advent of highly monochromatic light sources, using variable photon energies, and with better means to control the sample surfaces the XPS method is likely to develop to the most universal method for studying the function $\rho(E)$.

The most interesting and promising feature of XPS in valence band studies is the possibility it provides for studying the influence of the valence electrons on the core levels. In this way the XPS method can be used directly as an analytical tool, to study the elementary composition of the sample surface, and to some degree also to determine the physical and chemical state of a surface contaminant.

The most productive studies in XPS have to-date been made with the purpose of correlating core level shifts with an effective charge on the respective atoms. However, relatively little has been done in this area on metals and metal alloys, and probably this aspect should be pursued more actively especially with the recent access of ultra-high vacuum instruments. The core levels may not only be shifted in energy, but multiplet splitting may also occur due to unpaired spins of the valence electrons. The measurement of these multiplet effects therefore gives information on the electron configuration of an atom in a solid. These two types of measurements on core levels - shifts and splitting - are thus complementary. The general conclusion may be drawn that XPS spectra contain a wealth of information about the valence electrons in both the valence band region of the spectrum and in the core-level region.

REFERENCES

Baer, Y., Hedén, P.F., Hedman, J., Klasson, M. and Nordling, C. (1970); Solid State Comm. 8, 1479.

Baer, Y., Hedén, P.F., Hedman, J., Klasson, M., Nordling, C. and Siegbahn, K. (1970); Phys. Scripta 1, 55.

Dieke, G.H. (1968); "Spectra and Energy Levels of the Rare Earth Ions in Crystals", Interscience, New York.

Eastman, D.E. (1972); in "Techniques of Metal Research" VI, ed. Passaglia, P.E., J. Wiley, New York.

Eastman, D.E. (1971); Phys. Rev. Lett. 26, 1108.

Fadley, C.S. and Shirley, D.A. (1971); in "Electronic Density of States", ed. Bennet, L.H., P.163; NBS Spec. publ. 323.

Hagström, S.B.M. (1972); in "Electron Spectroscopy", ed. Shirley, D.A., p.515; North-Holland, Amsterdam - London.

Hedén, P.-O., Löfgren, H. and Hagström, S.B.M. (1972); Phys. Stat. Sol. B49, 721.

Melera, A., Hewlett-Packard Co., Palo Alto., Calif., unpublished communication.

Steinhardt, R.G., Hudis, J. and Perlman, M.L. (1972); Phys. Rev. 5, 1016.

Wertheim, G.K., Cohen, R.L., Rosencwaig, A. and Guggenheim, H.J. (1972); in "Electron Spectroscopy", ed. D.A. Shirley, p.813; North Holland, Amsterdam - London.

Wilson, L.A., Varian, Palo Alto, Calif., unpublished communication.

X-RAY PHOTOELECTRON STUDIES OF ALUMINIUM AND ALUMINIUM-GOLD ALLOYS

D.J. Fabian, J. Fuggle and L.M. Watson

University of Strathclyde, Glasgow, Scotland.

and

A. Barrie and D. Latham

Vacuum Generators, East Grinstead, England.

1. INTRODUCTION

X-ray photoelectron spectroscopy, as we have noted already in this Volume, is fast becoming an important tool for investigating the electronic structure of metals. Recently it has been used for measuring the effect of alloying on valence-electron and core-level binding energies (Hedman *et al* 1971). The detailed electronic description of the bonding between the component atoms of an alloy is a complex problem, and one of considerable importance in the development of a theory of alloying. Because soft x-ray emission spectra probe electron densities 'local' to the emitting atom (see for example Fabian 1971, and Watson this Volume p.125), while x-ray photoelectron spectra examine valence-band electron densities averaged through a region of the emitting alloy, these two methods complement each other well. Their combined use can be expected to provide a new field of attack on the many outstanding problems in the electronic structure of alloys.

In this report we describe measurements of x-ray photoelectron spectra (XPS) of pure and of oxidised aluminium, and of a series of aluminium-gold alloys, using a high-vacuum instrument (a VG-ESCA3) in which the sample can be prepared and examined in vacuum of better than 5×10^{-10} torr. On deliberate oxidation, doublet peaks involving 'chemical shifts' are observed in the spectra, and clearly demonstrate the necessity for extremely good vacuum conditions in XPS measurements on solids when the data are used for studying electronic structure. This is particularly important with measurements on alloys where shifts observed in core-levels are to be interpreted in terms of bonding effects between the component atoms.

Previous XPS data for one gold alloy, the strongly coloured (purple) $AuAl_2$, have been reported by Chan and Shirley (1971). These investigators concluded that their results for the valence-electron emission from $AuAl_2$ agreed reasonably well with the band structure calculation by Switendick (1971), particularly with reference to the position of the gold 5d-band states in the valence band of the alloy.

2. EXPERIMENTAL

In all the spectra reported, the samples were irradiated with Al Kα-radiation operating the aluminium x-ray anode at 12kV and 20-40mA. The VG-ESCA3 instrument operates with a varying retarding voltage, and constant electron analyzer energy which, for the measurements reported here, was usually selected to be 50eV.

Samples can be either evaporated onto a stainless-steel holder, under high vacuum ($\sim 5 \times 10^{-10}$ torr), or - in the case of alloys - prepared beforehand and suitable shaped specimens clamped to the target mount. Solid specimens can be cleaned by argon-ion bombardment, and by mechanical scraping with a tungsten carbide tool under vacuum. A high-vacuum gate valve permits the sample preparation chamber to be isolated from the analyzer chamber until preparation and cleaning are complete;

a linear-drive system then allows the sample and mount to be introduced to the analyzer and x-ray beam.

3. RESULTS AND DISCUSSION

A. Pure Aluminium

In the case of aluminium metal, successive evaporations of the pure element achieved an oxide-free layer - determined by the absence of an oxide peak in the spectrum - as a result of each evaporated layer performing a 'gettering' action inside the sample preparation chamber, purging it of oxygen and other contaminants.

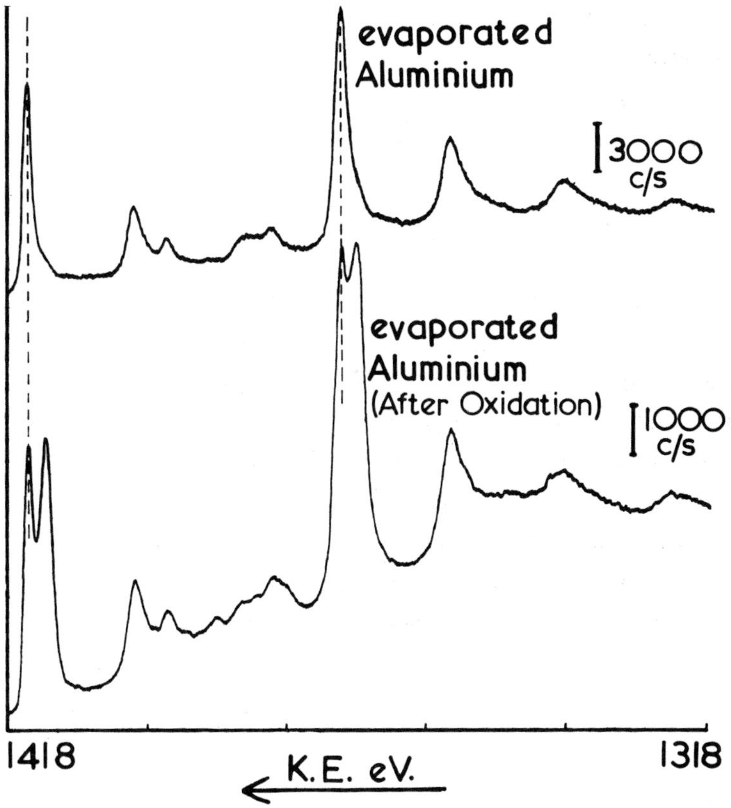

Figure 1 Al 2p and 2s photoelectron peaks for pure evaporated aluminium, and for the same sample oxidised by exposure to oxygen at ∿1 atmosphere for some hours.

Figure 1 illustrates the results achieved. Here the Al 2p and Al 2s core-level peaks are recorded for pure evaporated aluminium, purged of oxide after successive evaporations, compared directly with the spectra obtained for the same sample after deliberate oxidation brought about by exposure to pure dried oxygen at $\sim$1 atmosphere for some hours. The Al 2p peak ($2p^{1/2}$ and $2p^{3/2}$) and Al 2s peak, both become doublets on oxidation; the second peak is the result of electron emission from the same aluminium core-levels but from Al atoms in Al_2O_3. The distinct chemical shift observed for this peak results from the bonding of Al to O atoms in the oxide; transfer of electron charge from Al to O ions results in less electron screening at the Al-ion sites and stronger binding energies for Al atoms in the oxide. Binding energies are to be read in the opposite direction to kinetic energy for the photoemitted electrons.

In both cases a series of distinct and repeating aluminium 'plasmon' energy-loss peaks are observed to the lower kinetic-energy side of the aluminium core-level peaks. These are indicative of the enhanced electron-electron scattering effects recognised (Raether 1965) to be prominent with lighter metals.

Similar plasmon energy-loss bands are observed in the valence-band region of the spectrum. Figure 2 shows the XPS valence band measured for pure evaporated aluminium again compared with that obtained after oxidation. In the spectrum obtained after oxidation the valence band of aluminium oxide is observed superposed on the pure aluminium band, in very good agreement with the results obtained by soft x-ray spectroscopy using the Al K emission (Cauchois 1968). A large oxygen 2s peak appears on the first plasmon band

B. Aluminium-Gold Alloys

Samples of the three alloys Au_2Al, AuAl and $AuAl_2$ were prepared by melting weighed quantities of pure gold and pure

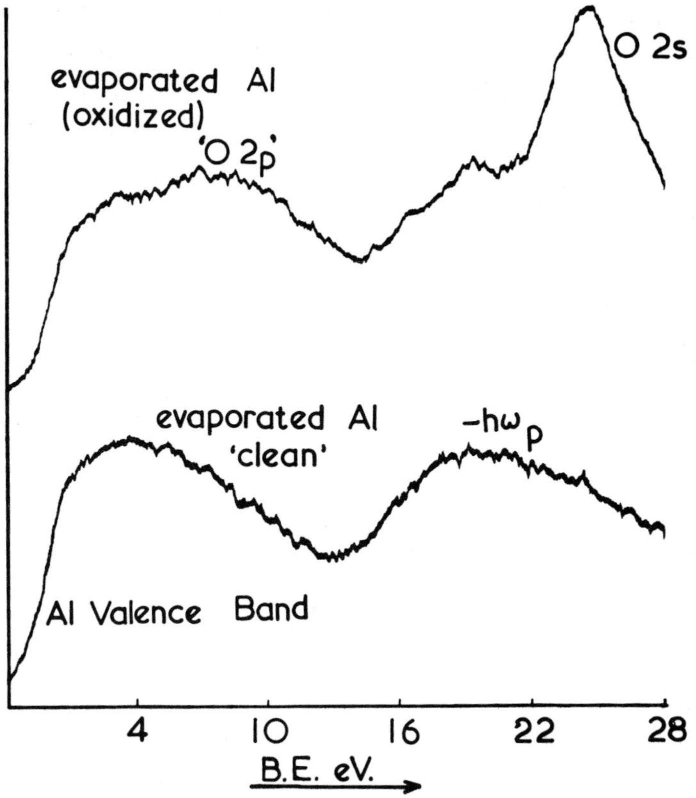

Figure 2 Measured photoelectron valence band for pure evaporated aluminium, and for the same sample after oxidation.

aluminium in a tungsten-arc furnace under argon atmosphere. The samples were subsequently annealed for several days in vacuum. Electron probe microanalysis indicated no trace of a second phase in the alloys $AuAl_2$ and Au_2Al; however, traces of a second phase were observable in the AuAl sample.

Suitable portions of the samples were in turn clamped to the XPS specimen holder and introduced, after scraping of the surface in a stream of dry nitrogen, to the sample preparation chamber. Further cleaning was performed by heating in vacuum ($\sim 5\times10^{-10}$ torr) and by argon-ion bombardment. In later experiments scraping of the specimen surface under vacuum was employed. Some oxide contamination of the alloys was observable from their photoelectron spectra – indicated by the presence of an oxygen 2s peak which was entirely absent in the spectra for pure aluminium and pure gold – and traces also of carbon (possibly carbide).

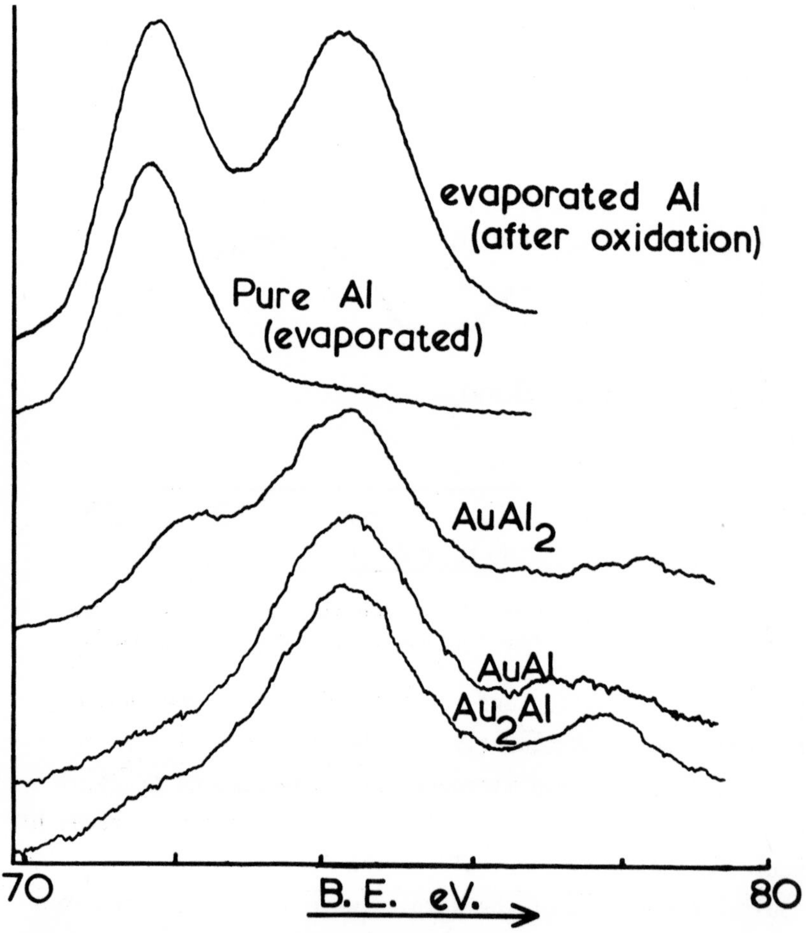

Figure 3 Photoelectron spectra recorded for gold-aluminium alloys after overnight baking in high vacuum followed by prolonged argon-ion bombardment and heating to 300°C (but before fresh surface was exposed by scraping sample under vacuum). Only with $AuAl_2$ is a 'genuine' Al 2p peak for the alloy detectable.

Figure 3 shows the Al 2p peak for the three gold-aluminium alloys, obtained in the early experiments before the technique of scraping a specimen surface under vacuum had been adopted. The spectra are compared with those for the pure aluminium and oxidized aluminium. It became clear for all three specimens that the surface, at least, was oxidized; the Al 2p peak appeared to be caused chiefly by aluminium oxide, since the peak coincided with the Al 2p for the deliberately oxidized pure aluminium. The same was observed for the Al 2s; only an Al 2s peak arising from the oxide was detectable for the alloys.

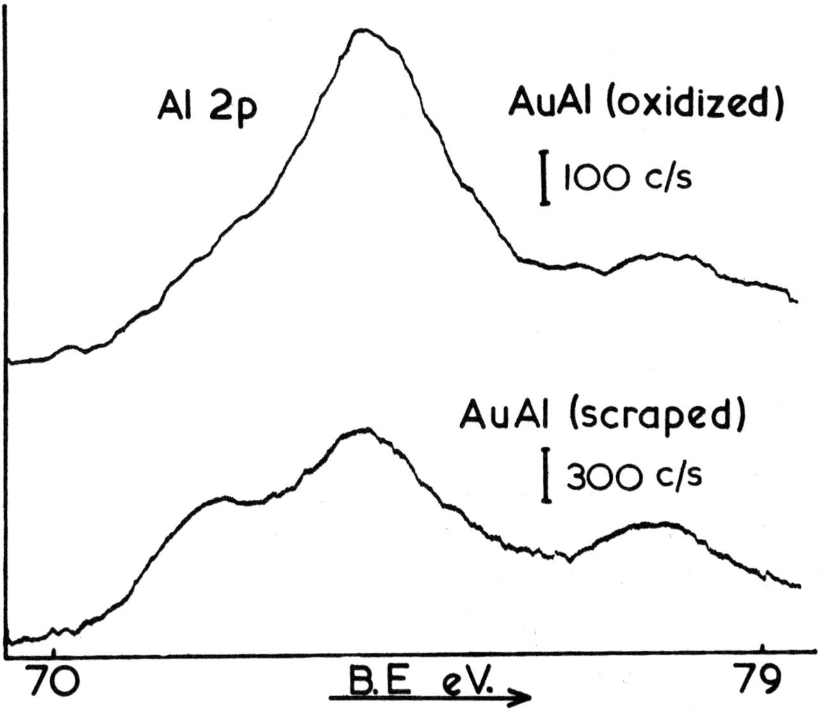

Figure 4 Photoelectron Al 2p peaks recorded for AuAl before and after scraping the specimen surface in high vacuum.

These results were found to persist even after overnight

baking, in high vacuum, followed in some instances by argon-ion bombardment and heating to 300°C. The question that naturally at first arose was whether the Al 2p and 2s peaks observed were in fact the 'genuine' Al peaks for the alloys. This would have indicated a large chemical shift - of the same order as that for the oxide - which is most unlikely. Therefore, in later experiments, after the system had been modified to mount a scraping tool within the sample preparation chamber, the samples were further investigated. Figures 4 and 5 indi-

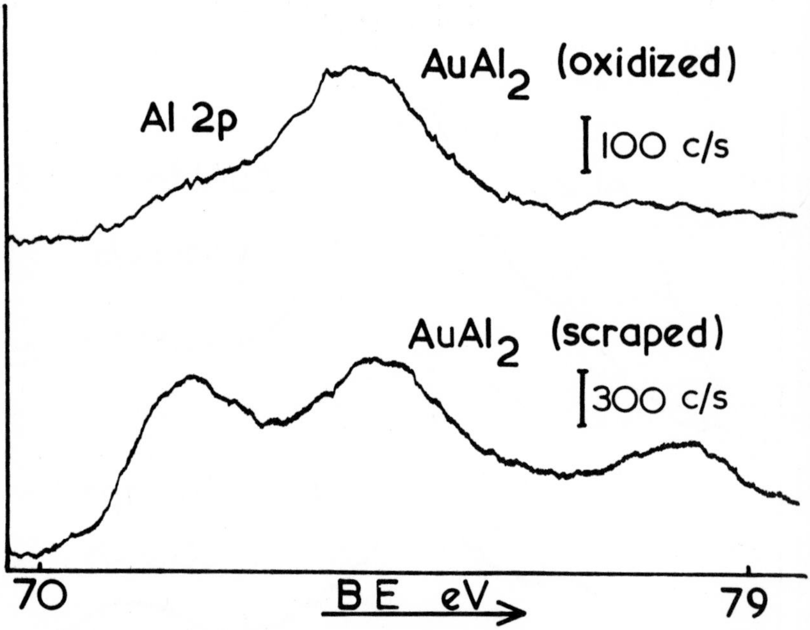

Figure 5 Photoelectron Al 2p peaks recorded for $AuAl_2$ before and after scraping the specimen surface in high vacuum.

cate the results then obtained for the Al 2p peaks in the spectra of AuAl and $AuAl_2$. The upper spectra in each case were recorded after overnight baking and argon-ion bombardment of the specimens; only the oxide Al 2p peak is observable. On scraping to expose a clean surface, at 5×10^{-10} torr, the overall photoelectron intensity is increased by two to three times, and the genuine Al 2p photoelectron peak for the alloy is also then observed alongside that for the aluminium oxide.

Figure 6 shows the results for the Al 2p peaks for all three alloys, after surface scraping, compared with those for oxidized aluminium. Energy alignment of the spectra necessarily assumes no intrinsic change in workfunction of the specimen with alloy composition, but the alignment here appears to be correct from the position of the oxide peak, which we can expect to be independent of the alloy and its composi-

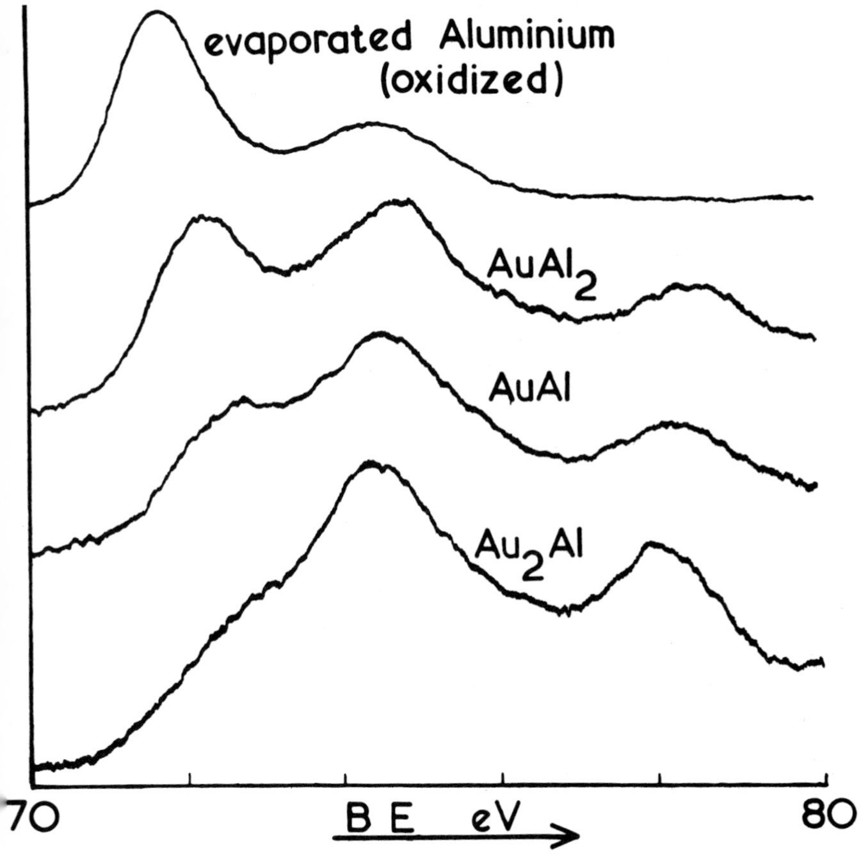

Figure 6 Photoelectron Al 2p peaks recorded for gold-aluminium alloys, after scraping in high vacuum, compared with spectrum for oxidized aluminium.

tion; although this comparison of the spectra is complicated by the gold N_6 and N_7 photoelectron satellite peaks, which arise from excitation by the Al $K\alpha_{3,4}$-emission satellites and overlap in energy with the oxide Al 2p peaks. An apparent chemical shift of the 'genuine' Al 2p peak is observed with increasing gold content, on going from pure aluminium through the alloys. It would be difficult to estimate the shift quantitatively, due to the uncertainty of alignment of the spectra arising from possible workfunction changes in the alloy (an entirely unknown factor); however, there is a distinct indication of the 2p peak shifting to increased binding energies, measured with respect to the Fermi level, as the gold content of the alloy is increased.

If the simple charge-screening model (see for example, Siegbahn *et al* 1967) is accepted as applicable in the case of alloys, this shift in the Al 2p peak suggests a depletion of valence electrons from the aluminium atoms in the alloy, with increasing gold content. Relative shifts for the gold core-level peaks are not so easily recorded, due partly to the difficulty of energy alignment which in this case cannot be checked with respect to the aluminium oxide peaks because there are of course no Al photoelectron peaks from pure gold. Some indication was obtained, figure 7, of a slight shift with composition of the Au N_6 and N_7 photoelectron peaks for the three alloys, measured relative to the oxide Al 2p and 2s peaks for every case, which have been taken to be independent of the alloy composition. However, further careful investigation of the alloy series is necessary before too much significance can be attached to these 'chemical' shifts, since the remaining oxide contamination, and possible workfunction changes on alloying, considerably cloud the interpretation. Meanwhile, electron depletion from the aluminium atoms in these alloys appears to be probable from soft x-ray emission measurements (Källne, this Volume p.205) while depletion from the gold atoms also is indicated by soft x-ray data (Kapoor *et al* 1972) and by the present preliminary photoelectron results. Transfer of valence d-electrons from the gold atoms to aluminium atoms, particularly for $AuAl_2$, is argued by Pauling (1959) on the basis of relative electronegativities for the metallic atoms. Probably the concept of ionic charge transfer, with

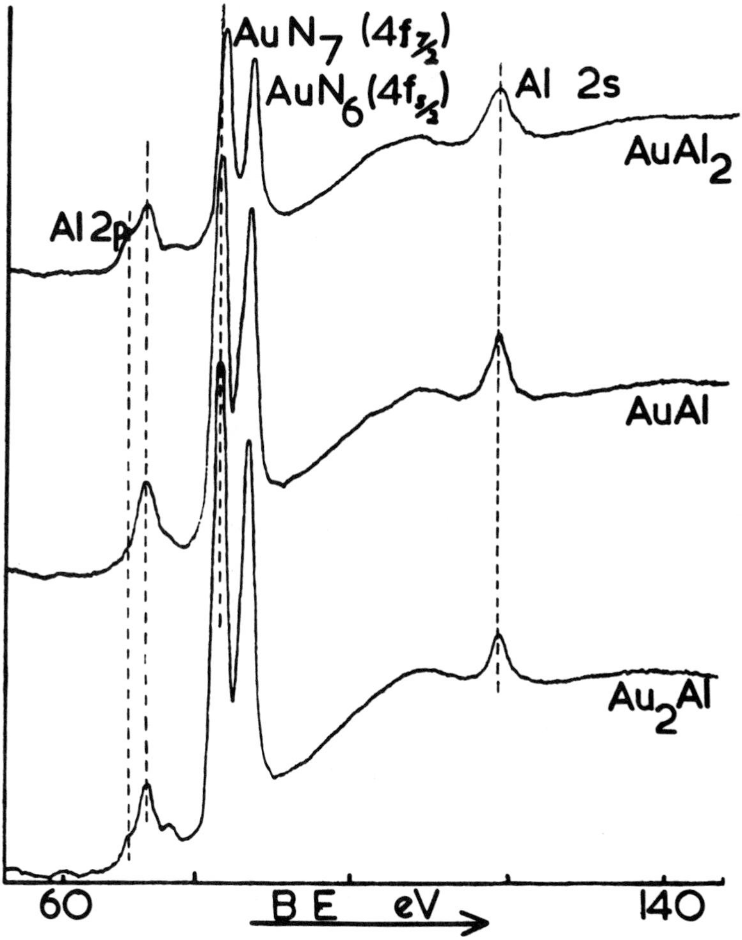

Figure 7 Photoelectron Au N_6 and N_7 peaks recorded for gold-aluminium alloys; energy alignment has been made arbitarily at the oxide-contamination Al 2s and 2p peaks in the spectra.

charge screening and 'chemical' shift interpretation, is an altogether oversimplified picture when applied to alloys. Further investigation, theoretical and experimental, is clearly desirable in this field.

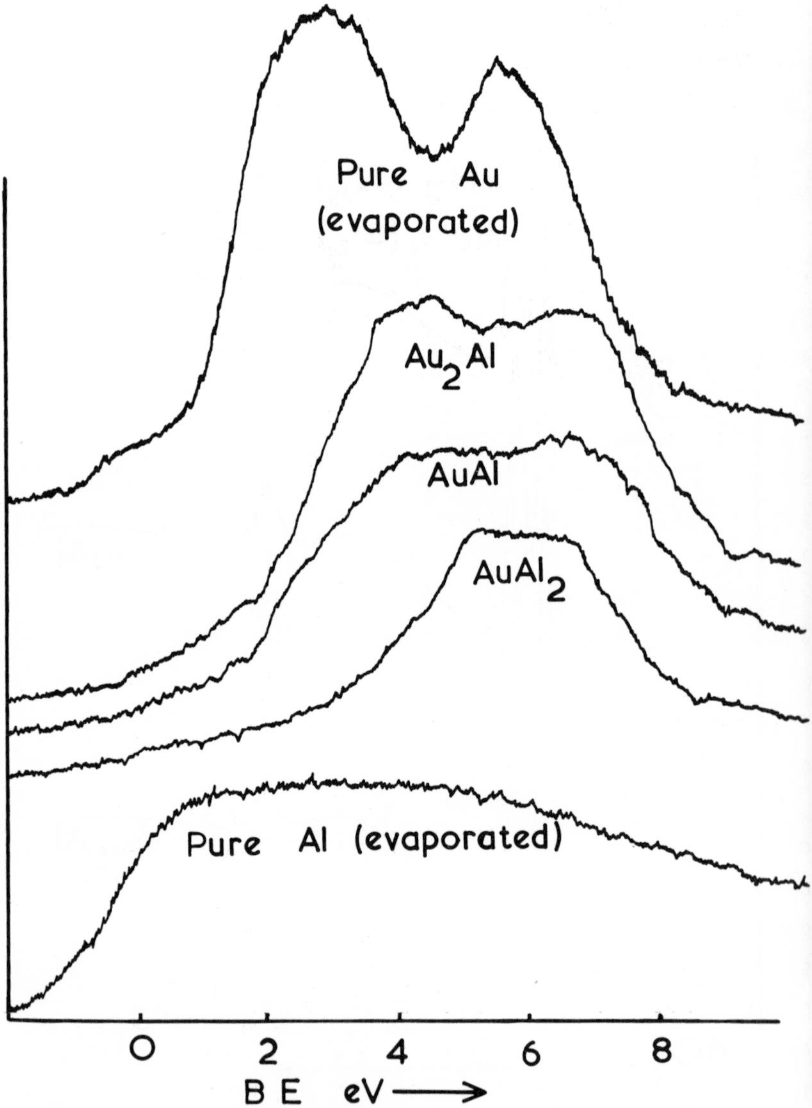

Figure 8 X-ray photoelectron valence band spectra for pure evaporated gold and aluminium, and for aluminium-gold alloys prepared as bulk specimens but cleaned under high vacuum.

A detailed examination was also made of the photoelectron valence-band spectra for the alloys and for pure evaporated gold and aluminium. These spectra are compared in figure 8. It is important to emphasize that such XPS valence bands do not necessarily reflect the overall density of electron states. In the present case, for example, structure observed in the photoelectron valence band for aluminium-gold principally reflects structure in the energy distribution of the Au 5d-like states; not only are there more d-electrons per atom, than p or s, but d-states generally appear to have a higher photoelectron cross-section.

The valence band recorded for pure gold agrees well with that observed by previous investigators (Fadley and Shirley 1971, Baer *et al* 1970); but also shows a distinct Fermi-edge usually only observable with higher resolution UPS and XPS studies (Shirley 1972). The photoelectron valence band for pure aluminium appears to reflect the well known free-electron-like character of the metal; the low-energy tail of the band is affected by the plasmon band, noted in figure 2, but an extrapolation of this tail to provide an estimate of the band-width leads to good agreement with the value $\sim$12eV obtained from soft x-ray emission measurements (see, for example, Fabian *et al* 1971).

The important feature with regard to the alloys is the decrease in splitting of the valence d-band observed with increasing aluminium content. The splitting which has been regarded as partly due to the effect of spin-orbit coupling persisting in the solid gold metal, appears too large to be explained completely in terms of this effect. Although its decrease on alloying can be reconciled with the explanation in terms of spin orbit effects, since we might expect the splitting to reduce to a 'free-atom' value as the gold atoms become separated by introduction of the aluminium atoms to the lattice, it is probable that relativistic effects cause the enhancement of the splitting for some heavy metals. Relativistic band-structure calculations appear to lead to a calculated energy spectrum (see, for example, Smith 1972) that agrees well with the photoelectron spectrum (Shirley 1972) for pure gold. The narrow valence d-band recorded for $AuAl_2$ agrees with that reported by Chan and Shirley (1971), giving for the position of the gold d-band – to which this peak is attributed – the value of $\sim$6eV below the Fermi level. However, the resolution and apparent counting-statistics in the present measurements appear to be much better and our estimated width of this valence d-band peak is $\sim$3eV, more closely agreeing with the value reported by Hüfner *et al* (1972) using Mg Kα radiation for excitation of the photoelectron spectrum.

The decrease of the splitting observed on going from pure gold to $AuAl_2$, with the consequent narrowing of the predominant Au d-band peak, also strongly supports the interpretation (Kapoor *et al* 1972) offered for the appearance of the two peaks recorded in the soft x-ray Au $N_{6,7}$-emission spectra of $AuAl_2$. The $N_{6,7}$-emission is caused by electron transitions from the valence band of the alloy to vacant N_6 and N_7 core-level states of constituent gold atoms. As the valence d-band sharpens into a narrower band, this becomes reflected, when due account is taken of transition probabilities and selection rules, in two peaks appearing in the $N_{6,7}$-emission spectrum, one for each of the well-separated N_6 and N_7 core-states. Valence-band splitting appears to persist with $AuAl_2$.

4. COMMENT

The results indicate beyond doubt that XPS measurements have an important place in the study of electron behaviour in alloys, especially when taken in combination with other forms of probing the valence band. It is also clear that contamination of the sample surface is an extremely important consideration in the interpretation of spectra. Moreover, the presence of oxide in the bulk cannot be ignored. In the present investigation samples were prepared in argon atmosphere, outside the instrument, but bulk oxide was still observable. Ideally the alloys should be prepared by controlled evaporation within the high vacuum of the sample preparation chamber. Further investigation, with equipment designed to achieve this, is already underway; see Fuggle *et al* 1973.

From the theoretical viewpoint, considerable doubt surrounds the cause of the splitting of the d-band for pure gold. The lessening of this splitting on alloying with aluminium should provide an interesting theoretical study; experimentally it is important to explore the effect of other metals.

Acknowledgments

This work was supported by a grant from the Science Research Council. The authors wish also to thank Professor G. Busch and Dr. Y. Baer for valuable discussions, and Professor K. Ulmer and Dr. G. Böhm for advice on related experiments.

REFERENCES

Baer, Y., Hedén, P.F., Hedman, J., Klasson, M., Nordling, C. and Siegbahn, K. (1970); Physica Scripta, 1, 55.

Cauchois, Y. (1968); in "Soft X-ray Band Spectra and Electronic Structure", ed. D.J. Fabian, p.73; Academic Press, London.

Chan, P.D., and Shirley, D.A. (1971); in "Electronic Density of States", ed. L.H. Bennett, p.791: NBS Spec. Publn. 323.

Fabian, D.J. (1971); Jnl. Physique, 32, (C4) 317.

Fabian, D.J., Watson, L.M. and Marshall, C.A.W. (1971); Rep. Prog. Phys. 34, 601.

Fadley, C.S. and Shirley, D.A. (1971); in "Electronic Density of States", ed. L.H. Bennett, p.163: NBS Spec. Publn. 323.

Fuggle, J., Fabian, D.J. and Watson, L.M., submitted to Sol. St. Comm.

Hedman, J., Klasson, M., Nilsson, R., Nordling, C., Sorokina, M.F., Kljushnikov, O.I., Nemnonov, S.A., Trapeznikov, V.A. and Zyryanov, V.G. (1971); Physica Scripta, 4, 1.

Hüfner, S., Wernick, J.H. and West, K.W. (1972); Sol. State Comm., 10, 1013.

Kapoor, Q.S., Watson, L.M., Hart, D. and Fabian, D.J. (1972); Sol. State Comm. 11, 503.

Pauling, L. (1959); in "The Nature of the Chemical Bond" 3rd Edn. p.394., Cornell U.P., New York.

Raether, H. (1965); Springer Tracts Mod. Phys. 38, 84.

Shirley, D.A. (1972); Phys. Rev. B, 5, 4709.

Sieghabn, K., Nordling, C. and others (1967); ESCA "Atomic, Molecular and Solid State Structure", Nova Acta. Reg. Soc. Sci. Upsal., Ser. 4, 20.

Smith, N.V. (1972); Phys. Rev. B, 5, 1192.

Switendick, A.C. (1971); in "Electronic Density of States", ed. L.H. Bennett, p.297: NBS Spec. Publn. 323.

X-RAY PHOTOELECTRON SPECTRA AND BAND STRUCTURE OF ALKALI HALIDE CRYSTALS

V.V. Nemoshkalenko, A.I. Senkevich and V.G. Aleshin

Institute of Metals, Academy of Sciences of the Ukrainian SSR, Kiev, USSR

1. INTRODUCTION

For long past x-ray emission and absorption spectra have served as the only sources of information on the electron states of solids over a wide energy range. The capabilities of x-ray spectroscopy (XS) as a tool for studying band structure have been amply demonstrated.

At present, due mainly to the works of Siegbahn and his co-workers (1967) a new technique is available for band-structure investigators, namely, x-ray photoelectron spectroscopy (XPS). This technique is capable of measuring electron binding energies precisely and for that reason may be used for both the determination of 'chemical' shifts, in core-electron energy levels, and for the study of valence-band structure.

The most complete information on band-structure can be obtained by combining together both the XS and XPS techniques. XS transitions, which lead to the emission or absorption of a photon, occur between two states of different kinds. One is always a core-electron state whose wavefunction in the region of each ion in the lattice resembles the atomic wavefunction; the second belongs to the valence or conduction band, and this state is described by a wavefunction with different mixed symmetries, depending on the particular state. Thus, for a full picture of electronic energy structure from XS it is necessary to measure the complete set of spectra bound to the scanning levels of all symmetries. For monoatomic crystals it is possible to determine from their x-ray spectra both the bandwidths and the energy distances between the bands. For crystals containing several elements, in XS there is a problem of how to combine the spectra of different constituent elements; the correct solution might be achieved by the use of XPS. Because in the x-ray photoelectric process an electron can be expelled from any level of any symmetry, a complete set of photoelectron spectra for all crystal states is received at once, and this makes it possible to determine with high precision the energy distance between the core states of a many-component compound and therefore accurately to correlate the x-ray spectra of different components. Another advantage of the XPS technique lies in the possibility of probing the closely-spaced energy bands with distances from several tenths to several tens of eV. A disadvantage of the method is its complexity, concerning the determination of bandwidths, due to the presence of a background of scattered photoelectrons.

The present work has been devoted to investigating photoelectron spectra for a group of alkali-halide crystals. Since the experimental XPS technique is comparatively new, it is useful to give a short description of the instrument as well as the principles involved in determination of the electron binding energy, with particular reference to insulating crystals.

2. EXPERIMENTAL

Measurements were made with a Varian (IEE-15) spectrometer with induced electronic emission. Figure 1 shows a schematic instrument. The pressure in the sample chamber during operation is $\sim 1 \times 10^{-7}$ torr.

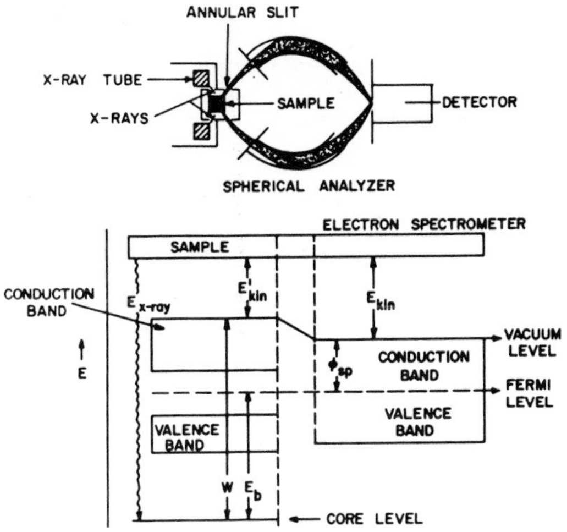

Figure 1 a schematic of the instrument.

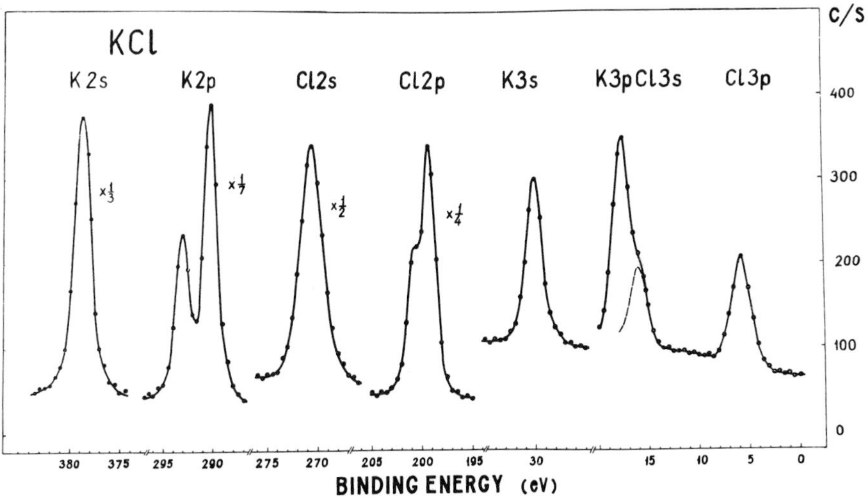

Figure 2 energy-level diagrams of the spectrometer (metal) and the sample (insulator).

In figure 2 we present the energy-level diagram of the spectrometer (metal) and the sample (insulator). The Al $K\alpha_{1,2}$-line (1486.6eV with half-width 0.8eV)serves as the source of photons. The kinetic energy of the emitted electrons is analyzed by a spherical electrostatic analyzer. Our measurements were carried out with an electron analyzer potential of 100V which yields an effective resolution of 1.85 eV. Peak positions are determined with a much greater accuracy, ±0.2 eV.

If the binding energy W of the electron is less than the energy of the x-ray radiation, and if inelastic interactions of the electron after it is excited can be neglected, then the kinetic energy E' of the electron emitted from the sample is given by the expression

$$E' = E_{x\text{-}ray} - W \tag{1}$$

Siegbahn *et al* (1967) have shown that in fact a dominant proportion of the electrons are emitted without inelastic scattering and that therefore the observed kinetic-energy spectra can be related to the structure of the electronic energy levels in the crystal. However, there are electrons which on leaving the sample do lose part of their energy as a result of inelastic interaction. Thus, in the fine-structure, to lower kinetic energies from the main emission peak, one can expect to detect satellite peaks corresponding to the plasmon losses of expelled electrons. In the present work we shall identify the principal peaks of our photoelectron spectra.

Despite the fact that the emitted electron leaves the sample with an energy E', with respect to the sample's vacuum level, it is the energy E with respect to the vacuum level of the spectrometer material that is actually measured by the spectrometer. If the work function ϕ_{sp} of the spectrometer material is known, then the binding energy of the electron E_b with respect to the Fermi-level of the material will be given by the following expression:

$$E_b = E_{x\text{-}ray} - E - \phi_{sp} \tag{2}$$

When the sample and the spectrometer are in a good electrical contact, as it is when the sample is a metal, the Fermi-levels of both the sample and the spectrometer materials will coincide, so that measurements of binding energies for the metals are actually made with respect to the Fermi-level of the sample. Since for dielectrics the notion of a Fermi-level does not have the same meaning as for metals, the binding energy in this case is measured with respect to the Fermi-level of the spectrometer material. The work function of the spectrometer can be determined by calibration with respect to a known binding energy. To this end we used the carbon 1s-level (284.0 eV).

Because during the process of measurement the sample is being charged, due to its poor conductivity and the emission of electrons, equation (2) should include an additional term ϕ to account for the effect of charge build-up. Finally, the expression for binding energy is of the form

$$E_b = E_{x\text{-ray}} - E - \phi_{sp} - \phi \qquad (3)$$

The surface of the sample in the spectrometer always contains a hydrocarbon film, and this can be used to determine the value $\phi_{sp} + \phi$ by comparing the energy of carbon 1s-level with its known value of 284.0 eV.

3. RESULTS

Figure 3 shows the photoelectron spectrum of KCl crystal. We see that the photoelectron lines from deep-lying electron states are narrow, their shape being close to Lorentzian. For different compounds the spectra are much alike, and for this reason only the valence-electron spectra are recorded here (figure 4) for the compounds LiF, LiCl, NaF, NaCl, NaI, KF, KBr, KI and CsI. These valence-band spectra are broader and asymmetric. The energies at which the important peaks occur in the spectra are indicated in Tables I and II.

Table I Energies in eV of the important photoelectron peaks in the spectra of the alkali halides.

		1S1/2	2S1/2	2P1/2	2P3/2	3S1/2	3P1/2	3P3/2	3d3/2	3d5/2	Valence Band
LiF	Li^+	58.8									11.1
	F^-	688.0	32.7								
LiCl	Li^+	57.4									6.1
	Cl^-		270.7	201.2	199.6	17.5					
NaF	Na^+	1074.2	66.2	31.4							10.3
	F^-	687.3	33.2								
NaCl	Na^+	1072.2	64.1	31.3							6.5
	Cl^-		270.3	201.0	199.3	17.3					
KF	K^+		378.7	296.7	294.0	34.1		17.9			7.7
	F^-	684.5	28.8								
KCl	K^+		378.7	293.0	290.2	29.8		17.7			5.8
	Cl^-		270.3	200.6	199.2	15.9					
KBr	K^+		378.8	297.2	294.4	30.1		17.8			5.0
	Br^-					257.0	190.0	183.4	71.3	70.0	

Table II Photoelectron peaks in spectra of alkali halides (continued).

		1S1/2	2S1/2	2P1/2	2P3/2	3S1/2	3P1/2	3P3/2	3d3/2	3d5/2
KI	K^+		378.5	296.8	294.0	27.4	19.3			
	I^-					1073.2	931.6	875.6	631.7	620.1
NaI	Na^+	1074.4	62.5	33.4						
	I^-					1074.4	933.1	877.0	633.1	621.5
CsI	Cs^+						1067.5	999.4	740.1	726.0
	I^-					1073.3	932.8	875.3	632.4	623.1

		4S1/2	4P1/2	4P3/2	4d3/2	4d5/2	5S1/2	Valence Band
KI	I^-	187.2	123.1		51.8	50.3	15.0	5.6
NaI	I^-	188.7	124.3		53.5	51.9	15.8	6.3
CsI	Cs^+	231.7	171.6	160.6	79.1	76,9	24.8	12.0
	I^-	187.5	123.3		51.9	50.3	16.1	5.5

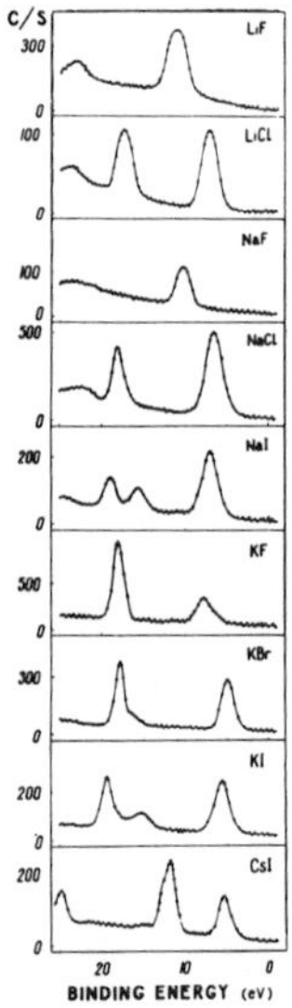

Figure 3 X-ray photoelectron spectrum from KCl.

4. DISCUSSION

Let us compare our values of electron binding energy, for the compounds NaI, KI and LiCl, with the calculated results obtained by Kunz (1969, 1970a, 1970b). We require a common energy reference, and we fix this by taking the binding energy of an electron at the bottom of the conduction band as equal to zero. The experimental values of binding energy are therefore shifted in such a way that the bottom of the conduction band is zero. The amount of shift was determined by comparison of the values obtained in the present study with the values corresponding to the onset of absorption in the ultra soft x-ray region. Such data are available for KI (the $N_{4,5}$-spectrum of I and from KI - Brown *et al* 1970), for NaI (the $L_{2,3}$-spectrum of Na^+ from NaI - Haensel *et al* 1968a), and for LiCl (the K-spectrum of Li^+ from LiCl - Haensel *et al* 1968b). The absorption edges for these compounds have been determined, and are 52.5, 32.2 and 59.4 eV respectively.

It has been shown by several investigators (Kunz 1969, 1970a, 1970b; Brown *et al* 1970; Lipari and Kunz 1971) that states

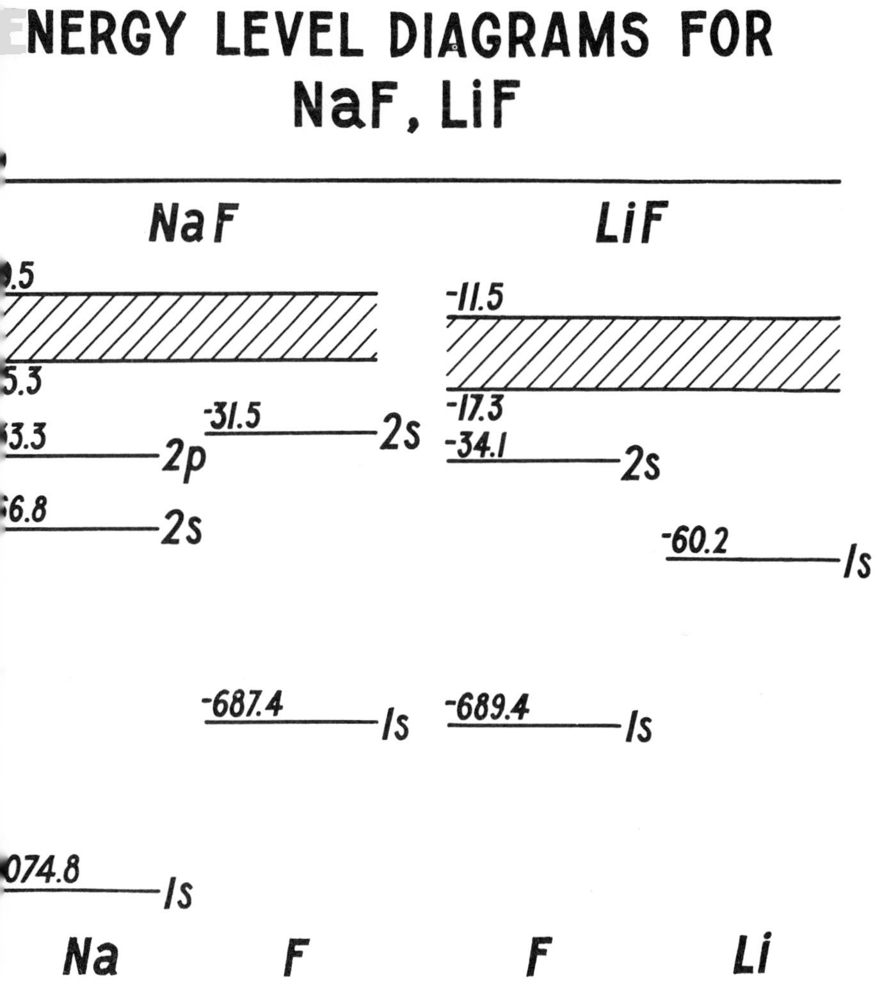

Figure 4 X-ray photoelectron spectra of valence bands of alkali halide crystals.

which lie at the very bottom of the conduction band are of Γ_1-symmetry for a wide group of alkali-halide crystals. Despite the fact that in the K and $N_{4,5}$ spectra transitions to the Γ_1 point are forbidden by the selection rules (applying the dipole approximation), the choice of a reference point aided by the x-ray spectra will not result in an appreciable increase in binding energy because starting from the energy that corresponds to the Γ_1 state, the density of states for the conduction band of the compounds under consideration quickly increases, and transitions in the *vicinity* of the point Γ_1 *are* allowed by the selection rules.

We see from Table III that the experimental and calculated results are in satisfactory agreement. However, we must consider a possible source of discrepancy. When calculating the NaI and KI band-structures, Kunz used Slater's approximation for the exchange potential. As shown by Lindgren (1965), the eigenvalues of the energy operator do not coincide in this case with the values of binding energy. Corrections that take account of the eigenvalue deviations from the values of binding energies are known as Koopmans correction. For example, the correction for atomic iodine is ∿30eV for the 3s-electrons) and ∿1.6eV for the 5s-electrons (Lindgren 1968). Better agreement with experiment was reached by Kunz; this probably arose because neglect of the Koopmans correction in the calculation is partly compensated by the neglect of correlation effects. For the LiCl and NaCl crystals, calculated in the Hartree-Fock exchange approximation, there is a considerable discrepancy in the energy positions of features originating from the 3s-states of Cl^-. For example, for LiCl the difference between theory and experiment is 15.6eV (Table III) which is inadmissible in such a calculation. A comparison of computed (Haensel *et al* 1968b Lipari and Kunz 1971) and experimental data for NaCl also leads to a significant difference in the energy position

Table III Experimental and calculated energies of photoelectron peaks.

		1S1/2	2S1/2	2P1/2	2P3/2	3S1/2	3P1/2	3P3/2	Valence Band
KI	K^+exp		380.7	299.0	296.2	29.6	21.5		
	K^+theor		368.6	299.3	295.9	36.6	19.6	19.9	
	I^-exp					1075.4	933.8	877.8	7.8
	I^-theor					1052.4	935.3	883.9	5.5
NaI	Na^+exp	1073.2	61.3	32.2					
	Na^+theor	1060.0	61.4	33.7	33.5				
	I^-exp					1073.2	931.9	875.8	5.1
	I^-theor					1051.8	934.8	883.1	5.9
LiCl	Li^+exp	59.4							
	Li^+theor	69.4							
	Cl^-exp		272.7	203.2	201.6	19.5			8.1
	Cl^-theor		290.5	221.5		35.1			10.8

		3d3/2	3d5/2	4S1/2	4P1/2	4P3/2	4d3/2	4d5/2	5S1/2
KI	I^-exp	633.9	622.3	189.5	125.3		54.0	52.5	17.2
	I^-theor	683.6	671.3	187.1	142.2	135.5	59.0	57.1	17.1
NaI	I^-exp	631.9	620.3	187.5	123.1		52.3	50.7	14.6
	I^-[illegible]	[illegible]	[illegible]	[illegible]	[illegible]	[illegible]	[illegible]	[illegible]	[illegible]

of Γ_1 (Cl^- 3s) and Γ_{15} (mainly Cl^- 3p) bands. The theoretical distance between these bands is 18.4eV, while experiment yields a much lesser value of 10.8eV. Thus the position is not satisfactory with regard to calculating states that are deep-lying, and do not belong to the uppermost part of the valence band nor to the region close to the conduction band bottom. To achieve better agreement of theoretical and experimental data it seems advisable to choose a more reasonable one-electron potential, as well as to take account of relaxation, correlation and polarization effects (Lindgren 1968).

The XPS results along with the XS data open the possibility of evaluating the band gap. For this purpose the absorption data in the x-ray region are required for determination of the position of energy bands with respect to the bottom of the conduction band. For some of the alkali-halide crystals such results were obtained in the present study. The evaluation was carried out from the peak in the photoelectron distribution nearest to the conduction band, since it is not easy to find an exact limit of the valence band from the spectrum, and because such an estimation will result in greater values than those of the corresponding band gaps. The data obtained in that way were compared with the values already reported (Table IV).

Table IV

Crystal	Δ_{XPS}	Δ^1_{OPT}	Δ^2_{OPT}	References
LiF	12.5	13.6	11.5	Rössler and Walker 1967
LiCl	8.1	9.2	8.6	Teegarden and Baldini 1967
NaF	10.9	11.5	9.5	Rössler and Walker 1967
NaCl	8.2	8.2	7.5	Miyota and Tomiki 1968
KCl	6.8	8.1	7.4	Tomiki 1967
KI	6.3	5.9	5.4	Hopfield and Worlock 1965

It can be seen, as a rule, that the magnitudes of band gaps obtained from optical absorption spectra in the far UV-region (third column) exceed the values evaluated with the help of XPS and XS (second column); this should not occur. However, the values cited by some of the authors (see for example Brown *et al* 1970) were obtained by attributing the first absorption peak, which is 1-2eV wide, to the Γ_1-exciton. The band gap, which makes allowance for the width of the first peak (assuming that it is caused by the band transitions), will become less than the quantity Δ_{xps}; the origin of disagreement for the crystals LiCl and KCl is not clear. Hence it is possible to conclude that at least for some alkali-halide crystals the first absorption peak to the long-wave side is a band structure effect. Consequently, in the x-ray absorption spectra the first absorption peak cannot originate from the excitation process (c.f. the reports of several investigators; Kunz 1970b, Brown *et al* 1970, Lipari and Kunz 1971).

Complementary measurements using both XPS and XS (together, desirably, with optical data) can yield a united energy for a crystal. Let us for example look at LiF and NaF. For these compounds Fischer (1965) obtained the x-ray emission band of fluorine, which agree quite well with our results. In fact, according to the x-ray data the density-of-states peak in the valence band is at 672.2eV for LiF, and at 676.8eV for NaF, while the photoelectron spectra give respectively 676.9eV (LiF) and 677.0eV (NaF). Remembering that the K-absorption of Li^+ in LiF sets in at 60.2eV, and the $L_{2,3}$-spectrum of Na^+ in NaF at 32.0eV, the energy diagrams for these crystals can be constructed, as indicated in figure 5 (the width of valence band is estimated from the emission spectra).

The photoelectron spectra from NaCl, KCl, KBr,and KI crystals have been reported some time ago by Siegbahn *et al* (1967) but their data differ considerably from ours with regard to

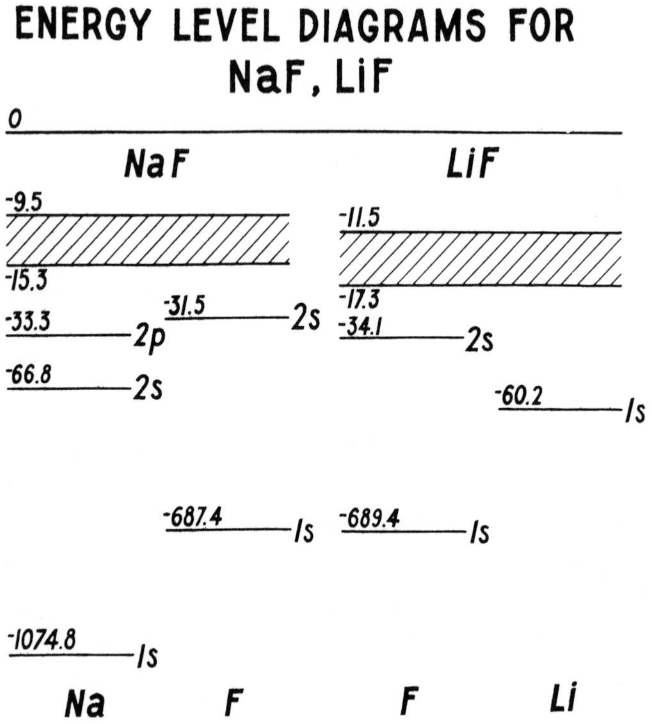

Figure 5 Energy level diagrams NaF and LiF.

the energies of the separate levels. On the other hand, our results are in much better agreement with more recent measurements by Siegbahn *et al* (1970) for NaCl. Probably, therefore, the earlier results for the alkali-halide crystals should be regarded as preliminary.

REFERENCES

Brown, F.C., Göhwiller, C., Kunz, A.B., Scheifley, W. and Carera, N. (1970); Phys. Rev. B2, 2126.

Fischer, D. (1965); J. Chem. Phys. 42, 3814.

Haensel, R., Kunz, C. and Sonntag, B. (1968); Phys. Rev. Lett. 24, 262.

Haensel, R., Kunz, C., Sosaki, T. and Sonntag, B. (1968b); Phys. Rev. Lett. 18, 1436.

Hopfield, J.J. and Worlock, J.M. (1965); Phys. Rev. 137, A1455.

Kunz, A.B. (1969); Phys. Rev. 180, 934.

Kunz, A.B. (1970a); J. Phys. Chem. Solids, 31, 265.

Kunz, A.B. (1970b); Phys. Rev. B2, 5015.

Lindgren, I. (1965); Phys. Lett. 19, 382.

Lindgren, I. (1968); Phys. Rev. 176, 114.

Lipari, N.O. and Kunz, A.B. (1971); Phys. Rev. B3, 491.

Miyota, T. and Tomiki, T. (1968); J. Phys. Soc. Japan 24, 1286.

Roessler, D.M. and Walker, W.C. (1967); J. Phys. Chem. Solids 28, 1507.

Siegbahn, K., Gelius, U., Siegbahn, H. and Olson, E. (1970); Physica Scripta, 1, 272.

Siegbahn, K., Nordling, C., and others (1967); ESCA "Atomic, Molecular and Solid State Structure studied by means of Electron Spectroscopy", Nova Acta. Reg. Soc. Sci. Upsal., Ser. 4, 20.

Teegarden, K. and Baldini, G. (1967); Phys. Rev. 155, 896.

Tomiki, T. (1967); J. Phys. Soc. Japan, 22, 463.

Part 2

SOFT X-RAY EMISSION FROM ALLOYS

A SURVEY OF CHARACTERISTICS OF ALLOY X-RAY EMISSION SPECTRA

L.M. Watson

Department of Metallurgy, University of Strathclyde, Glasgow, Scotland.

1. INTRODUCTION

Many properties of alloys such as phase stability, and metal to non-metal transitions, still await theoretical understanding. Their full interpretation will require detailed knowledge of the distribution of valence electrons throughout the valence bands for the pure components of the alloy, and - more difficult - a detailed understanding of how the valence electrons regroup to provide the bonding between constituent atoms and to form the valence band of the alloy. No unified theory of the behaviour of electrons in alloys exists; several theories have been proposed, supported in part by experimental work on, say, one alloy series, only to be rejected when considering another alloy series. Of the experimental techniques in current use, capable of providing information about the energy distribution of states throughout the valence band, ultra-violet photoelectron spectroscopy (UPS), x-ray photoelectron spectroscopy (XPS) and, the oldest established technique, soft x-ray spectroscopy (SXR) are the most widely used. This survey will partially review the present state of soft x-ray emission spectroscopy of binary alloys, with occasional recourse to soft x-ray absorption and UPS studies. These aspects of the spectroscopy of alloys are covered in detail elsewhere in this Volume.

2. THE MECHANISM OF SOFT X-RAY EMISSION

In a pure solid the emission of x-rays arises from transitions of electrons from a level of lower binding energy to one of higher binding energy in which a vacancy has previously been created. The vacancy is a prerequisite to the emission process, and it therefore follows that soft x-ray emission always results from an excited state of the solid. Parratt (1959) and later Fabian (1970) have reviewed the types of excitation state that are likely to exist and more theoretical studies by, for example, Hedin (1968), Nozières and de Dominicis (1969), Brouers (1967), Glick *et al* (1968), have examined the effect of the inner core vacancy on the soft x-ray emission process from simple metals. The general conclusion is that the effect of the inner core vacancy on the overall shape of the valence band emission spectrum is small, although it is significant at the Fermi threshold where a singularity in the case of L-emission, or a general 'broadening' in the case of K-emission, can occur. Other effects, caused by excitation states of the solid, include an extended tailing of the low-energy limit of the band and the possible simultaneous creation of a plasmon coupled with the emission of the x-ray photon resulting in attenuation of the emitted photon energy. This second effect causes the emission of low-energy plasmon satellite bands, which have been observed in the emission spectra of light metals; see, for example, Rooke (1963), Watson *et al* (1968), Neddermeyer and Wiech (1970). In the present survey we shall be describing briefly plasmon satellites observed also for alloys.

However, for the most part, we shall accept that the effect of the inner core vacancy does not modify greatly the measured spectra of alloys, and we consider the emission spectra as providing information about electron states characteristic of the unexcited solid.

The intensity of emission of x-rays, of energy $h\nu$, from singly ionized atoms is given by

$$I_{h\nu} = 2 \int_{E_{const}} \frac{d^2k}{\nabla_k(E)} T(E_k) \qquad (1)$$

where the integral is taken over the constant energy surface $E(k)$ or E_k in k-space, and $T(E_k)$ is the transition probability for the emitting transition, at every $E(k)$. This equation is a one-electron approximation and neglects many-body interactions. The first term in the integral is the density of states in the valence band $\rho(E)$, given by

$$\rho(E) = \int_{E_{const}} \frac{d^2k}{\nabla_k(E)} \qquad (2)$$

When studying alloys probably the most important factor is $T(E_k)$, given by

$$T(E_k) = \frac{4e^2}{3h^4c^2} (h\nu)^3 \left| \int_{all\ \mathbf{r}} \psi_f^*(\mathbf{r})\, \mathbf{r}\, \psi_i(\mathbf{r},k)\, d^3r \right|^2 \qquad (3)$$

where $\psi_i(\mathbf{r},k)$ is the initial state wavefunction and $\psi_f(\mathbf{r})$ is the final state wavefunction. It has been common to assume that $T(E_k)$ is constant throughout the valence band, so that the density of states term and the transition probability term can be separated. That is to say, the intensity profile of a soft x-ray band spectrum gives a direct reflection of the density of states in the valence band. It is now generally recognised that this assumption is far from valid, and that in many cases the transition probability, if not the dominant factor, is at least as important as the density of states.

An immediate consequence of equation (3) is that the integral must be non-zero only in regions of real space for which both the initial and final state wavefunctions are non-zero. Since the final-state wavefunction is a well defined inner core level, it will be non-zero only in a small region of space around the nucleus of the emitting atom and is essentially zero elsewhere in the solid. Therefore it follows that soft x-ray emission spectra arise from electron transitions only in a region localized around the emitting atom, and can not provide information about electron states throughout the metal. Equation (3) also takes account of the well-known dipole selection rules whereby, when the inner core vacancy is created in a level of s-symmetry, only valence-band states of p-symmetry about the emitting atom can undergo radiative transitions, and when the inner core vacancy is a state of p-symmetry, then valence electrons of s or d symmetry can undergo radiative transitions. The possibility of a valence electron localised in the region of one atom or ion making a transition to a vacancy in a neighbouring atom can not be ruled out. This possibility gives rise to the so-called crossover transitions, for which the selection rules are different. However, because of the small amplitude of the overlapping wavefunction for such transitions, the probability of their occurrence is small.

3. SOFT X-RAY EMISSION FROM ALLOYS

A. Component Spectra

In the binary alloys considered here, the bonding between the two constituent atoms varies - according to their species - from almost entirely metallic to a strong admixture of metallic with covalent or ionic bonding. That is, we have a range from complete delocalization and uniform sharing of the valence electrons, to those cases in which charge transfer occurs from the Wigner Seitz cell of say atom A to that of atom B.

In cases such as the latter, some of the valence electron states become semi-localized in the region of atom B. The density of electron states in the region of the A atoms will then be different from the density in the region of B atoms, and for a complete investigation of an alloy using soft x-ray emission it is important that the series spectra, K, L, M, etc., of both component species be investigated. Unfortunately, this has been done for only a few alloy systems, due mainly to experimental problems such as those arising from insufficient intensity of overlapping of spectra; thus we are often restricted to spectra arising from only one of the atom species.

The electronic character of bonding in solids can be illustrated by figure 1 (from Nagel 1970) in which the covalent

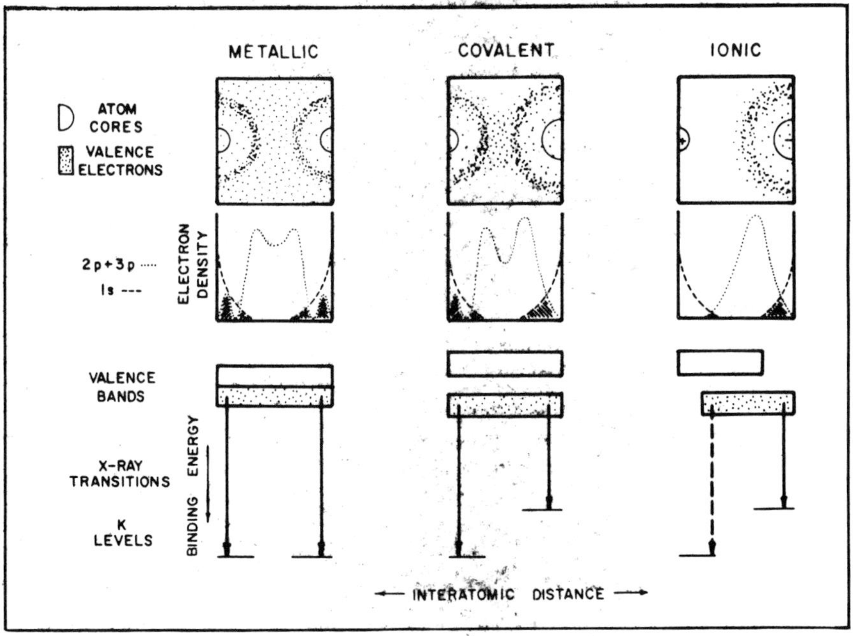

Figure 1 Schematic of valence electron distributions for the three major forms of bonding. The electron density distributions for third-row cations (Al, Si, etc.) and second-row anions (C,N,O etc.) are indicated. The hatched areas are the regions in which the valence 1s electron densities overlap.

picture is the one most closely approaching the bonding in an alloy. The conduction and valence bands are not in general, separated. Thus we can see, for example, that the valence L-emission spectra of atoms A and B for an alloy AB will give rise to two emission bands, usually of different shape due to the variation in the density of states around atoms A and B, and due also to the variation in extent of overlap of the valence-electron wavefunctions with the inner core-level wavefunctions (e.g. $2p_{1/2}$ or $2p_{3/2}$) of the emitting atoms.

Self absorption of the x-ray photons within the alloy can provide an added complication. If the x-ray photons emitted by atom A have energies a few electron volts in excess of the energy difference between the Fermi-level and an inner core level of atom B, there is a high probability that B atoms will strongly absorb the x-rays emitted from atom A, resulting in a distortion of the valence band spectrum of atom A. We shall discuss this effect for aluminium-magnesium alloys.

B. Bandwidths

One of the most quoted conclusions drawn from investigations of emission spectra of alloys is the apparent difference in measured bandwidths for the valence band, when observed by the emission from atoms A or B of the alloy AB. In many cases the bandwidths appear to be close to those measured for the pure metals A or B (Curry 1968, Fabian 1970). However, in many cases determination of the low-energy limits of the emission bands is extremely difficult due to extended tailing, to the effects of overlapping higher-order spectra, and to satellites and other causes of a non-linear variation in the background intensity. An important contribution to the study of alloy emission bandwidths has been made by Dimond (1970), who made extensive studies of the $L_{2,3}$-emission spectra of a series of Al-Mg alloys and compared the emission intensities with those for the pure metal. A more complete study of this alloy system has been carried out by Neddermeyer (this volume p.153).

Direct comparison of the emission intensities from two different alloy samples is extremely difficult due to the difficulty of precisely reproducing exact experimental conditions in each case. Dimond overcame this problem to some extent, in his investigation of Al-Mg alloys, by examining the ratio $I(E)_{alloy}/I(E)_{metal}$ and graphing the logarithm of this ratio *vs* energy. $I(E)_{alloy}$ is the $L_{2,3}$-emission intensity at energy E for one constituent of the alloy, and $I(E)_{metal}$ the intensity of $L_{2,3}$-emission for the corresponding pure metal component at the same energy. To obtain the ratio the intensity was first normalized at a suitable point in the emission band.

The curves of $ln[I(E)_{alloy}/I(E)_{metal}]_{Al}$ *vs* energy were compared directly with the published absorption curves for magnesium in the same energy region (see figure 2). The comparison

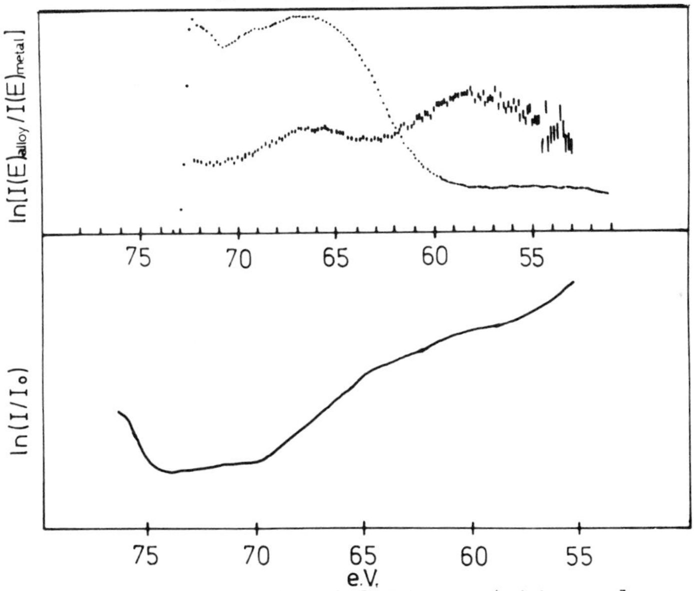

Figure 2 Upper curves show $ln[I(E)_{alloy}/I(E)_{metal}]$ *vs* energy for the Al $L_{2,3}$-emission from the alloy $Al_{25\%}Mg_{75\%}$, superposed on the spectrum for pure aluminium. The lower curve is the absorption coefficient (ln I/Io) in the same energy region for an 880Å magnesium foil (Townsend 1953).

is favourable indicating that self absorption has a marked effect on the shape of the aluminium emission band for the alloy. Other factors affecting the comparison are genuine changes in the density of occupied and unoccupied states due to alloying. We do not expect the 'ratio' curves of magnesium to be influenced greatly by self absorption since the photons do not have sufficient energy to promote electrons from the Al L-shell to the unoccupied states above the Fermi level. Departures from a linear curve must therefore be associated mainly with changes in the emission spectrum on alloying.

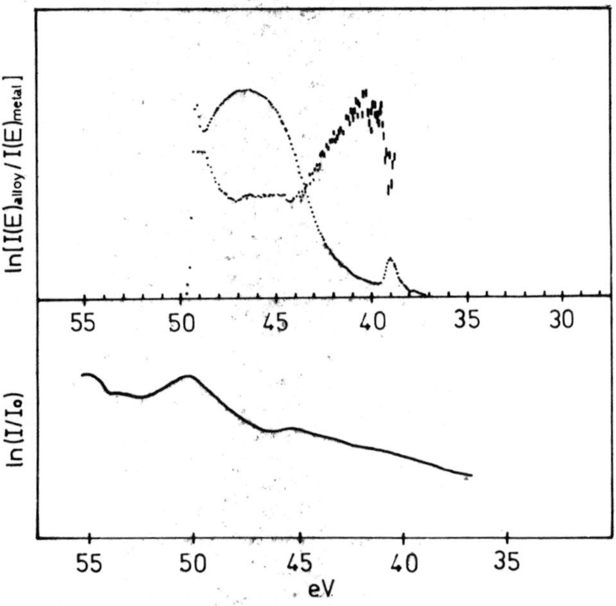

Figure 3 Upper curves show $ln[I(E)_{alloy}/I(E)_{metal}]$ *vs* energy for the Mg $L_{2,3}$-emission from the alloy $Al_{70\%}Mg_{30\%}$, superposed on the spectrum for pure magnesium. The lower curve is the absorption coefficient (ln I/Io) in the same energy region for a 5000Å aluminium foil (Tomboulian and Pell 1951).

Figure 3 shows an example of a 'ratio' curve, obtained by Dimond, compared with the absorption spectrum of aluminium in the same energy region (Tomboulian and Pell 1951). Here it can be seen that the curve shows a marked hump in the region from ∿43eV (which corresponds to the bottom of the emission band of pure magnesium) to ∿38eV. This indicates that the emission intensity for the alloy is greater in this region than it is for the pure metal, and in fact causes the alloy bandwidth for magnesium to be extended to ∿11eV - the bandwidth for pure aluminium. The evidence from this work lends support to the existence of a valence band of common width 'seen' from either the aluminium or the magnesium atoms. The weak emission intensity of the low-energy region of the magnesium band is attributed to the extremely low amplitude of the low-energy valence s-type wavefunctions in the region of the magnesium ion cores.

This conclusion was reached by Jacobs (1969) in a theoretical treatment of the soft x-ray emission spectra of a hypothetical 50% 50% ordered aluminium magnesium alloy. Neddermeyer reports elsewhere in this Volume (p.153) a more complete investigation of the Al-Mg alloy series, reaching similar conclusions, and extending the discussion to include the p-type wavefunctions involved in K-emission from these alloys. In general it can be said that the electron state density 'seen' by the constituent atoms is nearly always different in detail of shape and fine structure, but it is in fact possible that the width of the valence band observed is common to the constituent atoms; i.e. that the apparent differences in bandwidth measured by the constituent atom emission bands are simply explained by the experimental difficulty of measuring low intensities and of detecting the low-energy limit of the band. This would not be at variance with theories such as those employing the coherent potential approximation, in which two distinct bands are predicted for the constituent atoms, with density-of-states peaks or 'humps' at quite different energies, but nonetheless with the same overall energy bandwidths.

In addition to differences in the bandwidth observed on probing from the separate constituent atoms, we look for genuine changes of bandwidth expected with change in the electron atom ratio. In the simplest model for an alloy - the rigid band model (Jones 1934) - the density-of-states profile is assumed to remain constant with alloy composition, and the only change expected would be the shift of the Fermi level, to higher or lower energies as the electron atom ratio is respectively increased or decreased. Evidence of such shifts have been obtained by Donahue and Azároff (1967), studying the K-absorption spectra of cobalt-nickel alloys. Similar evidence has been reported by Eggs and Ulmer (1968), who investigated isochromat spectra of rhodium-palladium and palladium-silver alloys (see this Volume p.521). We might reasonably expect that an alloy of two metals with closely similar band structure will show little change in this band structure on alloying; i.e., to a first approximation the rigid band model should hold. Evidence that this is so is found in the results obtained by Nemnonov and Finkelshtein (1966) in their study of the x-ray emission spectra of alloys of neighbouring transition metals from titanium to chromium. More recently Nemnonov *et al* (1971) conclude that a common energy band occurs in these alloys with complete collectivization of the dsp-electrons.

Most often the band structures of the pure metals that constitute the alloy are not similar, and the rigid band model does not hold. Considerable modification of the density of states curves results from the regrouping of valence electrons on alloying. This results in changes in bandwidth not necessarily consistent with changes in electron atom ratio. However, as described above, the absolute measurement of emission bandwidths is difficult and often not reliable.

C. Charge Transfer

The electron cell model for alloys proposed by Bolsaitis and Skolnick (1968) and the two-band model proposed earlier by Varley (1954) have met with some success in explaining the heats of formation of binary alloys. These theories essentially consider the transfer of charge from the Wigner Seitz cell of the lower-valency ion to the Wigner Seitz cell of the higher-valency ion, and take account also of the change in dimensions of the cells due to the charge transfer. Although the models are oversimplified, and heats of formation must be influenced strongly by ion-ion exchange interactions, charge transfer will undoubtedly take place in many alloys. X-ray emission spectra should detect such charge transfer since each emitting species is 'sampling' the density of states in the neighbourhood of the emitting ion. The changes we expect to observe are an overall change of shape and a shift in energy of the emission bands. The higher-valence ion would be expected to show an increase in intensity at the low-energy end of the band due to the extra charge screening-out the ion and also a shift to lower energies. The lower-valence component would be expected to show the opposite trend. This is in fact the case for several alloys, as observed by Curry (1968). However, it is important to note that the energy shifts of the bands measured by soft x-ray spectroscopy are not absolute since redistribution of the valence electrons is known to cause energy shifts of the inner core levels. Therefore, for absolute measurement, it is necessary to combine inner core-level shifts measured by x-ray photo-electron spectroscopy with relative shifts obtained from x-ray band spectra.

The above comments describe only qualitatively the changes in shape of the emission spectra expected due to charge transfer. To obtain quantitative measurement of the charge transfer it is necessary to measure the change in absolute intensity of emission, from the component atoms, suitably corrected for self-absorption effects. This is difficult experimentally. However, Wenger *et al* (1971) have performed a careful series of such experiments, in which they measured the changes, in alloys of transition metals with aluminium, of the number of 3d-electrons on the transition-metal atom and the number of 3p-electrons on the aluminium atom. This has been done for alloys of aluminium with manganese, iron cobalt, nickel and copper. No attempt was made to measure the shape of the emission bands, only the integrated intensity of - for example - the Lα band from nickel and the Lℓ line from nickel. The Lα band results from transitions of the valence electrons to the nickel 2p inner core level while the Lℓ line is due to 3s$\rightarrow$2p inner core-level transitions. It is assumed that alloying will have minimal effect on the 3s level so that the intensity of the Lℓ line serves as a measure of the number of Ni atoms ionized at the 2p level, and can thus be used to normalise the intensities of the Lα band from alloys of various concentrations. In this way Wenger *et al*, assuming that the changes in the intensity of the Lα band are due to changes in the d-electron concentration in the valence band around the transition metal ion, show that the number of 3d electrons on the transition metals increase with increasing concentration of aluminium. In a similar manner, by measuring the Al K$\alpha_{1,2}$-line and the Al Kβ-band, these investigators show that the number of p-electrons on the aluminium atoms increases with the concentration of the transition metal (see figure 4). The number of d-electrons on the transition-metal atom, for alloys of the same concentration, increases from manganese to iron, reaching a maximum at cobalt, and decreases through nickel to copper. The results obtained from such investigations greatly enhance the knowledge to be derived from soft x-ray emission spectra, and emphasises the

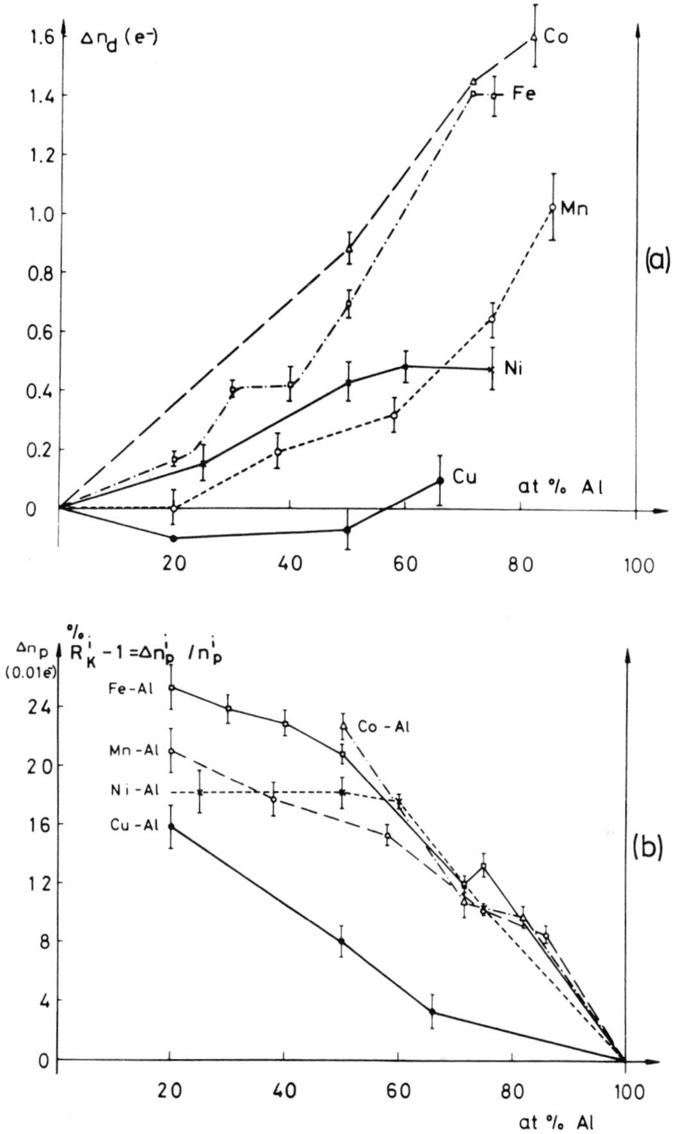

Figure 4 (a) Number of 3d electrons, n_d, on Mn, Fe, Co, Ni and Cu atoms in binary alloys of these metals with aluminium. (b) Relative change $\Delta n_p^i/n_p^i$ with alloying of the number of p electrons inside the aluminium ion, and estimated change Δn_p of the number n_p of p electrons on an Al atom (Wenger *et al* 1971).

need for measurements of the absolute change in emission intensity as well as changes of shape of the emission bands.

D. Hybridisation of States

The theories outlined above take no account of the effects of hybridisation of the valence-electron states. In many alloy spectra extremely sharp low-energy peaks have been observed. For example Marshall *et al* (1969) explained the low-energy peak observed in the Al $L_{2,3}$-emission spectra from aluminium-silver alloys in terms of the hybridisation of the silver d-bands with the aluminium s-p band. The alloy $AuAl_2$ is another example. Switendick and Narrath (1969) have made a detailed calculation of the band structure of this alloy, and have shown that hybridisation of the d bands of gold with the s, p band of aluminium, give rise to a greatly enhanced density of s-states at the bottom of the valence band of the alloy in the region of the aluminium atoms, while the states in the region of the gold atoms remain predominantly d-like. Consequently a peak is observed in the Al $L_{2,3}$-emission spectrum for the alloy, attributed to the increase in the density of states around aluminium coupled with a contribution from the overlapping d-states of gold. The density of states calculated by Switendick and Narrath with the spectrum obtained by Williams *et al* (1971) are shown in figure 5. A further investigation by Kapoor *et al* 1972, of the Al $L_{2,3}$-emission and Au $N_{6,7}$-emission from $AuAl_2$, together with the spectra for two additional alloys in the same series AuAl and Au_2Al, also support the general interpretation of the spectrum of $AuAl_2$ in terms of hybridisation and the bandstructure calculated by Switendick and Narrath.

A pronounced low-energy peak occurs in the Al $L_{2,3}$-emission spectra of alloys of aluminium with noble-metals (Fabian *et al* 1971, Kapoor *et al* 1972), and with second transition-series

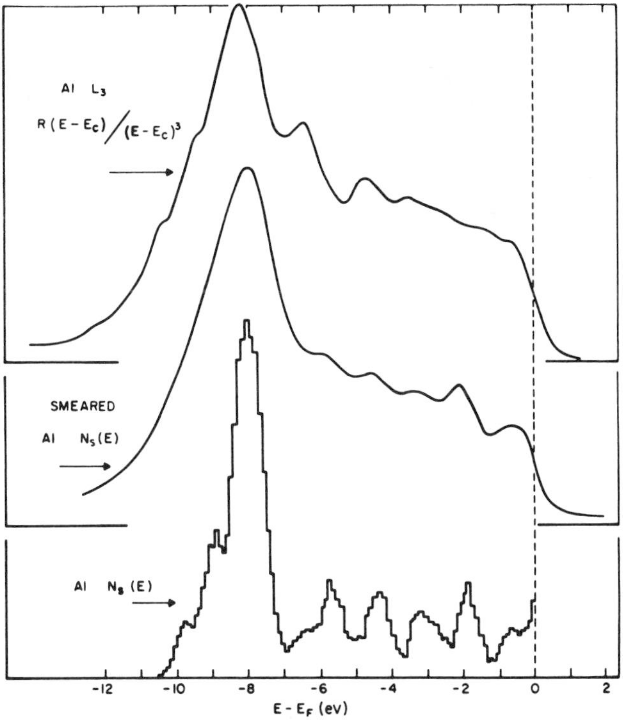

Figure 5 Comparison of density of states for $AuAl_2$ with the soft x-ray emission spectrum. Lower curve: the s-orbital state density at aluminium sites, estimated by Switendick (1971). Middle curve: approximate effect of final-state lifetime and instrumental broadening on the density of states. Upper curve: measured Al L_3-spectrum (Williams *et al* 1971).

metals (Al-Nb and Al-Pd; Watson *et al* 1971). The appearance of this peak strongly suggests the formation of semi-localised states of s-d symmetry around the aluminium atoms. In the case of the noble metals with aluminium, Fischer and Baun (1966) have shown that the Al K-emission spectra also exhibit a low-energy peak. In general peaks in the K-emission spectra - due to valence-band p-states - occur at higher energies than the s-d peaks and have a broader structure. Figure 6 demonstrates this for the aluminium noble metal alloys. With Al-Nb (figure 9) where a strong peak occurs in the Al $L_{2,3}$-spectra, no second peak is observed in the Al K-spectra; whereas in

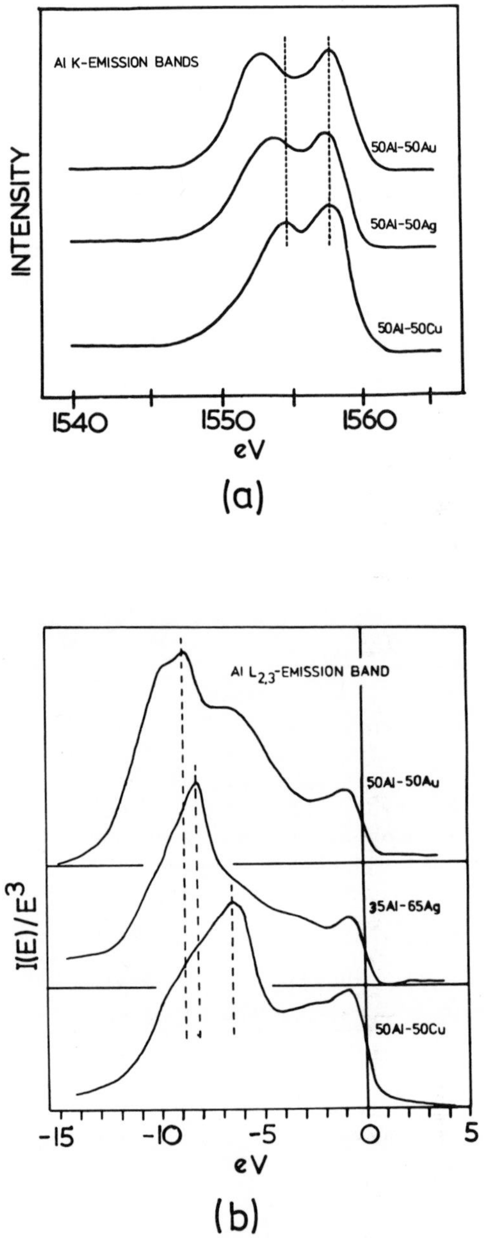

Figure 6 (a) Al K-emission bands from aluminium noble-metal alloys (Baun and Fischer 1967).
(b) Al $L_{2,3}$-emission bands from aluminium noble-metal alloys.

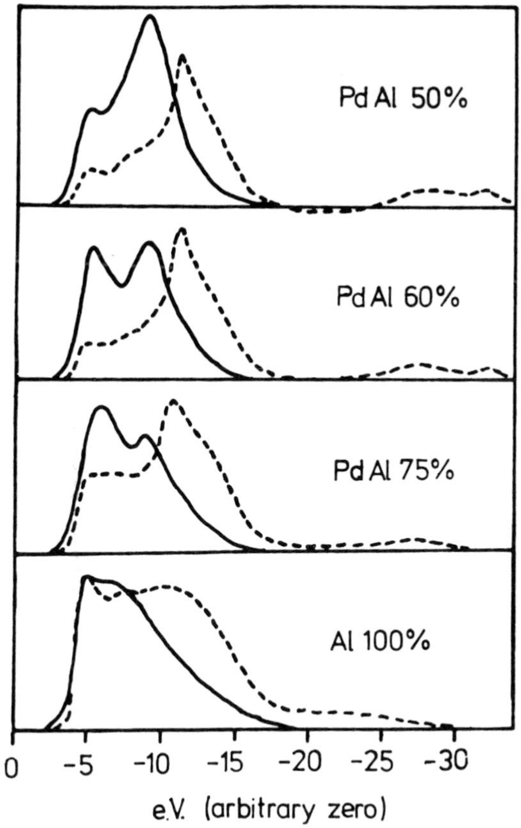

Figure 7 Soft x-ray emission spectra of aluminium in aluminium-palladium alloys. Solid curves are the Al K-emission spectra, dashed curves are the Al $L_{2,3}$-emission spectra.

Al-Pd (figure 7) the Al K-emission band is split, the low-energy peak becoming predominant at high palladium concentrations. Nemnonov *et al* (1971) have investigated the $K\beta_5$ bands of alloys of vanadium with cobalt, rhodium, iridium, nickel, palladium and platinum. They observed a splitting of the spectrum into two peaks. This becomes more pronounced as the atomic number of each series is increased, and is even more so on going to the higher transition series (figure 8). The splitting of the bands has been related to the energy separa-

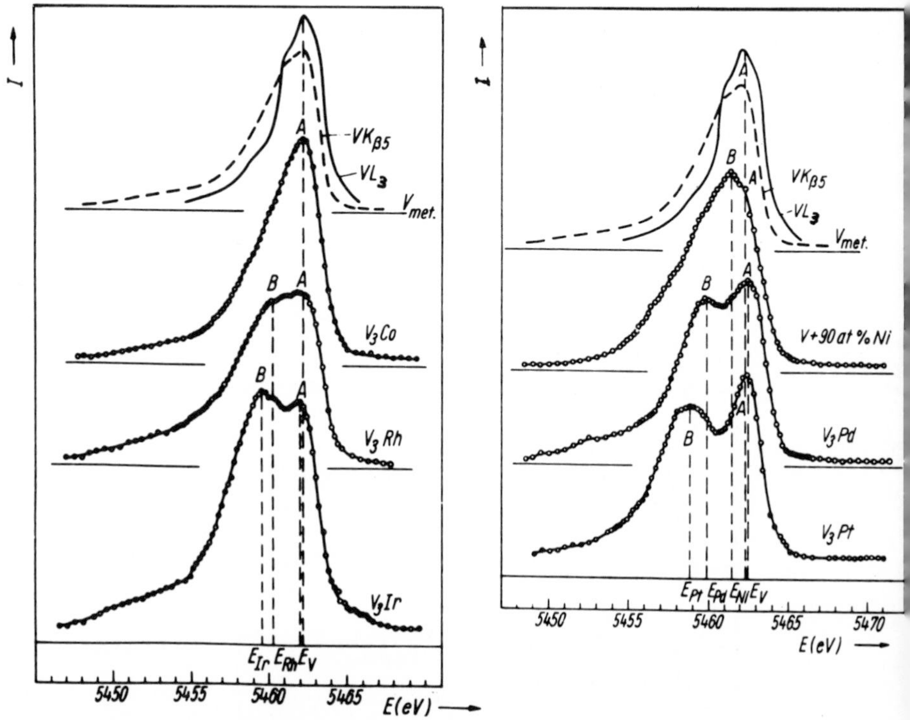

Figure 8 Vanadium $K\beta_5$-emission bands for pure vanadium and alloys with cobalt, rhodium, iridium, nickel, palladium and platinum (Nemnonov *et al* 1971).

tion of the component d-bands; the p-states, involved in the Kβ-emission, arise from hybridization with the d-bands.

In general, aluminium alloyed with transition or noble metals shows an enhancement of the intensity at the low energy region of the $L_{2,3}$ emission band with respect to that at the Fermi energy. This enhancement becomes sharper as the atomic number of the second component is increased, and for aluminium noble and second transition series alloys assumes the shape of a semilocalized bound state at the bottom of the band. The aluminium K-emission spectra show a single peak where only the lower d-bands of the pure metal are filled and show a double peak where the upper d-bands of the pure metal are filled.

E. The Density of States for Different Components of the Alloys

Velicky *et al* (1968) and Soven (1969), using the coherent potential approximation have predicted that the density of states in concentrated disordered alloys will vary greatly from the region of atom A to the region of atom B. The degree of variation will depend on the similarity of the alloying components. We have seen that for neighbouring transition-metal alloys the rigid-band model appears to hold to a first approximation. With alloys of aluminium with transition metals, the rigid-band model certainly does not apply, and the predictions of Soven and Velicky *et al* provide a more plausible interpretation for the spectra. Figure 9 shows the Al

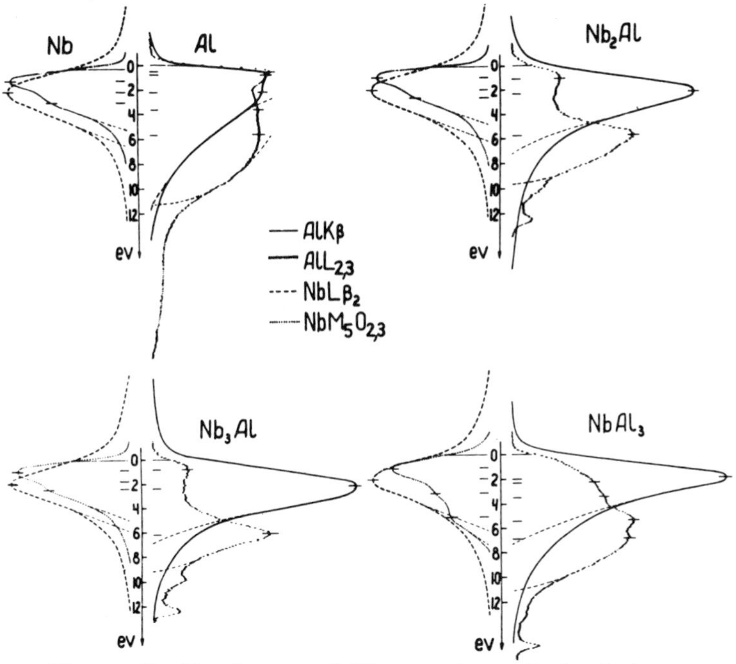

Figure 9 The $L_{2,3}$ and $K\beta_x$ spectra of aluminium and the $L\beta_2$ and $M_5O_{2,3}$ spectra of niobium from a series of aluminium-niobium alloys.

$L_{2,3}$ and K emission bands, and the Nb $L\beta_2$ and $M_5O_{2,3}$ emission spectra for a series of aluminium-niobium alloys (Watson *et al* 1971, V.V. Nemoshkalenko unpublished). We can see that the spectra from the various components show features that are certainly not common to both species. The bandwidths may in fact remain the same - as indicated above the low-energy limit in many of the spectra is difficult to determine - but the density of states in the region of the component atoms differs considerably. It is important to note that relative intensities, such as in the spectra of figure 9, are meaningless; only the shapes of the emission bands have significance.

An important aspect of the aluminium-niobium results is the gross change in shape of the Al $L_{2,3}$-emission spectra of Al_3Nb and $AlNb_2$, suggesting a strong dependence on crystal structure. In general it has been found that metallic alloy spectra are more concentration-dependent than they are crystal-structure dependent. This conclusion has been reached by Baun and Fischer (1967) for Al-Cu alloys, and by Fabian *et al* (1971) for Al-Ag alloys. Baun and Fischer found that the peak separation in the double-peaked Al K-emission spectra of a series of aluminium-copper alloys varies linearly with composition.

F. Bound States

We find little evidence from x-ray emission spectra to confirm the prediction by Friedel (1958) that resonant bound states are formed in the valence band. This is hardly surprising since these bound states arise from isolated impurities in an electron sea, and the alloys generally examined by soft x-ray emission are too concentrated. To obtain accurate spectra of dilute alloys is extremely difficult. The question arises whether such states will be formed in concentrated alloys. Curry and Harrison (1970) report Al $L_{2,3}$-emission spectra for alloys of aluminium with first transition-series, of roughly 50%-50% composition, and relate the spectral changes at the high-energy limit of the bands to the contribution from the transition-metal d-states which, they suggest, may be in resonant bound states. Evidence supporting the existence of resonant bound states is supplied by experiments more sensitive to small amounts of impurities, such as optical properties and ultraviolet photoemission spectroscopy (see for example Spicer, this Volume p. 7). Virtual bound states at the impurity atoms in a dilute alloy have been predicted by Stott (1969), but again the experimentalist is faced with measuring the solute spectrum in every dilute alloy and no convincing evidence exists from soft x-ray emission to support the prediction. However, as outlined above, there exists strong evidence of semi-localised states in a number of concentrated aluminium alloys. The Mg $L_{2,3}$-emission spectra of magnesium noble-metal alloys exhibit similar behaviour (Curry and Harrison 1970).

4. PLASMON SATELLITE EMISSION BANDS OF ALLOYS

Extremely weak low-energy satellites of the main emission band for pure light metals have been reported by a number of investigators (Rooke 1963, Fomichev 1967, Watson *et al* 1968, Neddermeyer and Wiech 1970). These are interpreted as arising from an attenuation of the main emission band photons by the creation of a plasmon within the metal. The resulting satellite is observed at an energy $\hbar\omega_p$ below the parent band, where $\hbar\omega_p$ is the bulk plasmon energy. Because of their low intensity (typically 3-5% of the parent band) it is possible to measure plasmon satellites only when the nature of the background radiation is known. The latter is complex both for heavy elements and for alloys and very few plasmon satellite measurements for alloys have therefore been reported.

Dimond (1970) describes the measurement of $\hbar\omega_p$ from plasmon satellite bands for a number of aluminium-magnesium alloys.

The shapes of the satellite bands are distorted due to overlap: with the magnesium $L_{2,3}$ spectrum in the case of the plasmon satellite for aluminium, and with the Mg L_1-L_3 line and second-order of the Al $L_{2,3}$ spectrum in the case of the satellite for magnesium. Nonetheless, Dimond reports bulk plasmon energies for the alloys in good agreement with characteristic energy loss data obtained by Spalding and Metherell (1968), indicating that the plasmon energies vary almost linearly from 15.2eV for pure aluminium to 10.6eV for pure magnesium (see figure 10). Many of the alloys studied by both

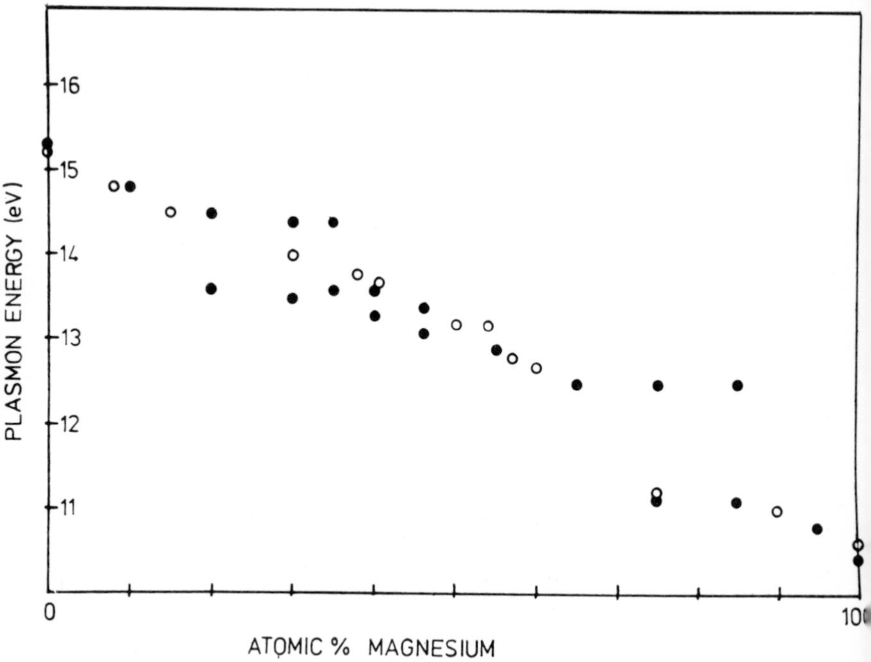

Figure 10 Bulk plasmon energies for Al-Mg alloys: measured by Dimond (1970) from soft x-ray spectra (open circles), and by Spalding and Metherell (1968) from characteristic energy loss measurements (filled circles).

Dimond and Spalding and Metherell contain two phases. In soft x-ray emission the difference in plasmon energies from the two phases cannot be distinguished and just one plasmon energy for these alloys has been quoted by Dimond. On the other hand, Spalding and Metherell were able to measure the characteristic energy loss from each phase; the incident electron beam in their case was the well-defined beam of an electron microscope. This results in the two plasmon energies for a particular alloy, shown in figure 10.

Lindsay (1969) has measured the aluminium $L_{2,3}$ plasmon satellite for alloys of copper, silver and zinc. In all cases the alloys were sufficiently dilute to allow the effects of self absorption and overlap of higher-order spectra to be neglected. Three alloy satellite bands are shown in figure 11, compared with the pure aluminium satellite; the intensities can be measured as a percentage of the parent band intensity as noted below the figure. We observe the remarkable result that the addition of 5-10at.% of the second component has a large effect on the intensity of the satellite.

5. SUMMARY

The information that can usefully be derived from the soft x-ray emission spectra of alloys has been briefly examined, and some examples discussed, of the changes in the spectra of pure metal on alloying. There is little doubt that much useful information can be obtained from alloy spectra; while combining soft x-ray emission with x-ray photoemission studies, and making absolute intensity measurements, can greatly increase our knowledge of the behaviour of electrons throughout the valence band. However, interpretation of the experimental results is difficult; the measurements involve the density of states, the transition probability and several complicated many-body effects. Therefore, the experimentalist must rely heavily on the results of theoretical calculations in order to eliminate one or more of the above variables, and to derive a reliable density-of-states curve from the data. Unfortunately

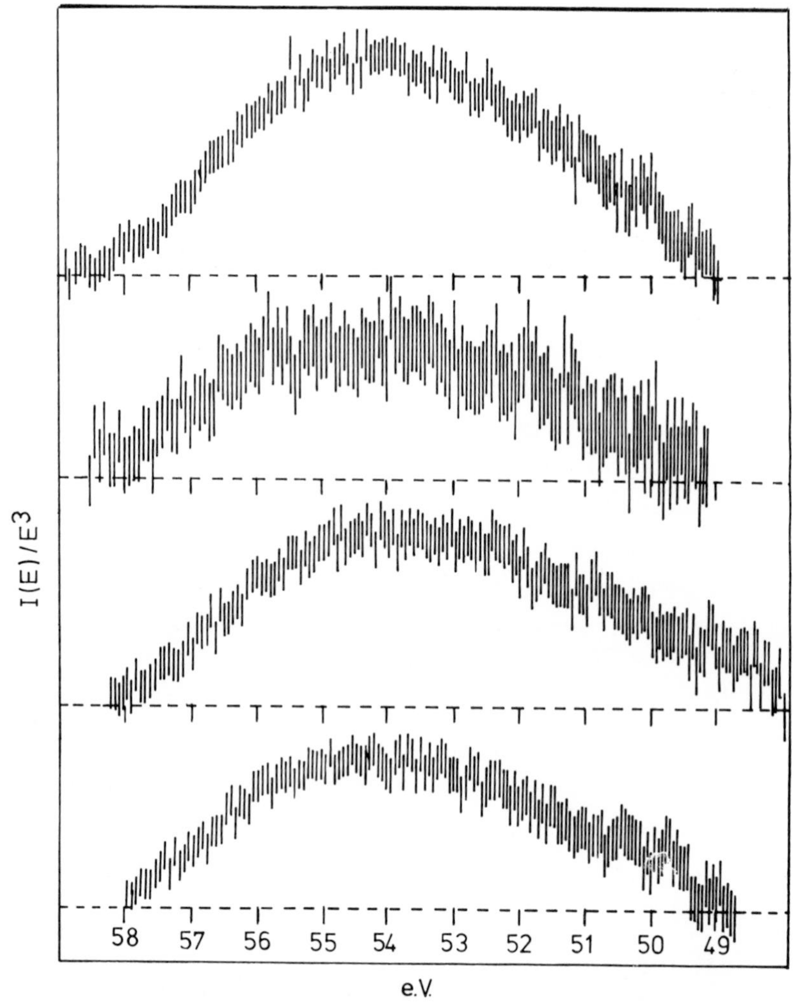

Figure 11 Al $L_{2,3}$-emission plasmon satellite bands for Al-Zn alloys (Lindsay 1969). (a) pure aluminium; intensity 3.9% of parent band. (b) Al-5at% Zn; intensity 6.3% of parent band. (c) Al-10at% Ag; intensity 3.2% of parent band. (d) Al-10at% Cu (2-phase alloy); intensity 7.4% of parent band. Error bars indicate the 67% certainty limits.

few full theoretical treatments of specific alloy systems have been done, due to the complexity of the alloy problem. However, improvements in alloy theory are being made, the coherent potential approximation has provided theoretical support for observed differences in component spectra from alloys, and we can expect that scattering theories currently being developed (see for example Gyorffy, this Volume p.641) will advance the interpretation of alloy spectra. Compounds for which a great deal of experimental information exists, have not been included in this discussion. These materials lend themselves better to interpretation in terms of standard chemical bonding theory, a subject which is adequately covered elsewhere in this Volume by Fischer (p.669) and by Slater and Urch (p.655).

Acknowledgments

It is a pleasure to record my indebtedness to Dr. D.J. Fabian, as a friend and colleague, for his advice and constructive criticism, to the late Professor E.C. Ellwood for his encouragement and support, and to the Science Research Council for a Fellowship and funds in support of this research.

REFERENCES

Baun, W.L. and Fischer, D.W. (1967); J. App. Phys. 38, 2092.

Bolsaitis, P. and Skolnick, L. (1968); Trans. Met. Soc. AIME, 242, 215.

Brouers, F. (1967); Phys. Stat. Sol. 22, 213.

Curry, C. (1968); in "Soft X-Ray Band Spectra" ed. D.J. Fabian (Academic Press, London) p.173.

Curry, C. and Harrison, R. (1970); Phil. Mag. 21, 659.

Dimond, R.K. (1970); Thesis, University of Western Australia.

Donahue, R.J. and Azároff, L.V. (1967); J. Appl. Phys. 38, 2813.

Fabian, D.J. (1970); Mater. Res. Bull. 5, 591.

Fabian, D.J., Lindsay, G.McD. and Watson, L.M. (1971); in "Electronic Density of States", NBS spec. publ. 323, p.307.

Fischer, D.W. and Baun, W.L. (1966); Internal Research Publ. AFML-TR-66-191, Airforce Materials Laboratory, Ohio.

Eggs, J. and Ulmer, K. (1968); Z. Phys. 213, 293.

Fomichev, V.A. (1967); Fiz. Tverd. Tela, 9, 1833.

Friedel, J. (1958); Nuovo Cimento 7 (suppl.), 287.

Glick, A.J., Longe, P. and Bose, S.M. (1968); in "Soft X-Ray Band Spectra" ed. D.J. Fabian (Academic Press, London) p.319.

Hedin, L. (1968); in "Soft X-Ray Band Spectra" ed. D.J. Fabian (Academic Press, London) p.337.

Jacobs, R.L. (1969); Phys. Letts 30A, 523.

Jones, H. (1934); Proc. Roy. Soc. A144, 225.

Kapoor, Q.S., Watson, L.M., Hart, D. and Fabian, D.J. (1972); Sol. State Comm. 11, 503.

Lindsay, G.M. (1969); Ph.D. Thesis, University of Strathclyde.

Marshall, C.A.W., Watson, L.M., Lindsay, G.M., Rooke, G.A. and Fabian, D.J. (1969); Phys. Letts. 28A, 579.

Nagel, D.J. (1970); Advances in X-Ray Analysis, 13, 182.

Neddermeyer, H. and Wiech, G. (1970); Phys. Letts. A31, 17.

Nemnonov, S.A. and Finkelshtein, L.D. (1966); Fiz. Metallov i Metallovedenie 22, 538.

Nemnonov, S.A., Kurmaev, E.Z., Ishmukhametov, B.K. and Belash, V.P. (1971); Phys. Stat. Sol.(b) 46, 77.

Nozières, P. and de Dominicis, C.T. (1969); Phys. Rev. 178, 1097.

Parratt, L.G. (1959); Rev. Mod. Phys. 31, 616-45.

Rooke, G.A. (1963); Phys. Letts. 3, 234.

Soven, P. (1969); Phys. Rev. 178, 1136.

Spalding, D.R. and Metherell, A.J.F. (1968); Phil. Mag. 18, 41.

Stott, M.J. (1969); J. Phys. C. 2, 1474.

Switendick, A.C. and Narath, A. (1969); Phys. Rev. Lett. 22, 1423.

Switendick, A.C. (1971); in "Electronic Density of States" NBS Spec. publ. 323, p.297.

Tomboulian, D.H. and Pell, E.M. (1951); Phys. Rev. 83, 1196.

Townsend, J.R. (1953); Phys. Rev. 92, 556.

Varley, J.H.O. (1954); Phil. Mag. 45, 887.

Velicky, B., Kirkpatrick, S. and Ehrenreich, H. (1968); Phys. Rev. 175, 747.

Watson, L.M., Dimond, R.K. and Fabian, D.J. (1968); in "Soft X-Ray Band Spectra" ed. D.J. Fabian (Academic Press, London) p.45.

Watson, L.M., Kapoor, Q.S. and Nemoshkalenko, V.V. (1971); J. de Phys. 32, C4-325.

Wenger, A., Burri, G. and Steinemann, S. (1971); Sol. State Comm. 9, 1125.

Williams, M.L., Dobbyn, R.C., Cuthill, J.R. and McAlister, A.J. (1971); in "Electronic Density of States" NBS spec. publ. 323, p.303.

X-RAY EMISSION BAND SPECTRA AND ELECTRONIC STRUCTURE OF ALLOYS OF LIGHT ELEMENTS

H. Neddermeyer

Sektion Physik der Universität München

1. INTRODUCTION

In recent years the interpretation of soft x-ray emission spectra has made good progress. With a detailed knowledge of the electronic band structure, transition probabilities, and many-body-effects it has been possible to calculate the shape of emission spectra of a few pure elements; as demonstrated for silicon by Klima (1970), and for aluminium by Smrčka (1971). However, the situation is much more complicated in the case of alloys, where the problems are far from being solved.

In the present paper we restrict ourselves to discussion of binary alloys of light elements, and especially to the system aluminium-magnesium, where atomic d-electrons are not involved in the electronic band structure. A survey will be given of recent measurements of x-ray emission band spectra of a series of Al-Mg alloys and an interpretation of the spectra in the light of recent theoretical work will be attempted.

The complete x-ray emission spectrum for an Al-Mg alloy consists of the Kβ and $L_{2,3}$ emission bands of both aluminium and magnesium. A study of the Al and Mg Kβ-emission bands for Al_3Mg_2 and Al_2Mg_3 by Farineau (1938) was one of the first investigations of an alloy system. Farineau found for each alloy that the Al and Mg Kβ-emission bands were of similar form and quite different from those for pure Al and pure Mg; with the bandwidths of the Al and Mg Kβ-bands lying between those for the pure elements. These results seem to support the rigid-band model introduced by Jones (1934).

However, the results of subsequent investigations of the Al and Mg $L_{2,3}$-emission bands of a number of Al-Mg alloys by Das Gupta and Wood (1955), Gale and Trotter (1956), Crisp and Williams (1960), and Appleton and Curry (1965) conflict with the simple rigid-band model. As is well known, the $L_{2,3}$-emission bands of both constituents prove to have different shapes and bandwidths. Appleton and Curry found that the bandwidths of the spectra of Al and Mg for a certain Al-Mg alloy remain about equal to those of the corresponding pure metals, in sharp contrast to the simple rigid-band model which requires the same bandwidths and similar shapes for the Al and Mg emission bands.

The conclusion was that the rigid-band model and even a common-valence-band model is not applicable for Al-Mg alloys and attempts have been made to explain the different shapes of the emission band spectra with some form of the two-band model proposed by Varley (1954). It is assumed that separate localized valence bands must be associated with each type of atom (Rooke 1969, Fabian 1970).

The failure of the simple rigid-band model as a basis for the interpretation of the x-ray emission band spectra is not surprising; the reason was first given by Mott (1936) who proposed the concept of localized screening of the charge surrounding a dissolved atom of valence higher than that of the solvent metal. This leads to a non-uniformity of the valence-electron distribution within the crystal. In the case of Al-Mg alloys this charging effect is expected to be large, since the difference in valence of the two constituents is of the same magnitude as the smaller of the two valences; in such cases the rigid-band model does not apply, as discussed by Stern (1966).

The question now arises whether we have to reject also a common valence band model as did Appleton and Curry. It is true there is some evidence for a two-band model which could possibly explain the different shapes of the $L_{2,3}$ emission bands. On the other hand the results of Farineau for the Kβ-emission bands do support a common valence band. Thus the situation concerning the K and L emission is contradictory. Therefore because experimental techniques have been considerably improved we decided to remeasure systematically the x-ray emission band spectra for a complete series of Al-Mg alloys including alloys with well-defined phases.

More recent measurements by Fischer and Baun (1966) and by Nemoshkalenko and Gorskii (1968) of the Al Kβ-emission only for Al-Mg alloys will not be discussed here, since only consideration of the spectra of both constituents can give full insight into the electronic structure of an alloy system.

2. EXPERIMENTAL

A. Samples

In addition to the pure metals Al (99.99% purity) and Mg (99.99% purity) the following Al-Mg alloys were investigated: $Al_{5\%}Mg_{95\%}$ (*i.e.* 5at.% Al and 95at.% Mg), $Al_{10\%}Mg_{90\%}$, $Al_{30\%}Mg_{70\%}$; and $Al_{12}Mg_{17}$, Al_3Mg_2, and Al_2Mg. Metallographic and x-ray diffraction studies were made to check the crystallographic order. According to the phase diagram (Eickhoff and Vosskühler, 1953) $Al_{12}Mg_{17}$ (γ-phase) and Al_3Mg_2 (β-phase) are single-phase alloys. The dilute alloys $Al_{5\%}Mg_{95\%}$ and $Al_{10\%}Mg_{90\%}$ are single-phase (δ) at 430°C, while at room temperature they slowly decompose into two phases; in the case of $Al_{10\%}Mg_{90\%}$ this decomposition could be detected during the x-ray emission studies. Finally, $Al_{30\%}Mg_{70\%}$ and Al_2Mg are binary phases.

From the theoretical point of view it is important to know something about the degree of order. The dilute alloys $Al_{5\%}Mg_{95\%}$ and $Al_{10\%}Mg_{90\%}$ are typical examples of disordered systems. The x-ray diffraction studies showed these alloys to have the hexagonal crystal structure of pure magnesium, with reduced lattice constants relating to the concentration of aluminium. It can be assumed that the Al atoms are statistically distributed within the hexagonal lattice and that clustering of the Al atoms is negligible. The alloys $Al_{12}Mg_{17}$ and Al_3Mg_2 should be ordered; in the case of Al_3Mg_2 some disorder could not be excluded. The crystal structure of the pure β-phase is very complicated (Samson 1965) and complete order in all parts of the sample is improbable.

B. $L_{2,3}$-Emission Bands

The Al and Mg $L_{2,3}$-emission band spectra were obtained using a concave grating spectrometer (constructed by Wiech 1966) with ultra-high vacuum conditions and photoelectric recording (Neddermeyer 1969, 1971). The x-ray tube was operated at 1-2 x 10^{-8} torr. The radiation was excited directly by bombardment voltages of 2kV and emission currents of 1-3mA. The spectra were recorded either continuously with a strip-chart recorder or step-wise using digital equipment. The influence of contamination and self-absorption could be neglected. The conclusion reached by Dimond (1967) that self-absorption could strongly influence the band shapes in case of alloys was not confirmed (Neddermeyer 1969). The spectra were corrected for the known reflectivity of the grating. The influence of the quantum efficiency of the multiplier photocathode seems to be small in this case. No correction has been made for instrumental line-width broadening, which amounts to 0.14eV and 0.07eV for the Al and Mg $L_{2,3}$-emission bands respectively.

C. Kβ Emission Bands

The Kβ-emission band spectra we remeasured with a Johann-type bent crystal spectrometer (constructed by Läuger 1968). For the Al K-emission we used a quartz crystal cut in the (10$\bar{1}$0)-plane and for the Mg K-emission an ADP (Acid di-phthalate) crystal cut in the (101)-plane. The x-rays were excited in fluorescence using the radiation of a tungsten target. The x-ray tube was run at a voltage of 10kV and emission currents up to 360mA.

The spectra were recorded using a gas-flow proportional counter moving steadily along the Rowland focal circle. The counts were stored in a specially designed interface of a Honeywell computer. After constant time intervals the number of counts was read into the memory, punched on paper tape and plotted under computer control by a Hewlett-Packard X-Y recorder. The distance travelled by the continuously moving proportional counter during one time interval was small compared to the slit width, so deterioration of resolving power due to this method could be neglected. Target contamination and systematic changes of spectra during the measurements were not observed. The effect of self-absorption could not be evaluated, self-absorption certainly has an influence on the high-energy side of the spectra. The instrumental line-width broadening was about 5 times larger than that for the corresponding L-emission bands. This means that it was of the same magnitude as the inner-level broadening.

D. Evaluation of the Spectra

The measured $L_{2,3}$-emission bands consist generally of a pronounced peak at the high-energy side and a smooth part at lower energies. The intensity distribution in the peak region was taken from step-by-step measurements, and in the low-energy region from averaged strip-chart recordings.

The final Kβ-emission spectra were obtained by summing approximately eight single scans for each sample.

The energy scale for the Kβ emission from one constituent, relative to that for its $L_{2,3}$ emission, was adjusted with the energy of the maximum of the $K\alpha_{1,2}$ line, independently measured for each alloy. The energy scale of the Mg spectra, relative to that of the Al spectra, was shifted so that the middle of the emission edges of the L spectra coincided. From the Kβ bands a constant background was subtracted, and for the $L_{2,3}$ bands a linearly varying background was assumed.

3. RESULTS

A. The Pure Metals Aluminium and Magnesium

In figure 1 the complete x-ray emission spectra of the pure elements Al and Mg are shown. The $L_{2,3}$ spectra start at the high-energy side with the emission edge, corresponding to electrons occupying states near the Fermi-level. Because of lower instrumental resolving power and broader inner levels the high-energy side for the Kβ spectra is less pronounced. The shapes of the Al spectra are now understood in detail. For magnesium a full calculation has not been performed (for an estimate see Watson *et al.* 1968); however, a quantitative interpretation of the Mg spectra should in principle be possible without difficulty. To the low-energy side of the Mg $L_{2,3}$-spectrum the interband transition $L_{2,3} \to L_1$ is observed. This line has been removed from the Mg L-spectra of the alloys. The assumption that the background is energy-independent is poor, at least in the case of the Kβ spectra. Because of self-absorption the portion of the background corresponding to the continuous bremsstrahlung varies strongly within a Kβ band, and in fact one should subtract the background indicated by the dashed lines B.

The widths of an emission band plays an important role in

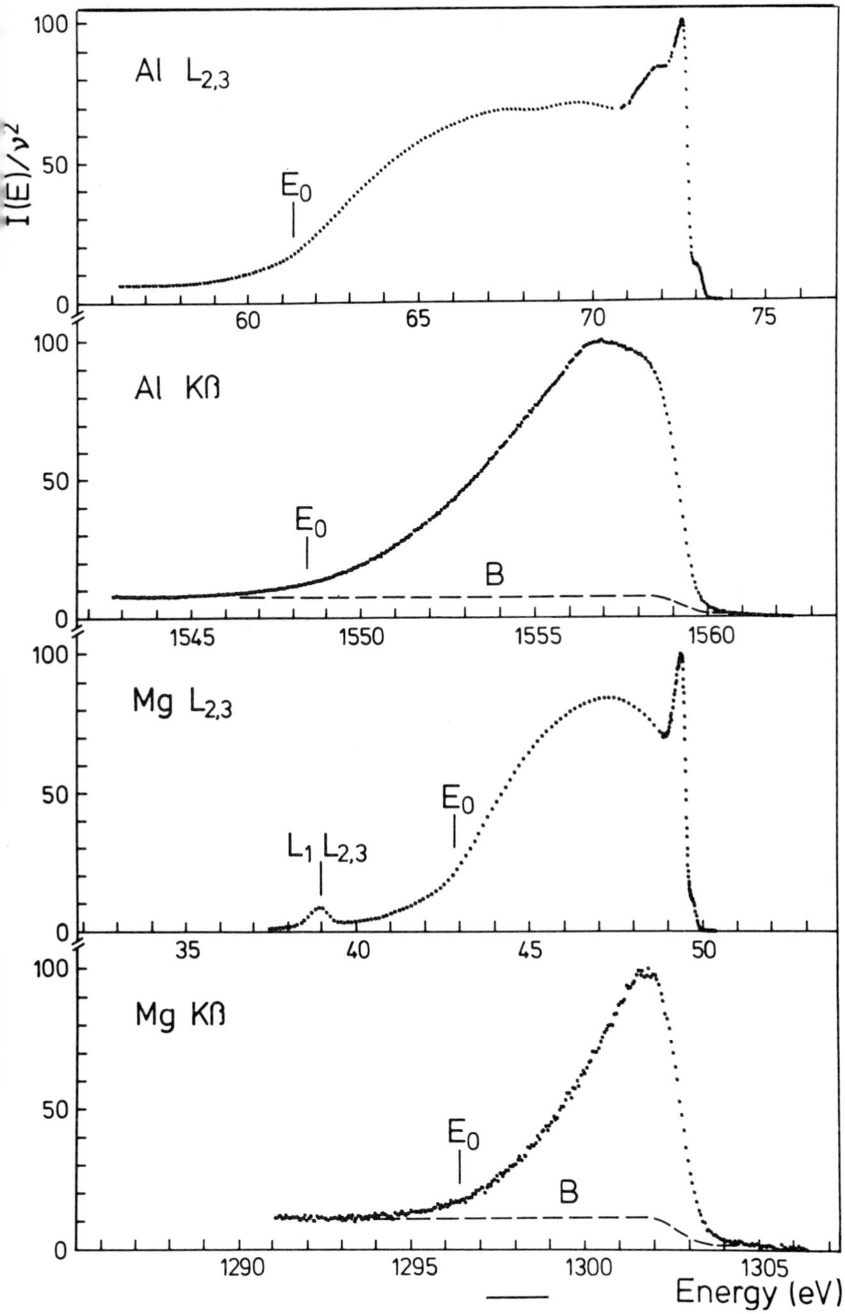

Figure 1 Kβ and $L_{2,3}$ emission band spectra of aluminium (above) and magnesium (below).

the interpretation. We determined the beginning of an emission band in the following way: according to Jones *et al.* (1934) for simple metals, with free-electron-like behaviour, the low-energy region the spectra should follow $I(E)/\nu^4 \sim (E-E_0)^{3/2}$ for K emission and $I(E)/\nu^4 \sim (E-E_0)^{1/2}$ for L emission, E_0 being the bottom of the band. We calculated $(I(E)/\nu^4)^{2/3}$ and $(I(E)/\nu^4)^2$ for the K and L emission bands respectively and extrapolated these values linearly. In fact this procedure appeared to be a good approximation for the low-energy region of the spectra of the pure elements, apart from very low energies where tailing effects due to many-body interactions caused deviations from this simple behaviour. The bottom of the band E_0 is indicated in each spectrum. The experimental bandwidths (*i.e.* the differences between the middle of the emission edges and E_0) are in good agreement with the free-electron bandwidths of 11.6eV and 7.1eV for aluminium and magnesium respectively.

B. Single Phase Non-dilute Alloys Al_3Mg_2 and $Al_{12}Mg_{17}$

In figure 2 the results for the non-dilute single phase alloy Al_3Mg_2 are presented. The general shape of the $L_{2,3}$ spectra is in rough agreement with measurements reported by Appleton and Curry (1965), only at low energies are differences observed. However, the situation is quite different in case of the Al and Mg Kβ-emission bands. We obtained the very surprising result that the two Kβ-emission bands are identical, thus confirming the measurements of Farineau. Both spectra show previously unreported characteristic fine-structure, indicated by the vertical lines. For the Kβ-emission spectra the bottom of the bands were found to lie between those for the pure metals. For the $L_{2,3}$ spectra we did not try to construct the bottom of the band, as the low-energy region of the Mg $L_{2,3}$-emission band has an entirely concave

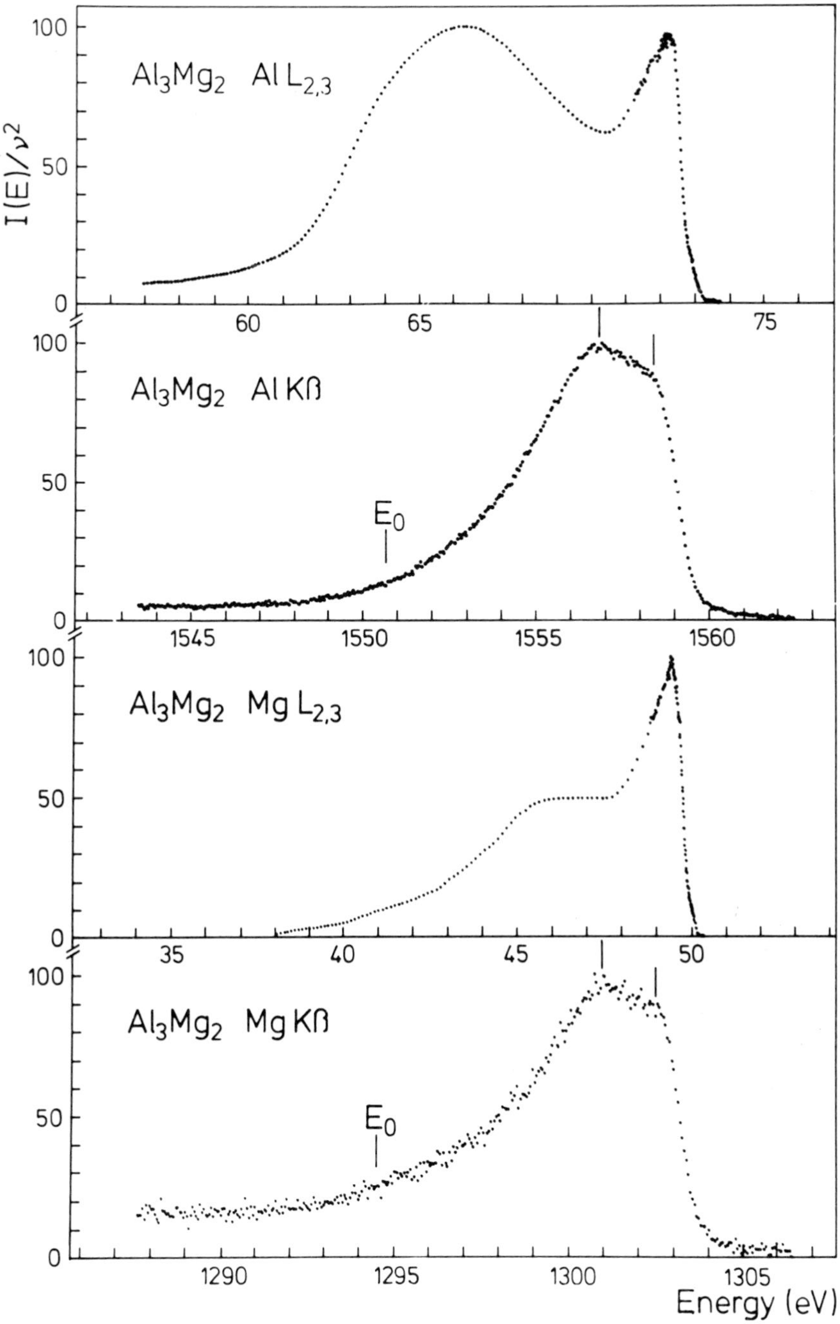

Figure 2 Kβ and $L_{2,3}$ emission band spectra of Al_3Mg_2.

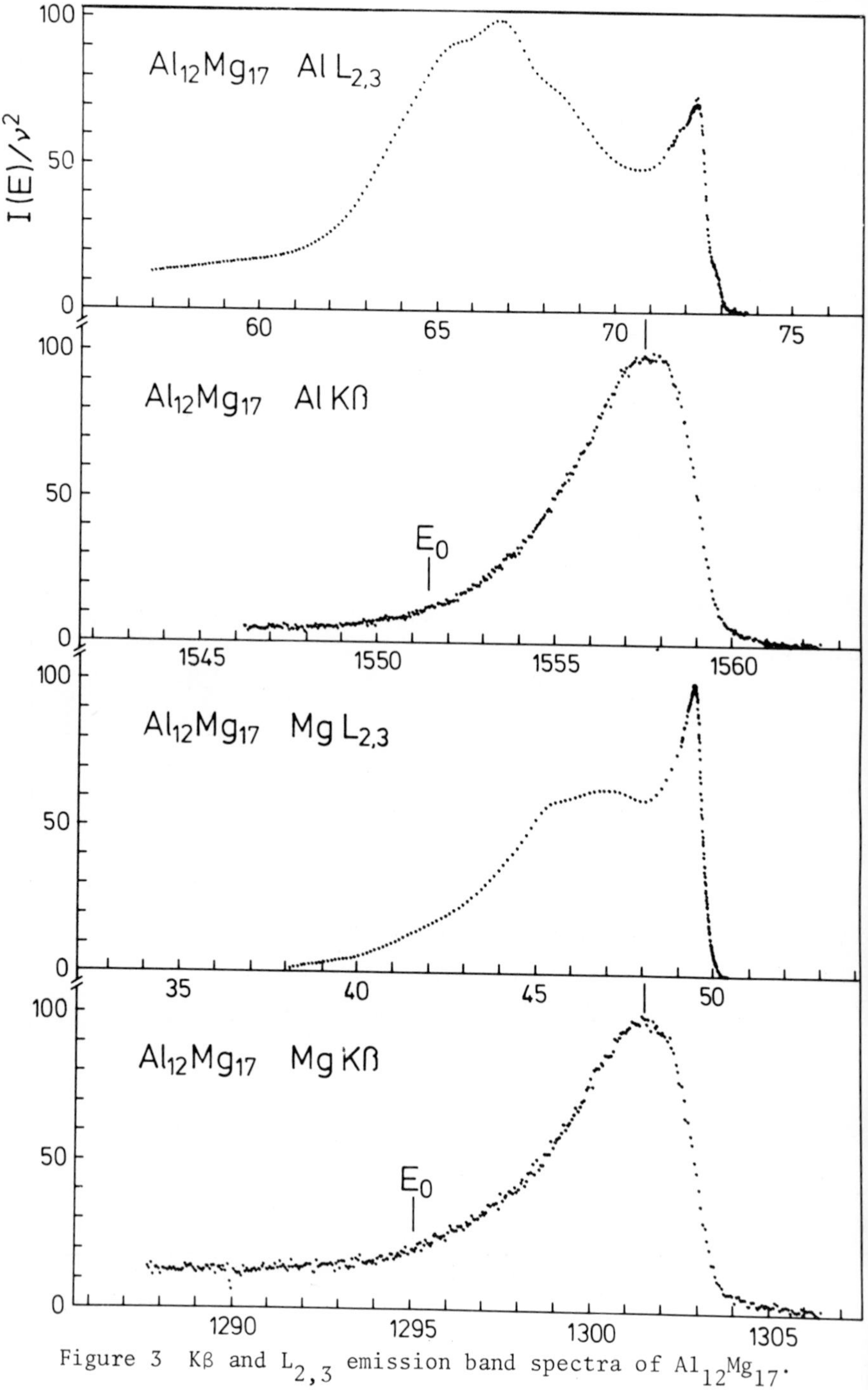

Figure 3 Kβ and $L_{2,3}$ emission band spectra of $Al_{12}Mg_{17}$.

shape and a parabolic extrapolation is not possible. A simple theoretical model by Jacobs (1969) appears to provide an explanation for the difference between the Al $L_{2,3}$ and Mg $L_{2,3}$ spectra.

The spectra for $Al_{12}Mg_{17}$ show a quite similar behaviour (figure 3). Again the $L_{2,3}$ emission bands reveal a different form and the Kβ emission bands are practically identical. On the other hand, differences from the K-emission spectra of Al_3Mg_2 are clearly visible: the two fine-structure features in Al_3Mg_2 have disappeared, and the bottom of the band has moved to higher energies indicating a reduction of the bandwidth. However, it is obvious that for alloys a bandwidth determined by this method is a satisfactory criterion only for a comparison of Kβ spectra. Solely in the case of the pure elements does it seem possible to make conclusions about the actual bandwidth of occupied electron states.

Now the question is are the Al $L_{2,3}$ and Mg $L_{2,3}$ as different as they appear on first view. In fact, we observe certain relationships in the shape of the $L_{2,3}$ spectra for both Al_3Mg_2 and $Al_{12}Mg_{17}$ (Neddermeyer 1969, 1971):

1) The general shape of the emission bands is about the same; the $L_{2,3}$ spectra of Al_3Mg_2 are smooth whereas the $L_{2,3}$ spectra of $Al_{12}Mg_{17}$ have pronounced fine structure.

2) The widths of the peaks agree fairly well.

3) The emission edges have the same width when instrumental broadening (the spectral window) is taken into account.

Thus a closer examination shows that the L-spectra for the concentrated alloys indeed have some common features.

C. Single Phase Dilute Alloys $Al_{5\%}Mg_{95\%}$ and $Al_{10\%}Mg_{90\%}$

Figures 4 and 5 show the results for the dilute alloys $Al_{5\%}Mg_{95\%}$ and $Al_{10\%}Mg_{90\%}$. No remarkable differences between the Mg emission spectra for the alloys and for pure magnesium could be found. In contrast to the single-phase non-dilute alloys, the Al emission spectra for both $Al_{5\%}Mg_{95\%}$ and $Al_{10\%}Mg_{90\%}$

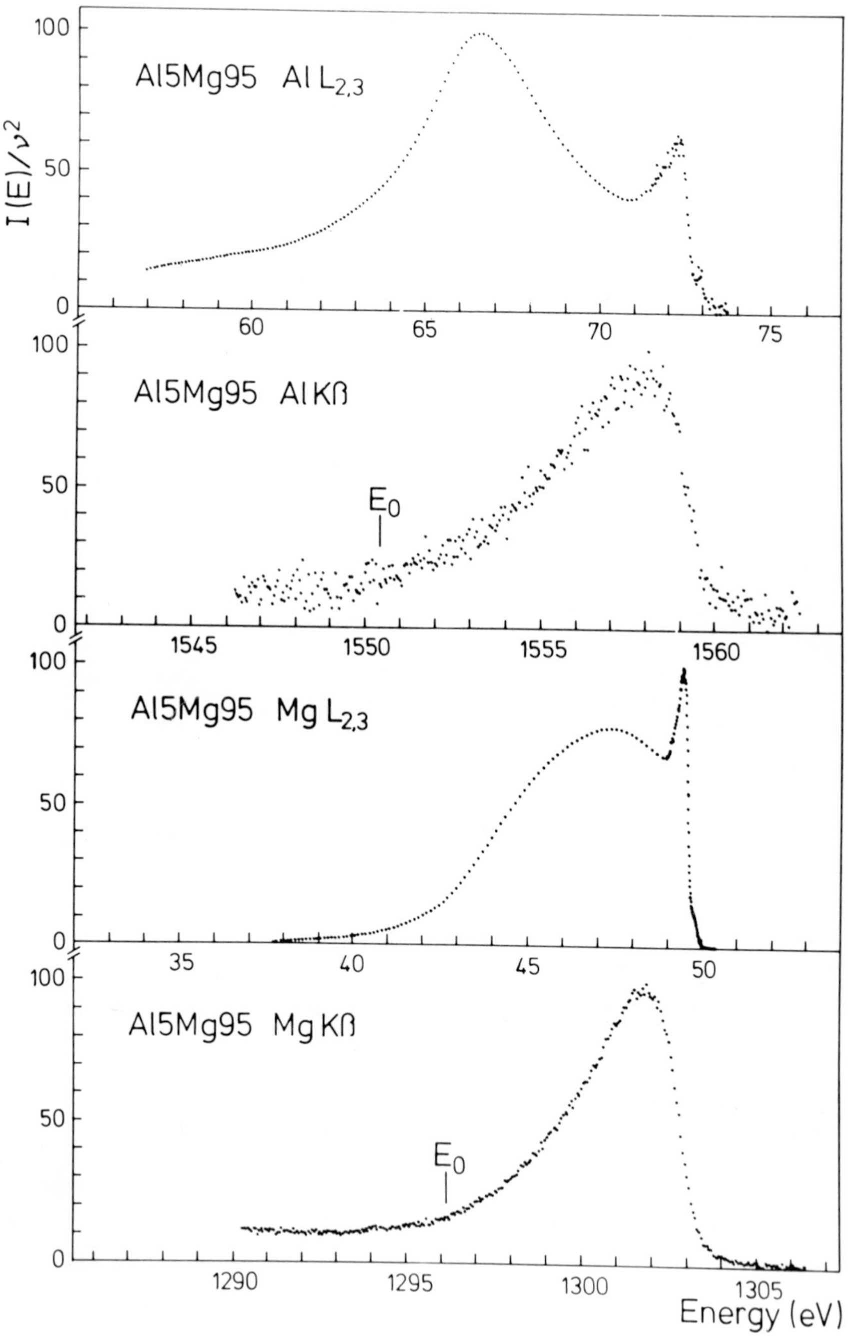

Figure 4 Kβ and $L_{2,3}$ emission band spectra of $Al_{5\%}Mg_{95\%}$.

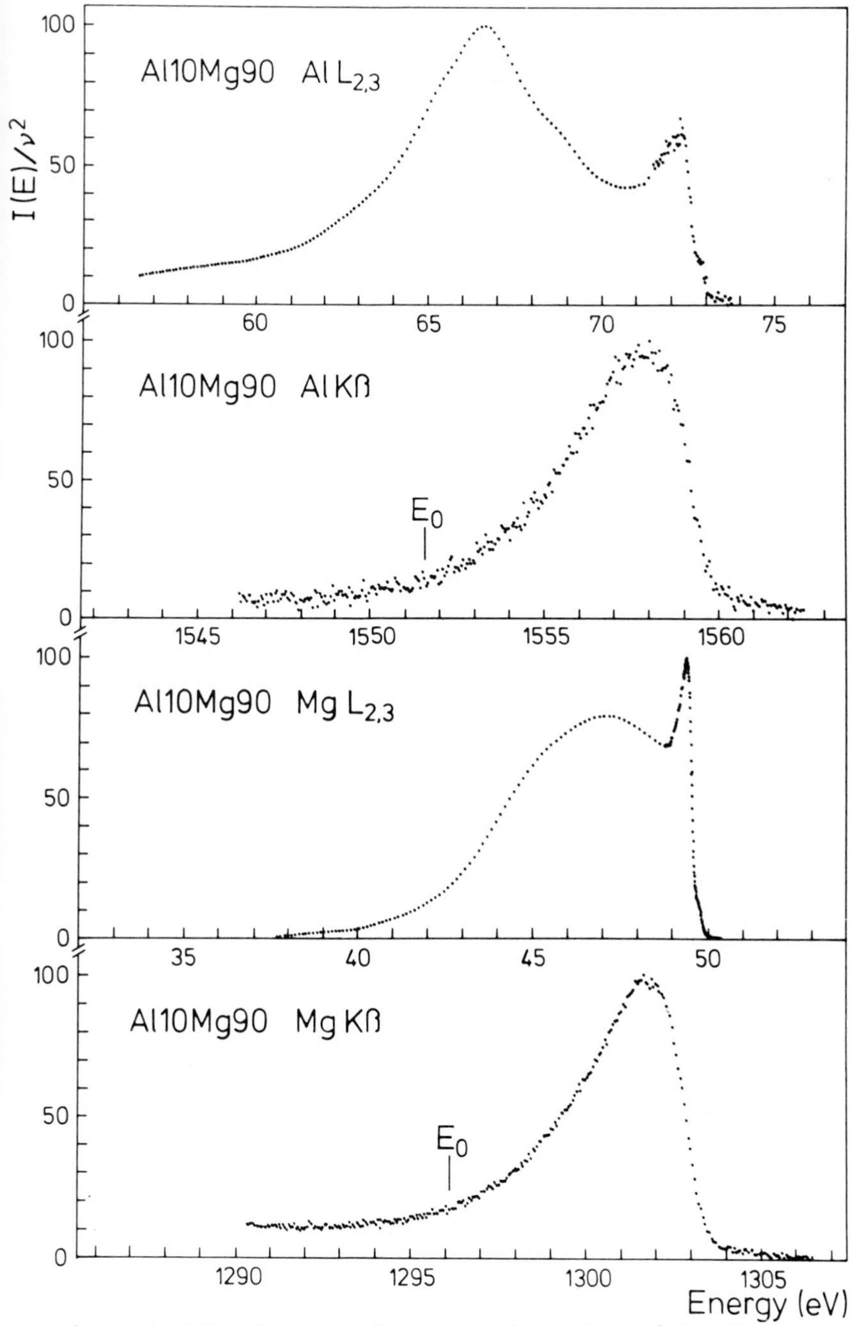

Figure 5 Kβ and $L_{2,3}$ emission band spectra of $Al_{10\%}Mg_{90\%}$

bear practically no relation to the corresponding Mg spectra, but relate more to the Al spectra of $Al_{12}Mg_{17}$. However, characteristic differences exist for the L spectra: the peak-to-maximum ratio is decreased and the low-energy side moves to higher energies compared with $Al_{12}Mg_{17}$. These differences allow us to conclude that in the main the Al $L_{2,3}$-spectra originate in emission from aluminium atoms that are statistically distributed within the dilute alloys, and not from clusters of Al atoms or from the Al atoms in clustered arrangements of Al and Mg atoms similar to the γ-phase.

D. Two-phase Alloys Al_2Mg and $Al_{30\%}Mg_{70\%}$

The experimental results for the binary phases Al_2Mg (figure 6) and $Al_{30\%}Mg_{70\%}$ (figure 7) appear to be more complicated. One would expect the emission spectra to be the superposition of the spectra of the pure phases according to their quantity ratio. This does not occur for the L spectra. In fact, the spectra of Al_2Mg and $Al_{30\%}Mg_{70\%}$ show similarities respectively to those of Al_3Mg_2 and $Al_{12}Mg_{17}$ with the exception of the Mg $L_{2,3}$-emission band of $Al_{30\%}Mg_{70\%}$ which resembles more that of pure magnesium. As for the case of the non-dilute single-phase alloys the $L_{2,3}$-emission band spectra are different from one another while the Kβ-emission band spectra of one alloy are identical.

4. DISCUSSION

The most striking properties of the x-ray emission band spectra of Al-Mg alloys are:

1) The Kβ-emission band spectra of both constituents are equal their bandwidths being dependent on the relative concentration of the alloy components, as expected from a simple rigid-band model; 2) The $L_{2,3}$-emission band spectra of both constituents are different in shape (*i.e.* in the gross shape of the inten-

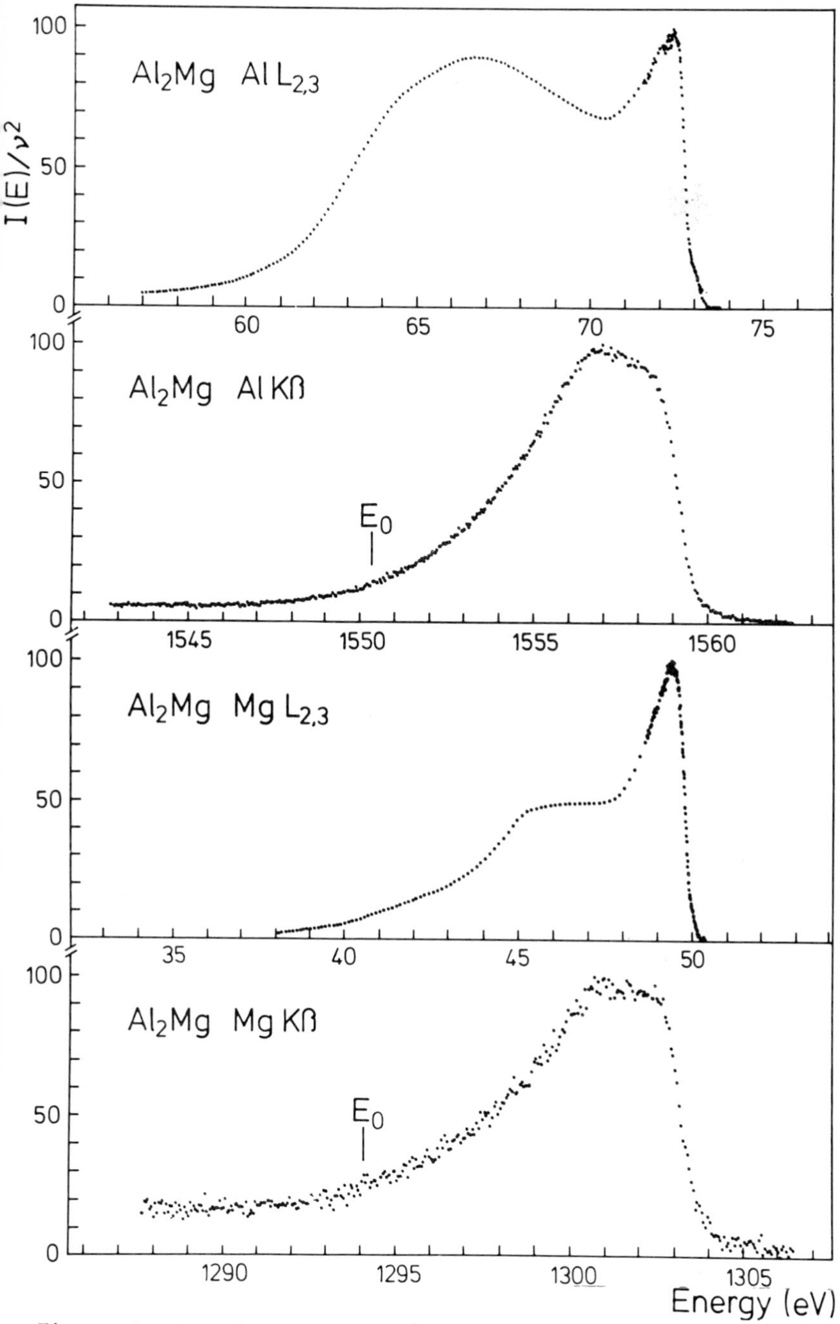

Figure 6 Kβ and $L_{2,3}$ emission band spectra of Al_2Mg.

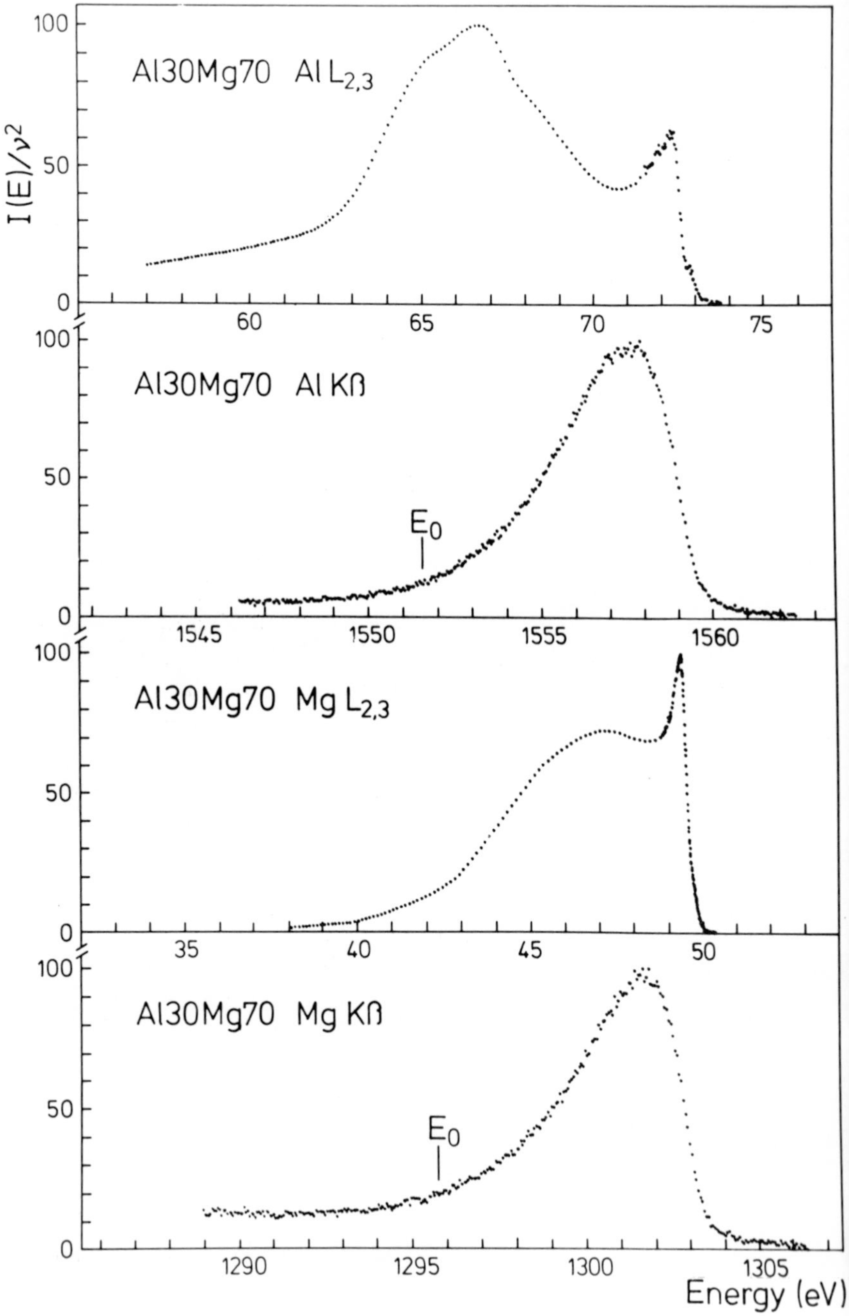

Figure 7 Kβ and $L_{2,3}$ emission band spectra of $Al_{30\%}Mg_{70\%}$

sity distribution) however, similarities are observed especially in the high-energy region of the spectra.

The statements of Appleton and Curry concerning the general behaviour of the $L_{2,3}$ spectra of aluminium and magnesium on alloying are confirmed to a certain extent by the present measurements, but there are also some essential discrepancies. The most important of these is the following: the widths of the $L_{2,3}$ emission bands were found by the above investigators to be equal to those of the pure metals to within 0.5eV. Our measurements show that the usual parabolic extrapolation of the low-energy side of the Mg $L_{2,3}$ bands is not possible for Al_2Mg, Al_3Mg_2, and $Al_{12}Mg_{17}$. Therefore, the bottom of these bands cannot be determined with accuracy, and it cannot be concluded from the experimental data that for an Al-Mg alloy two different valence bands of different widths exist, or in other words that a common-valence-band model cannot be applied.

The main difficulty in our understanding the difference in shape of the $L_{2,3}$ spectra, in terms of a common-valence-band model, seems to be as follows: to-date it has always been considered that any theoretical model yielding one density-of-states curve, should automatically yield similar forms of the spectra of the components provided the symmetry of the inner level is the same, and that such similar forms should be found experimentally. However, it must be always borne in mind that x-ray emission from a certain atom is produced by electronic transitions from the local environment to the core-state hole of this atom. In the one-electron description the intensity distribution of x-ray emission band spectra can be written in a simplified form as a product of the density of states and transition probability. In this picture the difference in shape of the L spectra can easily be understood as a consequence of different transition probabilities between transitions to the Al and Mg atoms following from the charging effect mentioned above. Indeed the similarities in the high-energy region of the $L_{2,3}$ spectra and the resemblance of the respective Kβ-emission spectra appear to prove that a common valence band does exist and that one single density-of-states curve might be the basis for an interpretation of the two $L_{2,3}$ spectra for the concentrated alloys.

It is also possible to explain the experimental results in terms of local densities of states. This theoretical model does not so much mean a difference in a physical sense, but rather a different mathematical description. Local densities of states are used in recent alloy theories, one of the most promising being the coherent potential approximation introduced by Soven (1967) and examined carefully by Velický *et al.* (1968). The total density of states N(E) can be calculated as a weighted sum of local densities of states $N_A(E)$ and $N_B(E)$ corresponding to the type of atom A or B: $N(E) = x.N_A(E) + (1-x).N_B(E)$, where x is the concentration of atoms A in the alloy system AB.

At present we cannot explain the difference in behaviour of the K and L emission, since the wavefunctions of the valence electrons are not known. The only thing one can do is to conclude from the experimental data some statements regarding their general character: the p-like electrons involved in the K emission must be described by rather extended wavefunctions, therefore participating to the same extent in transitions to the K holes of both aluminium and magnesium atoms. On the other hand the s-like and d-like electrons are more localized and give rise to different Al and Mg $L_{2,3}$ spectra for the alloy components, a conclusion supported by Jacobs (1969).

We can further state that the coherent potential approximation seems to be a good basis for interpreting the $L_{2,3}$ spectra. This approximation, originally introduced with the assumption of the same bandwidths for both components, may be generalized to alloys in which the pure components have different bandwidths (Soven 1970, Blackman *et al.* 1971) as is the case for Al-Mg alloys. Soven gives some numerical results for such a system with a simple model density-of-states curve. At low concentrations of one component (B) a sub-band belonging to B is observed, whereas the local density of states of the second component (A) in the alloy is practically the same as that for the pure component A. These model calculations are supported by the results found for the dilute alloys $Al_{5\%}Mg_{95\%}$ and $Al_{10\%}Mg_{90\%}$. We believe that the x-ray emission of the concentrated alloys can be qualitatively understood in the same way.

Acknowledgment

The author is grateful to G. Breidenbach, who developed the interfaces for the Honeywell computer. The author wishes also to thank Professor A. Faessler for his current interest and support of this work and many colleagues for helpful discussions.

REFERENCES

Appleton, A. and Curry, C. (1965); Phil. Mag. 12, 245.

Blackman, J.A., Esterling, D.M. and Berk, N.F. (1971); Phys. Letters, A 35A, 205.

Crisp, R.S. and Williams, S.E. (1960); Phil. Mag. 5, 1205.

Das Gupta, K. and Wood, E. (1955); Phil. Mag. 46, 77.

Dimond, R.K. (1967); Phil. Mag. 15, 631.

Eickhoff, K. and Vosskühler, H. (1953); Z. Metallkunde 44, 223.

Fabian, D.J. (1970); Mat. Res. Bull. 5, 591.

Farineau, J. (1938); Ann. Phys. Paris 10, 20.

Fischer, D.W. and Baun, W.L. (1966); in "Advances in X-Ray Analysis" Vol. 10, (edited by J.B. Newkirk and G.R. Mallett), p374: Plenum Press, New York.

Gale, B. and Trotter, J. (1956); Phil. Mag. 1, 759.

Jacobs, R.L. (1969); Phys. Letters 30A, 523.

Jones, H. (1934); Proc. Roy. Soc. London A144, 225

Jones, H., Mott, N.F. and Skinner, H.W.B. (1934); Phys. Rev. 45, 379.

Klima, J. (1970); J. Phys. C - Solid State Phys. 3, 70.

Läuger, K. (1968); Thesis, University of München.

Mott, N.F. (1936); Proc. Cambridge Phil. Soc. 32, 281.

Neddermeyer, H. (1969); Thesis, University of München.

Neddermeyer, H. (1971); in Proc. 3rd IMR Symp., 'Electronic Density of States', Nat. Bur. Stand. Spec. Publ. 323: in press.

Nemoshkalenko, V.V. and Gorskii, V.V. (1968); Phys. Stat. Sol. 28, K 15.

Rooke, G.A. (1969); in Proc. Conf. X-Ray Spectra and Electronic Structure of Matter, II, p.64: Inst. of Metal Physics, Ukr. Acad. Sci. Kiev.

Samson, S. (1965); Acta Cryst. 19, 401.

Smrčka, L. (1971); Czech. J. Phys. B 21, 683.

Soven, P. (1967); Phys. Rev. 156, 809.

Soven, P. (1970); Phys. Rev. B2, 4715.

Stern, E.A. (1966); Phys. Rev. 144, 545.

Varley, J.H.O. (1954); Phil. Mag. 45, 887.

Velický, B., Kirkpatrick, S. and Ehrenreich, H. (1968); Phys. Rev. 175, 747.

Wiech, G. (1966); Z. Physik 193, 490.

Watson, L.M., Dimond, R.K. and Fabian, D.J. (1968); in "Soft X-Ray Band Spectra and Electronic Structure", ed. Fabian, D.J., p.45; Academic Press, London.

ELECTRONIC PROPERTIES OF ALUMINIUM AND SILICON INTERMETALLIC COMPOUNDS FROM X-RAY SPECTROSCOPY

G. Wiech and E. Zöpf

Sektion Physik der Universität München, West Germany

1. INTRODUCTION

One of the main factors governing electronic band structure is the arrangement of the atoms in the solid. Since x-ray emission and absorption band-spectra are closely related to the electronic structure, there is also a connection between the intensity distribution of the x-ray spectra and the crystal structure. In this paper we wish to illustrate this connection by presenting new measurements of x-ray emission bands of intermetallic compounds. Intermetallics are suitable for this investigation for several reasons. There exists a large number of intermetallics with a variety of crystal structures, and many are quite stable and do not decompose under electron bombardment.

To interpret the x-ray emission spectra of binary phases in terms of their electronic structure one must measure the complete x-ray spectrum of both components. However, for compounds of elements with high atomic number this is a difficult sometimes insoluble problem, due to experimental difficulties and to the large inner core-level widths. Therefore, all possible x-ray emissions can only be investigated in the case of compounds containing elements of low atomic number. Nonetheless, investigations of binary compounds for which one component is a heavy element show that extensive information about the electronic structure can be derived by studying the complete x-ray spectrum of the light element only.

If we restrict our interest to elements of the third period the complete x-ray emission spectrum - neglecting satellites - consists of only two x-ray emission bands ($K\beta$ and $L_{2,3}$), and the $K\alpha_{1,2}$ doublet. In the binary and ternary phases we have investigated, always one component has been either aluminium or silicon. For these elements we have measured the $K\alpha$-doublet and the $K\beta$ and $L_{2,3}$ emission bands.

2. EXPERIMENTAL

For investigation of the $L_{2,3}$ emission bands a 2-metre grazing incidence concave grating spectrometer was used. The measurements of the $K\alpha_1$ lines and the $K\beta$-emission bands were performed with a spectrometer of the Johann-type. Both instruments have been described in detail elsewhere (Wiech 1964, 1966, Läuger 1968). Therefore, only the most salient features are mentioned here.

Two different gratings, each with 600 lines/mm, were used to measure the L-emission bands of aluminium (blaze angle $3^{0}31'$, gold coating) and silicon (blaze angle $1^{0}31'$, aluminium coating). The compounds under investigation were pressed onto a water-cooled copper anode and excited by bombarding with a 0.3-4.0mA beam of 3keV electrons. The x-ray tube was operated at a pressure of about 10^{-8} torr. All compounds proved to be stable during the course of the measurements. The x-ray photons were counted using an open multiplier of the Bendix (M306) type. The intensity varied from 10^3 to 10^5 counts/min. The calibration of the spectrometer was based on the wavelengths of the $M\xi$ lines of ^{38}Sr to ^{47}Ag, which have recently been re-measured (Dannhäuser and Wiech 1971a, 1971b).

With the Johann-type spectrometer the $K\alpha_1$ lines and the $K\beta$-emission bands were obtained in first-order reflection from the $(10\bar{1}0)$ planes of a quartz crystal, bent to a radius of 1m, with a reflecting area of 8 x 30 mm^2. K-spectra were obtained using fluorescence excitation. The operating conditions of the x-ray tube were 10kV and 100-400mA at a pressure of about 10^{-6} torr. The x-ray photons were counted with a Geiger-Müller detector. The intensity of the $K\alpha_1$ lines was about 6 x 10^4 counts/min, while the intensity of the emission bands varied (with the material studied) from $\sim$1 x 10^3 to $\sim$4 x 10^3 counts/min. Data for Bragg angle ϕ, the average dispersion and the reference lines used are listed in Table I.

Table I Bragg angle ϕ, the average dispersion, and wavelength of reference lines (Bearden 1964) for the $K\alpha_1$ lines and $K\beta$-emission bands of Al and Si.

	Bragg-angle ϕ	Average dispersion (XU/mm)	Reference lines	Wavelength (XU)
Al $K\alpha_1$	78.5°	1.6	Ag $L\alpha_1$ (II)	4145.82±0.03
			Cr $K\beta_{13}$ (IV)	2080.55±0.02
Al $K\beta$	69°	2.7	Cd $L\alpha_1$ (II)	3948.15±0.04
			La $L\alpha_1$ (III)	2660.18±0.05
Si $K\alpha_1$	57°	4.35	Fe $K\beta_{13}$ (IV)	1752.92±0.02
			Co $K\alpha_1$ (IV)	1785.259±0.009
Si $K\beta$	53°	5.0	Ni $K\alpha_1$ (IV)	1654.457±0.008
			Cr $K\alpha_2$ (IV)	2288.854±0.003

With the concave grating spectrometer resolution was about 0.35eV for the silicon $L_{2,3}$-emission bands and 0.20eV for the aluminium $L_{2,3}$-emission bands. For the Johann-type spectrometer the resolution, including the width of the K levels ($\sim$0.5eV), amounts to 0.75eV and 0.65eV for the silicon $K\beta$- and aluminium $K\beta$-emission bands respectively.

Before the investigation all specimens were tested for homogeneity by x-ray diffraction.

3. RESULTS AND DISCUSSION

A. The aluminium Kβ and $L_{2,3}$ emission bands of the Laves phases $CaAl_2$, $CeAl_2$ and $ErAl_2$

Laves phases comprise a large group of related intermetallic compounds, AB_2, with one of the three following structure types $MgCu_2$, $MgZn_2$ and $MgNi_2$. $CaAl_2$, $CeAl_2$ and $ErAl_2$ belong to the Laves phases of the $MgCu_2$ type.

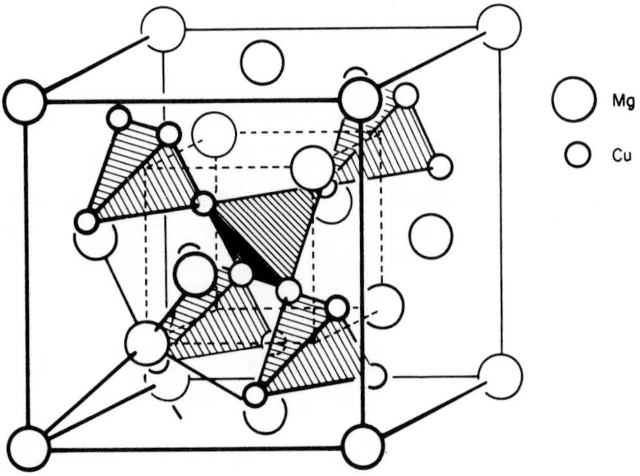

Figure 1 The $MgCu_2$ structure

The $MgCu_2$ structure is cubic (with 8 formula units per cell) and can be regarded as built up of two interpenetrating lattices of A and B atoms (figure 1). The B atoms lie at the corners of tetrahedrons which are joined at their apexes. Between the tetrahedrons are placed the larger A atoms. The array of A atoms is cubic as in silicon; each A atom is surrounded by four other equidistant A atoms and twelve B atoms at a rather smaller distance; each B atom is surrounded by six B atoms and six A atoms at a rather larger distance. Thus the lattice is distinguished by high symmetry, large coordination number and great compactness. Crystals showing this structure tend to exhibit metallic bonding.

In connection with the investigation of intermetallic compounds, the Kβ-emission band of pure aluminium was measured (figure 2). The spectrum is found to be in good agreement with that obtained by Läuger (1968). For the Al $L_{2,3}$-emission band of pure aluminium we refer to Neddermeyer and Wiech (1970).

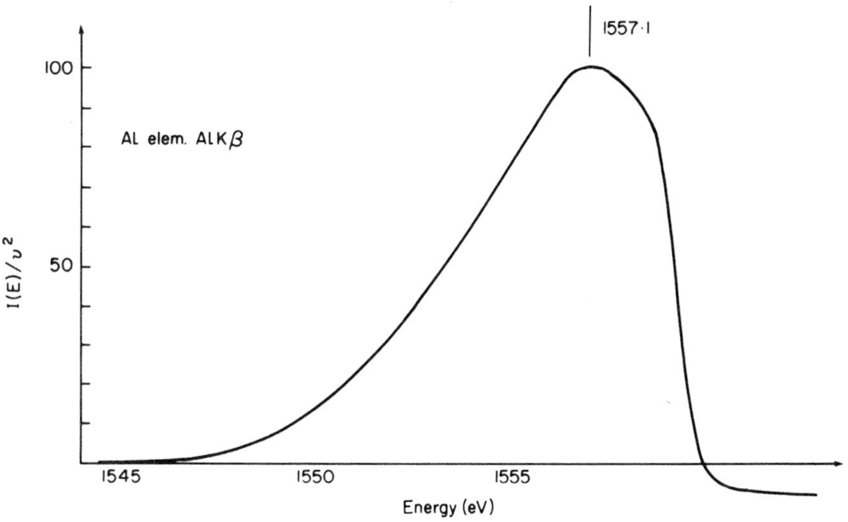

Figure 2 Kβ-emission band of pure aluminium

The Al Kβ and $L_{2,3}$ emission bands of XAl_2 (X = Ca, Ce, Er) are shown in figures 3, 4 and 5. For all spectra a constant background was assumed and subtracted, and no corrections for self-absorption were made. The K and L emission bands here and in the following figures are correlated in energy by the $K\alpha_1$ lines. In Table II the wavelengths and energies of the $K\alpha_1$ lines, including the wavelength for the pure metal, are listed; and in table III the energies of the most pronounced

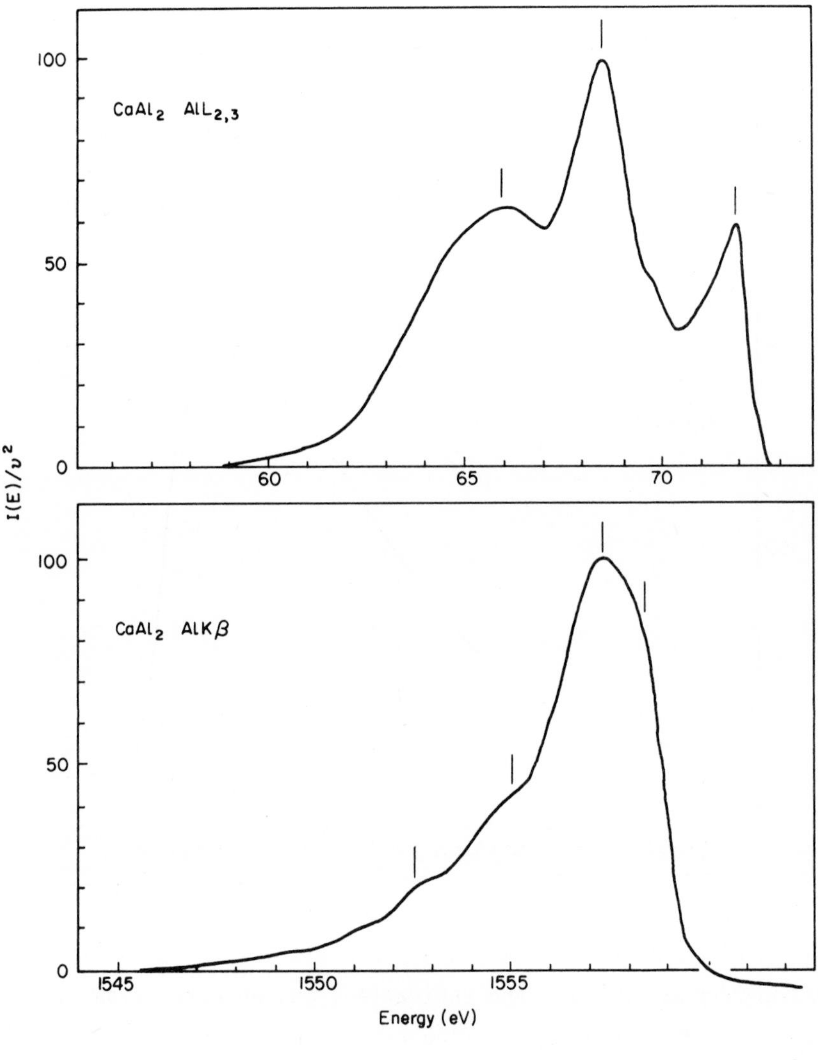

Figure 3 Al $L_{2,3}$ and Kβ emission bands of $CaAl_2$

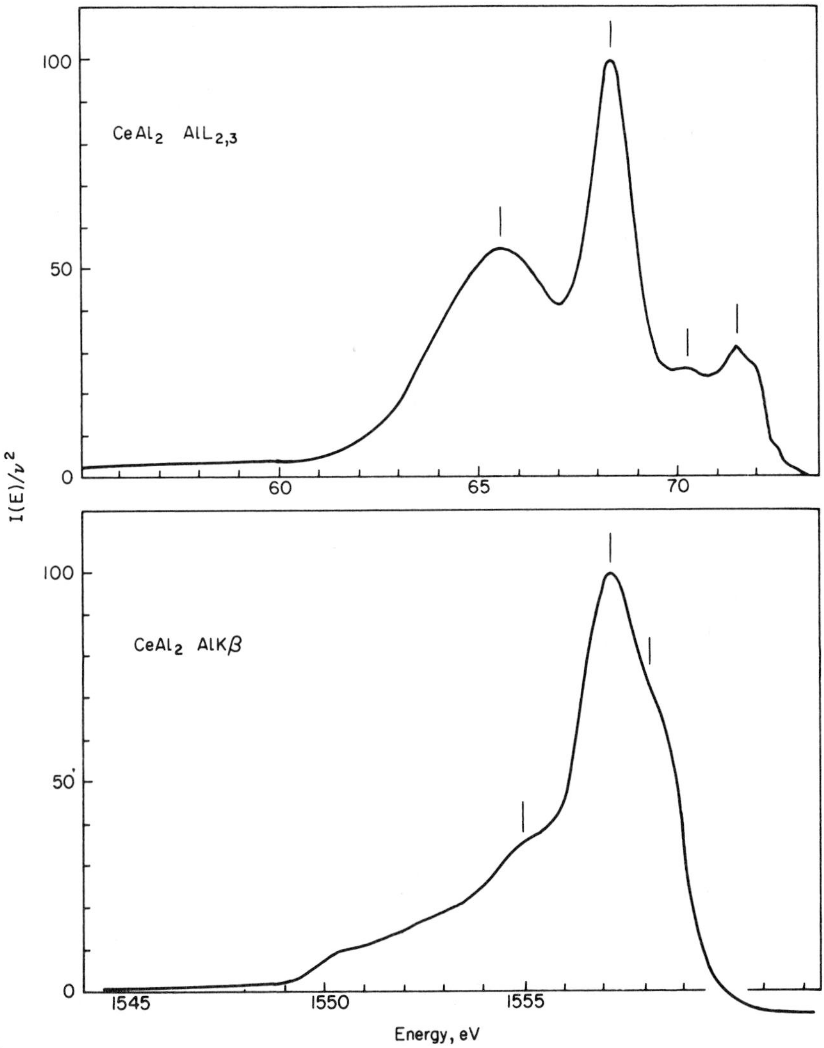

Figure 4 Al $L_{2,3}$ and Kβ emission bands of $CeAl_2$

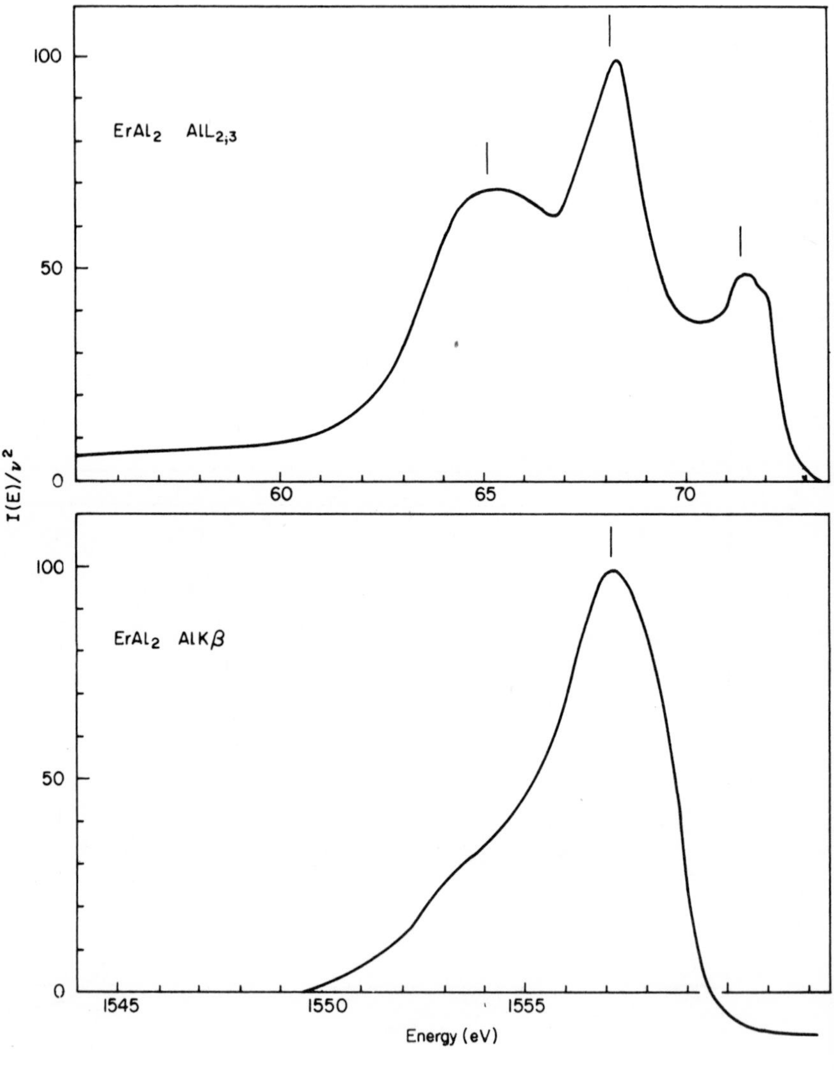

Figure 5 Al $L_{2,3}$ and Kβ emission bands of $ErAl_2$

characteristics are given.

Table II Wavelengths and shifts of $K\alpha_1$ lines of aluminium and silicon compounds.

Substance	Wavelength		Shift	
	λ(XU)	E(eV)	Δλ(XU)	ΔE(eV)
Al elem.	8322.35	1486.65		
$CaAl_2$	8322.56	1486.61	0.21	- 0.04
$CeAl_2$	8322.59	1486.60_5	0.24	- 0.045
$ErAl_2$	8322.65	1486.59_5	0.30	- 0.055
Si elem.	7110.95	1739.91		
$CaSi_2$	7110.77	1739.95_5	-0.18	0.04_5
$Ca_{0.75}Sr_{0.25}Si_2$	7110.80	1739.95	-0.15	0.04
$BaSi_2$	7110.67	1739.98	-0.28	0.07

Table III Energy position of peaks in the Al Kβ and Al $L_{2,3}$ emission bands of aluminium intermetallic compounds (eV).

Compound	$L_{2,3}$-emission band			Kβ-emission band
$CaAl_2$	65.9	68.4	71.8	1557.3
$CeAl_2$	65.7	68.4	71.6	1557.2
$ErAl_2$	65.2	68.4	71.5	1557.2
$Ce_8Al_{13}Si_3$	66.5		71.3	

The K and L emission bands of these intermetallics show a marked sililarity but differ considerably from the emission bands of pure aluminium. The main intensity of the K bands is concentrated in a relatively narrow energy range, the half width being only about one half of that of the pure metal. This means that p-electrons are predominant at the top of the valence band.

The L bands show three characteristic peaks, indicating that s and d electrons form comparatively sharp sub-bands with little overlap. The peak near the high-energy edge of the L-emission band correlates with a slight shoulder in the K-emission band, while the maximum of the Kβ emission coincides with the minimum of the $L_{2,3}$ emission at about 70eV. The relatively rapid increase of intensity at the K and L emission edges indicates the metallic character of the compounds. By considering the intensity ratios of the three peaks of the L-emission bands we find that $CaAl_2$ is more similar to $ErAl_2$ than to $CeAl_2$.

The Al K and L emission bands of $LaAl_2$, $PrAl_2$, $NdAl_2$, $GdAl_2$, $DyAl_2$, which are all Laves phases of the $MgCu_2$ type have also been measured (Wiech and Zöpf, to be published). The aluminium Kβ and $L_{2,3}$ emission bands for all these compounds are similar to those shown in figures 4 and 5, but there are small systematic differences on going from one compound to another. Since the atomic arrangement in all of these compounds is the same, these systematic differences are probably due to the influence of the various rare-earth metals. It appears that the increasing number of 4f-electrons from lanthanum to erbium is to some extent reflected in the band structure.

The Al $K\alpha_1$ lines of all the aluminides mentioned above are shifted towards longer wavelengths with respect to pure aluminium (see table II), indicating a charge transfer from the rare-earth atom to the aluminium atom.

B. Al $L_{2,3}$-emission band of $Ce_8Al_{13}Si_3$

The compound $Ce_8Al_{13}Si_3$ is obtained by substituting some silicon atoms for aluminium atoms in $CeAl_2$; it has the hexagonal AlB_2-type structure as is shown in figure 6. The cerium atoms lie between planes of hexagonally arranged aluminium atoms; each aluminium atom has three aluminium atoms as nearest neighbours and is surrounded by six cerium atoms.

Figure 7 shows the Al L-emission band of $Ce_8Al_{13}Si_3$.

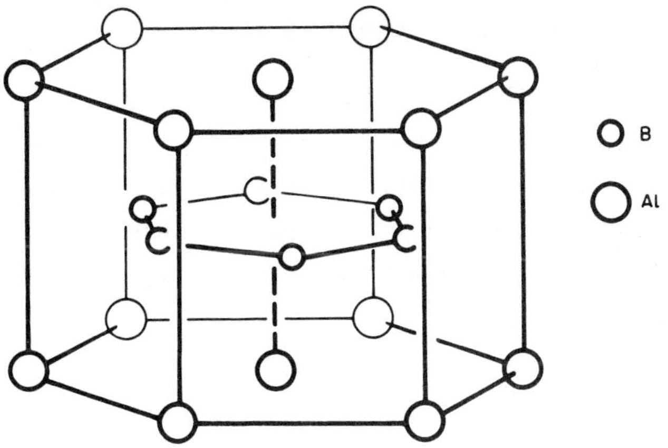

Figure 6 The AlB_2 structure of $Ce_8Al_{13}Si_3$

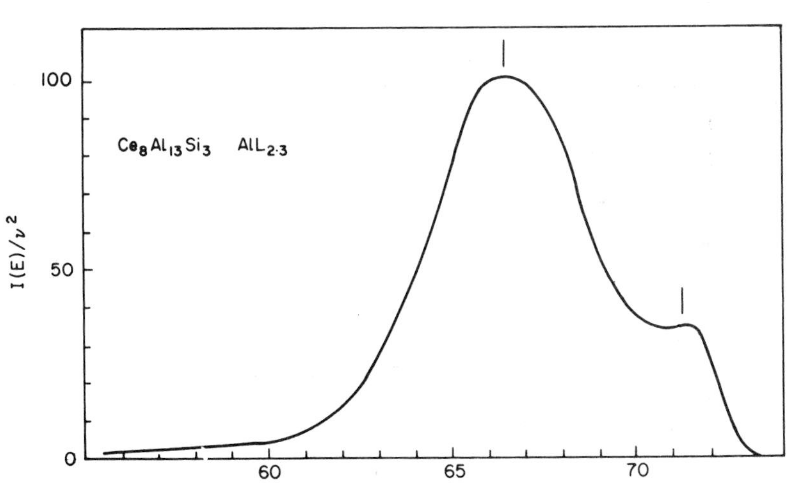

Figure 7 Al $L_{2,3}$-emission band of $Ce_8Al_{13}Si_3$

The marked structure found in the Al L band of the aluminides (figures 3, 4 and 5) has disappeared and instead of two peaks there is only one broad hump, followed by a shoulder with a flat maximum near the high-energy edge.

C. Si-emission bands of $BaSi_2$

According to Gladischewskij (1959) $BaSi_2$ has the AlB_2-type structure (figure 6), but more recent investigations (Axel *et al* 1968, Janzon *et al* 1970) indicate that $BaSi_2$ has an orthorhombic structure with isolated Si_4 tetrahedrons; the atomic arrangement being similar to that of white phosphorus.

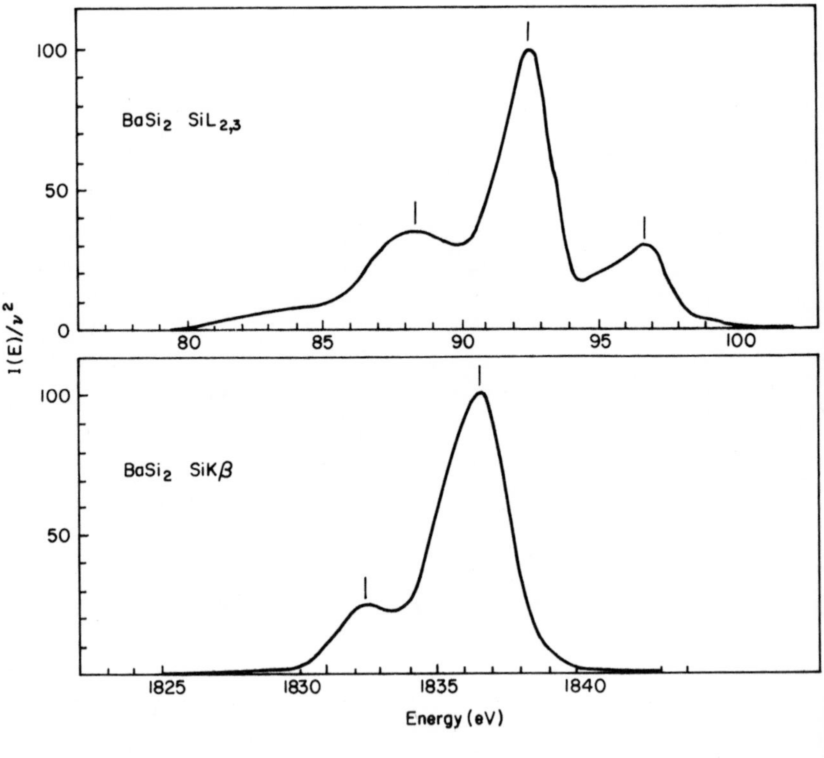

Figure 8 Si $L_{2,3}$ and Kβ emission bands of $BaSi_2$

Figure 8 shows the Si Kβ and Si L emission bands of $BaSi_2$. The wavelength of the Si $K\alpha_1$ line and the energy position of the peaks are listed in tables II and IV respectively. The Si L-emission band has three pronounced peaks; the main peak and the peak at the high-energy side corresponding closely to the two peaks of the Si Kβ-emission band.

Table IV Energy position of peaks in the Si K and Si $L_{2,3}$ emission bands of Si Compounds (eV)

Compound	$L_{2,3}$-emission band			Kβ-emission band		
$CaSi_2$	89.7	92.1			1836.1	1837.1
$Ca_{0.75}Sr_{0.25}Si_2$	89.8	92.05			1835.9	1837.1
$BaSi_2$	88.4	92.5	96.7	1832.4		1836.7
WSi_2	90.15		96.7			
$MoSi_2$	90.45		96.95			

In connection with the measurements of $BaSi_2$ we have investigated the Si Kβ and $L_{2,3}$ emission bands of the compound Er_3Si_5 (Zöpf, unpublished) which has the AlB_2-type structure. If $BaSi_2$ had the AlB_2-type structure one would expect a close similarity between the silicon emission bands of $BaSi_2$ and Er_3Si_5. However, the silicon emission bands of $BaSi_2$ differ strongly from those of Er_3Si_5.

Also the Al $L_{2,3}$-emission band of the ternary compound $Ce_8Al_{13}Si_3$ (figure 7) whicn has the same crystal structure as Er_3Si_5, is similar to the Si $L_{2,3}$-emission band of Er_3Si_5.

On the other hand, both the Si Kβ and $L_{2,3}$ emission bands of $BaSi_2$ show a striking similarity with the aluminium emission bands of the aluminides shown in figures 3, 4 and 5. This similarity may be caused by the related atomic arrangement of the aluminium and silicon atoms; in both compounds these atoms form tetrahedrons. Apparently the x-ray emission bands are essentially determined by the tetrahedral arrangement of the atoms and depend to a lesser degree on whether the emitting atom in question is a silicon or an aluminium atom. In the light of these results it appears that $BaSi_2$ has an orthorhombic structure.

D. Si-emission bands of $CaSi_2$ and $Ca_{0.75}Sr_{0.25}Si_2$

In the solids $CaSi_2$ and $Ca_{0.75}Sr_{0.25}Si_2$ (Eisenmann *et al* 1967, Janzon *et al* 1970), as well as in Er_3Si_5 (AlB_2-type), there is a hexagonal arrangement of Si layers. In the case of Er_3Si_5 the hexagons are plane, and in Ca-Si compounds they are corrugated. Figure 9 shows the atomic arrangement of the two Ca-Si compounds. The calcium atoms lie in planes between the corrugated silicon layers. The position of the calcium atoms relative to the silicon atoms is indicated by the atoms 1, 2 and 3 for $CaSi_2$, and by the atoms 3, 4 and 5 for the ternary compound. This means that in $CaSi_2$ the calcium atoms are each surrounded by 8 silicon atoms, and in $Ca_{0.75}Sr_{0.25}Si_2$ by 6 silicon atoms. In both compounds each silicon atom has three silicon atoms as neighbours.

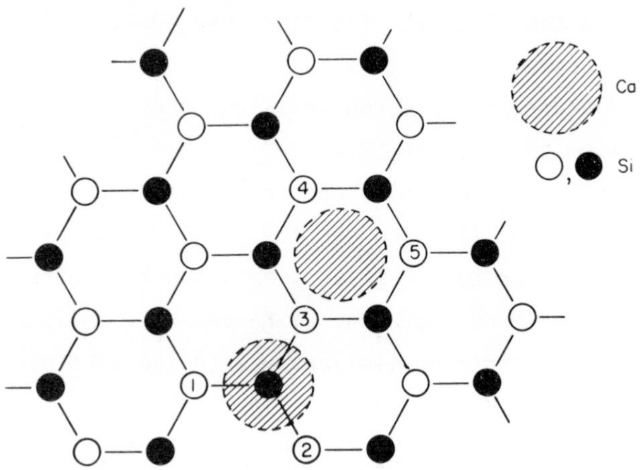

Figure 9 Orthorhombic arrangement of calcium-silicon compounds (from Eisenmann *et al* 1970).

The Si K and L emission bands of both compounds are presented in figures 10 and 11. The wavelengths of the $K\alpha_1$ lines and the energy position of characteristic points of the emission bands are listed in tables II and IV. Due to the fact that the compounds have a related crystal structure, similar emission bands are observed. All spectra show marked structure, the structural features in the L bands corresponding to structural features in the K bands. For both compounds partial heteropolar bonding is assumed (Klemm and Busmann, 1963), silicon being charged negatively. However, the observed shifts of the $K\alpha_1$ lines (table II) suggest that silicon is positively charged.

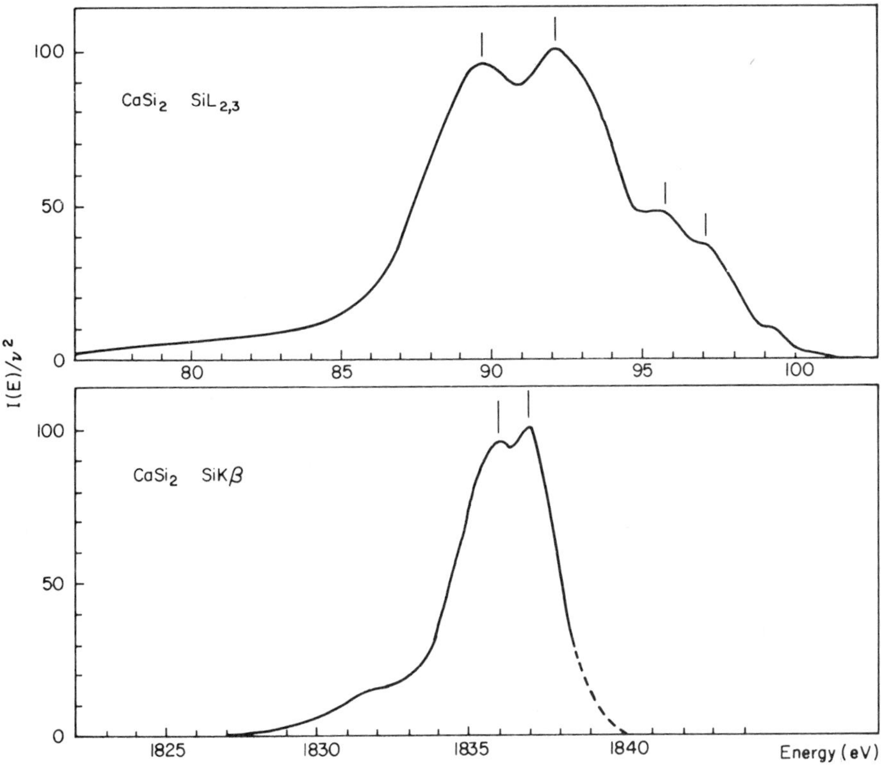

Figure 10 Si $L_{2,3}$ and Kβ emission bands of $CaSi_2$

E. Si $L_{2,3}$-emission bands of WSi_2 and $MoSi_2$

Both WSi_2 and $MoSi_2$ have $MoSi_2$-type structure. The atomic arrangement of this structure is quite different from those

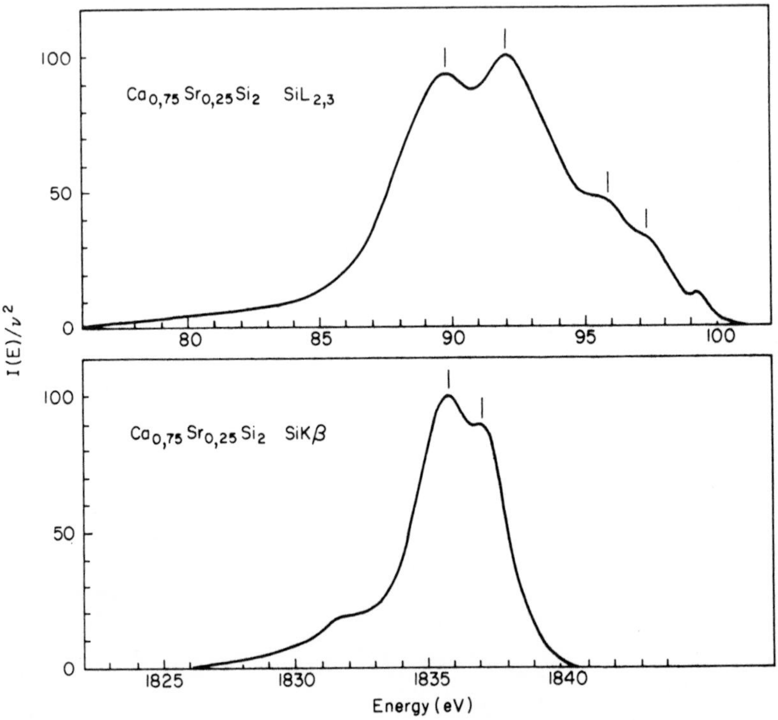

Figure 11 Si $L_{2,3}$ and Kβ emission bands of $Ca_{0.75}Sr_{0.25}Si_2$

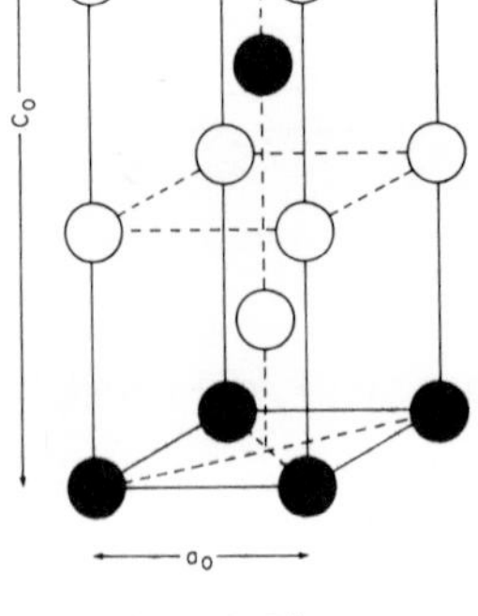

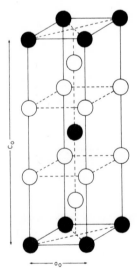

Figure 12 The $MoSi_2$ structure

discussed above and is shown in figure 12. Each silicon atom is surrounded by five tungsten or molybdenum atoms and five silicon atoms. In figure 13 the Si $L_{2,3}$-emission bands of WSi_2 and $MoSi_2$ are presented. The energy position of the peaks

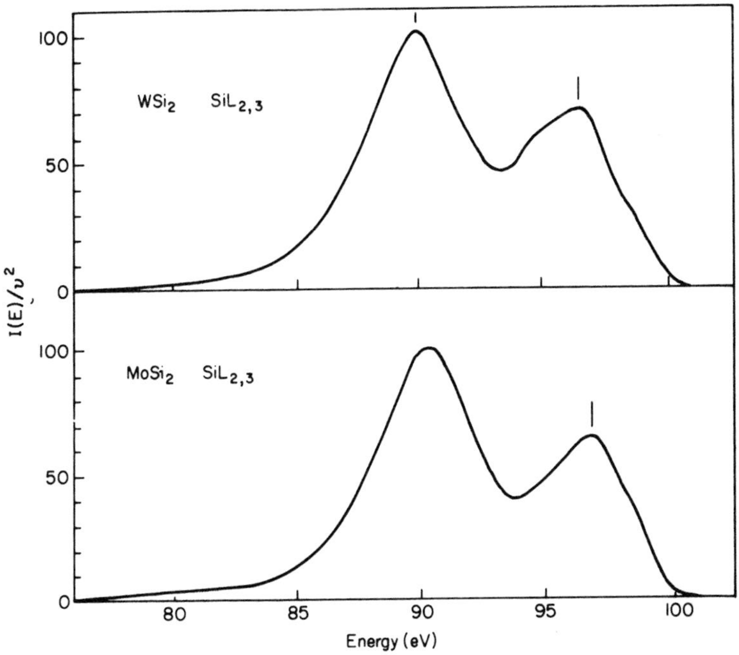

Figure 13 Si $L_{2,3}$-emission bands of WSi_2 and $MoSi_2$

of the bands are listed in table III. As expected, the intensity distributions of both spectra are similar and differ remarkably from the intensity distribution of all the other spectra presented here.

Acknowledgments

Thanks are due to Dr. H. Schäfer of the Institut für Anorganische Chemie der Universität München for supplying the Ca-Si compounds. The authors are also indebted to

Mr. G. Melchart and Mr. F. Widera for assistance during the course of the measurements.

REFERENCES

Axel, H., Janzon, K.H., Schäfer, H. and Weiss, A. (1968); Z. Naturforsch, 23b. 108.

Bearden, J.A. (1964); X-ray wavelengths, U.S. Atomic Energy Commission; Oak Ridge Tennessee.

Blokhin, M.A. and Saschenko, W.P. (1957); Jzv. Akad. Nauk SSSR, Ser. Phys., 21, 1343.

Dannhäuser, G. and Wiech, G. (1971a); Phys. Letts., 35A, 208.

Dannhäuser, G. and Wiech, G. (1971b); Z. Physik, 244, 415.

Eisenmann, B., Janzon, K.H., Riekel, Ch., Schäfer, H. and Weiss, A. (1967); Z. Naturforsch. 22b, 102.

Eisenmann, B., Riekel, Ch., Schäfer, H. and Weiss, A. (1970); Z. anorg. allg. Chemie, 372, 325.

Gladischewskij, E.I. (1959); Dopovid, Akad. Nauk. Ukr. SSR, 3, 294.

Janzon, K.H., Schäfer, H. and Weiss, A. (1970); Z. anorg. allg. Chemie, 372, 87

Klemm, W. and Busmann, E. (1963); Z. anorg. allg. Chemie, 319, 297.

Läuger, K. (1968); Dissertation, Universität München.

Neddermeyer, H. and Wiech, G. (1970); Phys. Letts., 31A, 17.

Schäfer, H., Janzon, K.H. and Weiss, A. (1963); Angew. Chem., 75, 451.

Wiech, G. (1964); Dissertation, Universität München.

Wiech, G. (1966); Z. Physik, 193, 490.

SOFT X-RAY STUDY OF THE d-BANDS IN $AuAl_2$

A.J. McAlister, J.R. Cuthill, R.C. Dobbyn, and M.L. Williams

Institute for Materials Research, National Bureau of Standards, Gaithersburg, Md. 20760

1. INTRODUCTION

The series of isostructural, isoelectronic intermetallic compounds, AuX_2 (X = Al, Ga, In) is of considerable current interest, largely because of the unusual behaviour of the magnetic susceptibility and gallium Knight-shift in $AuGa_2$. Specific heat (Rayne, 1963) and de Haas-van Alphen studies (Jaccarino *et al*, 1968) reveal Fermi surfaces that are fairly well described by the nearly free electron model, and show no strong differences between the compounds. Yet while the aluminium and indium Knight-shifts are positive and essentially temperature independent, the gallium Knight-shift displays a large and unusual temperature dependence ranging from -0.13% at 4°K to +0.45% at 230°K (Jaccarino *et al*, 1968). The magnetic susceptibilities of $AuAl_2$ and $AuIn_2$ are temperature independent, but $AuGa_2$ displays a temperature dependence (Jaccarino *et al*, 1968). Switendick and Narath (1969) have reported non-relativistic augmented plane wave (APW) band calculations for the three compounds. They found a feature of the $AuGa_2$ bands not present in the others, a flat s-like band just below the Fermi level. Such a feature could result in a temperature-dependent band repopulation effect, and thus explain the observed Knight-shift behaviour, though not necessarily the susceptibility. Another feature of the calculations by these investigators was the predicted location (6-7eV below the Fermi level, E_F) and width (1-1.5eV full width at half maximum) of the Au d-bands. Their positioning of the d-bands is in sharp contrast to the 2eV or so below E_F suggested by interpretation (in terms of d-band to E_F optical transitions; Wernick *et al* 1969) of the optical properties (Vishnubhlata and Jan 1967) of these unusually coloured materials. Strong support has been given to the $AuAl_2$ calculation by measurements (Williams *et al* 1971; Curry and Harrison, 1970; Kapoor *et al*, this Volume p215) of the Al $L_{2,3}$ soft x-ray (SXS) emission spectrum from $AuAl_2$.

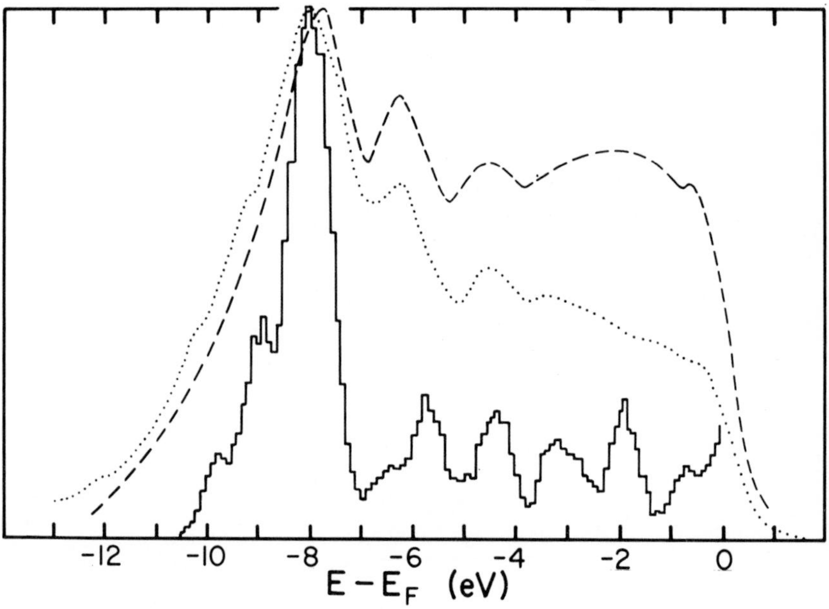

Figure 1 Comparison of the measured (SXS) Al $L_{2,3}$-emission spectrum with a theoretical estimate of the distribution in energy of s-like orbital character at aluminium sites. (This is the leading term in a one-electron estimate of the spectrum; omitting such factors as the smooth variation in the radial matrix element, and various broadening mechanisms.) Solid curve: theory, after Switendick (1971). Dotted curve experiment, after Williams *et al* (1971); dashed curve: experiment, after Curry and Harrison (1970). The difference between the experimental curves appears to arise from different instrumental frequency response.

As shown in figure 1 the measured SXS profiles are in close structural agreement with the distribution in energy of s-like charge within Al APW-spheres (the leading term in a one-electron estimate of the Al emission spectrum; see Bennett *et al* 1971), calculated by Switendick (1971) from the band structure results.

We report here a study of the Au $N_{6,7}$ SXS emission (5d to 4f transition) from $AuAl_2$. The results, together with the x-ray photo-emission (XPS) data obtained by Chan and Shirley (1971) indicate a double peaked d-band complex, the maxima occurring at 4.7 and 6.5eV below E_F. The details of the measurements, and their bearing on other studies - both experimental and theoretical - of the electronic structure of $AuAl_2$, are discussed in the following sections.

2. EXPERIMENTAL DETAILS

Measurements were made using a vacuum spectrometer with glass grating and photoelectric detection as described elsewhere (Cuthill *et al* 1967; Cuthill 1970). Prior to these measurements the instrument was reset: grazing angle 4.5^o, entrance and analyzer slits 0.05mm. The spectrum was scanned repetitively, total counts being recorded over successive short time intervals during each scan. Successive scans were summed to enhance signal-to-noise ratio. The relative counting error in the raw data, $(N)^{-1/2}$ where N is the accumulated count per channel, ranged from 0.9 to 1.2%. To achieve this confidence level rather long counting intervals were used, resulting in a degradation of instrumental resolution which we estimate at 0.4eV for this experiment. The average sample chamber pressure was 4×10^{-7} torr, and the sample temperature was maintained at approximately 600^oC. Electron beam excitation was used at an energy of 2.5keV with the beam striking the sample surface at grazing angle (approximately 20°).The x-ray takeoff angle was 90°. The sample, a polycrystalline rod, was given a light abrasive polish and washed in acetone and absolute alcohol before mounting in the instrument. It was prepared from 99.999% pure gold and aluminium, by slowly adding the gold, in stoichiometric proportions, to an aluminium melt whose temperature was gradually raised above the melting point of the compound. The final melt was then drawn into a graphite lined quartz tube and the resulting rod zone-refined. Optical metallographic examination of slices from the ends of the rod showed that traces of free aluminium (much less than 1%) occurred at the grain boundaries in the interior of the sample, but not at the surface. The pure Al $L_{2,3}$-spectrum, which differs strongly from that in the compound, was not observed in our earlier studies of the Al L-emission from this same sample (Williams *et al* 1971). X-ray diffraction studies on the actual sample employed in the measurement showed only the $AuAl_2$ pattern. An XPS scan of a portion of the original zone-refined rod (not the sample studied here) showed only a single Au-phase, chemically shifted from pure Au, and some (presumably occluded) carbon. The latter was not observed in our measurements, indicating either its absence in the portion of the original rod actually studied in the SXS measurements, or that it was deposited on the sample during the XPS measurements.

3. RESULTS AND INTERPRETATION

The upper curve of figure (2) shows the raw data obtained for the $N_{6,7}$ emission spectrum of gold in $AuAl_2$; the vertical bars indicate the 70% confidence level. The lower curve was obtained by drawing a smooth line through the raw data in a manner consistent with the standard counting error, and then subtracting off an estimated background. Four distinct humps appear in the spectrum. This structure is consistent with a

4f splitting of 3.6 ± 0.2eV and a 5d band with two peaks separated by 1.8 ± 0.2eV. This interpretation is supported by available XPS data.

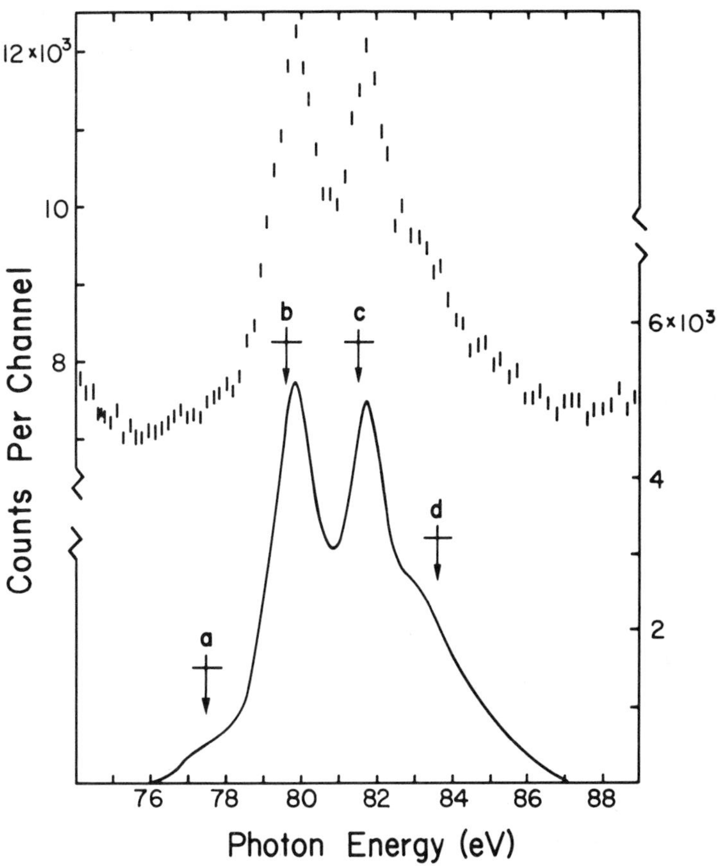

Figure 2 The $N_{6,7}$ soft x-ray emission spectrum of gold in $AuAl_2$. The upper curve shows the raw data. Bar lengths represent the 70% confidence level. The lower curve is the smoothed spectrum, corrected only for background. The arrows show the locations of structure predicted from x-ray induced photoemission data. The left ordinate applies to the raw data, the right to the smoothed curve.

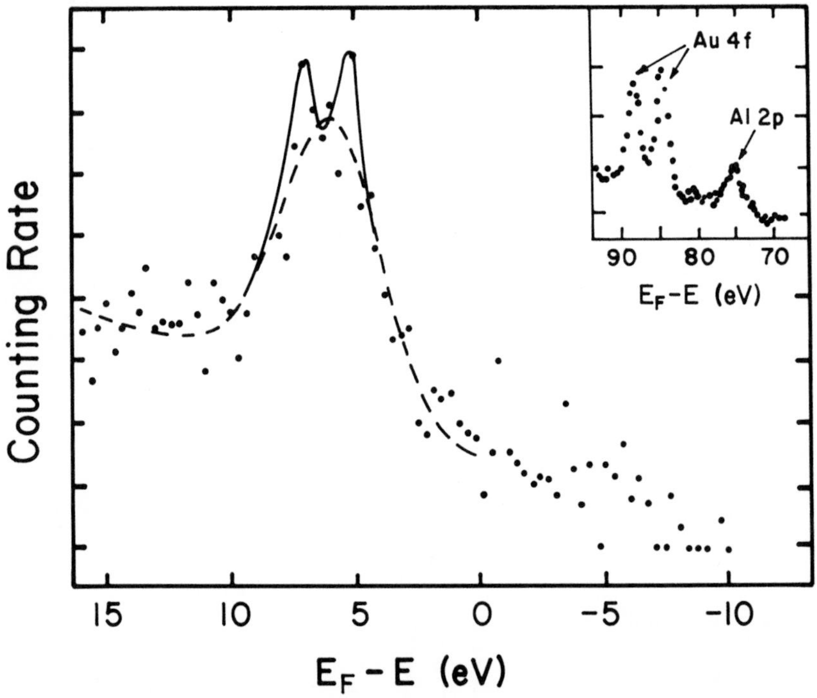

Figure 3 The x-ray induced (Al $K_{\alpha_{1,2}}$ radiation) photoemission from $AuAl_2$ (Chan and Shirley, 1971).

The XPS results obtained by Chan and Shirley (1971) are shown in figure 3. These yield binding energies for the 4f states of gold in the alloy of 84.6$\pm$0.2 and 88.6$\pm$0.2eV. Also shown in figure 3 are the results obtained by these investigators for the valence band. The latter are rather 'noisy' and possibly are open to interpretation. Chan and Shirley fitted a single Gaussian curve centred at -6.4eV to the spectrum (dashed in figure 3). However, they noted that these data can be represented also by a two-peaked curve. We have drawn the solid curve in figure 3, with maxima at -7.1 $\pm$ 0.2 and -5.0 $\pm$ 0.2eV. Using this two-peak curve, and assuming that it results mainly from the d-electrons, one expects structure in the $AuAl_2$ gold $N_{6,7}$-emission spectrum at 77.5$\pm$0.4, 79.6$\pm$0.4, 81.5$\pm$0.4 and

83.6±0.4eV. These photon energies are indicated with arrows in figure 2, marked a, b, c, and d respectively. Agreement is quite good, so far as the location of structure is concerned. However, the intensity variation in the spectrum merits comment. After correction for the variation in dispersion of the spectrometer and the dependence of the dipole matrix elements on photon energy, the relative intensities of the features a, b, c, and d have been estimated, using a simple four-band fitting scheme, to be 0.12, 1.00, 0.56, and 0.35 respectively. As for the case of the $N_{6,7}$ spectrum of pure gold which we have discussed elsewhere (McAlister *et al*, 1971), these numbers are closer to the expected atomic values of 0.00, 1.00, 0.70, and 0.05 (Condon and Shortley 1935) than to estimates of orbital quantum weights alone (*i.e.* to the fractions of initial and final state electronic charge residing in ℓ-partial waves coupled by the $\Delta\ell$ dipole selection rules). The latter depends on the assumption that conduction band states can be characterized approximately by their orbital angular momentum admixtures. A rough quantum weight estimate, made by assuming that the two-peaked curve of figure 3 reflects mainly d-orbital character at gold-atom sites, and combining the relative amplitudes of the two peaks with the population numbers of the 4f-shell, predicts relative intensities of 0.92, 0.69, 1.00, and 0.75. The better agreement of the observed intensities with predictions for the atom, where the 5d states are well described by their total angular momenta and the spin-orbit splitting is large (almost 1.5eV, Herman and Skillman, 1963) suggests that the spin-orbit interaction plays a significant rôle in the splitting of the gold d-bands in the alloy, and that the upper and lower humps of the d-bands can be described approximately by values of $j = 5/2$ and $3/2$ respectively.

In view of the statistical error in the XPS valence band

results, it seems desirable to estimate the binding energies of the 5d peaks indirectly using the more precise XPS values of the 4f binding energies, and the 4f-5d energy differences given by the SXS data. Chan and Shirley's value for the binding energy of the gold N_7-level in $AuAl_2$ is 84.6 ± 0.2eV. For the same level, Watson, Hudis, and Perlman (1972) report 84.5 ± 0.1eV, relative to a value of 306.6eV assigned to the simultaneously monitored Rhodium M_5-line. Our positioning of SXS spectral features a,b,c, and d (based in part on the above four-band fitting scheme) are 78.1 ± 0.2, 79.9 ± 0.2, 81.7 ± 0.2eV, and 83.5 ± 0.2eV respectively. These data place the 5d humps at 4.7 ± 0.3 and 6.5 ± 0.3eV below the Fermi level. This observation of a double peaked d-band complex spread over some 4eV is at variance with the non-relativistic band theory prediction (Switendick and Narath, 1969) of a single, rather narrow ($\sim$1.5eV width at half maximum) d-band complex centred at 7.0eV below the Fermi level. This raises questions concerning the interpretation of the Al L-emission spectrum from $AuAl_2$ (Williams *et al* 1971; Curry and Harrison, 1970). The gold and aluminium bands have been overlaid in figure 4. (Only the structural features c and d of figure 2 are retained for this comparison; features a and b reflect the same d-band structure shifted down by the amount of the spin-orbit splitting of the f-levels.) The large prominent peak at -8.0eV in the Al L-spectrum was interpreted by Switendick (1971) as a peak in the s-charge of aluminium in a region of strong 'Al s' with 'Au d'-hybridization, marking the lower d-band edge. The same effect has been noted in calculations for pure transition metals (Hodges *et al*, 1966) and transition metal diborides (McAlister, unpublished). In view of the gold d-peak observed at -6.5 ± 0.3eV - nearly the same energy at which the non-relativistic band calculation centres a single narrow d-band complex - this hybridization interpretation seems likely to be correct. The origin of weaker structures at higher energies in the Al L-spectrum is now open to question. The non-relativistic calculation associates them mainly with zone-face contacts of s-like bands. The comparison in figure 4 suggests that one or more may be associated with 'Al s and Au d' hybridization effects.

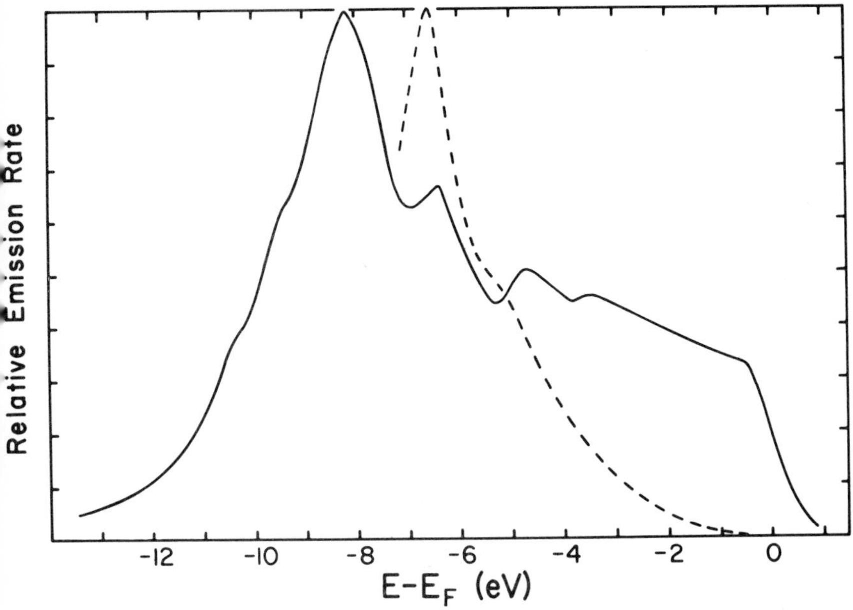

Figure 4 Comparison of the Al $L_{2,3}$ (solid curve) and Au $N_{6,7}$ (dashed curve) soft x-ray emission spectra of $AuAl_2$. Only peaks c and d of figure 2 are shown here, since a and b reflect the same d-band structure shifted down in energy by the amount of the f-level spin-orbit splitting.

There is a discrepancy in the energy scale assignments of figure 4. The spectrum shows a distinct 'metallic' high-energy emission edge at 73.3eV. Therefore, in figure 4 we have placed the Fermi level at the edge. Chan and Shirley's data in figure 3 place the Al 2p-levels at 75.0 ± 0.2eV below the Fermi level. The reasons for this discrepancy are not clear to us. The absolute error in the soft x-ray photon energy is certainly not greater than 0.2eV. (The dispersion, which is the spectroscopist's major concern in this sort of study, is known to greater accuracy.) Moreover, identification of the edge as the Fermi-threshold appears correct because both Knight-shift measurements and soft x-ray measurements on aluminium and $AuAl_2$ indicate a decrease in the height of the emission edge by a factor of 0.4 in the alloy (Bennett *et al*, 1971). On the other hand, if as seems likely the XPS measurements locate the Au 4f-levels to within a few tenths of a volt, a 1.7eV-error in placing the Al 2p-levels seems improbable. Of course, the two measurements are not exactly equivalent. The position of the SXS Al L-emission edge is determined almost solely by the Al $2p^{3/2}$ binding energy, while the XPS Al 2p-line depends on the widths and splitting of the $2p^{3/2}$ and $2p^{1/2}$ levels. This point requires further study.

The present data also have possible implication for the optical properties of $AuAl_2$. Both the XPS and SXS curves show a very slow decrease in intensity between the Au d-peak at -4.7eV and the Fermi level. Considerable experimental broadening exists in both measurements, and the inherent width of the d-bands cannot be estimated precisely. Nevertheless, the presence of broad structure so near the Fermi level suggests that hybridized d-states may exist in this energy range. This suggestion is strengthened by the comparison of the intensities of the SXS spectra from the compound and pure gold shown in figure 5. (The Au N_7 binding energies in Au and $AuAl_2$ from Watson *et al* (1971) have been used in constructing figure 5 since they are consistently calibrated. The SXS spectrum for pure gold is from McAlister *et al* (1971). Experiments with better resolution in the critical range between the upper peak and E_F, and a relativistic, self-consistent calculation of the band structure of $AuAl_2$, are required to resolve these problems.

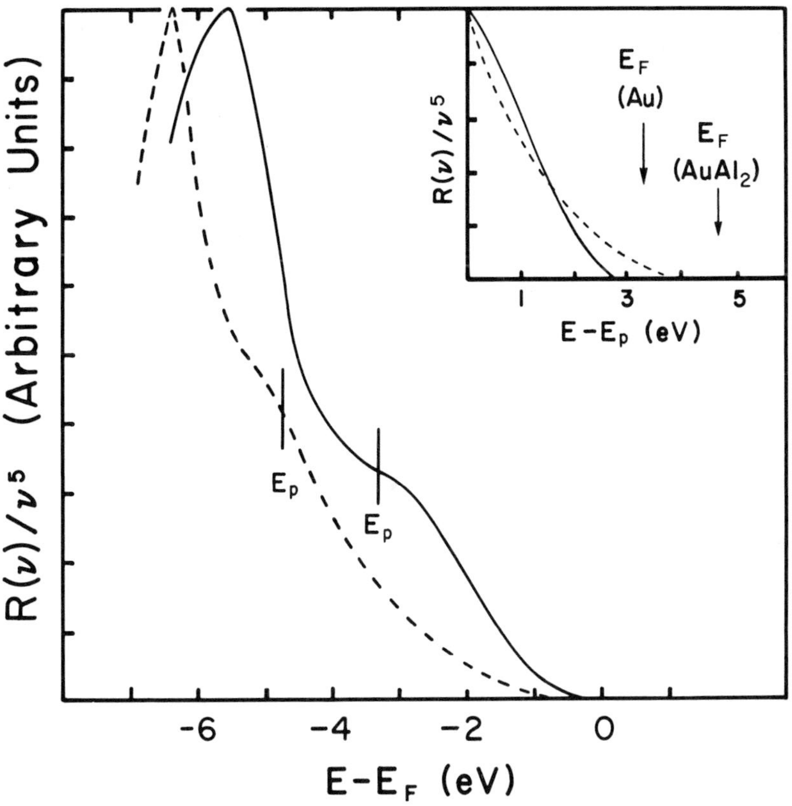

Figure 5 Comparison of the SXS Au $N_{6,7}$-emission spectra for pure gold (solid curve) and for $AuAl_2$ (dashed curve). The curves are normalized to peak intensity, and the zero of energy is the Fermi-level E_F, as estimated from the SXS data and the XPS N_7 binding energies (Watson *et al* 1971). In the inset, where attention is focused on the region between the upper d-peak and E_F, the curves are positioned to match, in energy, at the estimated location of their upper d-band peaks (denoted by E_p) and are normalized in intensity there.

Summary

Measurements of the SXS Au $N_{6,7}$-emission from $AuAl_2$, together with XPS measurements from this compound (Chan and Shirley, 1971) show the d-bands to be distributed over an energy range ∿4eV, with distinct maxima at 4.7 and 6.5eV below the Fermi level. This result is at variance with the predictions of non-relativistic band calculations. It raises questions about the interpretation of the Al L-spectrum from the compound, (Williams *et al*, 1971, Curry and Harrison, 1970) which appeared to confirm band theory (Switendick and Narath, 1969; Switendick, 1971). Further, while offering no definitive test, owing to experimental broadening, it once again raises the possibility of d-band participation in the strong colouring of the compound, an effect which the aluminium $L_{2,3}$-emission studies appeared to preclude.

Acknowledgments

We thank Drs. L.H. Bennett and R.E. Watson for many helpful conversations during the course of this work. Thanks are also due D.P. Fickle for preparing the sample, C.H. Brady for metallographic studies, C.J. Bechtoldt for x-ray diffraction measurements, and J.F. Rendina for XPS measurements.

REFERENCES

Bennett, L.H., McAlister, A.J., Cuthill, J.R. and Dobbyn, R.C. (1971); ed. L.H. Bennett, p.665, Electronic Density of States Nat. Bur. Stand. (U.S.) Spec. Publn. 323.

Chan, D. and Shirley, D.A. (1971); ed. L.H. Bennett, p.791, Electronic Density of States, Nat. Bur. Stand. (U.S.) Spec. Publn. 323.

Condon, E.U. and Shortley, G.H. (1935); "The Theory of Atomic Spectra", Cambridge U.P., Cambridge, England.

Curry, C. and Harrison, R. (1970); Phil. Mag. 21, 659.

Cuthill, J.R., McAlister, A.J., Williams, M.L. and Watson, R.E. (1967); Phys. Rev. 164, 1006.

Cuthill, J.R. (1970); Rev. Sci. Instr. 41, 422.

Herman, F. and Skillman, S. (1963); "Atomic Structure Calculations", Prentice-Hall, Englewood Cliffs, N.J.

Hodges, L., Ehrenreich, H. and Lang, N.D. (1966); Phys. Rev. 152, 505.

Jaccarino, V., Weber, M., Wernick, J.H. and Menth, A. (1968); Phys. Rev. Letts. 21, 1811.

McAlister, A.J., Williams, M.L., Cuthill, J.R. and Dobbyn, R.C. (1971); Sol. State Comm. 9, 1775.

Rayne, J.A. (1963); Phys. Letts. 7, 114.

Switendick, A.C. and Narath, A. (1969); Phys. Rev. Letts. 22, 1423.

Switendick, A.C. (1971); ed. L.H. Bennett, p.297, Electronic Density of States, Nat. Bur. Stand. (U.S.) Spec. Publn. 323.

Vishnubhlata, S.S. and Jan, J.P. (1967); Phil. Mag. 16, 45.

Watson, R.E., Hudis, J. and Perlman, M.L. (1971); Phys. Rev. B, 4, 4139.

Wernick, J.H., Menth, A., Geballe, T.H., Hulle, G. and Maita, J.P. (1969); J. Phys. Chem. Solids 30, 1949.

Williams, M.L., Dobbyn, R.C., Cuthill, J.R. and McAlister, A.J. (1971); ed. L.H. Bennett, p.303, Electronic Density of States, Nat. Bur. Stand. (U.S.) Spec. Publn. 323.

X-RAY K-EMISSION SPECTRA OF ALUMINIUM-GOLD ALLOYS

Elisabeth Källne
Institute of Physics, University of Uppsala, Sweden.

1. INTRODUCTION

Aluminium-gold alloys have received much interest from different investigators (see, for example, this Volume pp 91 and 191). The present investigation was initiated by some x-ray photoelectron experiments on electron escape depths, where a thin evaporated gold layer on aluminium was used. After moderate heating of the sample, the gold signal in the electron spectrum became weaker and it was noted that the sample had become purple. To study the electronic structure of the thin gold layer we then recorded the x-ray emission band spectrum of this sample and of the alloys : $Al_{88\%}Au_{12\%}$, Al_2Au and AlAu. Comparison of the recorded aluminium K band spectra showed that the thin gold layer produces a spectrum with similar features to the alloy band spectra, but with one extra peak not present in the alloy spectra. This appears to indicate that a sound aluminium-gold alloy had not been formed.

We report here the Al K-emission band, the Al $K\alpha_1\alpha_2$ lines, and the Al $K\alpha_3\alpha_4$ high-energy satellite lines for the alloys $Al_{88\%}Au_{12\%}$, Al_2Au and AlAu. The alloys Al_2Au and AlAu have earlier been studied both by x-ray spectroscopy (Fischer and Baun 1966 and 1967, Curry and Harrison 1970, Nemnonov *et al* 1971, McAlister *et al* 1971, Kapoor *et al* 1972) and by electron spectroscopy (Chan and Shirley 1971, Kapoor *et al* this Volume p215). The present measurements are compared with the results reported from these investigations.

2. EXPERIMENTAL

The alloys were prepared by melting pure aluminium and pure gold in stocheiometric proportions in a boron-nitride tube. The samples were examined micrographically and by x-ray diffraction. The Al_2Au sample was found to be cubic single-phased, while the AlAu sample consisted of a mixture of two phases, Al_2Au and $AlAu_2$. The AlAu sample was therefore annealed under controlled atmosphere for about 30 hours; subsequent x-ray diffraction examination showed it to be single-phased.

The spectrometer, a Johann-type instrument with quartz crystal $(10\bar{1}0)$ of 2m curvature, has been described previously (Nordfors 1961). The water-cooled anode sample holder was fitted with two targets, pure aluminium as reference and the investigated alloy. Spectra from the two targets were then recorded alternately, three times for each spectrum. The purity of the aluminium anode, with respect mainly to oxidation, was continuously checked by recording the Al $K\alpha_3\alpha_4$ satellite emission lines (see Nordfors 1956).

The power at the x-ray anode was kept low (8-10mA and 6kV) to minimize the heating of the sample. The recording times for the alloy spectra were 1-2 hours, with operations pressure in the spectrometer tank and the x-ray tube $\sim 1\text{-}2\times 10^{-5}$torr. The Al $K\alpha_1\alpha_2$ lines, the Al $K\alpha_3\alpha_4$ satellite lines, and the Al K-emission band were recorded for each alloy. No gold spectra from the alloys were recorded.

3. RESULTS AND DISCUSSION

A. Aluminium K-emission bands

Figure 1 shows the Al $K\beta$ bands obtained for aluminium, and

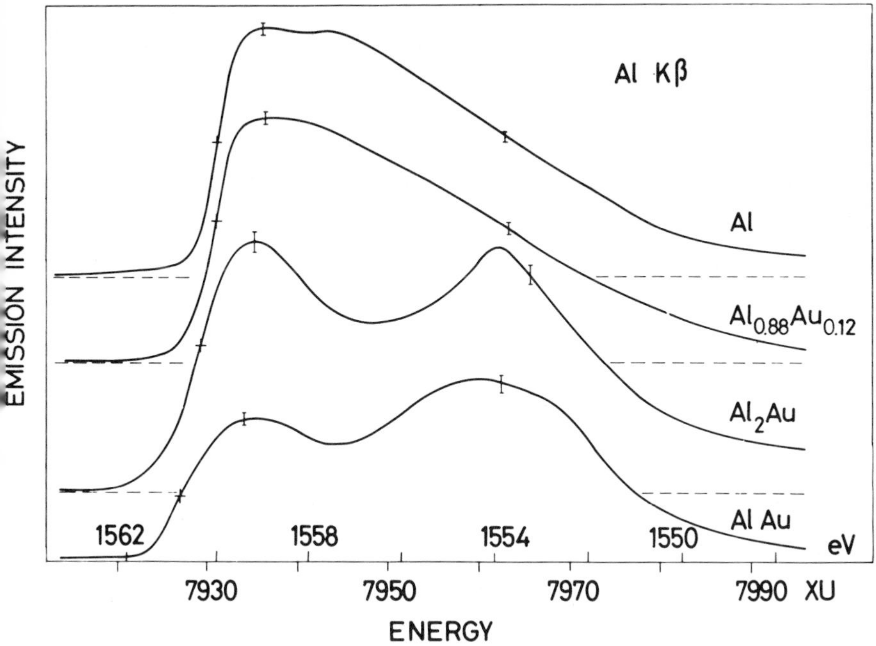

Figure 1 Al K-emission bands for Al, $Al_{88\%}Au_{12\%}$ Al_2Au and AlAu. The error bars indicate one standard deviation. The midpoint of the linear part of the Fermi edge is denoted with a cross.

$Al_{88\%}Au_{12\%}$ Al_2Au, and AlAu. The wavelength scale is computed from the position of the centroid of the Al $K\alpha_1\alpha_2$ lines, 8322.94 XU, Noreland *et al* 1967), and from the linear dispersion of the spectrometer, 4.232 XU mm^{-1} (Nordfors 1961). We can see that for Al_2Au and AlAu the Al K band is shifted towards higher energy. The energy shifts of the Fermi level, taken as the midpoint of the linear part of the emission edge, are shown in Table 1. The shifts of the Al $K\alpha_1\alpha_2$ lines for the different alloys are also indicated. The table shows that the Al $K\alpha_1\alpha_2$-line shifts are very small, indicating that the 1s and 2p levels shift by closely similar amounts. These results agree well with those for the Al $L_{2,3}$-emission bands of Al_2Au and AlAu recorded by Kapoor *et al* (1972).

Table 1 Energy shifts, in eV, of the aluminium K-emission edge (Fermi level → 1s) and of the centroid of the Al $K\alpha_1\alpha_2$ lines ($2p_{3/2,1/2} \rightarrow 1s$) relative to pure aluminium. The errors are estimated from the maximum deviation.

Alloy	Al$K\alpha_1\alpha_2$ centroid	Al K band edge
$Al_{88\%}Au_{12\%}$	0.00±0.04	0.0±0.2
Al_2Au	0.06±0.04	0.33±0.17
AlAu	0.08±0.04	0.78±0.17

It is noted also that the shift of the Al K-band of Al-Au agrees with the measurements reported by Fischer and Baun (1966, 1967). It is interesting to compare the shifts shown in Table 1, all towards higher energy, with the results obtained by Fischer and Baun for AlCu, AlNi and AlFe. For AlFe and AlNi the Al K-emission edge shifts are towards lower energy, while for AlCu the band position remains constant. For all these alloys the Al $K\alpha_1\alpha_2$ lines shift towards higher energy. The shifts of inner lines, for example the $K\alpha_1\alpha_2$, have widely been theoretically interpreted in terms of charge transfer (Nefedov 1962, Lindgren 1966, Manne 1967). In particular for aluminium a positive shift of the $K\alpha_1\alpha_2$ lines means a loss of 3p electrons. This is also in agreement with the electronegativity of the two alloy constituents, aluminium and gold. Interpretation of shifts of inner levels is more difficult, as noted by Manne (1967). However, regarding the Fermi level as fixed relative to the vacuum level, a removal of electrons from an atom is always connected with increased binding energy of the inner levels (Siegbahn *et al* 1967). The shifts of the Al K band shown in Table 1 are all towards higher energy. Thus the measured shifts show increased binding energy of the 1s level, in agreement with the present interpretation. Kapoor *et al* (1972) have found positive shifts for the Al $L_{2,3}$ bands, also in agreement with the present results.

The shape of the Al K band changes considerably upon increasing the gold content. The most significant feature is the appearance of a low-energy peak, which increases rapidly with increasing gold in the alloy. A comparison of the relative intensities of the two peaks in the Al K band for the alloy Al_2Au shows that the low-energy peak is more intense in the spectrum obtained in the present investigation than in the measurements by Nemnonov *et al* (1971). Since this peak increases rapidly with increasing gold content, the discrepancy may be due to different sample compositions.

The separation between the peaks in the Al K band spectrum for Al_2Au, 5.4eV, agrees well with the results on the same alloy reported by Nemnonov *et al* (1971). The Al $L_{2,3}$-emission spectra for aluminium-gold alloys, studied by Kapoor *et al* (1972), show similar changes with increasing gold content. The separation between the peaks in the Al $L_{2,3}$ spectrum is 1-2eV larger than the separation in the Al K spectrum in both the alloys Al_2Au and AlAu. Augmented-plane-wave calculations for Al_2Au, by Switendick and Narath (1969), predict a set of gold d-bands $\sim$7eV below the Fermi level. This is in agreement with the Al $L_{2,3}$ spectrum, as noted by Kapoor *et al*; however, the present K-emission measurements show a low-energy peak $\sim$6.5eV below the Fermi level for the alloys Al_2Au and AlAu. This indicates that the states contributing to this low-energy peak have also some Al 3p-symmetry. Finally, it is also noted that the position of the low-energy peak with respect to the Fermi edge remains constant with changing content of the alloy. The position of this peak agrees with the energy separation of the d-band from the Fermi level for pure gold metal.

It is interesting to compare the Al K bands of these alloys with earlier measurements on titanium compounds (Ramqvist *et al* 1969). A low-energy peak appearing in the Ti K-emission band could successfully be explained in terms of transitions from non-metal 2s states. Likewise we can interpret the low-energy peak in the alloy Al K band as cross-transitions, remembering only that the intensity in the Al K band spectrum comes from admixing Al 3p-states. Marshall *et al* (1969) interpreted results for Al $L_{2,3}$-emission from the alloy AlAg in terms of transitions from virtual bound states.

B. Al $K\alpha_3\alpha_4$ high-energy satellite lines

The Al $K\alpha_3\alpha_4$ satellite lines for pure aluminium and for the three Au-Al alloys are shown in figure 2. The wavelength

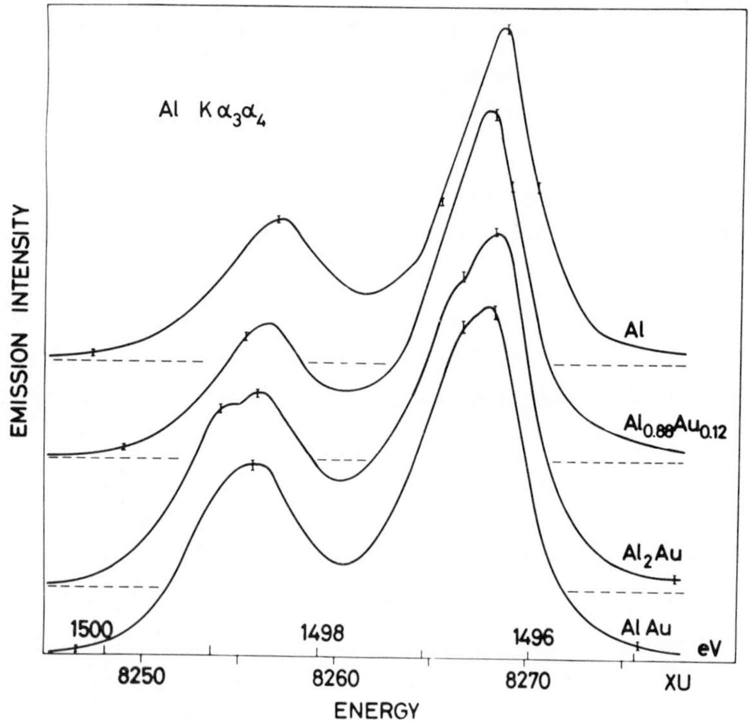

Figure 2 Al $K\alpha_3\alpha_4$ satellite lines for Al, $Al_{88\%}Au_{12\%}$, Al_2Au, and AlAu. The error bars indicate one standard deviation.

scale is computed in the same manner as for the Al K bands. We note that both the $K\alpha_3$ and $K\alpha_4$ satellite lines shift, in energy, and show also a multiplet structure in the case of the alloys. We do not resolve the structure into components, because we do not have information regarding the number of components, the relative spacing, nor the relative intensities. Therefore, no numerical values for the shifts of the separate lines are given. As shown by Nordfors (1956), the Al $K\alpha_3\alpha_4$ satellite lines are strongly asymmetrical even for pure aluminium, thereby making it difficult to make a unique deconvolution of the lines. As far as we are aware, splitting of the Al $K\alpha_3\alpha_4$ satellite lines has not previously been experimentally reported. Fischer and Baun (1966, 1967) reported the Al $K\alpha_3\alpha_4$ lines for AlFe and AlNi alloys, and their spectra indicate that the shape of the satellites changes upon alloying.

We have compared the present results, for aluminium-gold alloys, with the results obtained for Al_2O_3 by Nordfors (1956), and we cannot exclude the possibility of traces of aluminium oxide in our samples. However, the Al K bands obtained for the alloys (figure 1) show no peak in the energy region where the Al K band for Al_2O_3 is known to exhibit a peak. This tends to indicate that oxidation in our samples is unlikely.

The relative intensities of the $K\alpha_3$ and $K\alpha_4$ satellite lines change on alloying; this has previously been carefully studied by Fischer and Baun (1967a). In our measurements the voltage on the x-ray target was kept constant at 6kV, and the relative intensity of $K\alpha_3/K\alpha_4$ should have reached saturation value (Baun and Fischer 1964). The relative peak intensities for the alloys varies from 0.43 to 0.54. This variation is much less than reported by Fischer and Baun for other aluminium compounds and alloys.

Utriainen *et al* (1968) have interpreted the satellite structure $K\alpha_3\alpha_4\alpha'$ in terms of the sudden approximation. They then calculated the relative integrated intensities of the $K\alpha_3\alpha_4\alpha'/K\alpha_1\alpha_2$ satellites for different elements and compounds. In the present investigation the integrated intensity Al $K\alpha_3\alpha_4$/Al $K\alpha_{1\ 2}$ does not change significantly upon alloying; the ratio varies between only 0.11 and 0.15, and possible changes in the integrated intensity-ratio are thus too small to be examined.

The multiplet structure of the Al $K\alpha_3\alpha_4$ satellite lines has also been studied theoretically by Demekhin and Sachenko (1967) in terms of multiple ionization. The theory predicts six components for the $K\alpha_3$ satellite line, and three components for the $K\alpha_4$. In the present investigation the multiplet structure is much more pronounced for the alloys Al_2Au and AlAu than for the pure metal. The reason for this is not understood. It would be interesting to study more aluminium alloys, and to investigate further the multiplet structure of the satellite lines.

4. CONCLUSION

The Al K-emission band shifts towards higher energy for the alloys Al_2Au and AlAu. Also the lines Al $K\alpha_1\alpha_2$ show small shifts towards higher energy, indicating a charge transfer from the aluminium metal to the gold metal in these alloys. The Al K band changes in shape on alloying, in a similar manner to the Al $L_{2,3}$ emission band, showing a low-energy peak in the spectrum. This peak can be interpreted in terms of transitions from gold 4d states with admixed Al 3p-states to the vacancy in aluminium 1s core level. The high-energy Al $K\alpha_3\alpha_4$ satellites shift in energy, and show a pronounced multiplet structure for the alloys Al_2Au and AlAu.

Acknowledgments

I wish to thank Doc. S. Rundqvist of the Institute of

Chemistry, Uppsala, for kindly preparing and analysing the alloy samples, and also Doc. E. Noreland and Prof. R. Manne for valuable comments on the manuscript. This investigation received financial support from the Swedish Natural Science Research Council.

REFERENCES

Baun, W.L. and Fischer, D.W. (1964); Phys. Lett. 13, 36.

Curry, C. and Harrison, R. (1970); Phil. Mag. A21, 659.

Demekhin, V.F. and Sachenko, V.P. (1967); Bull. Acad. Sci. USSR 31, 913, 921.

Fischer, D.W. and Baun, W.L. (1966); Phys. Rev. 145, 555.

Fischer, D.W. and Baun, W.L. (1967); J. Appl. Phys. 38, 229, 2092.

Fischer, D.W. and Baun, W.L. (1967a); J. Appl. Phys. 38, 2404.

Kapoor, Q.S., Watson, L.M., Hart, D. and Fabian, D.J. (1972); Solid State Comm. *in press*.

Lindgren, I. (1966); in 'Röntgenspektren und Chemische Bindung', Leipzig, p.182.

Manne, R. (1967); J. Chem. Phys. 46, 4645.

Marshall, C.A.W., Watson, L.M., Lindsay, G.M., Rooke, G.A. and Fabian, D.J. (1969); Phys. Lett. 28A, 579.

McAlister, A.J., Cuthill, J.R., Dobbyn, R.C. and Williams, M.L. (1971); in "Electronic Density of States", ed. Bennet, L.H., 3rd IMR Symp. p.303; NBS Special publn. 323, Washington.

Nefedov, W. (1962); Phys. Stat. Solidi 2, 904.

Nemnonov, S.A., Zyryanov, V.G., Minin, V.I. and Sorokina, M.F. (1971); Phys. Stat. Solid. b 43, 319.

Nordfors, B. (1956); Arkiv Fysik 10, 279.

Nordfors, B. (1961); Arkiv Fysik 19, 259.

Noreland, E., Ekstig, B., Källne, E. and Chetal, A.R. (1969); Metrologia 5, 80.

Siegbahn, K., *et al*, (1967); Atomic, Molecular and Solid State Structure Studied by means of electron spectroscopy, p.76; ff. Uppsala.

Switendick, A.C. and Narath, A. (1969); Phys. Rev. Lett. 22, 1423.

Ramqvist, L., Ekstig, B., Källne, E., Noreland, E. and Manne, R. (1969); J. Phys. Chem. Sol. 30, 1849.

Utriainen, J., Linkoaho, M., Rantavuori, E., Åberg, T. and Greaffe, G. (1968); Z. Naturforschung 23a, 1178.

SOFT X-RAY EMISSION SPECTRA AND ELECTRONIC STRUCTURE OF ALLOYS OF ALUMINIUM WITH FIRST TRANSITION-SERIES METALS

Q.S. Kapoor, L.M. Watson and D.J. Fabian

Department of Metallurgy, University of Strathclyde, Glasgow, Scotland.

1. INTRODUCTION

The phase-equilibrium diagrams of alloys of aluminium with transition metals show the existence of intermetallic compounds with only small solubility ranges. The intermetallics have a marked tendency to form ordered structures, even at high temperatures prior to melting, and many are brittle. These and many additional properties indicate that the bonding between component atoms is strongly covalent in character, and that considerable interaction occurs between unlike atoms.

Atomic interaction in the alloys is manifest in a redistribution of the valence-band electron states between the component atoms, and soft x-ray emission spectroscopy is powerfully suited to studying this effect: it provides information on the distribution of valence electrons and also on the symmetry of the states involved. The electron states are sampled locally to the emitting atom, and emission bands for each component can be examined while the symmetry of the electron states can be explored because, for example, K-emission spectra reflect the density of p states and L-emission spectra the density of s and d states in the band.

To obtain maximum information, it is necessary to measure,

for a given alloy, all possible emission bands (K, L, M, etc.) for each component. It is also desirable to have a means of normalizing the total integrated emission intensity of the spectra for a given alloy. Comparison of the integrated intensities for the two components also provides useful information on the electron transfer from one atom to the other. The necessary normalization has, for example, been achieved by Wenger *et al* (1971) who measured, at the same time as the band emission, an emission line involving an inner core-level transition, and used the intensity of this line as a measure of the rate of initial-state vacancy creation.

Sampling of local electron densities is a consequence of the transition probability matrix element, which has a significant value only in regions of space where the initial-state wavefunction (a valence-band state) overlaps appreciably the final-state wavefunction (an inner core-state). The final-state wavefunction is localized and therefore it follows that we measure the electron density in a region local to the emitting atom.

The spectra reported in the present investigation are the aluminium $L_{2,3}$-emission bands for the alloys of aluminium with some first transition-series metals. Since only the emission from the aluminium atoms is recorded, the measurements constitute only a partial examination of the alloys investigated. However, comparisons and correlations with the complementary spectra, measured for the same alloys in other laboratories, contribute to a general study of the electronic structure of these alloys and intermetallics.

2. EXPERIMENTAL

The alloys were prepared from pure metals by melting in a tungsten arc furnace, under argon atmosphere. The samples were placed in alumina boats, sealed under vacuum in silica capsules, and homogenized by annealing at suitable temperatures for several days. Metallographic study, and in some cases examination by electron probe, showed most samples to be single-phase. With some alloys traces of a second phase were observed, but in all cases the quantity and composition of the precipitate was such that emission from it would be weak and unlikely to cause distortion of the spectrum of the phase under investigation.

The procedure for recording and processing the spectra has been described in detail previously (Watson *et al* 1968). A platinum coated one-metre radius concave grating with 600 lines/mm and a blaze angle of $2^o4'$ was used for all the measurements. The alloy specimens were mounted on a water cooled anode, using a conducting silver-based cement, and the x-ray tube was operated at 5-7mA and 3kV in vacuum of better than 2×10^{-6}torr. The detector was a Bendix M306 photomultiplier with copper-beryllium photo-cathode. Successive scans of each spectrum were recorded, until sufficient counts had been accumulated at each step in the main band to reduce the statistical counting error to less than 1%. Between scans the sample was scraped under vacuum, using a tungsten-carbide-tipped tool, to expose fresh alloy surface and to reduce the build-up of contaminants.

The background bremsstrahlung radiation has been subtracted from each spectrum, estimating its effect by assuming it to vary linearly over the wavelength region of the recorded emission band. With alloys this may not be an entirely satisfactory approximation; however, it should not affect the main feature of the spectra. The results are presented in the form of $I(E)/E^3$ *vs* E where I(E) is the intensity, after subtraction of background, at energy E. Division by E^3 takes account approximately of the energy-dependent factor of the transition probability.

3. RESULTS

A. Aluminium-Vanadium Alloys

Figure 1 shows the Al $L_{2,3}$-emission bands from a series of aluminium-vanadium alloys. The energies of the principal

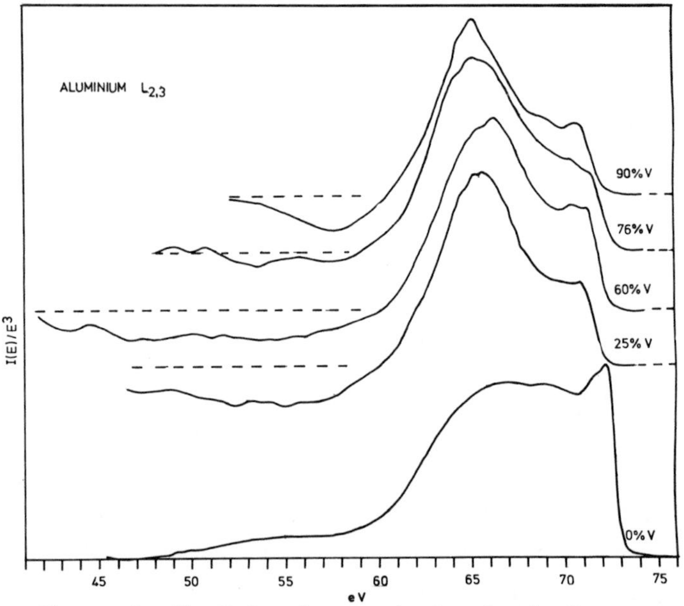

Figure 1 Aluminium $L_{2,3}$-emission bands for a series of aluminium-vanadium alloys.

features of the spectra are listed in Table I. In common with all measured aluminium $L_{2,3}$-emission bands for alloys aluminium with transition or noble metals, the intensity ratio of the low-energy to high-energy regions of the band shows an increase over that for the pure metal. As the vanadium content of the alloy is increased, the Fermi-level is lowered and the low-energy hump shifts — although not uniformly with the vanadium concentration. In the spectra shown in figure 1, the intensity beyond the bottom of the band appears to be negative; this follows subtraction of the known background Bremsstrahlung radiation, and is due to self absorption within the anode — in particular vanadium $M_{2,3}$-absorption.

Table I.

Alloy	Crystal Structure	Fermi-energy (eV)	High-energy peak (eV below Fermi-level)	Low-energy peak (eV below Fermi-level)	Fermi-edge breadth (eV)
		±0.05	±0.2	±0.5	±0.05
Al	Face-centred cubic	72.76	0.56	-	0.29
Al_3V	Tetragonal (Al Ti-structure)	72.1	1.0	6.9	0.94
$Al_{40}V_{60}$	Solid Solution	71.8	0.8	5.6	0.9
AlV_3	Cubic (βW-structure)	71.65	0.85	5.95	0.85
$Al_{10}V_{90}$	Solid Solution	71.5	0.7	6.6	0.85
$AlFe_3$	Cubic (BiF -structure)	71.8	0.5	7.7	0.9
AlFe	Cubic (CsCl-structure)	72.0	0.7	7.0	0.9
Al_3Fe	Monoclinic	72.5	-	6.2	0.52
AlCo	Cubic (CsCl-structure)	71.9	0.9	6.6	1.0
Al_9Co_2	Monoclinic	72.3	0.4	5.3	0.46

The aluminium Kβ-emission bands of the alloys Al_3V and AlV_3 have been measured by Nemoshkalenko *et al* (unpublished); they each show a single peak at ∼2eV below the Fermi-level. The Kβ band becomes more symmetrical as the vanadium content of the alloy is increased. Hence, in these alloys, the density of p-states around the aluminium atoms is concentrated in the high-energy part of the band, while the s,d states are more concentrated in the low-energy region.

B. Aluminium-Iron Alloys

Figure 2 shows the Al $L_{2,3}$-emission and the Fe $M_{2,3}$-emission bands for the alloys $AlFe_3$, AlFe and Al_3Fe. The intensity of

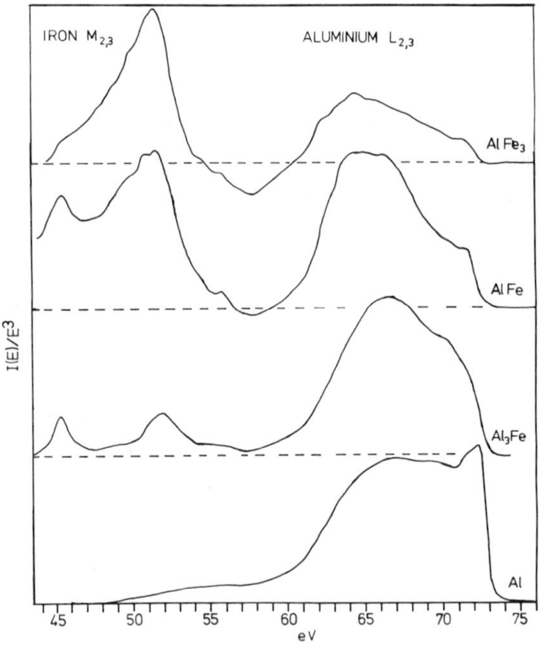

Figure 2 Aluminium $L_{2,3}$ and iron $M_{2,3}$ emission bands for aluminium-iron alloys.

the low-energy regions of the Al $L_{2,3}$ band has increased with respect to the high-energy region. The Fermi-edge shifts to lower energies as the concentration of iron is increased and the low-energy hump also moves progressively to lower energies (see Table I). The effect of self absorption in the anode is observed in the region between the Al $L_{2,3}$ and Fe $M_{2,3}$ bands (i.e. the apparent intensity goes negative after subtraction of background). The high intensity at ∿44eV in the spectrum for FeAl is not understood and requires further investigation.

Nemoshkalenko and Gorskii (1968) have measured the Al $K\beta_x$ and the Fe $K\beta_5$ spectra for aluminium-iron alloys. Their results show that the intensity maximum of the Al $K\beta_x$ band shifts progressively to lower energies with increasing content of iron (∿ 2.32eV for $AlFe_3$), and that the high-energy edge also shifts to lower energies, though by a lesser amount. The latter shift is in good agreement with the Fermi-edge shifts in the L-spectra noted in Table I. The Fe $M_{2,3}$-emission spectra, although of lower intensity and therefore not as reliable as the Al L and K spectra, show little change on alloying.

The same comments apply to the Fe $K\beta_5$-emission and the Fe L_3-emission bands of AlFe (Fischer and Baun 1967). The principal changes in the spectra, on alloying, occur in the Al $L_{2,3}$-emission, demonstrating a considerable redistribution of s,d states through the band in the vicinity of the aluminium atoms coupled with a shift in the density of p states away from the Fermi-level.

C. Aluminium-Cobalt Alloys

The Al $L_{2,3}$-emission spectra for Co_2Al_9 and CoAl are shown in figure 3. The Co $M_{2,3}$-emission band overlaps the low-energy region of the Al $L_{2,3}$ band, and accounts for the hump at about 58eV in the aluminium spectrum for the CoAl sample. The concentration of cobalt in Co_2Al_9 is too low to give a significant intensity of the Co $M_{2,3}$ band from this alloy. The features of the spectra are listed in Table I, where again it is noted that the Fermi-edge moves to lower energies.

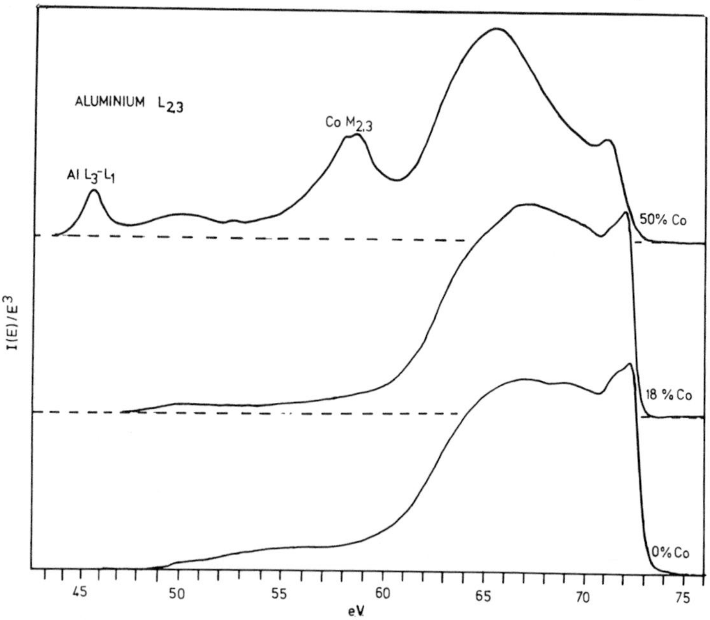

Figure 3 Aluminium $L_{2,3}$-emission bands for aluminium-cobalt alloys. The cobalt $M_{2,3}$ emission band overlaps at the low-energy end, and is observed in the tail of the spectrum for Al-Co.

D. Aluminium-Copper Alloys

Figure 4 shows the Al $L_{2,3}$ spectra for a series of aluminium-copper alloys, and the Cu $M_{2,3}$ spectrum of pure copper. The Cu $M_{2,3}$-emission band overlaps the Al $L_{2,3}$ band at the Fermi-level, and distorts the spectra for the copper-rich alloys. The alloys Al-10at%Cu and Al-20at%Cu are two-phase, consisting of α-Al and θ-$CuAl_2$. The edge shifts for these alloys are not reliable and therefore are not listed in Table I. In the Al $L_{2,3}$ spectra it is again noted that the low-energy region has increased in intensity with respect to the intensity at the Fermi-level and this becomes more pronounced as the copper concentration is increased. The Cu $M_{2,3}$ spectrum for the pure metal is in good agreement with the measurements reported by Dobbyn *et al* (1970).

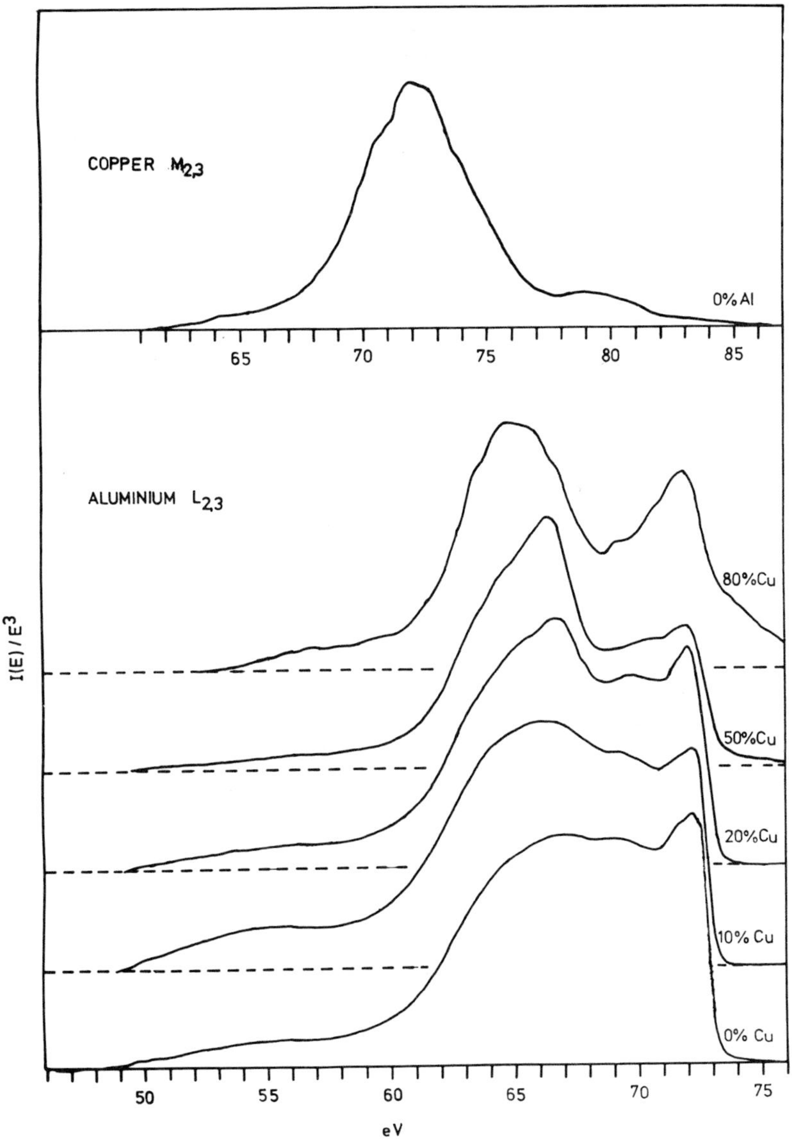

Figure 4 Aluminium $L_{2,3}$-emission bands for aluminium-copper alloys. The copper $M_{2,3}$-emission band, shown for the pure metal, overlaps the high-energy region of the aluminium emission from the alloys.

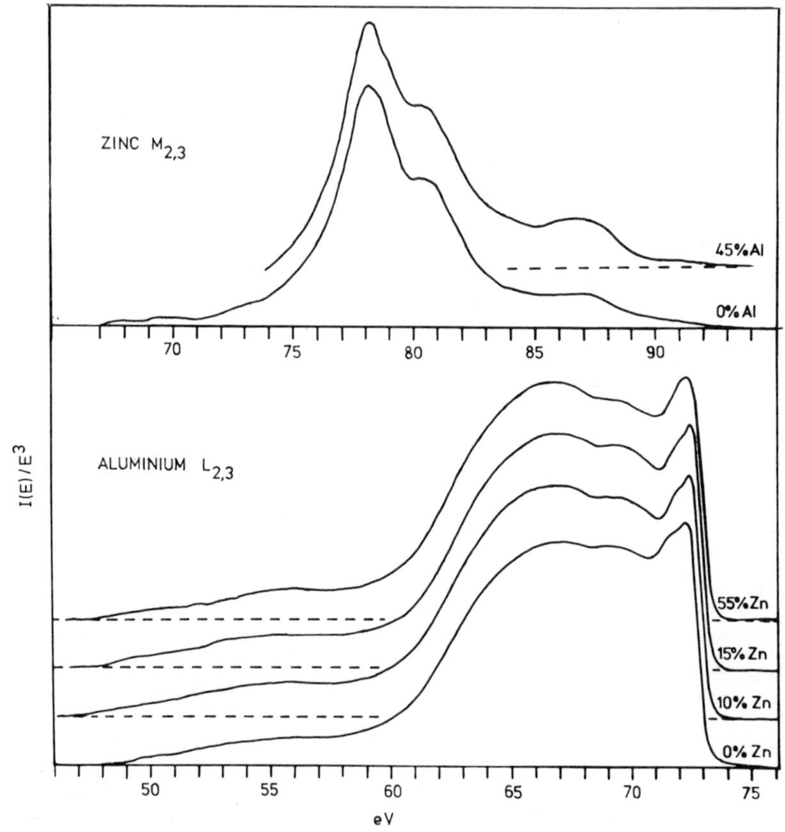

Figure 5 Aluminium $L_{2,3}$ and zinc $M_{2,3}$ emission bands from a series of aluminium-zinc alloys.

E. Aluminium-Zinc Alloys

The Al $L_{2,3}$ and Zn $M_{2,3}$ spectra from the pure metals and for Al-Zn alloys are shown in figure 5. Little change in the spectra takes place on alloying. It is important to note that this alloys system forms no intermetallic compounds and has an extended solubility range of zinc in aluminium, particularly at high temperatures. The alloys were prepared by quenching from elevated temperatures, and the resulting 55%atZn-alloy showed segregation into two phases, almost pure zinc and aluminium-zinc solid solution of varying composition. The other alloys were single-phase but are known to form *Guinier-Preston* zones indicating a preference for Al-Al and Zn-Zn bonds over Al-Zn.

3. DISCUSSION

The Al $L_{2,3}$-emission spectra for these alloy series show large changes in shape on alloying, which must be associated at least in part with the tendency to form covalent or covalent-ionic bonds. Investigation of the Al Kβ spectra for some of these alloys (Krivitsky *et al*, to be published) shows that the density of p-states around the aluminium atoms tends to move to lower energies, and that the spectra assume a form more characteristic of the Al_2O_3 spectrum, suggesting that some of the p-states are involved in the bonding.

Wenger *et al* (1971) have shown that the atomic d-states for transition metals towards the end of the first series, in these alloys, become progressively filled as the aluminium concentration is increased. The largest change in the number of 3d electrons is observed for the Co-Al alloys, and progressively decreases for the Fe-Al, Mn-Al, Ni-Al and Cu-Al systems. They also show that the density of p-states inside the aluminium ion increases with the concentration of the second component, being least for the Cu-Al system and greatest for the Co-Al alloy. This result is not in agreement with the results reported by Nemoshkalenko and Gorskii (1968), who consider that the integrated intensity of the aluminium $K\beta_x$ band remains almost constant for iron-aluminium alloys. Wenger *et al* claim that their results support the explanation of Hume-Rothery and Coles (1954) for the congruent melting points of aluminium transition-metal alloys in terms of increased cohesion due to increasing ionicity of the bond. However, without accurate measurements of the integrated intensities of all the valence-band spectra, it is not valid to assume that charge transfer takes place.

X-ray photoelectron-emission measurements for these alloys would provide further information. From the soft x-ray emission spectra we can conclude that there is a considerable redistribution of electron states between the atoms in the alloys. It is probable that the 3d-states of the transition-metal ions will tend to fill, at least for the transition metals up to nickel. However, in the alloys we can not assume that the 3d-states retain the same characterisitcs as in the pure transition metal. They tend to hybridize with the s,p states of the aluminium, and to produce the strong bonding that occurs; there must be a build-up of charge in the region between the transition-metal and aluminium atoms. The problem remaining is the correct interpretation of the changes in shape and intensity of valence-band emission spectra of the alloys, with particular reference to the nature of the bonding between the atoms; here we must rely on theoretical studies, and most probably on the molecular-orbital approach of the theoretical chemist.

Acknowledgments

The authors are indebted to the Science Research Council for funds in support of this investigation, and to Professor V.V. Nemoshkalenko and the Ukraine Academy of Sciences for supplying some of the alloys.

REFERENCES

Dobbyn, R.C., Williams, M.L., Cuthill, J.R. and McAlister, A.J. (1970); Phys. Rev. B2, 1563.

Fischer, D.W. and Baun, W.L. (1967); J. Applied Phys. 38, 229.

Hume-Rothery, W. and Coles, B.R. (1954); Adv. Phys. 3, 149.

Nemoshkalenko, V.V. and Gorskii, V.V. (1968); Ukr. Phys. J. 13, 1022.

Watson, L.M., Dimond, R.K. and Fabian, D.J. (1968); in "Soft X-Ray Band Spectra and the Electronic Structure of Metals and Materials", ed. Fabian, D.J., p.45: London, Academic Pres

Wenger, A., Bürri, G. and Steinemann, S., (1971); Sol. State Comm. 9, 1125.

TRANSITIONS INVOLVING SOLUTE d-ELECTRONS IN MAGNESIUM-BASED ALLOYS

P.R. Norris*, R.S. Crisp and R.K. Dimond
University of Western Australia, Nedlands, W.A.

1. INTRODUCTION

Recently a large number of investigations have been reported on the solvent soft x-ray emission band from alloys of noble and transition metals with either magnesium or aluminium. An almost universal feature of these spectra is a sharp peak within the band, at an energy whose value relative to the Fermi-edge varies with the particular alloy. To explain this peak Marshall *et al* (1969) postulated a mechanism involving hybridization of the flat d-bands with s-states of the conduction band; their interpretation gives good agreement with experiment in the case of aluminium-silver alloys. Curry and Harrison (1970) also found a correlation between the separation of the peak from the emission edge and the pure metal d-band energy, measured by other methods, consistent with an explanation involving initial states composed of a high proportion of atomic d-states. However, such transitions are quite sensitive to the spatial orientation of one atomic species with respect to the other in the alloy, and a complete test of the hypothesis would involve a variety of alloys of differing constituents and crystal structures. The present paper presents the results

*P.R.N. is now at: University of Strathclyde, Department of Metallurgy, Glasgow, Scotland.

of measurements on four alloys of magnesium, for which the sharp peak is also observed.

2. EXPERIMENTAL

The spectrometer is a 1-metre grazing incidence instrument with photoelectric detection, and has been described by Fisher, Crisp and Williams (1968).

One intermetallic was chosen from each alloy system. Where the phase diagram allowed a choice to be made, the alloy was selected to minimize the possibility of admixture of a magnesium-rich phase. Contamination of specimens with magnesium-rich material must be carefully avoided because of the anomalous weakness of the magnesium emission band from the majority of alloys.

The alloys were prepared by melting the weighed components under a suitable flux in recrystallized alumina crucibles. The melts were allowed to solidify in these crucibles, and were annealed. The compositions were then examined by powder x-ray diffraction. The crystal structures were found to be as expected from the phase diagram, and in no case was it possible to detect any trace of an impurity phase.

Targets were carefully cut from the ingots and clamped to the water-cooled target holder, with as brief exposure as possible to the atmosphere. The magnesium L-spectra were excited by currents of 4mA at 4kV, and the targets were frequently scraped clean within the spectrometer vacuum. Operating pressures were maintained between 1 and 2×10^{-6} torr, and examination of the fourth order of the carbon K-spectrum showed that carbon contamination was virtually absent.

3. RESULTS AND DISCUSSION

The emission spectra for the four alloys are shown in figure 1, where the only corrections to the original data are a linear

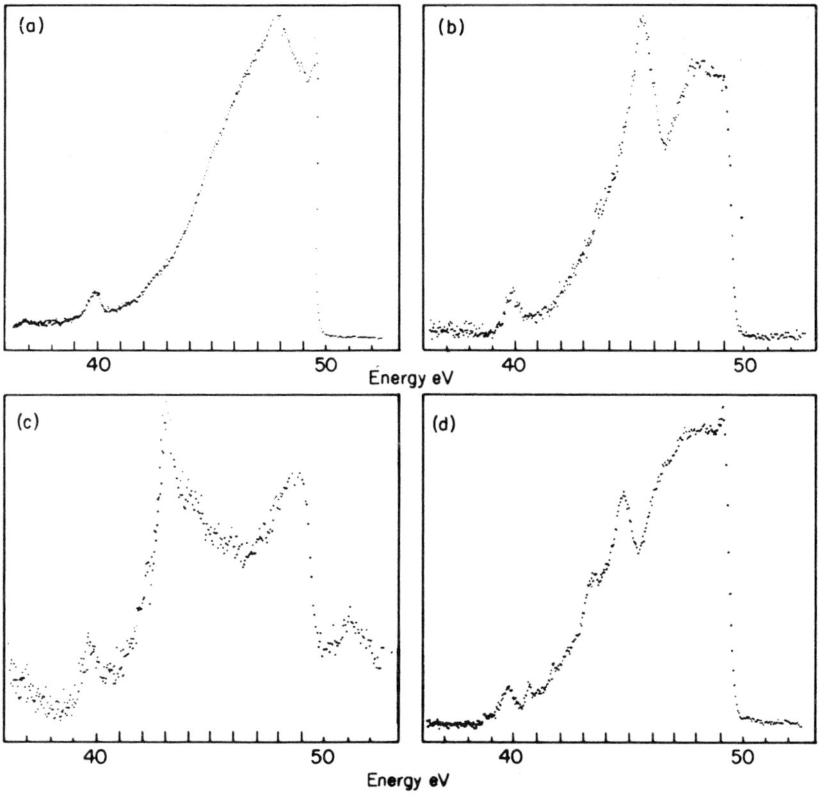

Figure 1 L-emission bands of (a) Mg_2Ni, (b) Mg_2Cu, (c) MgAg, and (d) Mg_3Au. The experimental data have been corrected to remove background radiation and instrumental effects.

interpolation of the intensities to allow for the effects of contamination (see for example, Watson, Dimond and Fabian 1968) and an adjustment of scales to correct for the slight change in energy width of the spectrometer 'window'.

Table 1 indicates the position and width of the emission edge for each spectrum, and also the separation between the edge and the characteristic peak mentioned above.

Table 1

Features of the spectra. Values in parentheses are from Curry and Harrison (1970).

Alloy	Composition	Peak (eV)	Edge (eV)	Edge-width (eV)
Mg	pure	-	49.35	0.4
Mg-Cu	Mg_2Cu	4.2 (4.3)	49.4	0.6
Mg-Ag	$Mg_{50\%}Ag_{50\%}$	6.6 (5.9)	49.6	0.8
Mg-Au	Mg_3Au	5.0 (large peak) 6.5 (small peak)	49.5	0.6
Mg-Ni	Mg_2Ni	2.3	49.5	0.5_5

It would be unrealistic to attach too much importance to the absolute energy-values of the edges, because of an inherent limit in accuracy of measurement, but it is worth noting that the shift observed in Mg_2Cu, relative to pure magnesium, is in the opposite direction to that reported by Curry and Harrison (1970); this may be related to the fact that the spectra themselves show pronounced disagreement in the region of the edge.

In no case is any extreme edge broadening observed; the edge-widths remain only a few times the resolution of the instrument. In particular, the highly distorted edge observed for Mg_2Ni by Appleton and Curry (1967) is not confirmed by the present results; the only broadened edge found in our series of alloys was with MgAg and this may be connected with the relatively indistinct appearance of the superlattice lines obtained in the powder x-ray diffraction photograph of this alloy; the degree of order in the specimen may have been quite low.

With all four alloys the peaks in the emission bands are quite distinct, and in general resemble those found for alloys of aluminium with various heavy metals. This correspondence even extends to the multiplicity of the Mg_3Au peak, which matches that found by Curry and Harrison in the spectrum of Al_2Au.

Because the shapes of the bands are greatly modified from parabolic, no attempt has been made to fit $E^{\frac{1}{2}}$-curves to the tails of the bands, nor to estimate the bandwidths. However, it is clear that with the alloy spectra there is a relatively higher emission intensity near the bottom of the band than with pure magnesium. This and other effects of alloying are shown clearly in figure 2, where the ratio of the intensity of the alloy to that of pure magnesium at each measured point is plotted as a function of energy. A logarithmic scale has been used for this intensity ratio, because the shape of the curve is then independent of the way in which the spectra are normalized; a change in normalization produces merely a translation of the whole curve in the ordinate direction. This method of presenting the results was first used by Dimond (1969); peaks in the spectra appear as steps in the intensity ratio relating them to pure magnesium, and this can assist in their interpretation.

It is of interest to compare the separations from the emission edge of these peaks with the calculated depth of the d-bands below the Fermi-level for each of the pure solute metals. While a certain amount of disagreement exists in the theoreti-

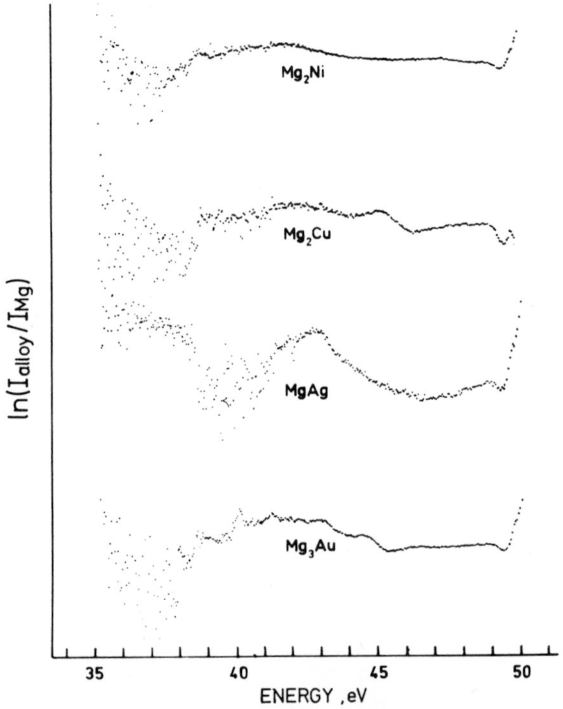

Figure 2 Logarithmic intensity ratio for L-spectra of the alloy with respect to pure magnesium. The scatter at the low-energy end is due to low counting ratio and has no significance, while the structure near the emission edge results from edge shifts in the alloy spectra.

cal data, there can be little doubt that the peaks lie consistently lower in energy than the pure metal d-bands. This is consistent with a rigid-band model for the alloys, assuming an effective valence of 1 for the noble metals, but such an interpretation cannot be anything like adequate considering the gross changes in the shapes of the spectra. In addition, it is certain that the d-levels themselves would be considerably modified, relative to their position in the pure metals, by the changes in screening due to the modified valence band, and by the s-d interaction itself, which is necessarily not small if a large overlap of the d-like states with magnesium cores is to occur.

REFERENCES

Appleton, A. and Curry, C. (1967); Phil. Mag. 16, 1031.

Curry, C. and Harrison, R. (1970); Phil. Mag. 21, 659.

Dimond, R.K. (1969); Thesis, University of Western Australia.

Fisher, P., Crisp, R.S. and Williams, S.E. (1958); Optica Acta, 5, 31.

Marshall, C.A.W., Watson, L.M., Lindsay, G.M., Rooke, G.A. and Fabian, D.J. (1969); Physics Letters 28A, 579.

Watson, L.M., Dimond, R.K. and Fabian, D.J. (1968); in "Soft X-Ray Band Spectra", : Academic Press, London, 45.

X-RAY SPECTRA AND ENERGY BAND STRUCTURE OF BINARY ALLOYS OF VANADIUM WITH THE ELEMENTS AT THE END OF THE TRANSITION SERIES

S.A. Nemnonov, E.Z. Kurmaev, B.Kh. Ishmukhametov and V.P. Belash

Institute of Metal Physics of the Ural Scientific Centre of the Academy of Sciences of the USSR, Sverdlovsk, USSR.

V.A. Fomichev and A.V. Rudnev

Institute of Physics, Leningrad State University, Leningrad, USSR.

1. INTRODUCTION

The theory of electronic structure of ordered and disordered alloys is presently under development.

The general quantitative examination of the band structure of an ordered binary transition-metal alloy has been reported by Stern (1965). He investigates a crystal built-up of atomic cells or spheres of different radii, for which the multi-electron problem for determining the energy spectrum of the atom within such a cell or sphere has already been solved. The single-electron wavefunctions are considered also to be known. In this way energy bands from individual atomic levels are formed. Stern (1965), using a tight-binding calculation, investigated the energy band-structure of an ordered 50%-50% alloy. He showed that the valence-electron spectrum of such an alloy consists of two bands of equal width Δ, which is determined - using the nearest neighbour approximation - by the exchange integral β. The reciprocal location of these bands is determined by the parameter $\delta=\varepsilon'_{21}/\Delta$, where ε'_{21} -

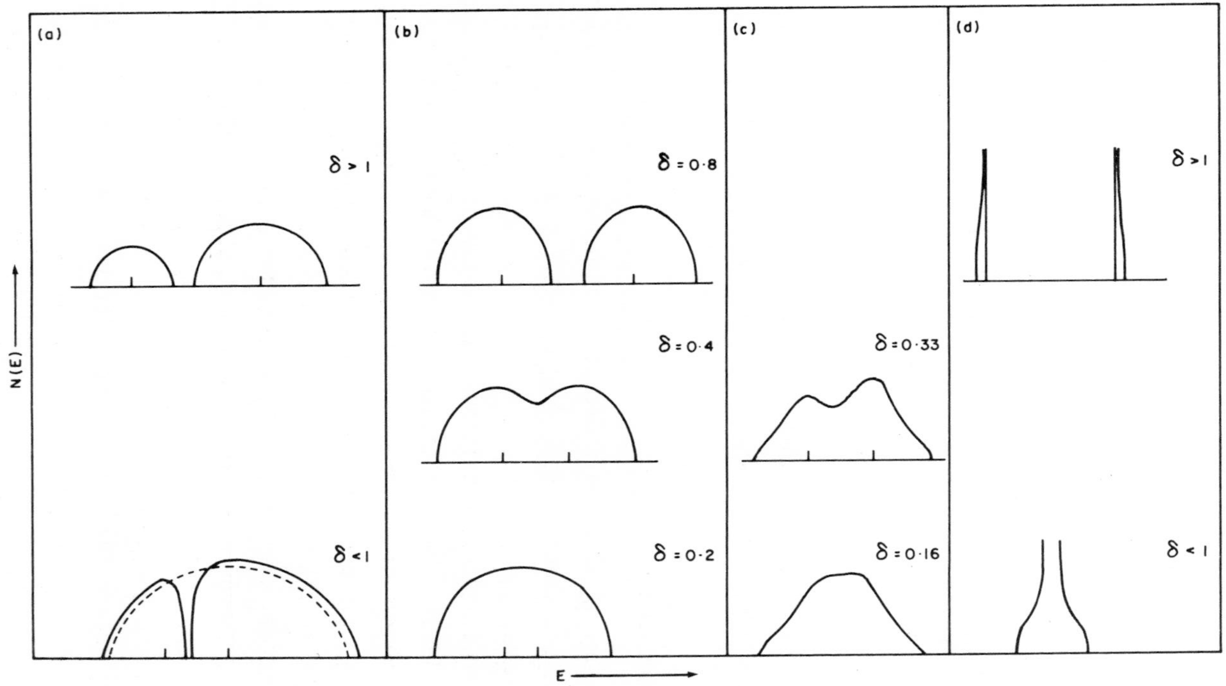

Figure 1 Schematic energy spectra of binary alloys, depending on the parameter $\delta=\varepsilon_{21}'/\Delta$; according to a) J.L. Beeby (1964), b) Y. Onodera and J. Toyozava (1968), c) P. Soven (1969), d) E.A. Stern (1965).

is the energy difference for the valence electrons of the different atoms. If the parameter $\delta = 0$ we have two overlapping energy bands; with $\delta > 0$ but small, the bands are separated by a narrow energy-gap; with $\delta \geqslant 1$ we have two well-separated energy bands (figure 1d).

However, while ordered alloys remain amenable to calculation any attempt to describe the energy spectrum of a disordered alloy meets with difficulty because the periodic structure is destroyed. We must therefore resort to different simplified models. One of the simplest for an alloy is the 'rigid-band' model (Mott and Jones 1936, 1958). The potential for the alloy is constructed from the periodic potential of the matrix atoms, and is perturbed by the potential due to the different fields of the solvent and solute atoms. Zero-order wavefunctions for electrons are considered identical to those for electrons in the crystal matrix, and the energy spectrum for electrons in the alloy is determined to a first approximation by perturbation theory. The alloy energy bands, and density of states, resemble those of the pure matrix, with the exception of a rigid shift in energy. Thus only the Fermi-energy is affected by the distortion of the crystal matrix caused by the impurity. However, the application of the rigid-band approximation can be justified only when the concentration of the solvent greatly exceeds the concentration of the solute. If the concentrations are comparable, the definitions 'solvent' and 'solute' are no longer sensible, in which case the theoretical calculation cannot be based upon the wavefunction and energy structure of only one constituent, and it is also impossible to use perturbation theory.

The next development in the theory of disordered alloys was a model suitable for any concentration of the alloy constituents (Nordheim 1931, Parmenter 1955, Schoen 1969). This model assumes the alloy electrons behave like electrons in a hypothetical ordered system; where the crystal potential is an average of the potentials of the components weighted according to their concentrations. A more complicated technique, based on the multiple-scattering formalism of Lax (1951), has been developed by Soven (1969), Velicky *et al* (1968), Beeby (1964) and Onodera and Toyozava (1968).

These investigations show that one of the main factors determining band structure in disordered alloys, as for the case of ordered structures, is the relation $\delta=\varepsilon'_{21}/\Delta$. Accordingly, if the parameter δ is close to zero there are two overlapping energy bands. With $0<\delta<1$ the bands are separated by a narrow energy-gap, and with $\delta\geqslant 1$ there are two well-separated energy bands. (See figure 1 a,b,c).

Thus it follows that superposition of constituent bands, and the formation of hybridized bands, occurs when the constituent bandwidths are considerably greater than their energy separation. This might arise, for example, when neighbouring transition elements (from titanium to chromium) are alloyed. It is therefore not surprising that for such alloys a common energy-band exists, with complete spd-hybridization. This is supported by experiments (Nemnonov and Finkelstein 1966, Nemnonov 1967 In such cases the rigid-band approximation is a sufficient description.

On the other hand, there is a large number of alloys with distinct valencies of their constituent elements and with different degrees of valence-electron localization. Component bandwidths can vary over wide limits. Generally, for such alloys the model representations are not sufficient, and the energy spectrum cannot be explicitly predicted "a priori". A rigorous calculation of the electronic structure for such systems would require preliminary knowledge of the pure metal bandwidths, as well as tentative data on the mutual position and the extent of separation of component bands in the alloy.

Such data can be obtained from experiments that give direct evidence about the energy spectrum of electrons in a solid.

The purpose of the present study was to investigate the energy-band structure of binary alloys of vanadium with the transition metals at the end of the first transition series (Fe, Co, Ni, Ru, Rh, Pd, Os, Ir, Pt), applying the technique of x-ray spectroscopy, to obtain information on the distribution of electron states with different types of symmetry.

2. EXPERIMENTAL CONDITIONS

With the exception of the alloys V-90at%Ni (fcc structure), V-80at%Fe and V-25at%Ru (bcc structure), all the investigated alloys were of the same crystal structure (β-W: V_3Co, V_3Rh, V_3Pd, VOs, V_3Ir and V_3Pt).

The fluorescent x-ray $K\beta_5$-emission bands of vanadium were measured with a Johann-type vacuum spectrograph (Trapeznikov and Nemnonov 1955). The quartz crystal (rhombohedron plane, bent to 900mm curvature) served as analyser. The energy resolution in the second order of reflection was 0.25eV. The secondary emitter was mounted on a copper water-cooled assembly. To protect the specimens from contamination the primary and secondary anodes were separated by a beryllium plate of 0.15mm thickness. Exposure times were from 60 to 90 h, with an excitation voltage of 20kV and a current of 50mA. Several spectra were photometered step-by-step, in 3 to 4 sections, and the results were averaged. The energies of spectral features are determined with an accuracy of ±0.2eV.

The $M_{2,3}$-emission spectra of vanadium were investigated using an ultrasoft x-ray spectrometer of the PCΛ-type (Lukirsky *et al* 1970). The instrument was equipped with a blazed "echelette" diffraction grating coated with gold, with 600 lines per mm and a radius curvature of 1m. The detector was a secondary-emission multiplier of the open type with gold photocathode. The higher-order diffractions were suppressed with the aid of a special filter, composed of two parallel glass plates covered with polystyrene. The take-off angle of emission from the anode was approximately 40°. All spectra were recorded in vacuum $\sim 10^{-6}$ of mercury, using an anode potential of 3kV and current 10-100mA. Samples were heated to achieve clean surfaces, monitored by disappearance of the K -emission line of carbon. The spectrometer slits were 200μ (Δλ = 4Å; ΔE = 0.4eV) Multiple recording of spectra showed good reproducibility of the emission fine-structure.

3. EXPERIMENTAL RESULTS

The measured $VK\beta_5$ and $VM_{2,3}$ emission bands for the pure metals, and their alloys with transition metals, are shown in figures 2 to 4. The open circles are the $VK\beta_5$-spectra; the dashed dotted lines are the $VM_{2,3}$ spectra. In figure 2 the curve traced through closed circles is the $VK\beta_5$-emission line for the alloy V-50at%Os corrected for inner-level width and instrumental broadening.

The emission intensity is in arbitrary units; spectra for each series are normalized to a single wavelength scale, with account taken of the energies of the inner $K\alpha_1$ and $K\beta_1$ lines.

The following features are noted:

(1) The K-emission and M-emission of vanadium in the alloys V-25at%Ru, V_3Rh, V_3Pd, VOs, V_3Ir and V_3Pt show a splitting (which is more pronounded in the K-spectra) into two sub-bands, indicated "A" and "B" in the figures. With the exception of the alloy V_3Pt, the intensity peaks for the K and M spectra sub-bands come at closely similar energies. With the V_3Pt alloy the K-emission sub-band A comes close in energy to the peak in the M-spectrum of vanadium. However, the splitting for the K-spectrum sub-band B is greater than for the M-spectrum.

(2) The short-wave length sub-band peaks in the K and M spec-

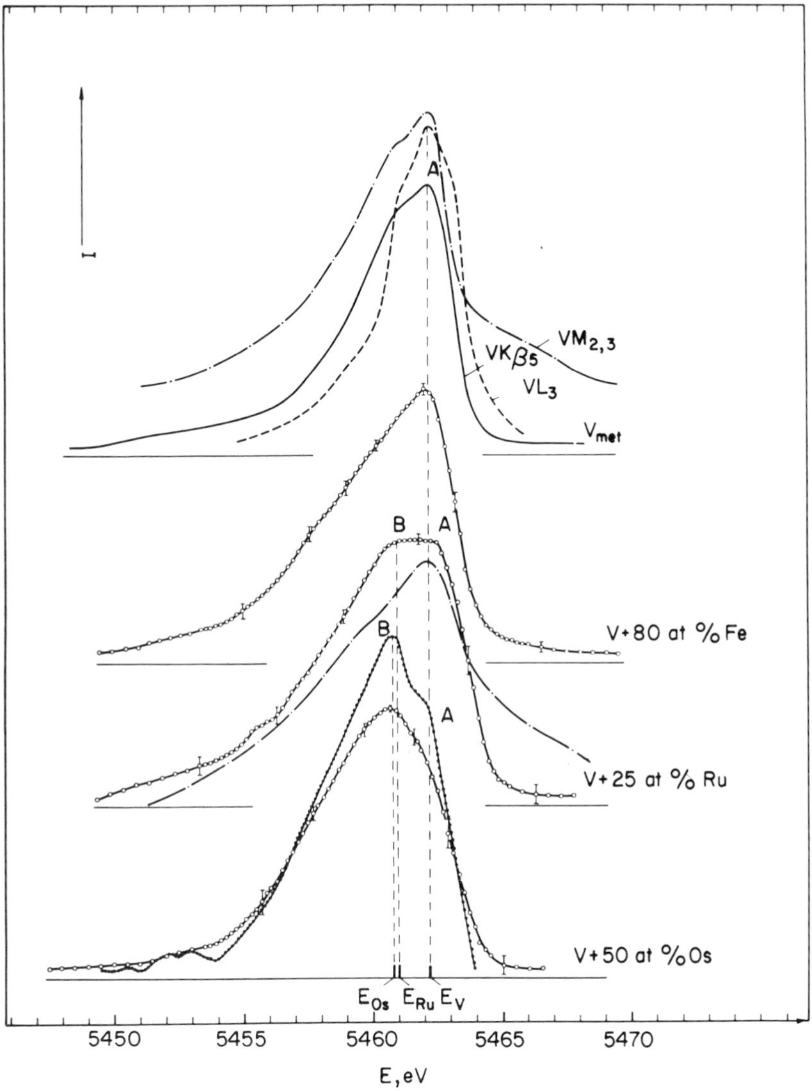

Figure 2 X-ray $K\beta_5$ and $M_{2,3}$ emission bands of vanadium for the pure metals and alloys V-80at%Fe, V-25at%Ru and V-50at%Os.

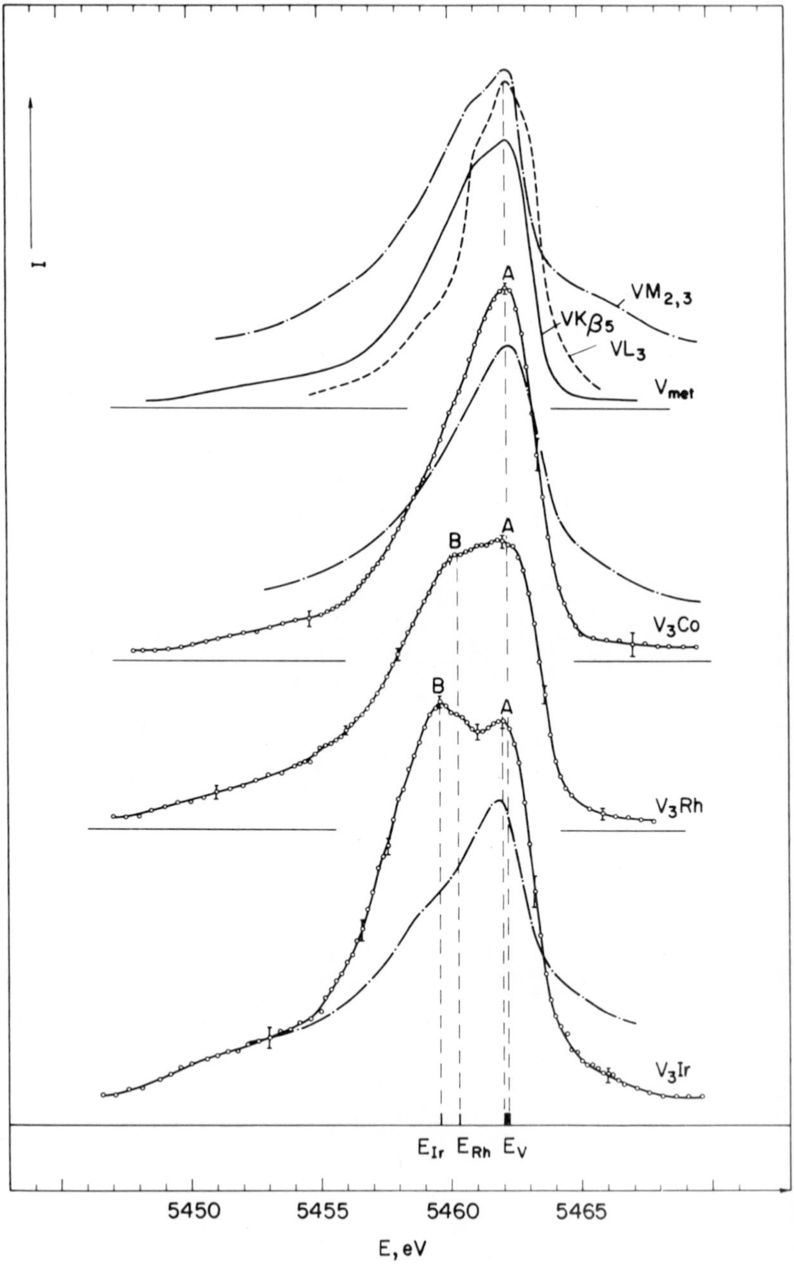

Figure 3 X-ray $K\beta_5$ and $M_{2,3}$ emission bands of vanadium for the pure metals and alloys V_3Co, V_3Rh and V_3Ir.

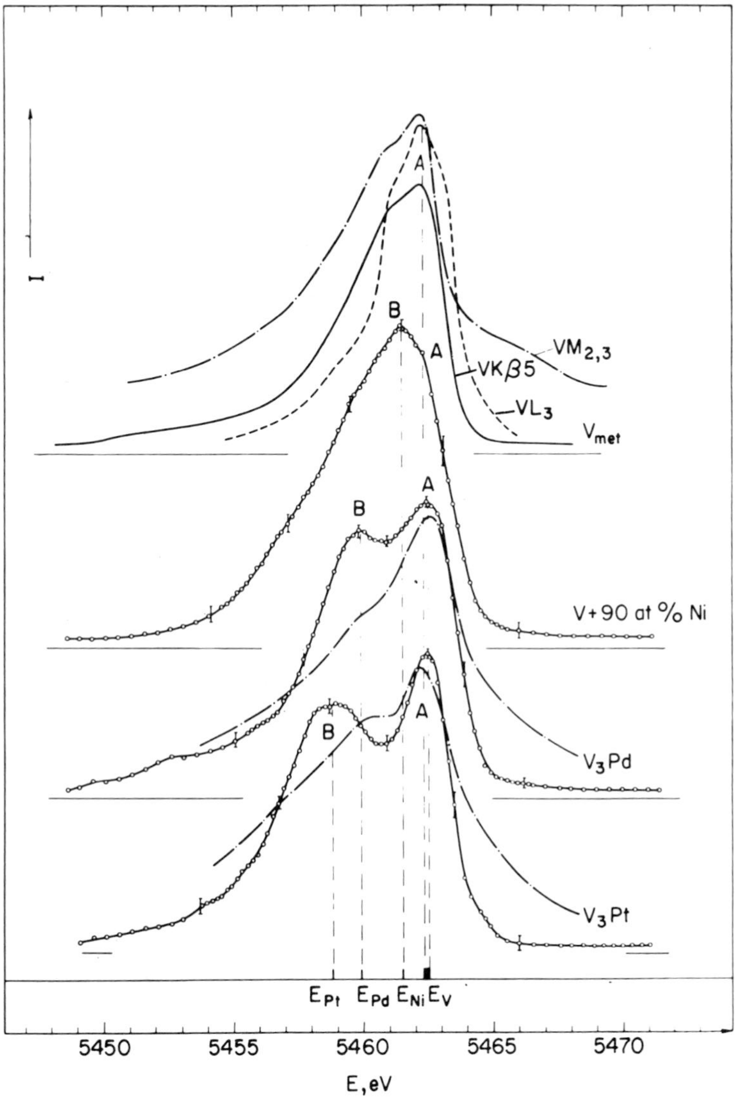

Figure 4 X-ray Kβ5 and $M_{2,3}$ emission bands of vanadium for the pure metals and alloys V-90at%Ni, V_3Pd and V_3Pt.

tra for the alloys coincide in energy with the peaks in the K and M emission bands for pure vanadium.

(3) The degree of B sub-band splitting (ΔE) depends on the second-constituent atom. We note, in spectra of alloys involving elements of the same Group, that ΔE usually increases with atomic number; and also that ΔE increases with increasing atomic number when crossing a given series.

4. DISCUSSION

The $VK\beta_5$-emission occurs with transition from the valence band to a vacancy in the 1s-level; the intensity of this emission band represents, in accordance with the dipole selection rules, the energy distribution of the 4p-states in the valence band arising from dsp-hybridization.

The $M_{2,3}$-spectrum corresponds to $3p\rightarrow 3d$ transitions. Measurement of the M-emission intensity gives information on the distribution of 3d-states in the occupied part of the conduction band.

We have already noted that K and M spectra of vanadium in the alloys have in most cases a two-band structure. The A-band peak in the K-spectrum of the alloys is close in energy to the main peak in the $M_{2,3}$-emission band for the same phases, leading to the conclusion that the sub-band A in the K-spectra of the alloys has its energy associated mainly with the 3d-electron states of vanadium, which in these alloys are concentrated chiefly in the upper part of the filled region directly below the Fermi surface.

Generally speaking, x-ray investigation of a binary alloy system makes it possible to obtain information about the electron-state distribution of the second component atoms, from the spectra of the first, provided the condition is met that the wavefunctions of the first differ from zero in that part of space where the wavefunctions of valence states for the second component have their maximum values. For this reason it becomes possible to observe, in the K-emission of titanium and vanadium from their carbides, nitrides and oxides, the energy bands originated from 2s (Kβ") and 2p ($K\beta_5^{II}$) states of C, N, and O (Nemnonov and Kolobova 1966, Kurmaev *et al* 1967). Similarly, the effect of energy bands of the second component was observed in the Al $L_{2,3}$ and Al$K\beta_x$ emission bands from aluminides of noble and transition metals (Marshall *et al* 1969, Nemnonov *et al* 1971, Fabian *et al* 1971, Switendick 1971).

Considering also approximate estimates of the binding energy for electron levels of the transition atoms, according to which the d-levels are lowered in energy on crossing the transition period (Herman and Skilman 1963, Beeby 1964, Clementi 1965), it is possible to interpret the low-energy band B in the alloy spectra as reflecting states arising from the d-states of the second constituent atoms.

Sokolov and Babanov (this Volume p431) have shown that the shape of the emission spectrum for one of the components, using the same calculation parameters as Soven (1969), is strongly dependent on the degree of overlapping of the wavefunctions of the different atoms forming a disordered alloy. When the wavefunctions strongly overlap (this is the case in the alloys considered) a second peak is observed in the emission band for the second component (e.g. for vanadium in the alloys illustrated in figures 2-4), associated with the energy states for the component.

The V_3Co and V-80at%Fe spectra exhibit simpler V$K\beta_5$-emission bands compared with other investigated phases. It is likely that the relationship among the energy differences of the V, Fe and Co 3d-states, and among the bandwidths is such that the energy spectrum of these alloys is characterized by two overlapping d-bands. Theoretical energy-bands for the alloy V_3Co (Mattheiss 1965) also indicate this.

Despite the fact that the series of elements Fe-Ru-Os, Co-Rh-Ir and Ni-Pd-Pt in each case possess the same configuration of valence electrons, for the free atoms, the energy spectra of their binary alloys with vanadium show certain distinctions. For the alloys of vanadium with transition elements belonging to the same Group, the degree of separation between the constituent energy bands generally grows with increasing atomic number of the transition element (with the exception of the alloys V-25at%Ru and V-50at%Os, for which the $K\beta_5$-band splitting is roughly equal). This might be due completely to the change of the crystal potential on going from one alloy to the next.

Similar results were reported from band structure calculations for hexagonal semiconductors (Maschke and Rossler 1968, Rossler 1969). In the compounds ZnO, CdO, ZnS and CdS, one of the constituents (O or S) is invariable, while the two others (Zn and Cd) are elements of the same group with similar configurations of valence electrons in the free atoms. The electronic structure of these compounds consists of two separated energy bands. The low-energy band has d-character, and the high-energy band originates from the p-states of oxygen or sulphur atoms. It is characteristic that transitions from ZnO to CdO or from ZnS to CdS indicate the lowered energy of the d-bands (by $\sim$1.3-1.5 eV).

Another interesting result is observed in the series of alloys V-25at%Ru, V_3Rh, V_3Pd and VOs, V_3Ir, V_3Pt; with increased atomic number of the second component along the period. Here energy separation of the sub-bands is also increased (see figures 2-4) and this is evidence that the energy of d-electrons falls with rising atomic number.

Consequently, the x-ray spectroscopic results presented here lead us to conclude that the electronic structure of binary alloys of vanadium with the elements at the end of the first transition series characterizes the energy separation of valence bands associated with the different constituents. The high-energy bands arise from the 3d-states of vanadium atoms and the low-energy bands from the d-states of the heavier constituent.

A similar result has been observed in the investigation of aluminium $K\beta_x$ and $L_{2,3}$-emission from metallic compounds of nickel, palladium and platinum with noble metals (Nemnonov *et al* 1971). In these compounds aluminium appears to behave like vanadium in the alloys, since these materials differ only in the energy separation between the corresponding sub-bands and in the symmetry of the high-energy electron states.

The results for the energies the component d-bands in this group of alloys are of considerable interest for an understanding of some of their physical properties; we believe that they can serve as a basis for rigorous calculation of the band structure.

Acknowledgments

It is a pleasure to express our gratitude to T.B. Shashkina and S.N. Petrov for translation of this paper into English.

REFERENCES

Beeby, J.L. (1964); Phys. Rev. 135, A130.

Clementi, E. (1965); IBM, J. Res. Development (Suppl), 9, 2.

Fabian, D.J., Lindsay, G.M. and Watson, L.M. (1971); in Electronic Density of States, 3rd IMR Symp. pp. 307-313, NBS. Spec. Publn. 323.

Herman, F. and Skilman, S. (1963); Atomic Structure Calculations, Prentice-Hall, Inc., Englewood Cliffs, New Jersey.

Kurmaev, E.Z., Nemnonov, S.A., Menshikov, A.Z. and Shveikin, G.P. (1967); Izv. Akad. Nauk SSR, Ser. fiz. 31, 996.

Lukirsky, A.P., Fomichev, V.A., Rudnev, A.V. (1970); in "Apparatura i metod rentgenovskogo analiza", vyp.6, str.89, SKB RA, Leningrad.

Lax, M. (1951); Rev. Mod. Phys. 23, 237.

Marshall, C.A.W., Watson, L.M., Lindsay, G.M., Rooke, G.A. and Fabian, D.J. (1969); Phys. Letts. (Netherlands) 28A, 579.

Maschke, K. and Rossler, U. (1968); Phys. Stat. Sol. 28, 577.

Mattheiss, L.F. (1965); Phys. Rev. 138, A112.

Mott, N.F. and Jones, H. (1958); Metals and Alloys (Dover Publications, Inc. New York.

Nemnonov, S.A. and Finkelstein, L.D. (1966); Fiz. Metallov i; Metallovedenie 22, 538.

Nemnonov, S.A. and Kolobova, K.M. (1966); Fiz. Metallov i Metallovedenie 22, 680.

Nemnonov, S.A. (1967); Fiz. Metallov i Metallovedenie 24, 1016.

Nemnonov, S.A., Zyryanov, V.G., Minin, V.I. and Sorokina, K.F. (1971); Phys. Stat. Sol. 43, 319.

Nordheim, L. (1931); Ann. Physik 9, 607.

Nordheim, L. (1931); Ann. Physik 9, 641.

Onodera, Y. and Toyozava, J. (1968); J. Phys. Soc. Japan 24, 341.

Parmenter, R.H. (1955); Phys. Rev. 97, 587.

Rossler, U. (1969); Phys. Rev. 184, 733.

Schoen, J.M. (1969); Phys. Rev. 184, 858.

Soven, P. (1969); Phys. Rev. 178, 1136.

Stern, E.A. (1965); Physics 1, 255.

Switendick, A.C. (1971); in Electronic Density of States, 3rd IMR Symp. pp.297-302, NBS. Spec. Publn. 323.

Trapeznikov, V.A. and Nemnonov, S.A. (1955); Fiz. Metallov i Metallovedenie 1, 562.

Velicky, B., Kirkpatrick, S. and Ehrenreich, H. (1968); Phys. Rev. 175, 747.

SOFT X-RAY SPECTRA OF VANADIUM, NIOBIUM AND THE ALLOYS V_3Sn AND Nb_3Sn

C.F. Hague and C. Bonnelle

Laboratoire de Chimie Physique, de l'Université de Paris VI, Paris Vème, France.

1. INTRODUCTION

Alloys of the Nb_3X and V_3X type are of particular interest because of their superconducting properties and high critical temperatures. Clogston and Jaccarino (1961) were able to explain the temperature-dependent Knight shifts and susceptibilities of V_3X type intermetallic compounds by proposing that a very narrow d band is intersected by the Fermi level. The conduction band consists of vanadium 4s and 4p wavefunctions overlapping the 3d band. The *n*p wavefunctions belonging to the X sites would lie sufficiently close to the Fermi energy to be admixed with the vanadium bands, while the *n*s electrons lie well below the Fermi level.

Labbé and Friedel (1966) proposed a simple band-structure model for the β-W type alloys. It is supposed that only interactions along the closely linked transition metal atoms need be considered since the interchain distances are large. The schematic density-of-states curve obtained on this assumption does indeed suggest a narrow d band with a high density peak close to the Fermi level.

More recently Weger (1970) has carried out a detailed analysis of the interactions involved, and Goldberg and Weger

(1971) have been able to estimate the density-of-states curve of such an alloy. Their results also give a narrow peak at the Fermi level and they find a distinct minimum some 1.6eV below this peak.

As part of a systematic study of the emission bands of transition elements, obtained from pure metals and their compounds (Hague 1972), we have studied the V Lα-emission band for the metal and for the V_3Sn alloy, as well as the Lβ band of Nb and Nb_3Sn. The Lα-emission band for vanadium arises from $3d \rightarrow 2p_{3/2}$ transitions, while the $4d \rightarrow 2p_{3/2}$ transitions give the niobium $L\beta_2$ band. Thus, in both cases the spectra give a measure of the distribution of electrons in the d bands.

2. EXPERIMENTAL PROCEDURE

It is well known that self-absorption of the emitted x-rays in the target can cause important distortions in the observed spectra (Bonnelle 1964, 1966). With this in mind, a special x-ray tube has been constructed in which the x-ray take-off angle is normal to the target surface, while the electron beam strikes the anode at almost grazing incidence (Hague 1971). In this way the distance travelled in the specimen by the observed x-rays is reduced to a few Å provided a suitable excitation potential is used (in the present investigation 1.5 to 2 times the threshold potential was used).

A further problem that is often overlooked is surface contamination. This can result from tungsten or carbon deposits and more seriously, from oxidation produced under the action of electron bombardment. The former type of contamination can in general be dealt with by using high vacuum in the x-ray tube. All the direct excitation observations were carried out under high-vacuum conditions (10^{-8}-10^{-7} torr). Even under such conditions, oxidation is often unavoidable, and satisfactory results can only be obtained by using fluorescence spectroscopy; the action of a primary x-ray beam on a sample has virtually no influence on its chemical state. However, it should be mentioned that the x-ray photons penetrate the target further than an electron beam of the same energy and that, for intensity reasons, it is often difficult to avoid using small take-off angles; thus, self-absorption distortions are again introduced. Nevertheless, this technique had to be employed to obtain the vanadium emission. The primary x-ray source was provided by the Cu Lα emission (∼1·keV) from a specially designed x-ray tube operated at 1A and 2kV.

A 25cm radius bent crystal reflection spectrograph was used. Both a KAP crystal (2d=26.4 Å) and a prochlorite crystal 2d=28.3 Å) were used to obtain the vanadium spectra while the niobium spectra were studied with a quartz crystal cleaved along the $10\bar{1}1$ plane (2d=26.67 Å). The dispersions on the Rowland circle were respectively 0.9 eV/mm, 1.25eV/mm and 7 eV/mm. The spectra were recorded on a highly sensitive film (KODAK DC3). We find that this technique is useful when fine details are important and absolute intensity measurements are not required. Moreover, the metal and alloy* spectra can be recorded one above the other on the same film giving a reliable indication of any displacements that may occur.

3. RESULTS

The vanadium Lα-emission band lies close to the O Kα band (23.6 Å), thus we could check on the presence of oxidation during the experiments. The O Kα band was never observed in the pure metal spectrum but was always present in the V_3Sn spectrum obtained by electron excitation. However, it should be noted that the V_3Sn sample remained completely unaltered in appearance after bombardment under normal vacuum conditions. The results obtained by electron excitation are shown in fig-

---

*Alloy samples were supplied by a specialized laboratory.

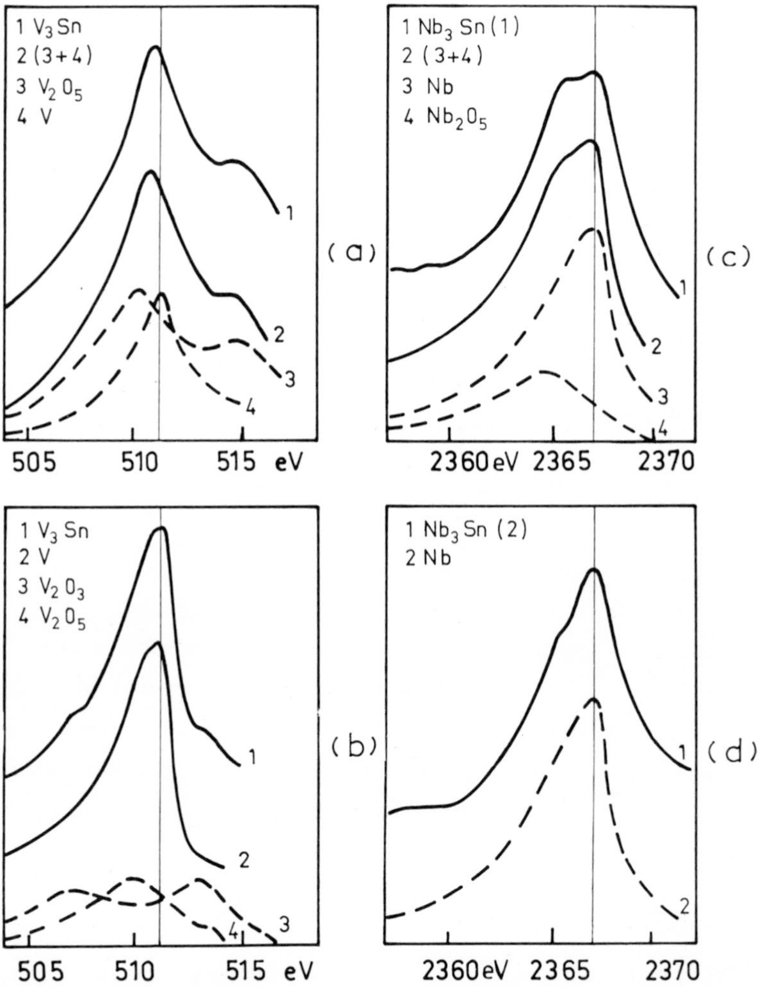

Figure 1 (a) Vanadium Lα-emission band obtained by electron excitation : (1) V_3Sn; (2) compound curve (3 + 4); (3) V_2O_5 (Fischer 1969); (4) Vanadium metal.

(b) Vanadium Lα-emission band obtained by fluorescence excitation : (1) V_3Sn; (2) Vanadium metal; (3) V_2O_3; (4) V_2O_5.

(c) Niobium $L\beta_2$-emission band obtained by electron excitation: (1) Nb_3Sn oxidized sample; (2) compound curve (3 + 4); (3) Niobium metal; (4) Nb_2O_5.

(d) Niobium $L\beta_2$-emission band obtained by electron excitation: (1) Nb_3Sn oxide-free sample; (2) Niobium metal.

ure 1a, while the x-ray excitation results are given in figure 1b. The bandwidths are given in Table I.

The niobium spectra were studied by electron excitation only. The excitation potential being much higher than for vanadium the electron penetration is a factor of 10 greater, so that despite the small electron incidence angle, the thickness of the target emitting the radiation is appreciable. The contribution from any surface contamination is thus reduced.

Two Nb_3Sn samples of different origin were studied. The results obtained with the first are given in figure 1c and with the second in figure 1d (see Table I).

4. DISCUSSION

The results obtained underline the rôle played by oxidation in this type of study. Curve 1 (figure 1a) is, perhaps, an extreme case where the presence of oxidation cannot be overlooked. We show in the same figure a compound curve obtained by adding the V_2O_5 curve (Fischer 1969) to that of the pure metal emission, both observed under electron excitation conditions. The similarity of this curve and the V_3Sn spectrum, obtained by direct excitation, is immediately obvious.

The O $K\alpha$ emission was never observed in the fluorescence spectra, so that the structure in the V_3Sn curve (figure 1b) is characteristic of the density of states of the alloy and suggests that the d band is somewhat different from that of the pure metal. For the alloy two peaks are observed, 0.5eV apart, representing structure in the d band; a shoulder is observed at 4eV below the peaks and is due to transitions from the Sn 4p states. The curve drops more sharply on both sides resulting in a narrower band at half-maximum than the equivalent curve for the pure metal; the steeper slope on the high energy side suggests a higher density of states at the Fermi level than for the pure metal.

The structure near the emission maximum and on the low-energy side of the spectrum from the oxide free Nb_3Sn sample (figure 1d) can be explained as for the alloy V_3Sn. Figure 1c shows the type of curve obtained with an oxidized sample. The pure niobium and the Nb_2O_5 results are taken from an earlier measurement (Hague 1965) under higher self-absorption conditions.

The experimental broadening has been carefully calculated and is almost negligible at the Bragg angles used here. The width of the vanadium L_3 level is taken to be 0.7eV while that of niobium L_3 level is about 1.3eV. These values, obtained by Callan (1960) from theoretical calculations of the width of the K level and experimental results of the $K\alpha_1$ lines, have been subtracted from the observed bandwidths; this is possibly an overestimation of the correction to be applied.

5. CONCLUSION

The excellent dispersion given by the prochlorite crystal and the higher resolution associated with the narrower inner excited level, were more favourable to the observation of structure in the vanadium spectra. Even so, the changes observed in the density of states from the metal to the alloy remain small.

Our results show the important part played by oxidation when the spectra of intermetallic compounds are obtained by electron bombardment. Such effects are virtually eliminated when fluorescence spectroscopy is employed.

Acknowledgments

The authors wish to thank Dr. M. Weger for the interest he has shown in this work, and for making available his unpublished results.

Table 1
Bandwidths at half-maximum (eV)[a]

	V	V_3Sn	Nb	Nb_3Sn
uncorrected	3.5±0.15	3.0±0.2	4.9±0.2[b]	4.5±0.2[b]
corrected[c]	2.8	2.3	3.6	3.2

(a) errors are root-mean-square values
(b) corrected for self-absorption
(c) corrected for instrumental errors and width of inner level

REFERENCES

Bonnelle, C. (1964); Thèse de Doctorat d'Etat, Paris; (1966) Ann. Phys. 1, 439.

Callan, E.J. (1960); private communication.

Clogston, A.M. and Jaccarino, V. (1961); Phys. Rev. 121, 5.

Fischer, D.W. (1969); J. Appl. Phys. 40, 4151.

Goldberg, I.B. and Weger, M. (1971); Bull. Israel Phys. Soc. 50.

Hague, C.F. (1971); J. Phys. E. 4, 119.

Hague, C.F. (1965); Diplôme d'Etudes Supérieures Univ. de Paris.

Hague, C.F. (1972); Thèse de Doctorat d'Etat, Paris.

Labbé, J. and Friedel, J. (1966); J. Phys. 27, 153.

Senemaud, C. and Hague, C.F. (1971); J. Phys. Colloque 32, C4-193.

Weger, M. (1970); J. Phys. Chem. Solids, 31, 1621.

NEW X-RAY EMISSION SPECTRA OF TRANSITION METALS IN THE REGION OF 100-1000Å

V.A. Fomichev, Tatiana M. Zimkina and A.V. Rudnev

Institute of Physics, Leningrad State University, USSR.

and S.A. Nemnonov

Institute of Metal Physics, USSR Academy of Sciences, Sverdlovsk, USSR.

1. INTRODUCTION

According to the data of Bearden and Burr (1965) all transition elements with unfilled d-subshells have inner core levels in the energy region of 15-50eV. These are the $M_{2,3}$ (3p-levels) for elements from scandium to copper, the $N_{2,3}$ (4p-levels) for elements from yttrium to silver, and the $O_{2,3}$ and $N_{6,7}$ (5p and 4f levels) for the elements from lutetium to gold.

The x-ray emission bands of these metals, arising from transitions of valence-band electrons to vacancies in the above core levels lie in the wavelength region 250-1000Å. Some of these emission bands have been investigated by Lukirskii *et al* (1966); the results indicate high intensities and peak-to-background ratios, and therefore offer information about the electronic structure of the transition metals.

2. EXPERIMENTAL

The x-ray spectrometer (an RSL-1500; described by Lukirskii *et al* 1970) has been developed for investigating x-ray spectra

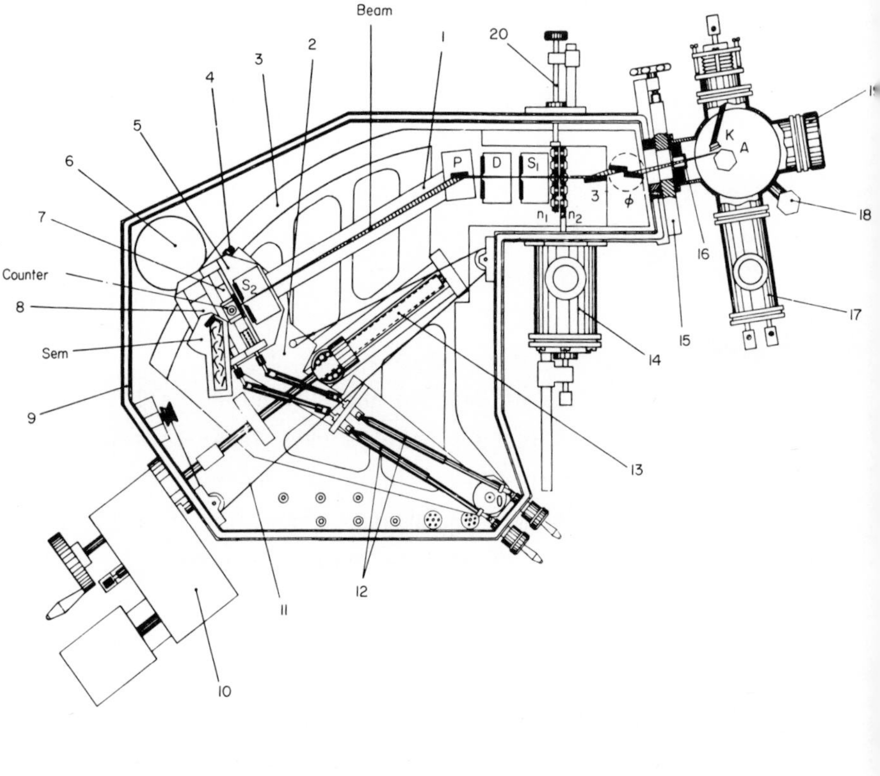

Figure 1 The spectrometer (RSL1500) : K cathode, A anode, 3 baseplate, 4 ball bearings, 6 diffusion pump, 7-8 adjustments for selecting counter or SEM, 9 '0' ring, 10 reduction gear, 11 cord, 12 manipulators, 13 screw mechanism, 17 evaporation chamber, 18 vacuum gauge, 19 inspection window. For additional designations see text.

in the region of 100-1000Å. Figure 1 shows the basic operating principles. The beam of x-rays from the anode A passes through a slit (16) which separates the x-ray tube from the spectrometer, to a reflecting filter Φ and a mirror (3) that focuses the beam onto the entrance slit S_1. The beam passes through a collimator D and is incident on a grating P at a grazing angle of $5^0 45'$. The diffracted x-rays are detected at the exit slit S_2. The grating and both slits are on the Rowland circle whose centre is located at the point O. The platform (5) and slit S_2 are attached to the rod (2) which rotates about the point O. The bar (1) ensures that the plane of the exit slit is always normal to the diffracted radiation.

The 'echelette' concave grating has 600 lines/mm and a radius of curvature of 1m. The surface is coated with a thin layer of gold (about 400Å). The detector is either a secondary electron multiplier (SEM) or proportional counter. Two sets of filters Π_1 and Π_2 are interposed between the entrance slit S_1 and the mirror. The Π_1 filters may be prepared outside the spectrometer and are placed in the beam by means of a pivot (20). The Π_2 filters are prepared by vacuum evaporation in the preparation chamber (14).

Spectra free from higher orders of diffracted radiation can be obtained by application of the reflecting filter (Φ) (Lukirskii and Omelchenko 1960). The principle of this filter is based on the phenomenon of the x-ray external reflection. In the ultra-soft region total external reflection takes place at grazing angles of incidence, and for every angle there is a critical wavelength λ_c whereby x-rays with $\lambda<\lambda_c$ penetrate into a medium and are absorbed completely and x-rays with $\lambda>\lambda_c$ are reflected from the surface. This property is utilized to cut off the short-wavelength radiation. The wavelength λ_c increases with increasing angle of incidence and in the ultrasoft x-ray region this angle ranges from 2^0 to 40^0.

In the RSL-1500 spectrometer the reflecting filter Φ is made up of two flat parallel plates. The operation of the filter is illustrated in figure 2.

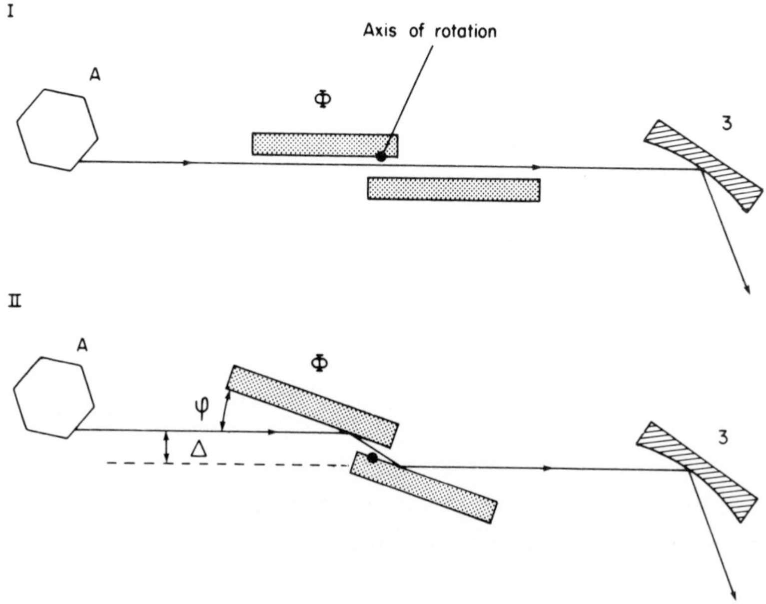

Figure 2 Operating principle of the filter Φ(see text).

In position I the x-rays from the anode A are incident directly on the focusing mirror (3). In position II the filter Φ has been rotated in a clockwise direction through an angle $(180 + \phi)^0$ and the x-ray tube has been displaced a distance Δ so that the emerging x-ray beam is undisplaced. The distance Δ varies from 3 to 5 mm, while the angle ϕ ranges from 5^0 to 60^0. The displacement of the x-ray tube is carried out using 'swallow-tail' slide mechanism (15 in figure 1). The whole operation can be performed without disturbing the vacuum. The filter plates are coated with polystyrene, which is ideally suited for the suppression of short-wavelength radiation Lukirskii *et al* 1965).

The spectral dependence of the x-ray continuum radiation for different angles is shown in figure 3 where it can be seen that λ_c=100Å for ϕ=10°, and λ_c=300Å for ϕ=40°. The intensity of the reflected radiation is small for ϕ=40° which is the useful limit of the angle of incidence.

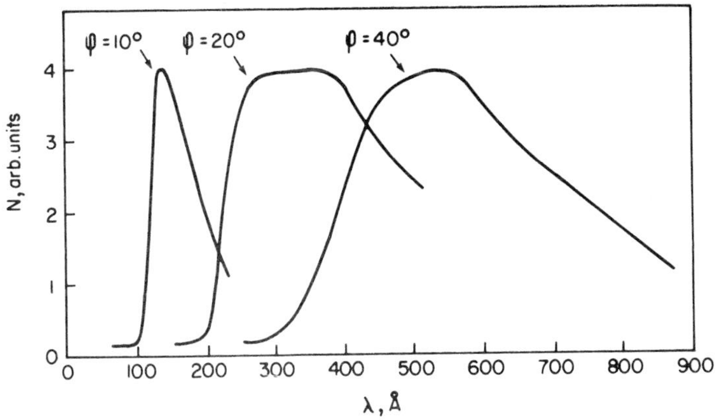

Figure 3 Spectral dependences of the bremsstrahlung for tungsten and copper anodes, at different angles of incidence ϕ on the filter Φ.

The action of the filter Φ is supplemented by the small reflection co-efficients of the grating for the second and third orders of wavelengths λ>150Å as illustrated in figure 4. The experimental points on the graph were obtained from measurements of the peak intensity of several monochromatic lines in the second and third orders of diffraction relative to the intensity in the first order. Using this grating and filter arrangement, with angle of incidence of about 20°, it is possible to obtain spectra in the region 300-800Å that are almost free from superposition of diffracted short-wavelength radiation.

The secondary electron multiplier, with a highly efficient dielectric photocathode of CsI or KCl (Savinov and Lukirskii 1967), has been used for the wavelength region 100-300Å. However, the photoefficiency of these photocathodes have a marked fine-structure in the region of λ>300Å (Samson 1967); for this reason photocathodes of gold, of copper and of tungsten, which have almost smooth quantum-yield spectral dependence, have been utilized to investigate the region above 300Å. The spectra reported here were measured using a gold photocathode.

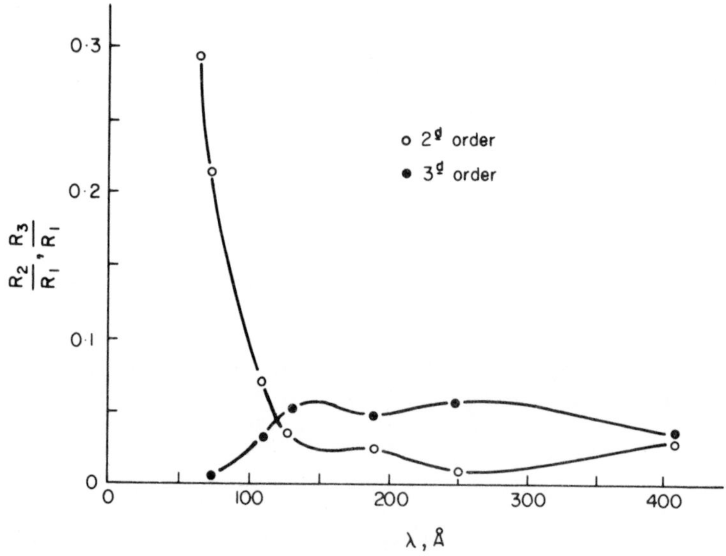

Figure 4 Spectral dependence of the grating reflection coefficient for second and third diffraction orders relative to the first diffraction order.

The samples were small slabs of high-purity transition metals, and the spectra were produced using direct excitation. The tungsten cathode was so constructed that the target surface was protected from tungsten contamination. The x-ray take-off angle was 40^0, and the x-ray tube was operated at 3kV and 10-100mA. To eliminate carbon contamination the target was heated until the carbon Kα line intensity became vanishingly small; an exception was made in the case of manganese due to its low temperature of evaporation, and here the spectrum obtained exhibited a significant carbon Kα line.

The width of the slits was varied from 20 to 300 microns and the energy resolution ranged from 0.1 to 0.6eV. Automatic recording of the spectrum took one hour during which time the count rate at the maximum intensity of the emission band was constant. All spectra were recorded several times with good reproducibility. The x-ray tube vacuum was 10^{-6}torr. The molybdenum M_ξ (λ=64.38Å) and the niobium M_ξ (λ=72.19Å) lines were used to calibrate the spectrometer by recording the lines up to and including the twentieth order. The accuracy of the energy calibration is estimated at 0.05eV.

3. RESULTS AND DISCUSSION

A. The $M_{2,3}$ Emission Bands of the First-series Transition Metals

The $M_{2,3}$ (3p) emission bands of the transition metals from titanium to copper have been investigated by Fomichev *et al* (1971). The vanadium $M_{2,3}$ emission band is shown in figure 5,

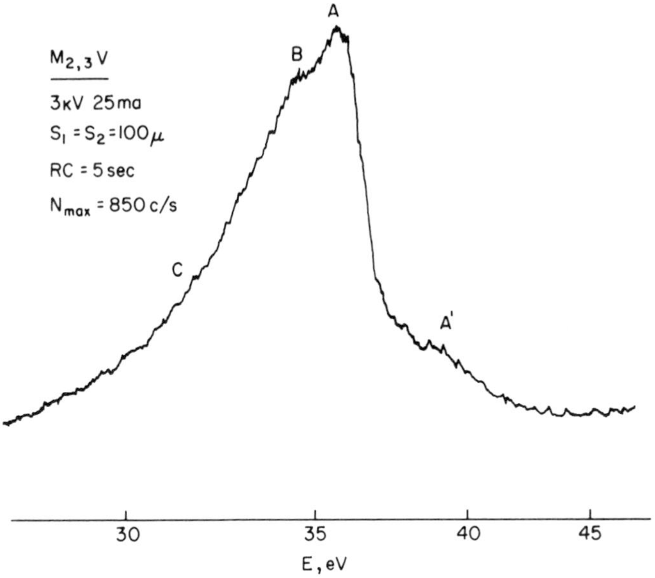

Figure 5 $M_{2,3}$ (3p)-emission band of vanadium (chart recording).

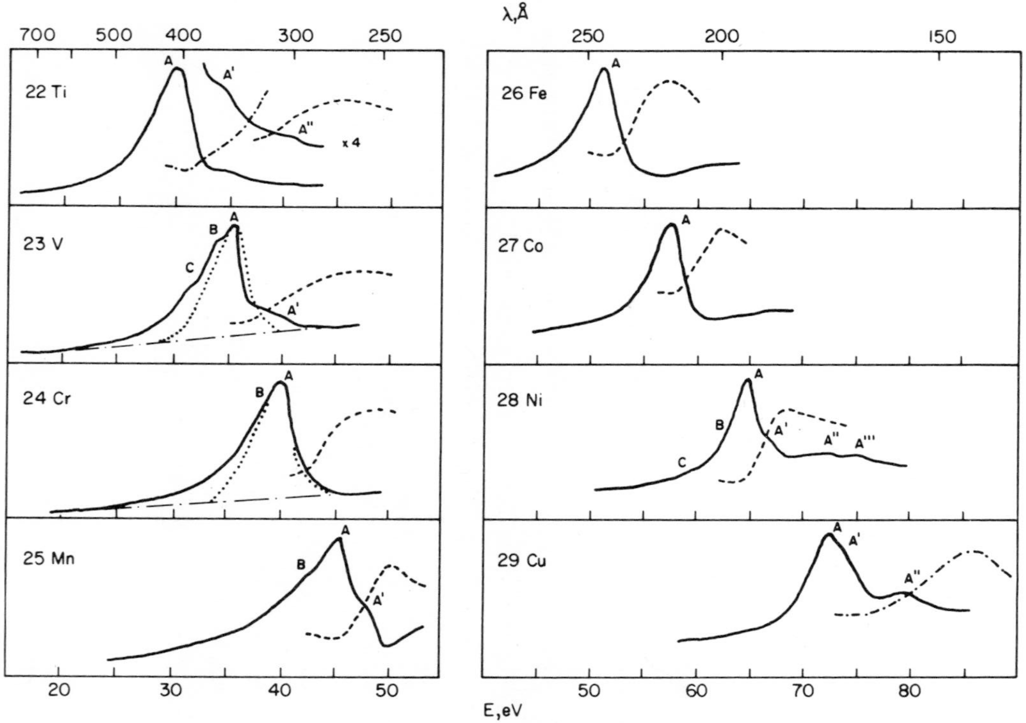

Figure 6 $M_{2,3}$ (3p) emission bands (solid curves) and absorption spectra of the first transition-series metals. Dashed curves - from Sonntag *et al* (1969). Dashed-dotted curves - Ti spectrum from Rustgi (1965) and Cu spectrum the present results. Dotted curves - $M_{2,3}$ emission bands of vanadium and chromium from Skinner *et al* (1954).

providing an example of an automatic chart-recorded spectrum obtained with the spectrometer. The $M_{2,3}$ emission bands obtained in the region of 150-700Å for the series of metals are shown in figure 6, together with the $M_{2,3}$ absorption spectra obtained by Rustgi (1965) and Sonntag *et al* (1969).

For comparison the $M_{2,3}$ emission bands of vanadium and chromium obtained by Skinner *et al* (1954) are also shown in figure 6. The bands are normalized at their maxima and are superimposed on the bremsstrahlung background emitted by these metals. The spectral dependence of the copper bremsstrahlung has been investigated in the range of 250-700Å to provide an estimate of the behaviour of the vanadium and chromium bremsstrahlung in the region of their emission bands. Comparison of these results shows that our $M_{2,3}$ bandwidths are larger than those obtained by Skinner *et al* (1954). The discrepancy is possibly due to inaccurate separation of the $M_{2,3}$ bands from the bremsstrahlung by these investigators.

In figure 7 our $M_{2,3}$ emission bands for nickel and copper are compared with those obtained respectively by Cuthill *et al* (1967) and Dobbyn *et al* (1970). Both results are in close agreement. However, we have observed feature C in the nickel spectrum at 59.0eV instead of the hump at 61.0eV recorded by Cuthill *et al* (1967). The M_2 emission bands of nickel and copper appear as the feature A'. The small peaks A" and A''' are attributed to high-energy satellites; A''' has been observed in the spectra of metals from copper to yttrium (Fomichev and Kuprijanov 1970). The shape of the M_3 emission bands of nickel and copper have been discussed in detail by Cuthill *et al* (1967) and by Dobbyn *et al* (1970).

It is of interest to compare the $M_{2,3}$ (3p) emission bands with the K(1s) and L_3 (2p) emission bands of the same metals. Figure 8 shows such a comparison for vanadium, where the K and L_3 bands are those reported by Nemnonov and Brytov (1968).

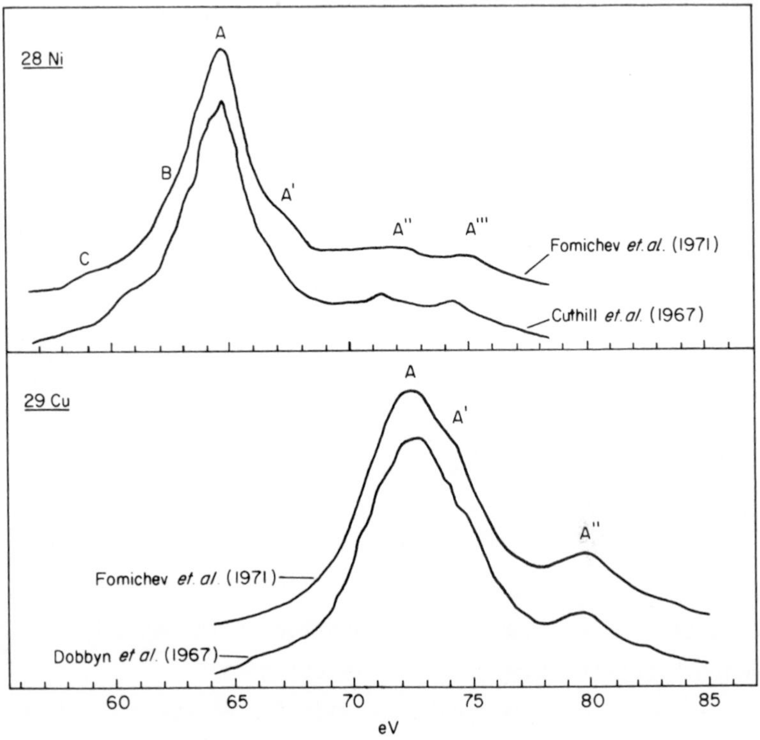

Figure 7 Comparison of the $M_{2,3}$ emission bands of nickel and copper. For a best fit the $M_{2,3}$ band of copper obtained by Dobbyn *et al* (1970) is shifted 0.5eV to higher energies.

The energy of the vanadium $K\beta_5$ and $K\beta_{1,3}$ lines (taken from the data collated by Bearden 1964) was used to position the K and $M_{2,3}$ bands correctly on a common energy scale. The spin-orbit splitting of the $M_{2,3}$ levels of the transition metals from titanium to cobalt is unknown. However, according to Herman and Skillman (1963), for the atoms of titanium, chromium and iron, this splitting is equal to 0.7, 1.06 and 1.6eV respectively, and this data may be used to estimate the spin-orbit splitting of the $M_{2,3}$ levels. The splitting for vanadium may be of the order of 0.85eV; but the $M_{2,3}$ band of vanadium shows no peak $\sim$0.85eV to the high-energy side of the peak A, justify-

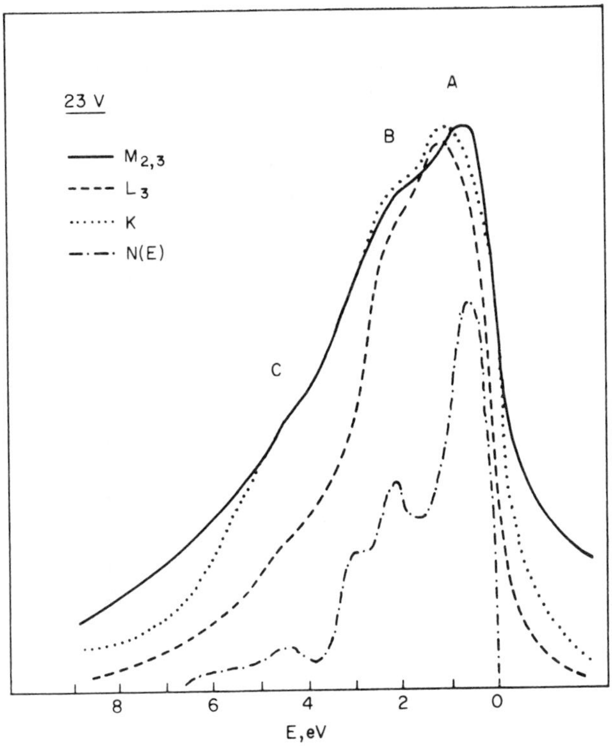

Figure 8 K(1s), L_3 (2p), and $M_{2,3}$ (3p) emission bands, and the density of states N(E) for vanadium. The position of the Fermi level is taken as the common zero of energy. The K and L3 bands are from Nemnonov and Brytov (1968); the density of states from Snow and Waber (1969).

ing our interpretation of this as the M_3 emission band.

It can be seen from figure 8 that the main features in the K, L_3 and M_3 emission bands occur at corresponding energies in the respective bands. Three maxima A, B and C are noted and are in good agreement with peaks in the theoretical density of states calculated by Snow and Waber (1969). The width and shape of the K and M_3 bands of vanadium are nearly the same while the L_3 band shows a similar shape but narrower bandwidth. For the first-series transition metals (Cr, Fe, Cu) the fact that the L_3 (2p) spectrum is always narrower than the M_3 (3p) spectrum has been explained by Cuthill *et al* 1967, by Dobbyn *et al* 1970, and by Sommer *et al* (1970). The probability of transition of 3d valence-band electrons to an L_3 (2p) final state has a different energy dependence from that of their transition to an M_3 (3p) final state; for the L-spectra there is a 60-80% increase in the transition probability towards the top of the valence band, whereas for the M-spectra the transition probability varies little throughout the band.

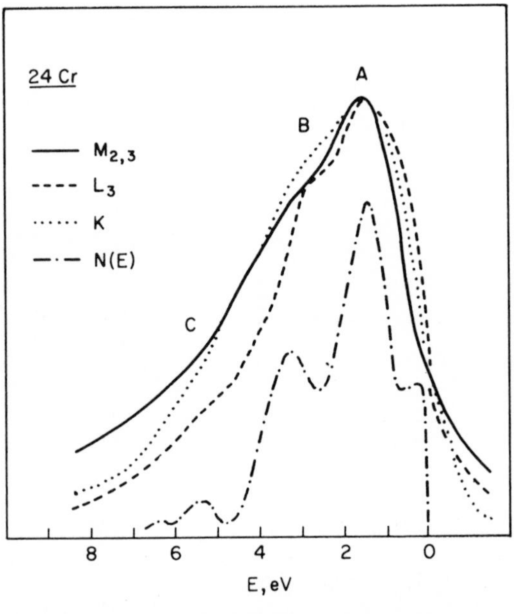

Figure 9 K(1s), L_3 (2p), and $M_{2,3}$ (3p) emission bands, and the density of states N(E) for chromium. The K and L_3 bands are from Nemnonov and Brytov (1968); the density of states from Snow and Waber (1969).

Figure 9 compares the same spectra for chromium and the same conclusions can be drawn. The shape of the emission bands is common to all spectra (K, L_3, M_3) with the exception of the feature at C, which was not observed in our M_3 spectrum.

As for vanadium the theoretical density of states agrees well with the shape of the emission bands, indicating that the s, p and d valence electrons are hybridised throughout the valence band as proposed by Nemnonov *et al* (1966 and 1968).

To summarise, the spectra show features that require more detailed investigation. Firstly, there are some weak fluctuations to the long-wavelength side of the $M_{2,3}$ emission bands of chromium, iron and cobalt. Secondly, in the manganese, iron and cobalt spectra, the minimum to the high-energy side of the emission bands coincides with the maximum of the absorption spectra. Hence we can expect the effect of self absorption to be large, which may change the slope of the emission edge. The $M_{2,3}$ spectra should perhaps be studied with various accelerating voltages and x-ray take-off angles (Liefeld 1968, Fischer and Baun 1968, and Fischer 1969).

B. The $N_{2,3}$ Emission Bands of the Second-series Transition-Metals

The $N_{2,3}$ (4p) emission bands of the metals from yttrium to palladium (with the exception of tellurium) have been investigated (Rudnev *et al* 1971a) and are shown in figure 10; they occur in the range of 200-1000Å.

The origin of the feature A', observed in all the spectra except for molybdenum, is difficult to associate with the N_2 emission band of these metals because the spin-orbit splitting of the $N_{2,3}$ levels is unknown. The calculated splitting, according to Herman and Skillman (1963), is 1.9, 2.5, 3.3 and 4.3eV for Zr, Mo, Ru and Pd respectively, whereas the separation A-A' is 2.7, 4.3, 10.7 and 4.2eV respectively. Nevertheless, we regard peak A for all the spectra as arising from the N_3 ($4p_{3/2}$) emission band alone.

In figure 11 the Zr N_3 emission band is superposed on the Zr M_5 (3d) band obtained by Holliday (1968) and the theoretical density of states calculated by Altmann and Bradley (1968).

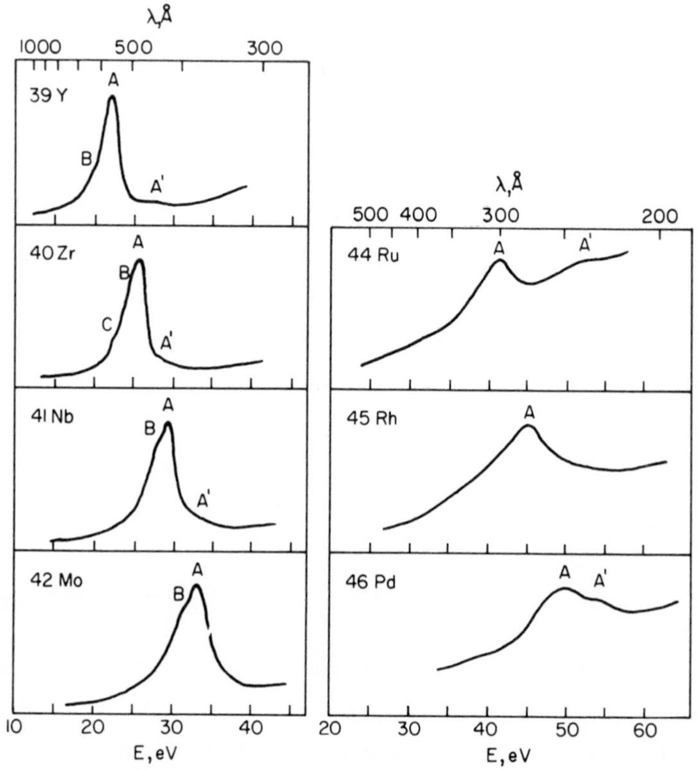

Figure 10 $N_{2,3}$ emission bands of the second transition-series metals.

The Zr M_ξ (M_5-N_3) line (151.4eV, according to the data by Krause and Wuilleumier 1971) was used to superimpose the M_5 and N_3 bands on a common energy scale. The peaks in both bands are separated by 0.6eV, while the emission edges coincide closely.

The position of the peak in the Zr M_5 band is 177.6eV (Zimkina *et al* 1964); the value of 179.2eV taken by Holliday (1968) is probably overestimated because this would not give a good superposition of the M_5 and N_3 emission bands onto a common energy scale.

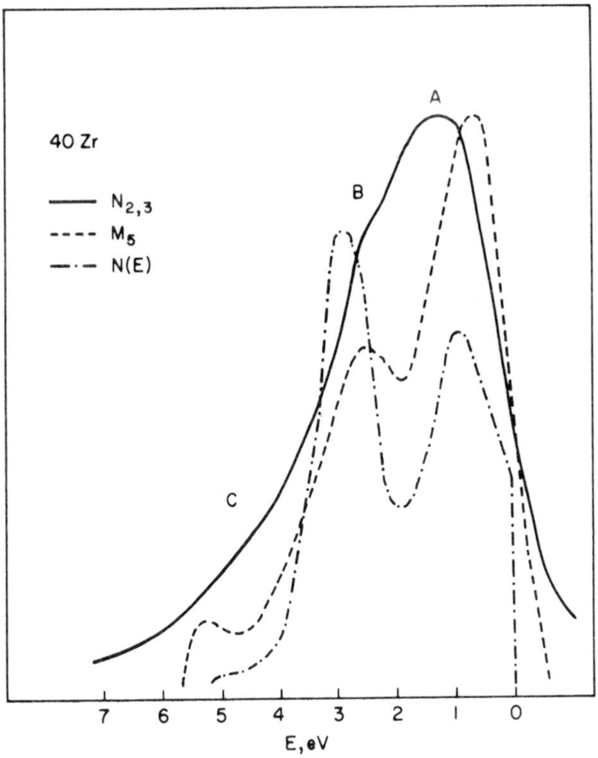

Figure 11 M_5 emission band (Holliday 1968), N_3 emission band (present results), and calculated density of states (Altmann and Bradley 1968) for zirconium.

The assumption that the common emission edge of the Zr N_3 and M_5 bands corresponds to the Fermi-edge provides a means of positioning the density-of-states curve in figure 11. The M_5 band has been corrected for instrumental and inner-level broadening, whereas the N_3 band has not. Taking this into account it can be seen that both bands have two peaks A and B at about 1.0 and 2.5eV from the Fermi-edge. We find qualitative agreement between the peaks in the emission bands and the two peaks in the density-of-states. According to Holliday (1968) the third feature at C is a satellite peak; however, the N_3 spectrum also shows a pronounced peak here, which therefore may be associated with the step in the density of states. Thus, the main features of the density of states for zirconium appear in the M_5 and N_3 emission bands, indicating that electrons with s, p and d wavefunction symmetry occur throughout the valence band. Similar comparisons may be made for yttrium, niobium and molybdenum.

C. $O_{2,3}$ and $N_{6,7}$ Emission Bands of the Third Transition-Series Metals

The $O_{2,3}$ (5p) and $N_{6,7}$ (4f) emission spectra of the metals from lutetium to gold (with the exception of osmium) have been studied (Rudnev *et al* 1971b). A chart recording of the tantalum spectrum is shown in figure 12. The threshold of the

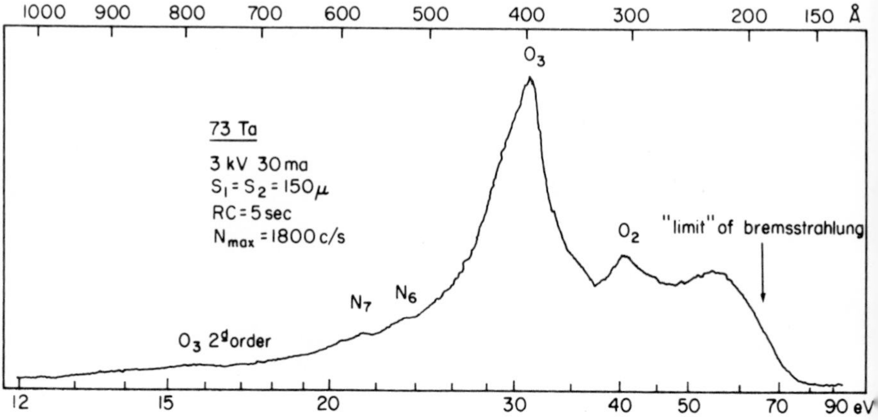

Figure 12 Chart recording of the emission spectrum of tantalum in the wavelength region 150-1000Å.

bremsstrahlung continuum is the result of the reflection filter Φ which cuts off the tantalum bremsstrahlung at $\lambda<150$Å (ψ 15°). As a result, the tantalum O_3 emission band is not distorted by the superposition of higher diffraction orders. The second and third orders of the bremsstrahlung maximum at $\lambda=220$Å have no effect on the O_3 emission band; this was confirmed by measuring the Ta O_3 emission band with the reflecting filter operating at $\phi>15^{\circ}$. It can be seen from figure 12 that the second order of the Ta O_3 emission band produces a weak peak on the low-

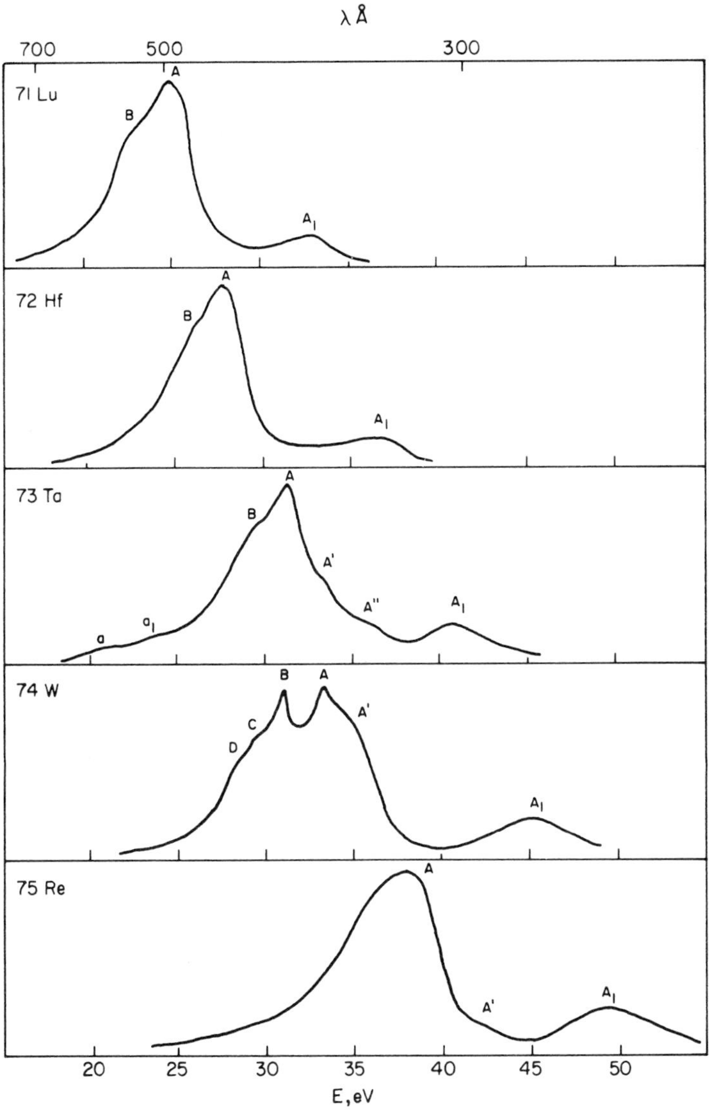

Figure 13 $O_{2,3}$ emission bands of the third transition-series metals.

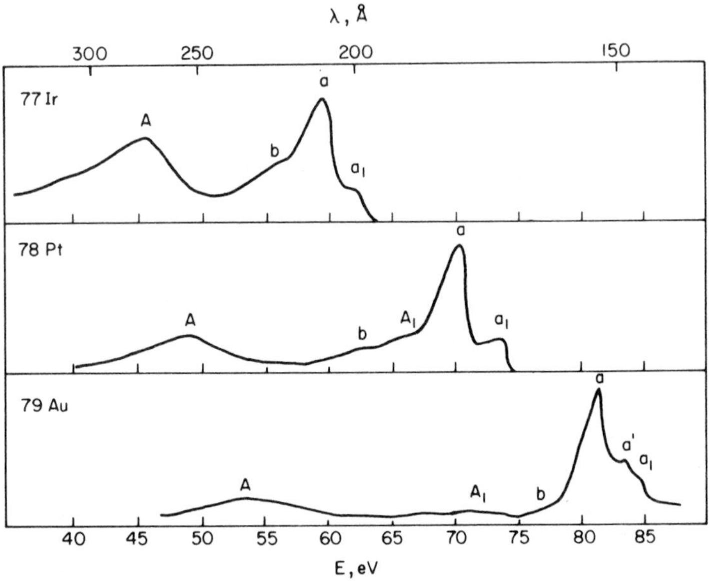

Figure 14 $O_{2,3}$ (5p) and $N_{6,7}$ (4f) emission bands of iridium, platinum and gold.

energy side of the O_3 band.

The emission bands of the third transition-series metals corrected for background, are shown in figures 13 and 14. A Mosely diagram (figure 15) identifies these spectra as the $O_{2,3}$ and $N_{6,7}$ emission bands. This enables us to associate the features A, A_1, a and a_1 with the electron transitions valence band to O_3 ($5p_{3/2}$), O_2 ($5p_{1/2}$), N_7 ($4f_{7/2}$) and N_6 ($4f_{5/2}$) respectively.

According to Hagström (1970) the 4f subshell is completely filled in atomic ytterbium but in the metal it comes at 2eV below the Fermi level. In lutetium and hafnium the $O_{4,5}$ (5d) subshell starts to fill in the metal, with the $N_{6,7}$ (4f) subshell full. According to Bearden and Burr (1965) and Siegbahn *et al* (1967) the energy of the 4f subshells in lutetium, hafnium and tantalum is respectively 7, 18 and 25eV below the Fermi energy. In the tantalum spectrum we observe the transitions valence-band

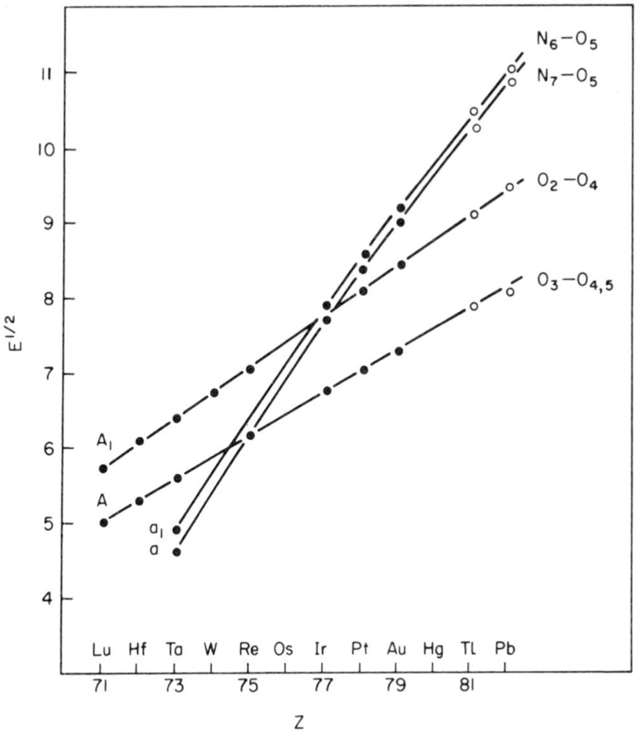

Figure 15 Mosely diagrams for the third transiton-series metals. Black circles - present results; open circles - data from Bearden (1964). The energy values for the $O_{2,3}$ and $O_{4,5}$ transitions for thallium (Z=81) and lead (Z=82) were calculated from those of the $L_1 \rightarrow O_{2,3}$ and $L_1 \rightarrow O_{4,5}$ transitions, using the data compiled by Bearden (1964).

to the $4f_{7/2}$ (N_7) and $4f_{5/2}$ (N_6) inner levels (a and a_1 in figure 13). In these transitions only d-valence electrons are expected to take part, since s→f transitions are forbidden by the dipole selection rules.

The features a and a_1 are 2.3eV apart, in good agreement with the value of the spin-orbit splitting of the $N_{6,7}$ levels of tantalum (2eV) reported by Siegbahn *et al* (1967).

As indicated by the diagram the $N_{6,7}$ and O_3 emission bands of tungsten and rhenium are superposed, and we can see from figure 13 that the emission spectrum of tungsten has much fine structure. Some of the peaks should be part of the N_6 and N_7 emission bands and, assuming that these give rise to the peaks A and B, we should expect to observe more fine structure in the rhenium spectrum. In fact the intensity of the $N_{6,7}$ emission bands increases with respect to the O_3 emission band with increasing atomic number (figure 14). However, the rhenium emission spectrum has a simple profile, therefore, we attribute the weak features D and C in the tungsten spectrum to the W N_6 and N_7 emission bands. In the rhenium spectrum the peak A may be due to the coincidence of the O_3 and N_7 emission bands and A' the N_6 band. From the Mosely diagram it follows that in the spectra for iridium, platinum and gold A is the O_3 emission band and A' the O_2 band. However, in the iridium spectrum the O_2 emission band is situated in the region of the $N_{6,7}$ bands and is not observed.

A' and A" in the tantalum spectrum are probably short-wavelength satellites since they lie beyond the O_3 absorption edge (at about 33.3eV according to Haensel *et al* 1969).

D. X-ray Spectra and Electronic Structure of Iridium, Platinum and Gold

Calculations of the energy band structure for iridium, platinum and gold have been reported by Mackintosh (1966), Jacobs (1968), Ballinger and Marshall (1969), Kupratakuln and Fletcher (1969), Kupratakuln (1970), Schlosser (1970) and Andersen (1970). Nilsson *et al* (1969) and Krolikowski and Spicer (1970) have studied the UV photoemission spectrum of gold and obtained the optical density of states (ODS). Baer *et al* (1970), and Fadley and Shirley (1970) have studied the valence band of these metals using x-ray photoelectron spectroscopy (XPS). The absorption spectra of platinum and gold in the region of the $O_{2,3}$ (5p) and $N_{6,7}$ (4f) absorption edges have been investigated by Haensel *et al* (1969). We use these results for comparison with the observed N_7 ($4f_{7/2}$) emission bands.

It is necessary to correct the N_7 emission band for background, and for overlap of the N_6 band. Also for the iridium and platinum spectra the O_2 band must be subtracted. This subtraction was carried out for platinum assuming that the O_2 and O_3 emission bands have the same width and that their intensity ratio is equal to 1.3 (as for gold).

The N_7 emission band of platinum was corrected for the N_6 band overlap by assuming that the N_6 and N_7 levels are separated by 3.3eV - a value which is in good agreement with the data of Bearden and Burr (1965), 3.2eV, Siegbahn *et al* (1967), 3.4eV, and Baer *et al* (1970), 3.4eV. Figure 16 shows the

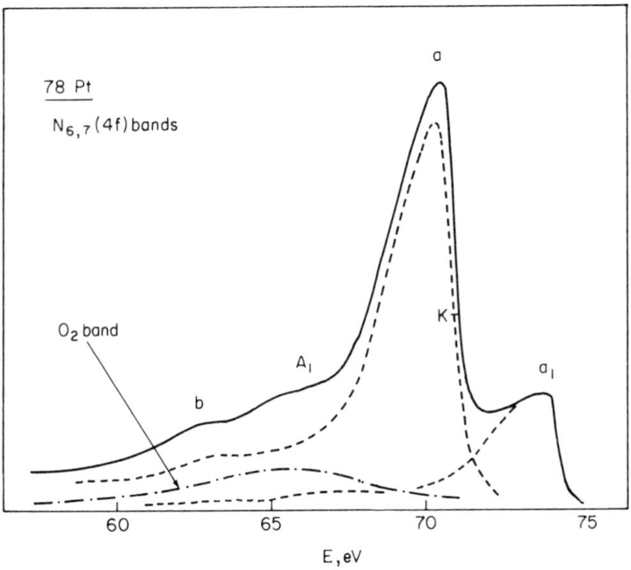

Figure 16 $N_{6,7}$ emission bands of platinum.

experimental spectrum of platinum, and the corrected N_7, N_6 and O_2 emission bands.

The inflection point K of the short-wavelength edge of the platinum N_7 emission band - taken to be the Fermi level - comes at 71.0eV. This agrees well with the Fermi level to N_7 energy separation of 71.0eV measured by both Baer *et al* (1970) and Fadley and Shirley (1970), using XPS, and also with the Pt N_7 absorption edge at 70.8eV reported by Haensel *et al* (1969).

The N_6 and N_7 emission edges were approximated by arctangent curves, and corrected for the spectrometer resolution (0.25eV). A value of 0.15eV was obtained for both edge widths. Considering the temperature broadening effect on the emission edge we can conclude that the N_6 and N_7 inner core levels have a width of less than 0.15eV.

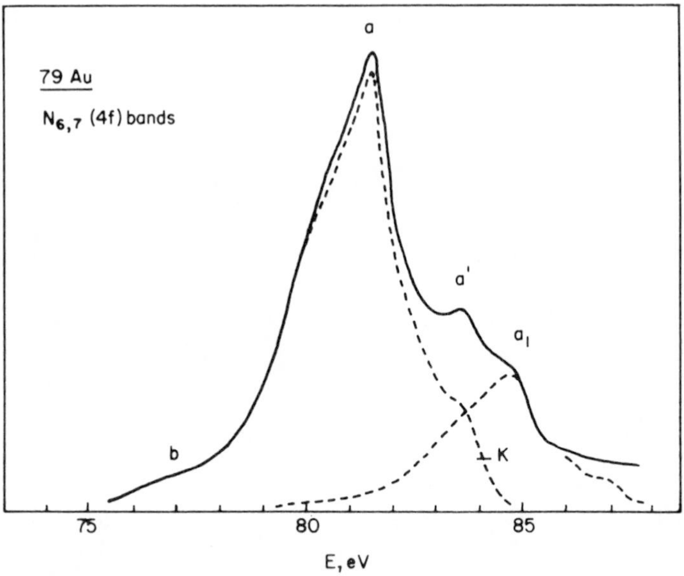

Figure 17 $N_{6,7}$ emission bands of gold.

The experimental spectrum of gold is shown in figure 17. The spin-orbit splitting of the Au N_6 and N_7 levels is 3.6eV, according to Bearden and Burr (1965) and Siegbahn *et al* (1967), and 3.7eV according to Baer *et al* (1970). The separation of a and a_1 (3.3eV) is in agreement with these values suggesting that these features are the N_7 and N_6 emission band peaks of gold. The A' is taken as part of the N_7 emission band. The intensity ratio of the N_6 and N_7 bands are taken to be 1:3.2 for resolving the spectrum into the N_6 and N_7 emission bands.

The K inflection point of the N_7 emission band edge is situated at 84.0eV which again agrees well with the XPS measurements by Fadley and Shirley (1970), 84.0eV, and Baer *et al* (1970), 83.8eV.

The theoretical density of states and the results obtained by Fadley and Shirley and Baer *et al* (1970) are compared with the Pt N_7 emission band in figure 18. Both XPS-curves are

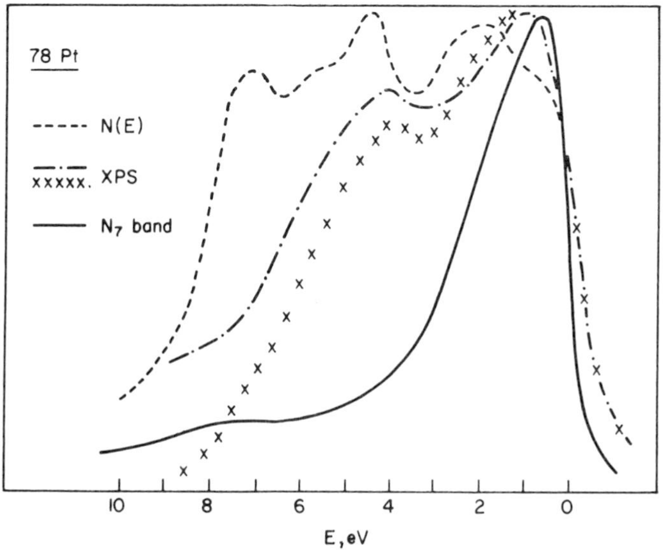

Figure 18 Comparison of data for the valence band of platinum. Dashed-dotted curve - XPS (Baer *et al* 1970); crosses - XPS (Fadley and Shirley 1970); dashed curve - density of states (taken from Fadley and Shirley 1970); solid curve - N7 (4f7/2) emission band. The position of the Fermi level is the common zero of energy. For a best fit the XPS-curve reported by Fadley and Shirley (1970) has been shifted 0.2eV to higher energies.

of similar shape, and have two maxima at a distance of 1.3 and

3.9eV from Fermi level. These may be attributed to the d-band structure of the N(E) curve. The peak at 7eV in the density of states curve is not observed in the XPS measurements, and the N_7 emission band differs strongly from both the density of states XPS curves. The N_7 band has its prominent maximum at 0.6eV and a second peak at 8eV from Fermi level; it also differs from the N(E) curve at the low-energy side of the band. Clearly this is due to the effect of transition probabilities.

The intensity of the x-ray emission band can be written as

$$I(E_i - E) \sim N(E) \cdot P'(E_i - E)$$

where E_i is the energy of the inner level, E is the energy in the valence band, N(E) is the density of states, and $P'(E_i-E)$, which we can write as P(E), is a suitably averaged transition probability. P(E) may be written

$$P(E) \simeq \left| \int \psi(r)\, r^3\, \Psi\,(r,E)\, dr \right|^2$$

where $\psi(r)$ is the wavefunction of the inner level and $\Psi(r,E)$ is the wavefunction of the valence band.

The N_7 emission arises from a 4f core state and, according to the selection rule $\Delta\ell=\pm1$, will reflect the density of d-states in the valence band. However, as seen from figure 17, the N_7 emission band does not follow the density of states curve, which - for platinum and for gold - may be regarded as almost entirely representing the density of d-states.

Theoretical calculations of the L_3 and M_3 spectra of iron, copper and chromium, by Cuthill *et al* (1967), by Dobbyn *et al* (1970) and by Sonntag *et al* (1970), show that for L-spectra there is a noticeable decrease in the transition probability at the bottom of the valence band. On the basis of these results we conclude that the difference in shape of the N_7 band from the N(E) curve may be associated with variation of the transition probability with spatial distribution of the d-

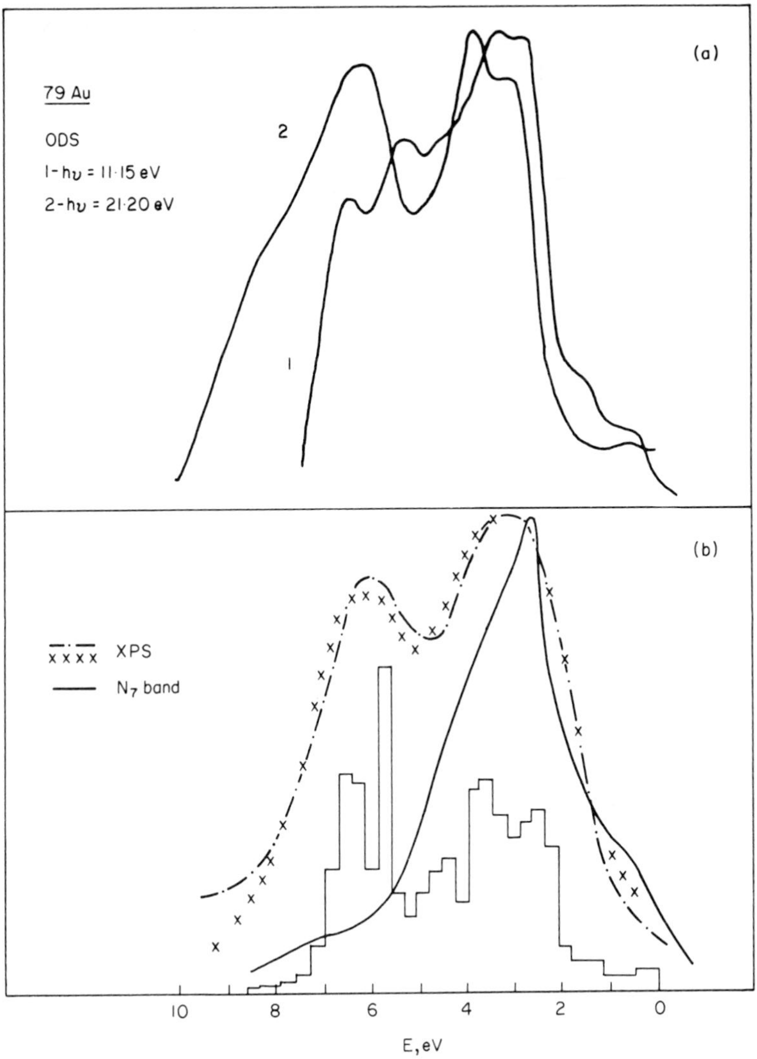

Figure 19 Comparison of data for the valence band of gold.
a) ODS-curves (EDCs) reported (1) by Nilsson *et al* (1969), and (2) by Krolikowski and Spicer (1970).
b) Dashed-dotted curve - XPS (Baer *et al* 1970); crosses - XPS (Fadley and Shirley 1970); density of states histogram calculated by Kupratakuln (1970); solid curve - N_7 $(4f_{7/2})$ emission bands.
The position of the Fermi level taken as a common zero. For a best fit the XPS-curve reported by Fadley and Shirley (1970) is shifted 0.9eV to higher energies.

electron wavefunctions.

Data for gold are shown in figure 19. Photoemission electron energy distribution curves (or optical densities of states - ODS) obtained by Nilsson *et al* using 11eV photons, and by Krolikowski and Spicer (1970) using 21eV photons, are shown in figure 19a; and the x-ray photoelectron spectra (XPS) obtained by Baer *et al* (1970) and by Fadley and Shirley (1970) are shown in figure 19b, together with the N_7 emission band and the calculated density-of-states histogram obtained by Kupratakuln (1970). It is interesting to compare the ODS and XPS curves. The ODS from low-energy photoemission measurements include structure from both the valence band and the conduction band. Furthermore the ODS curves can include structure which is not part of the valence or conduction bands but is due to direct transitions taking place between states that have the same slope of E(k) curve (Phillips 1966).

Increasing the photon energy makes it necessary to analyse the energy of the photoelectrons in a higher energy region where the density of states in the conduction band can be assumed to have an $E^{\frac{1}{2}}$ dependence. In this case the shape of ODS must approximate to the density-of-states curve in the valence band and to the shape of the XPS-curve. This can be seen clearly in figure 19. Also, increasing the photon energy (to >21eV) improves the agreement between the ODS and XPS curves in the region 7-9eV below the Fermi energy.

The common features of the XPS and ODS curves, which agree with the calculated N(E), are: (1) the small density of states at the Fermi level; and (2) two maxima in the d-band at 3.2 and 6.0eV from Fermi level.

The N_7 emission band of gold has a small step to the high-energy side of the peak, in agreement with the ODS and N(E) curves. The peak in Au N_7-emission band comes at 2.6eV from the Fermi level and coincides with the first broad peak in the

N(E) histogram. However, the N_7 (or $4f_{7/2}$) emission intensity falls in the region of the second peak in the density of states, as for platinum, and only a small step is observed (at 7.2eV from the Fermi level). This is attributed again to decrease of the transition probability at the bottom of the band.

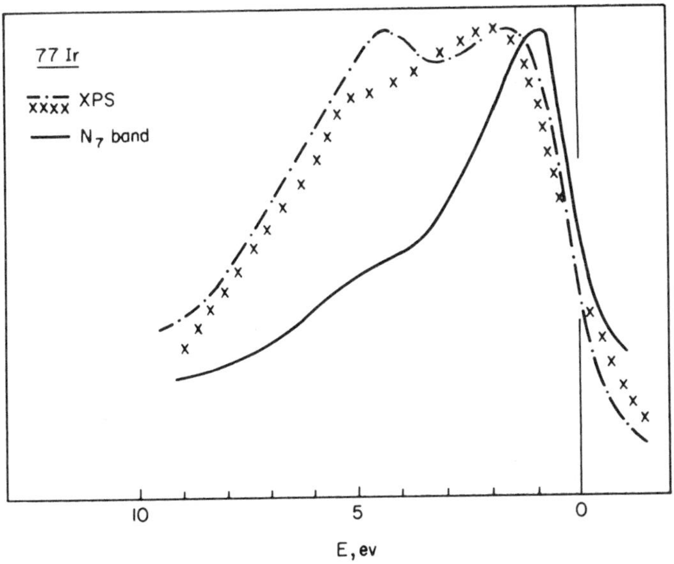

Figure 20 Spectra of iridium. Dashed-dotted curve - XPS (Baer *et al* 1970); crosses - XPS (Fadley and Shirley 1970); solid curve - 4f-emission band. The position of the Fermi level taken as common zero.

Figure 20 shows XPS data for iridium (Baer *et al* 1970, Fadley and Shirley 1970), and the measured $N_{6,7}$ emission spectrum uncorrected for overlap of the O_2 emission band. The spectrum is mainly the N_7 band since the O_2 emission and the lower-energy N_6 emission have low intensity. The inflection point in the band edge comes at 60.5eV, and is in a good agreement with the binding energy of the Ir N_7 level from XPS measurements (61.0eV, Baer *et al* 1970; 60eV, Fadley and Shirley

1970; and 60.5eV, Bearden and Burr 1965). This point was taken as the Fermi level in our spectrum. Comparison of the curves shows the same discrepancies for the iridium N_7 emission band as for platinum and gold.

Hence measurements of the $N_{6,7}$ (5d→4f) emission spectra of platinum, gold and iridium show that the second peak in the d-band is barely detectable. As noted above, this is an effect of transition probability which decreases rapidly with decreasing energy of the valence band state. The change in the transition probability for the $N_{6,7}$ emission band may be caused by the difference in the spatial distribution of the wavefunctions of the valence band states at the top and at the bottom of the d-band. On the other hand, Fadley and Shirley (1970) suggested that the two peaks in the d-bands of iridium, platinum and gold may be attributed to the atomic-like spin-orbit split 5d-levels, with j-value respectively 5/2 and 3/2 for the first and second peaks. If this is the case then the N_7 emission band ($5d_{5/2} \rightarrow 4f_{7/2}$) should have only one maximum, since the atomic selection rule forbids the transition $5d_{3/2}$ to $4f_{7/2}$. The intensity of the N_7 emission band agrees approximately with this atomic selection rule, the second peak, which may be related to the transition $5d_{3/2} \rightarrow 4f_{7/2}$, has very low intensity. This unexpected result is of considerable interest; confirmation should be sought from other experimental measurements and from theoretical studies.

The present results show that the shape of the x-ray emission band depends significantly on the transition probability. Earlier data obtained by Fomichev *et al* (1970) and Aleshin and Smirnov (1969) have also demonstrated the importance of the transition probability in determinating the shape of the x-ray emission band.

E. The X-ray Emission Bands of the V_3Au alloy

The vanadium $M_{2,3}$ (3p) emission and the gold $N_{6,7}$ (4f)

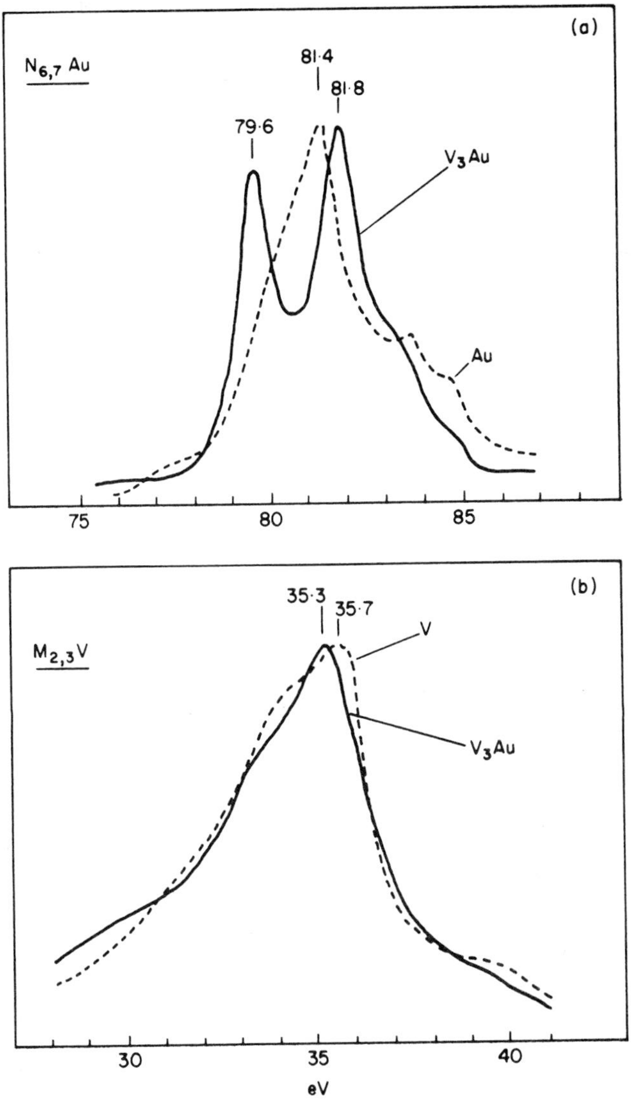

Figure 21 The Au $N_{6,7}$ and V $M_{2,3}$ emission bands for pure gold and for the alloy V_3Au. (For the alloy measurements were made using 15mA and 3kV in the x-ray tube and instrumental resolutions of 0.4eV and 0.5eV respectively).

emission from the alloy V_3Au have been measured and compared with spectra for the pure metals. Figure 21 shows the Au $N_{6,7}$ and V $M_{2,3}$ emission bands for gold and for V_3Au. The Au $N_{6,7}$ spectra differ markedly in shape; in place of one peak at 81.4eV there are two peaks at 79.6 and 81.8eV, in the V_3Au band. These have about the same intensity, and halfwidths of ∿1-1.5eV. The V $M_{2,3}$ spectra differ little in shape.

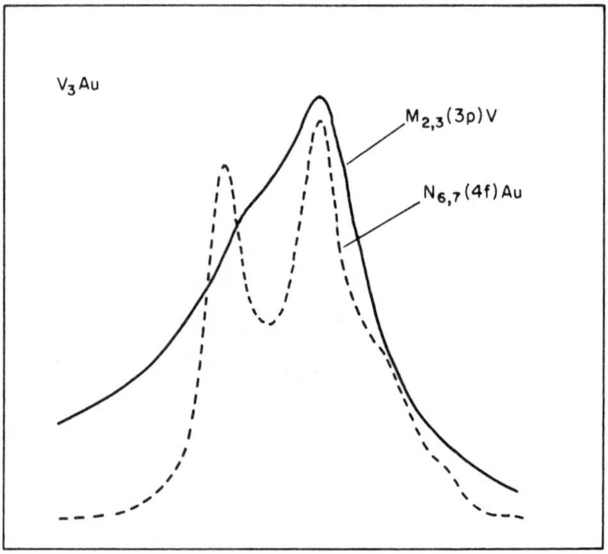

Figure 22 Comparison of the vanadium $M_{2,3}$ and gold $N_{6,7}$ emission bands for V_3Au. The superimposition shown is not necessarily correct, the main peaks of the spectra were made to conincide.

In figure 22 the gold $N_{6,7}$ and the vanadium $M_{2,3}$ emission bands for V_3Au are superimposed such that the V $M_{2,3}$ peak coincides with the first peak of the Au $N_{6,7}$ emission. We have no data for the energies of the vanadium 3p level and gold 4f level in V_3Au, and therefore cannot match these spectra precisely.

F. Comments and Discussion

The results show that the changes of emission intensity with increasing atomic number in the second and third series transition metals have features in common. To compare the intensity of different emission bands we use the ratio of the maximum intensity to the background. We find, for the $N_{2,3}$ (4p) and $O_{2,3}$ (5p) emission bands, that this relation decreases with increasing atomic number, as is clearly seen in figures 10, 13 and 14.

With indium we do not find an emission line corresponding to the 4p-4d transition (in the region $\sim 200\text{Å}$); and according to Bearden (1964) this line is missing in x-ray spectra of heavy elements. Similarly the 5p→5d lines were not observed in spectra of heavy elements with filled 5d-shells.

The $N_{6,7}$ (4f) emission intensity increases with increasing atomic number (figures 13 and 14), and an emission line arising from the 4f→5d transition has a high intensity in the spectra of elements with Z>79. According to a calculation of transition probabilities for all subshells of atoms with atomic numbers from 5 to 100 (Manson and Kennedy, unpublished) the transition probabilities for 4p→4d, 5p→5d and 4f→5d transitions increase with increasing atomic number. The experimental results show that the intensity of the measured x-ray spectra do not agree with these theoretical predictions when the transitions take place between subshells having the same principal quantum number. Possibly Auger transitions are responsible for the decreasing line intensities. It would be interesting to estimate the Auger transition probabilities for the 4p and 5p subshells. According to recent calculations by McGuire (1972) the probability of the Coster-Kronig transitions $M_{2,3} \rightarrow M^2_{4,5}$ increases with atomic number. This leads to the conclusion that the number of radiative transitions $M_{2,3} \rightarrow M_{4,5}$ will decrease with increasing atomic number.

The intensity ratios of the emission bands of the third transition-series metals appear to be irregular. The ratios for the O_2 ($5p_{1/2}$) to O_3 ($5p_{3/2}$) varies between 1:6 and 1:3 for elements from lutetium to gold; whereas, according to the statistical weights of the O_2 and O_3 levels, the ratios should be 1:2.

The relative intensities of the N_6 ($4f_{5/2}$) to N_7 ($4f_{7/2}$) emission bands for iridium, platinum and gold are about 1:3. This value differs also from the statistical weights ratio (1:1.3) (6:8). An explanation proposed by Tomboulian (1957) for similar results involves the Auger effect. For example, the N_6 vacancy may be filled with an electron from the N_7 level accompanied by the removal of a valence electron (the $N_6 \rightarrow N_7V$ transition). This process reduces the N_6 emission band intensity with respect to the N_7 and should lead to the Auger broadening of the N_6 level. However, as we have seen above, the N_6 and N_7 emission bands of platinum have almost the same edge-width (0.15eV) which suggests that the N_6 and N_7 levels have the same lifetime broadening. These results are not fully understood.

Acknowledgments

We are very grateful to Professor S.T. Manson and Dr. D. Kennedy for communicating their calculated atomic transiton probabilities.

REFERENCES

Aleshin, V.G. and Smirnov, V.P. (1969); in "X-Ray Spectra and Electronic Structure of Matter" Vol. I, p.314 : Institute of Metal Physics, Academy of Sciences of the Ukr.SSR, Kiev (in Russian).

Altmann, S.L. and Bradley, C.J. (1968); in "Soft X-Ray Band Spectra and the Electronic Structure of Metals and Materials" (ed. D.J. Fabian), p.265 : Academic Press, London.

Andersen, O.K. (1970); Phys. Rev. B 2, 883.

Ballinger, R.A. and Marshall, C.A.W. (1969); J. Phys. C 2, 1827.

Baer, Y., Hedén, P.F., Hedman, J., Klasson, M., Nordling, C. and Siegbahn, K. (1970); Physica Scripta 1, 55.

Bearden, J.A. (1964); "X-Ray Wavelengths" : U.S. Atomic Energy Commission, Oak Ridge, Tennessee.

Bearden, J.A. and Burr, A.F. (1965); "Atomic Energy Levels" : U.S. Atomic Energy Commission, Oak Ridge, Tennessee.

Cuthill, J.R., McAlister, A.J., Williams, M.L. and Watson, R.E. (1967); Phys. Rev. 164, 1006.

Dobbyn, R.C., Williams, M.L., Cuthill, J.R. and McAlister, A.J. (1970); Phys. Rev. B 2, 1563.

Fadley, C.S. and Shirley, D.A. (1970); J. Res. Nat. Bur. Stand. 74A, 543.

Fischer, D.W. and Baun, W.L. (1968); J. Appl. Phys. 39, 4757.

Fischer, D.W. (1969); J. Appl. Phys. 40, 4151.

Fomichev, V.A. and Kuprijanov, V.N. (1970); Fizika Tverdogo Tela 12, 2639.

Fomichev, V.A., Zhukova, I.I. and Rumsh, M.A. (1970); in "X-Ray Spectra and Electronic Structure of Matter" Vol. II, p.3 : Institute of Metal Physics, Academy of Sciences of the Ukr. SSR, Kiev (in Russian).

Fomichev, V.A., Rudnev, A.V. and Nemnonov, S.A. (1971); Fizika Tverdogo Tela 13, 1234.

Haensel, R., Radler, K. and Sonntag, B. (1969); Solid State Comm. 7, 1495.

Hagström, S.B.M., Hedén, P.O. and Löfgren, H. (1970); Solid State Comm. 8, 1245.

Herman, F. and Skillman, S. (1963); "Atomic Structure Calculations" : Prentice-Hall Inc., New Jersey.

Holliday, J.E. (1968); in "Soft X-Ray Band Spectra and the Electronic Structure of Metals and Materials" (ed. D.J. Fabian), p.101 : Academic Press, London.

Jacobs, R.L. (1968); J. Phys. C 1, 1296.

Krause, M.O. and Wuilleumier, F. (1971); Phys. Lett. 35, 341.

Krolikowski, W.E. and Spicer, W.E. (1970); Phys. Rev. B 1, 478.

Kupratakuln, S. and Fletcher, G.C. (1969); J. Phys. C. 2, 1886.

Kupratakuln, S. (1970); J. Phys. C. 3, S109.

Liefeld, R.J. (1968); in "Soft X-Ray Band Spectra and the Electronic Structure of Metals and Materials" (ed. D.J. Fabian), p.133 : Academic Press, London.

Lukirskii, A.P. and Omelchenko, Yu.A. (1960); Optika i spectroscopija 8, 563.

Lukirskii, A.P., Savinov, E.P., Ershov, O.A., Zhukova, I.I. and Fomichev, V.A. (1965); Optika i spectroscopija 19, 425.

Lukirskii, A.P., Brytov, I.A. and Fomichev, V.A. (1966); Fizika Tverdogo Tela 8, 95.

Lukirskii, A.P., Fomichev, V.A. and Rudnev, A.V. (1970); in "Apparatura i metodi rentgenovskogo analiza" no.6, p.85 : SCB RA, Leningrad.

McGuire, E.J. (1972); J. Phys. Chem. Solids 33, 577.

Mackintosh, A.R. (1966); Bull. Am. Phys. Soc. 11, 215.

Nemnonov, S.A. and Finkelshtein, L.D. (1966); Fizika Metallov i Metallovedenie 22, 538.

Nemnonov, S.A. and Brytov, I.A. (1968); Fizika Metallov i Metallovedenie 26,43.

Nilsson, P.O., Norris, C. and Wallden, L. (1969); Solid State Commun. 7, 1705.

Phillips, J.C. (1966); Solid State Phys. 18, 55.

Rudnev, A.V., Fomichev, V.A. and Nemnonov, S.A. (1971a); Fizika Tverdogo Tela 13, 2483.

Rudnev, A.V., Fomichev, V.A., Shulakov, A.S. and Nemnonov, S.A. (1971b); Fizika Tverdogo Tela 13, 2053.

Rustgi, O.P. (1965); J. Opt. Soc. Am. 55, 630.

Samson, J.A.R. (1967); "Techniques of Vacuum Ultraviolet Spectroscopy" : John Wiley and Sons Inc., New York.

Savinov, E.P. and Lukirskii, A.P. (1967); Optika i spectroscopija 23, 303.

Schlosser, H. (1970); Phys. Rev. B 1, 491.

Siegbahn, K., Nordling, C. and others, (1967); "Atomic Molecular and Solid State Structure Studied by Means of Electron Spectroscopy" : Nova Acta Reg. Soc. Sci. Upsal. Ser. IV, 20; Uppsala.

Skinner, H.W.B., Bullen, T.C. and Johnston, J.E. (1954); Phil Mag. 45, 1070.

Sommer, G., Volkov, V.F. Blokhin, M.A. and Nikiforov, I.Ya. (1970); Fizika Metallov i Metallovedenie 30, 669.

Sonntag, B., Haensel, R. and Kunz, C. (1969); Solid State Commun. 7, 597.

Snow, E.C. and Waber, J.T. (1969); Acta Metallurgica 17, 623.

Tomboulian, D.H. (1957); Handbuch. Phys. 30, 246.

Zimkina, T.M., Ershov, O.A. and Lukirskii, A.P. (1964); Izv. Akad. Nauk SSSR, Ser. Fiz. 28, 836.

Part 3

THEORY AND MANY-BODY EFFECTS OF SOFT X-RAY AND PHOTOELECTRON EMISSION FROM METALS AND FROM DISORDERED ALLOYS

THEORY OF SOFT X-RAY EMISSION AND MANY-PARTICLE EFFECTS

N.H. March

Department of Physics, The University, Sheffield, England

1. INTRODUCTION

At the present time we can usefully talk about the theory of soft x-ray emission in metals in three parts:

(1) Band structure

(2) Correlations between electrons

(3) Core-hole electron interaction

These three aspects of the problem are by no means independent; but the division is helpful, for the purposes of this review, and at the present stage of development of the theory.

Since the Strathclyde Conference on soft x-ray band spectra and electronic structure in 1967, there has been a resurgence of interest in calculating soft x-ray emission spectra from band theory: for lithium, sodium and potassium, for aluminium and for some non-metals silicon and germanium. Although we shall refer to some of these calculations briefly, no attempt will be made to give a complete review of such work. Rather, we shall focus on points where the theory appears still to need development: we shall be dealing therefore with those areas where questions still remain, and focusing on doubts and uncertainties.

This should not obscure the fact that great progress in the theory has occurred since 1967. Indeed the theory is now sufficiently well-developed to allow us to relate soft x-ray emission to some other properties, and in particular to link it with the shape of the Compton line in x-ray scattering.

2. BAND THEORY IN TERMS OF MOMENTUM EIGENFUNCTION

Let us begin with the usual one-electron discussion, in which we neglect the hole in the inner core state, and account for electron correlation only through an appropriate choice of the one-body periodic potential generating the energy bands.

As emphasized previously by Stott and March (1968), it is useful in soft x-ray emission to describe the conduction band by the momentum eigenfunction

$$v_{\mathbf{k}K_n} = v(\mathbf{k} + K_n)$$

where the K_n are reciprocal lattice vectors. This is related to the Bloch waves $\psi_{\mathbf{k}}(\mathbf{r})$ by

$$\psi_{\mathbf{k}}(\mathbf{r}) = \sum_{K_n} v(\mathbf{k} + K_n) e^{i(\mathbf{k} + K_n)\cdot\mathbf{r}} \qquad (1)$$

which is a statement of Blocn's theorem, since it represents a product of $\exp(i\,\mathbf{k}.\mathbf{r})$ with a Fourier series.

It is then readily shown for simple metals, and in particular for the alkalis, that the momentum eigenfunction is flat for values of $|\mathbf{k}|$ out to approximately the Fermi momentum k_f, with $|v|^2 \sim 1$ for $K_n = 0$; and that at large momenta, for example $p = \hbar(K+K_n)$, the atomic-like oscillations dominate. Thus we have a free-electron correspondence at small p, and an atomic correspondence at large momenta. However, the completeness relation

$$\sum_{K_n} |v(\mathbf{k} + K_n)|^2 = 1$$

shows a link between the two regions.

The transition probability $T(\mathbf{k})$, apart from a constant, is given by

$$T(\mathbf{k}) = |\int \psi^*_{\text{core}} \nabla_{\mathbf{r}} \psi_{\mathbf{k}}(r) dr|^2 \qquad (2)$$

where ψ_{core} is the core wavefunction and $T(\mathbf{k})$ is readily expressed in terms of v(**p**).

This method has been used recently by Borland and Cooper (1971) to calculate the soft x-ray emission spectrum of lithium using the usual Seitz potential as employed, for example by Ham (1962) in his energy band calculation on Li. We shall be referring to this below, together with the Compton studies by Lundqvist and Lydèn (1971), and briefly referring also to positron annihilation results.

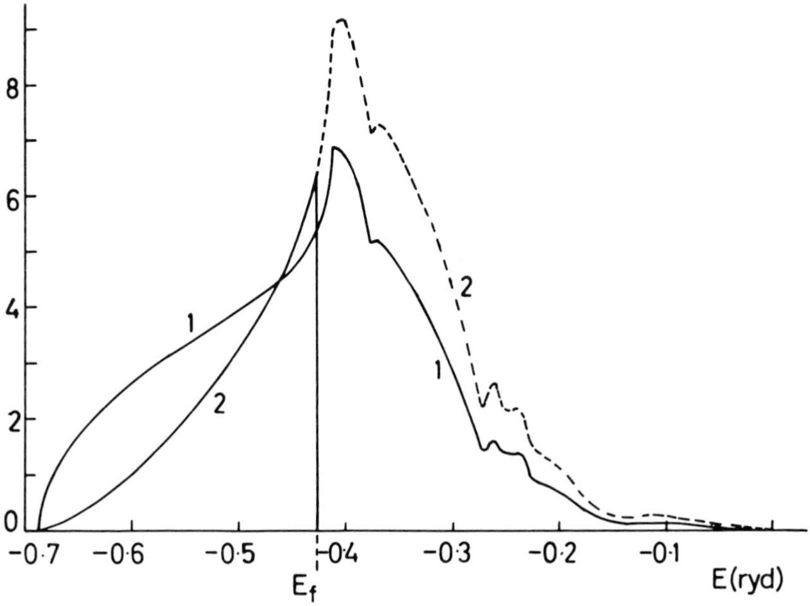

Figure 1 Theoretical results obtained by Borland and Cooper for density of states (curve 1) and for soft x-ray emission intensity (curve 2) for metallic lithium in bcc phase. Results obtained by calculating momentum eigenfunction v(**p**) from Seitz potential.

Figure 1 shows the theoretical results obtained by Borland and Cooper for both the density of states (A) and the soft x-ray emission spectrum (B) of metallic lithium. There is no sign of the premature peak observed in the soft x-ray emission at about ½ ev below the Fermi-energy E_f. Furthermore, using the Seitz potential, there is very little anisotropy in the transition probability T(**k**); the largest anisotropy amounting to a factor of about 1.2 at the Fermi surface.

However, McMullen (1970) has also calculated the soft x-ray spectrum for some alkali metals, including lithium. He used a self-consistent potential calculated by Lawrence (1969) based on Slater exchange. The intensity of soft x-ray emission as a function of energy, I(E), is such that its gradient begins to decrease at the Fermi-level, in contrast to the behaviour shown in figure 1. Furthermore, and clearly associated, the anisotropy in T(**k**) is much greater in McMullen's calculation, as emphasized by Borland and Cooper, and reaches a factor of 2 at the Fermi surface. The semi-empirical analysis by Stott and March (1968) indicates that if the anisotropy is sufficient a premature peak in I(E) can indeed result. This could well now hinge on the choice of potential, and as we discuss below this involves the correlation problem.

We wish to emphasize at this point that the soft x-ray experiment probes only that part of the electronic distribution which is within the region where the core wavefunction involved in the transition probability has appreciable amplitude (c.f. equation 2). The usual band structure results depend heavily, in the nearly free-electron-like metals, on the fact that the electron density is relatively constant over a large fraction of the atomic cell.

However, in soft x-ray emission, the core wavefunction is localized, and the flat region of electron density is therefore hardly explored at all. On the other hand, it is the atomic-like regions of the conduction band wavefunctions that are observed in soft x-ray emission. These short-wavelength components in **r** space are obviously reflected in the high momentum components of $v(\mathbf{k} + \mathbf{K}_n)$.

Therefore, in a one-electron framework, we must ask the question: 'What is the one-body potential that we should use to calculate the soft x-ray emission spectrum of metals?'.

The usual potentials have - underlying them - the philosophy that they should correctly give the charge density $\rho(\mathbf{r})$ of the metal, as measured by the intensities of x-ray scattering at the Bragg reflections. This can be expressed in terms of a one-body potential

$$V_{\text{Bragg}} = V_{\text{Hartree}} + \frac{\partial E_{xc}[\rho]}{\partial \rho} \qquad (3)$$

where V_{Hartree} is to be calculated using the correct density $\rho(\mathbf{r})$ in the metal; $E_{xc}[\rho]$ is the exchange and correlation contribution to the energy density of the inhomogeneous electron gas in the metal and its functional derivative, with respect to ρ, gives the contribution to the one-body potential in equation 3. It appears (see for example, Jones and March 1970) that the potential (eq.3) is the appropriate one for discussing the Fermi surface as mapped out by measuring the Kohn anomaly in lattice vibrations.

However, we should not be surprised if ambiguities arise when we attempt to fit the many-electron problem of a metal into a one-body mould. In particular, it is not physically reasonable to expect the potential (eq.3) to work equally well in describing the electron correlations in the whole of the inhomogeneous electron gas. Particularly the Dirac-Slater form

$$E_{xc}[\rho] \propto \rho^{4/3}$$

used in the calculations of Lawrence referred to above, is appropriate for describing the correlations in a relatively uniform gas. But in the region where the amplitude of the core wavefunction is appreciable, the electron density is varying rapidly in space, and the correlation problem is surely more atomic-like than free-electron like.

Our conclusion, then, is that it is <u>more</u> appropriate to find a potential that gives the high-momentum components correctly in this case than to use the potential $V_{Bragg}(\mathbf{r})$ defined in equation (3). We wish to stress again that the difficulties arise because of the practical necessity of dividing the theory into parts (1) and (2).

A. Momentum density and Compton line

It is therefore of considerable interest to enquire what sort of description the potential $V_{Bragg}(\mathbf{r})$, given by equation (3), would yield for the high-momentum components. To answer this question, we move attention from the charge distribution in $\mathbf{r}$-space, accessible through Bragg scattering of x-rays, to the Compton line in x-ray scattering, which measures more directly the electronic momentum distribution. In particular, in the usual dimensionless form, the intensity of the Compton line as a function of wavelength measured from the Compton wavelength is described by the function J(q) given by

$$J(q) = \tfrac{1}{2} \int_q^\infty \frac{I(p)}{p}\, dp \qquad (4)$$

where I(p)dp is the probability of an electron having momentum of magnitude between p and p+dp. The momentum density P(p) is related to I(p) through

$$I(p) = 4\pi p^2\, P(p) \qquad (5)$$

and in writing euqations (4) and (5) we have taken the simplest case of spherical symmetry. The point we wish to stress is that P(p) is accessible to experiment through the shape of the Compton line J(q).

A full discussion of the relation of $\rho(\mathbf{r})$ and P(p) has been given elsewhere by March and Stoddart (1971); refer therefore to this paper for details.

However, to make the discussion specific, let us again refer to metallic lithium. Available evidence, suggestive rather than conclusive, points to the fact that the Seitz potential is probably a useful starting approximation to V_{Bragg}, the potential used by Lawrence presumably being an even better estimate. But we can enquire whether the Seitz potential correctly generates the high-momentum components as measured by Compton scattering from lithium. This was discussed initially by Donovan and March (1956), and subsequent experiments have shown that their work gives a very reasonable account of the shape of the Compton profile but that the theory gives a line which is too narrow. Thus the Seitz potential is not giving a fully quantitative account of the high-momentum components, which - as we have seen - are very important for calculating the transition probability $T(\mathbf{k})$. Within the Bragg potential framework, this would lead naturally to a discussion of the electron correlations; that is to part (2) of the theory.

Kilby (1963) took up this question and enquired whether correlation effects, simulated by the uniform electron gas could resolve the discrepancy; and recently Lundqvist and Lydén (1971) have taken the matter further and have attempted to put together high-momentum components from Bloch waves and from electron correlations, as simulated through a uniform electron gas having the same density as that in metallic lithium.

The later calculations show, nonetheless, that too little account of correlation is being included, though the agreement between theory and experiment is obviously improved over that found by Donovan and March.

Therefore, if in the soft x-ray problem we wish to simulate correlations in respect of the high-momentum components, we could ask the question: 'Is there a potential that correctly generates the electronic momentum density $P(\mathbf{p})$ in the many-body problem?'

That an exact form exists for such a potential, say

$V_{Compton}(\mathbf{r})$, has been demonstrated explicitly by March and Stoddart (1971) only for a two-electron system; some of their results are briefly summarized in Appendix 1.

For many electrons, it is clear that a potential can be found that will correctly yield certain definite moments of the momentum distribution.

The point we wish to emphasize is that, until we have a means of carrying out fully-quantitative calculations in a genuine many-body framework, it seems clear that a one-body potential - constructed to get the high-momentum components approximately correct as revealed in Compton scattering - is *more* appropriate for calculating the soft x-ray emission than the Bragg potential.

We suggest, in view of the marked dependence of the anisotropy of the transition probabilities $T(\mathbf{k})$ on the choice of potential, that further calculations are needed on lithium to see whether a potential exists that can give the Compton line correctly. Its effect then on $I(E)$ will be of considerable interest: since it is not clear how much of this can be attributed to the electronic structure of the pure lithium metal, and how much to the core-hole—electron interaction as we discuss below in §4.

B. Relation to Positron Annihilation Experiments

It might be objected that we have singled out Compton scattering, whereas positron annihilation experiments also give us information on momentum space properties of the electrons in a metal. This is true, but we must recall that the positron annihilation is related to the Fourier transform of the product of the electron wavefunction with the positron wavefunction. And in simple metals it is known that the positron wavefunction is 'excluded' from the core region, which means that it has such a low amplitude that the high-momentum components of the electronic distribution are substantially reduced. Thus, it would *not* be appropriate to use the positron annihilation experiments to help directly in the construction of the one-body potential to be used in soft x-ray emission without taking full account of the detailed nature of the positron wavefunction. However, it will clearly be of interest to relate the wavefunctions generated by $V_{Compton}(\mathbf{r})$ with the positron measurements.

3. PHASE SHIFT DESCRIPTIONS OF ELECTRON BAND STRUCTURE

To lead into the discussion of the third aspect of the theory, the core-hole electron interaction, let us consider briefly two other sets of results on the electronic structure of lithium metal.

The calculation by Stocks, Young and Meyer (1968) is essentially a non-local pseudopotential treatment. Their work does not include an explicit treatment of the momentum eigenfunction, and so we shall deal with it only in terms of the density of one-electron states N(E). They treat the scattering of the conduction electrons off a Li^+ core, represented by a Herman-Skillman potential; the screening of the core is dealt with approximately by putting a shell of surface charge equal to one electronic charge at a distance from the lithium nucleus fixed by the Friedel sum rule

$$Z = \frac{2}{\pi} \sum_{\ell} (2\ell + 1)\eta_{\ell}(k_f) \qquad (6)$$

where Z is taken to be unity, and the $\eta_{\ell}(k)$ are the phase shifts which are evaluated at the Fermi momentum k_f in the sum rule.

Figure 2 shows the energy dependent phase shifts for scattering off such a potential. The main point of interest is that there is a very substantial p-wave phase shift which varies quite rapidly with energy at the Fermi level. This feature singles lithium out from the other alkalis and can be interpreted as due to the fact that there are no p-electrons in the lithium core.

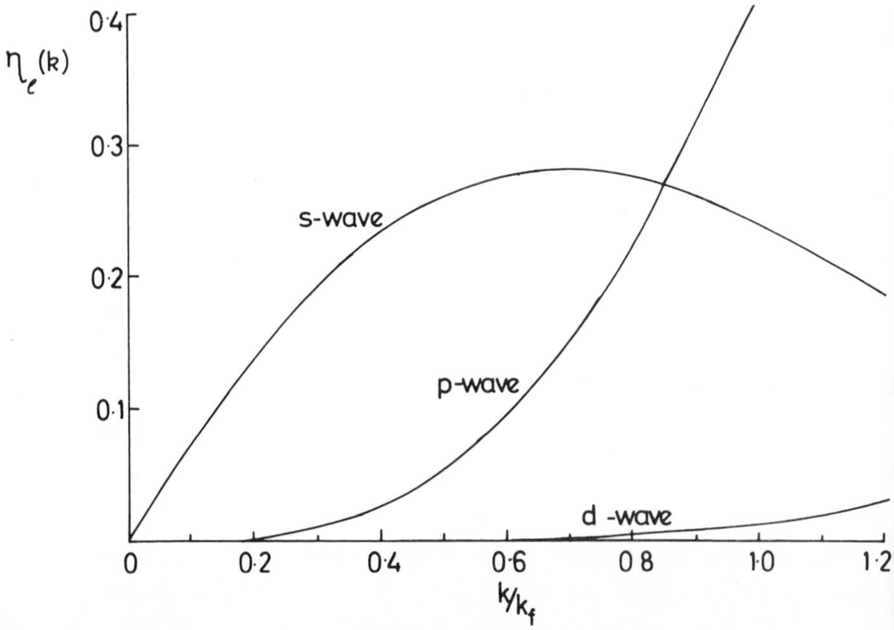

Figure 2 Phase shifts for scattering off Li^+ core (Stocks *et al* 1968) as function of energy. Substantial p-wave scattering at the Fermi-level is the principal feature to be noted.

The corresponding density of states obtained by Stocks *et al* (1968) is shown in figure 3. This curve agrees well with that deduced semi-empirically by Stott and March (1968) though their analysis neglected the core hole (see §4 below). There is remarkable agreement between the bandwidths, and both densities of states show a peak in N(E) just below the Fermi-level. However, this peak, in the calculations by Stocks *et al*, is somewhat sensitive to the precise values of the phase shifts.

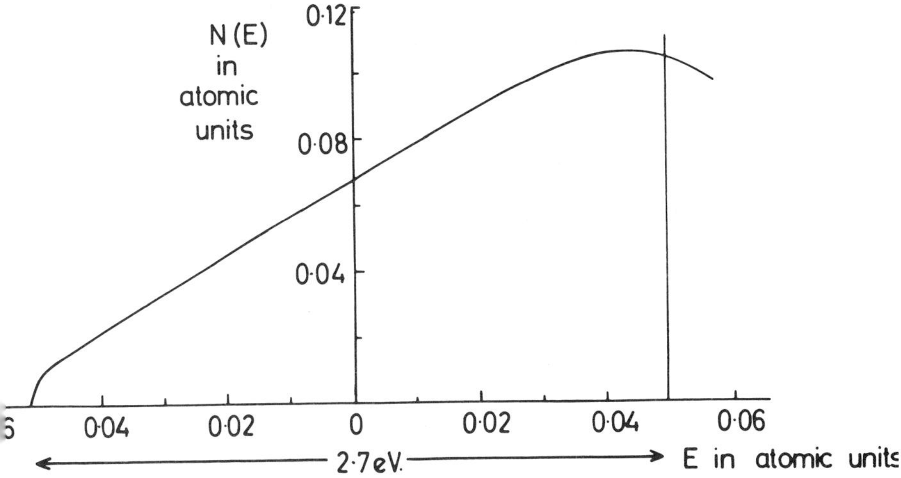

Figure 3 Density of states N(E) in bcc lithium, calculated from phase shifts shown in figure 2 (Stocks *et al* 1968).

We wish to make a further comment here on the narrow bandwidth obtained for lithium. Stocks *et al* argue from their phase shift analysis that the free-electron bandwidth is changed by a factor $[1/3 + (4/\pi)\eta_0(k_f)]$: according to this formula, the width of the band is therefore increased or decreased from the free-electron value according as $\eta_0(k_f) \lessgtr \frac{\pi}{6} =$ 0.52. Thus to the extent that large p-wave scattering at the Fermi-level reduces the s-wave component because of the sum rule (equation 6), Stocks *et al* argue that this narrow bandwidth, clear from the soft x-ray emission measurements of Crisp and Williams (1960), ties in with other (apparently diverse) properties of lithium, and in particular with its anomalous thermoelectric power and the increase in resistivity with pressure (see Dickey *et al* 1967). These latter properties again appear to depend on the relatively large p-wave phase shift at the Fermi-level.

While the method used by Stocks *et al*, based on the sum rule (equation 6) with Z = 1, may well lead to a very useful prescription for setting up a potential (the results of figure 3 are suggestive that this may be nearer V_{Compton} than V_{Bragg}!) we wish to draw attention to the analysis by Lee (1969) of the Fermi surface of lithium. Again he obtains a substantial p-wave phase shift, but the sum rule appears to be rather different from that used by Stocks *et al*.

We regard arguments as to which version of the sum rule is more appropriate for calculating electronic band structure as not fruitful; and we demonstrate in Appendix 2 the insensitivity of the band structure to the value of the Friedel sum.

The treatment of Stocks *et al* is not a structure sensitive theory. It may have a bearing on the results reported at this Conference, by Crisp for the soft x-ray emission from lithium in both the normal bcc and the close-packed phase which occurs when it is cooled to liquid helium temperatures. However, it will be necessary to calculate both N(E) and $T(\mathbf{k})$ in the close packed low-temperature phase before reliable conclusions can be reached on the interpretation of this new experiment.

The main conclusion from the work reported in this section is that the p-wave scattering is substantial in lithium metal.

We turn now to discuss the core-hole-electron interaction, again with particular reference to lithium.

4. EFFECTS ASSOCIATED WITH CORE-HOLE

The discussion so far has neglected the effect of the hole in the core on the soft x-ray emission spectrum. In the case of lithium, on which we shall focus particular attention, we are of course dealing with K-emission, though we shall make some more general remarks below.

Earlier Schiff (1954) proposed that the shape of the K-emission spectrum was strongly perturbed by the hole. Subsequent treatments dealt with the hole as a 'static impurity' (Goodings 1965, Shuey 1966, and Allotey 1967).

In the work by Stott and March (1968), attention was refocused on the experiment by Catterall and Trotter (1958) in which the satellite emission was studied and found to have largely the same shape as the principal emission. Stott and March argue that in terms of a static-hole description, it was hardly plausible, if one static hole had a large effect, that two holes would leave the spectrum with essentially the same shape.

Considerable progress has resulted since the last Strathclyde Conference in clarifying the fact that it is essential to treat the core-hole perturbation as 'switching off', and then transient effects lead to certain remarkable properties at the high-energy edge. That such considerations would be important in metals seems first to have been pointed out by Mahan (1967). A many-body formulation of this problem has been given by Nozières and his colleagues (Roulet *et al* 1969; Nozières *et al* 1969) but these studies involve a weak-coupling approximation. Of prime interest, for our consideration, is the work of Nozières and de Dominicis (1969) in which a theory valid for all coupling strengths is given in terms of a transient one-electron Green function; see in addition Hopfield (1969) for a helpful physical model allowing a simple derivation of the main results, and also the alternative derivations by Schotte and Schotte (1969), and by Langreth (1970) who uses an equation-of-motion method.

The principal result of the work by Nozières and de Dominicis is that, in deriving the intensity of soft x-ray emission near threshold, the transient effects referred to above introduce, for a particular orbital angular momentum quantum number l, a term of the form

$$\left(\frac{\varepsilon_o}{\varepsilon}\right)^{\alpha_\ell} \qquad (7)$$

where ε_o is a constant (probably of the order of the Fermi-energy), ε is the energy difference from the high-energy edge, and α_ℓ is given in terms of the phase shifts η_ℓ at the Fermi-energy for the scattering due to the electron-core-hole interaction. Explicitly, Nozières and de Dominicis show that

$$\alpha_\ell = \frac{2}{\pi}\eta_\ell - 2 \sum_j (2j + 1)\,(\eta_j/\pi)^2 \qquad (8)$$

A. Two-phase-shift Model

Before reviewing some specific phase-shift calculations, we shall construct an elementary model in which we retain only two phase shifts

$$\eta_o = \pi x_o \text{ and } \eta_1 = \pi x.$$

We shall assume

$$\eta_l = 0 \text{ for } l>1.$$

The Friedel sum-rule (equation 6) applied to the core-hole scattering then takes the form

$$2x_o + 6x = 1 \qquad (9)$$

We then find explicitly from equations (8) and (9) that, as a function of the p-wave phase shift πx,

$$\alpha_o = \frac{1}{2} - 24x^2 \qquad (10)$$

$$\alpha_1 = \frac{-1}{2} + 8x - 24x^2 \qquad (11)$$

and

$$\alpha_\ell = \frac{-1}{2} + 6x - 24x^2;\ \ell>1. \qquad (12)$$

These results equations (10) - (12) are plotted as functions of x in figure 4.

It is obvious, first of all, that if we retain only s-wave

and p-wave phase shifts, then from equation (8), $\alpha_\ell < 0$ for $\ell > 1$. It can be seen from figure 4 that

$$|\alpha_{\ell>1}| \geqslant \frac{1}{8}.$$

Similarly

$$\alpha_0 < \frac{1}{2} \quad \text{and} \quad \alpha_1 < \frac{1}{6}.$$

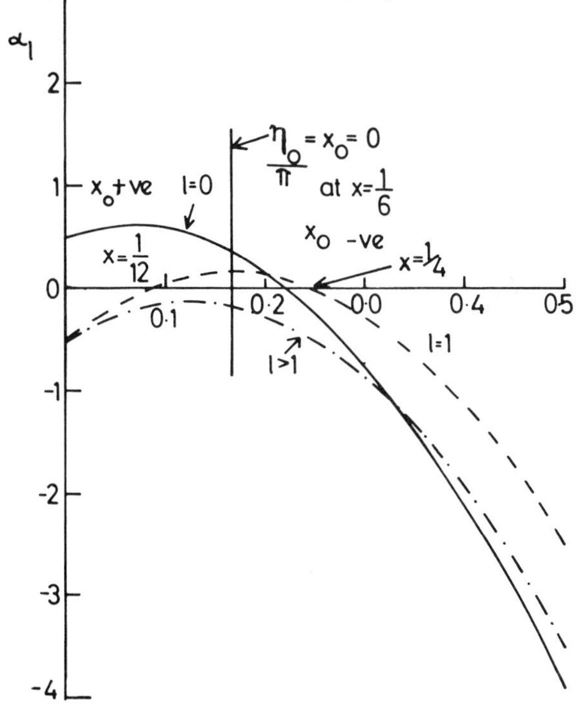

Figure 4 Two-phase shift model. Quantities α_ℓ shown are related to the phase shifts of core hole scattering by equation (8). Explicit plots are of α_0, α_1 and α_ℓ for $\ell>1$, as functions of p-wave phase shift $\eta_1 = \pi x$ at Fermi level. Case when x = 0 was already discussed by Nozières and de Dominicis.

Of particular relevance for the present discussion is the fact, as shown in figure 4, that α_1 can be positive in the range $\pi/12 < \eta_1 < \pi/4$. Only over a small part of this range is α_0 also positive.

B. Application to Lithium

In the case of lithium, we are dealing with K emission and are therefore concerned with $\ell = 1$, the p-wave. Since we know from experiment that the intensity goes to zero continuously at the high-energy edge, it is clear that α_1 must be negative.

This tells us, from the above two-phase-shift model, that the p-wave must have a phase shift less than $\pi/12$ or greater than $\pi/4$. We notice that in the latter case the s-wave phase shift must be negative.

To try to resolve this ambiguity, we shall consider briefly the direct calculations of phase shifts for the core-hole scattering, that have been made to-date. First, in the work of Ausman and Glick (1969) a screened Coulomb potential satisfying the Friedel sum-rule was used, with the result that the s-wave phase shift is large (0.95) and the p-wave phase shift is small (0.15). Therefore, Ausman and Glick obtain a small negative value of α_1 as seen immediately from the present two-phase-shift model (their d-wave and higher phase shifts are small).

To contrast this, let us choose another model potential satisfying the sum rule. We have referred to the phase shifts used by Stocks *et al* above, which satisfy the appropriate sum rule for the core hole. If we use their phase shifts we find α_1 to be small and positive, their p-wave phase shifts being quite large, but nonetheless substantially below $\pi/4$, as may be seen from figure 2.

As a third and final case, we refer to the work of Allotey (1971). Unfortunately, Allotey's model has not been constructed to satisfy the Friedel sum-rule, but with his potential he finds a large p-wave phase shift in excess of 0.38π. In fact, of course, it is not proved that a reasonable potential satisfying the sum rule can be found with such a large p-wave phase shift.

Thus, it is not to be regarded as finally established that the core-hole scattering in lithium leads to a p-wave phase shift $<\pi/12$, nor that this is the reason why α_1 is negative in lithium. The detail of the lithium ion core must evidently be properly built into the theory and, since this leads to the possibility of relatively strong p-wave phase shifts, the screened Coulomb potential used by some earlier workers may prove an inadequate approximation.

It is obvious, in cases where the d-wave phase shift is known - on physical grounds - to play a major rôle, that one could either rework the present model with s and d waves, or express α_ℓ as a function of p and d phase shifts, including s, p and d; we shall not go further into this here.

C. Other Alkalis and Noble metals

We have little doubt that for the other alkalis no large p-wave phase shifts will be found. Therefore, for the L spectra of sodium we can be sure that there is threshold enhancement, in agreement with experiment (see McMullen, 1970; especially figure 2 of this reference). However, it is the width of the threshold 'spike' in sodium, where presumably band-structure effects are very small, that is of prime interest. A refinement of the Nozières-deDominicis model is required to calculate the width of the threshold spike and this problem is currently being investigated (see also Longe, this Volume p341). The noble metals appear also to be interesting and from the phase shifts used by Stocks *et al* it seems that all the α_ℓ may turn out to be negative.

Thus, there is an obvious need now for careful construction of potentials to describe the electron-hole scattering. The starting point should be Herman-Skillman core potentials, or a semi-empirical potential, for the individual metals.

D. Satellite Emission in Lithium

Finally, we wish briefly to comment on the satellite emission for lithium, which Allotey discusses more fully in this Volume (p.361).

It seems that the essential point is that it is the 'switching off' of the hole which is crucial when we are at energies higher than the premature peak in the soft x-ray spectrum of lithium*. Although in the satellite emission the core is doubly ionized, it does not seem that the static hole introduces more than a rather weak energy-dependence up to the peak in the emission intensity. Thus the satellite emission would not be expected to differ very much in shape from the principal emission, and we appear to have a way of resolving what has been hitherto a very puzzling feature for lithium.

5. CONCLUSION

The present paper raises two basic problems in the theory of soft x-ray emission:

(1) Within a one-body framework, what potential should we use to calculate the soft x-ray intensity? We argue that we are in a region of the electron distribution with large density gradients, and that the correlation problem is therefore more 'atomic-like' than we generally encounter in the simple metals. We make the proposal that a one-body potential that fits the shape of the Compton line in x-ray scattering would be more appropriate for the soft x-ray calculation than those in current usage.

(2) How should we describe the core-hole scattering? That is, what potential should be used, and over what energy range from the threshold is the hole important? This will require a transcending of the Nozières-deDominicis solution. It may also require consideration of a core hole in a Bloch-

*The brief remarks made here on the satellite emission were stimulated by a discussion in Gothenburg with C. Kunz in May 1971, during a Seminar he gave there on synchrotron radiation.

electron system; in order to find the correct scattering potential to be used in the Nozières-deDominicis treatment.

We regard it as likely that in only a few cases (and lithium seems to be an outstanding exception) will the answers to questions (1) and (2) alter the situation qualitatively. However, there is a lot of work still required before we can be completely certain of the relative importance of band-structure effects, electron-electron interactions, and the core vacancy. The beautiful work on absorption spectroscopy using synchrotron radiation is obviously going to help considerably in unravelling these various contributions.

APPENDIX 1 : Bragg and Compton potentials for two-electron system

To make concrete, in a simple case, the concepts of the Bragg and the Compton potentials, we shall summarise here the main results of the calculation by March and Stoddart (1971) on a two-electron ion with atomic number Z for the case in which Z is sufficiently large for us to expand in 1/Z.

Here the electron density and the first-order density matrix are both known (for a review see March, Young and Sampanthar, 1967) and hence, by Fourier transform, the momentum density can be found. The Bragg potential is then obtained from the Schrödinger equation formed with a wavefunction for the one-electron orbital that is essentially the square root of the exact electron density $\rho(\mathbf{r})$. This potential may be written explicitly as

$$V_{Bragg}(r) + \frac{Z}{r} = \frac{3Z}{8}(1+\frac{0.67}{Z}) - \frac{Z^2r}{8}(1+\frac{0.67}{Z}) \quad (A.1)$$

to leading order in 1/Z.

Similarly, from the first-order density matrix the momentum density P(p) can be obtained, and this again is expressed in terms of the square of a one-body orbital in momentum space.

The corresponding potential reproducing this orbital then has the form

$$V_{\text{Compton}}(r) + \frac{Z}{r} = \frac{9Z}{8}\left(1+\frac{0.67}{Z}\right)^{\frac{1}{2}} - \frac{3Z^2r}{4}\left(1+\frac{0.67}{Z}\right)^{-\frac{1}{2}} \quad (A.2)$$

Naturally, because we have electron-electron interaction, these two potentials are not the same, though there are similarities of form between them. The difference gives us a quantitative measure of the effect of electron correlation.

We wish to stress that the two-electron system is a very special case; for when the two electrons have antiparallel spin, as in the ground state of the system considered here, there is only one orbital involved. We must refer the reader to the discussion by March and Stoddart (1971) for fuller details, but we must conclude by saying that it is not established, for N electrons, that a potential $V_{\text{Compton}}(\mathbf{r})$ exists that will represent the momentum density of the many-electron system exactly in terms of one-body orbitals. It is therefore possible that, for N>2, there is irreducible electron correlation in momentum space; whereas we know that the exact charge density $\rho(\mathbf{r})$ in the many-body problem, with N electrons, can always be exactly represented by the squares of N one-body orbitals generated by V_{Bragg}. However, it is certainly possible to obtain a potential that will correctly give certain moments of the momentum density.

APPENDIX 2 : Band structure in periodic crystal and Friedel sum rule

For a charged impurity in a metal crystal, then we must have the requirement of perfect screening, namely that the charge displaced by the impurity centre must be just enough to neutralize the excess charge on the centre itself.

In a uniform electron gas, this is expressed in terms of the phase shifts by equation (6), with Z the excess charge on the centre. The phase shifts of the different partial waves have to be known at the Fermi momentum k_f, in order to check the screening.

We wish to emphasize that while this is a perfectly proper condition for an impurity centre, there is a great deal of ambiguity when such a procedure is applied to a single ion in a perfect metal crystal. The argument presented below shows that there is no precise meaning in the Friedel sum, in the case of a perfect crystal. This does not mean that the prescription by Stocks *et al*, and indeed those by others with different Friedel sums, are not useful in setting up a periodic potential for calculating a band structure. It does mean, though, that there would be other ways of constructing exactly the same band structure, with a single-centre potential satisfying a different Friedel sum.

The reasoning hinges on the fact that a given periodic potential $V_p(\mathbf{r})$ is characterized entirely by its Fourier components V_{K_n}, through

$$V_p(\mathbf{r}) = \sum_{K_n} V_{K_n} \exp(iK_n \cdot \mathbf{r}) \qquad (A.3)$$

For a general periodic potential, there is no need for V_{K_n} to be the same for reciprocal lattice vectors of the same magnitude but different directions. However, in the example we take in this Appendix, this will in fact turn out to be the case.

It is obvious from equation (A.3) that knowledge of V_{K_n} at all reciprocal lattice vectors is sufficient to determine the band structure uniquely. But an alternative method, widely used in the theory of metal crystals, is to write $V_p(\mathbf{r})$ as a sum over direct lattice vectors R_n of a localized potential $V(\mathbf{r})$ say; that is

$$V_p(\mathbf{r}) = \sum_{R_n} V(\mathbf{r}-R_n). \qquad (A.4)$$

If we take the localized potential $V(\mathbf{r})$ to be spherical, then it is easy to confirm that $V_{\mathbf{K}_n} = V_{|\mathbf{K}_n|}$, and that, apart from a trivial multiplying constant,

$$V_{|\mathbf{K}_n|} = \int V(|\mathbf{r}|) \exp(i\mathbf{K}_n \cdot \mathbf{r}) d\mathbf{r}. \qquad (A.5)$$

However, while - through equation (A.5) - it is obvious that a given localized potential $V(r)$ generates $V_{\mathbf{K}_n}$ uniquely, the converse is evidently not true. Indeed, given $V_{|\mathbf{K}_n|}$ we can draw any continuous curve $V(|\mathbf{k}|)$ that passes through the same values $V_{|\mathbf{K}_n|}$ at the reciprocal lattice vectors, and each of these $V(k)$ will generate a different localized potential $V(r)$. But all such potentials will add up to yield the same periodic potential $V_p(\mathbf{r})$, because they all have the same Fourier transform at the reciprocal lattice vectors (although not elsewhere).

It is evident that if we now scatter electrons off these different localized potentials, they will yield different Friedel sums. But they lead to the same periodic potential, and hence to the same band structure.

To be sure on this last point, it is easy to work out the Friedel sum in the case when the localized potential is a weak screened Coulomb potential

$$V(r) = - \frac{ze^2}{r} \exp(-qr). \qquad (A.6)$$

In this case, to first order in z, the Friedel sum is determined completely by the Fourier transform at $k = 0$. But this is equivalent to a constant term in the periodic potential, and hence could not change the band structure. Needless to say, for larger z, the Friedel sum will depend to some extent on what the potential is in k-space at the first reciprocal lattice vector, but it will be quite clear from this argument that the band structure is insensitive to the Friedel sum.

As we said at the outset, the argument has nothing to do with the Friedel impurity problem, nor with the Nozières and de Dominicis core-hole problem discussed above in §4. In these cases, it is perfectly clear that the Friedel sum has a precise value; but in the case of a periodic crystal, there are, essentially, an infinite number of ways of decomposing a periodic potential into a sum of localized potentials placed on the lattice sites. Therefore, from the perfect crystal information alone, it is not possible to extract the localized potential V(r). Needless to say, displacing the ions from the perfect lattice sites by introducing a phonon, or more violently by melting the metal crystal, would allow a great deal more to be said about localized charge distributions and potentials.

Acknowledgments

Thanks are due to a number of colleagues for discussions of these problems over a long period, and especially to Drs. J.C. Stoddart, M.J. Stott and K.C. Williams.

REFERENCES

Allotey,F.K.(1967);Phys. Rev., 157, 467.

Allotey,F.K.(1971); Solid State Communications 9, 91.

Ausman, G.A. and Glick, A.J. (1969); Phys. Rev., 183, 687.

Borland, R.E. and Cooper, J.R.A. (1971); J. Phys. F, Metal Phys. 1, 237.

Catterall, J.A. and Trotter, J. (1958); Phil. Mag., 3, 1424.

Crisp, R.S. and Williams, S.E. (1960); Phil. Mag., 5, 525.

Dickey,J.M.,Meyer, A. and Young, W.H. (1967); Proc. Phys. Soc., 92, 460.

Donovan, B. and March, N.H. (1956); Proc. Phys. Soc., B69, 1249.

Goodings, D.A. (1965); Proc. Phys. Soc., 86, 75.

Ham, F.S. (1962); Phys. Rev., 128, 82; and 2524.

Hopfield, J.J. (1969); Comments on Solid State Physics, 2, 40.

Jones, W. and March, N.H. (1970); Proc. Roy. Soc., A317, 359.

Kilby, G.E. (1963); Proc. Phys. Soc., 82, 900.

Langreth, D. (1969); Phys. Rev., 182, 973.

Lawrence, M.J. (1969); Ph.D. thesis, University of Bristol.

Lee, M.J.G. (1969); Phys. Rev., 178, 953.

Lundqvist, B.I. and Lydên, C. (1971); Phys. Rev. B4 [10], 3360.

Mahan, G.D. (1967); Phys. Rev., 163, 612.

March, N.H. and Stoddart, J.C. (1971); IBM Wildbad Conference Proceedings.

March, N.H., Young, W.H. and Sampanthar, S. (1967); The Many-Body Problems in Quantum Mechanics (Cambridge U.P.).

McMullen, T. (1970); J. Phys. C., 3, 2178.

Nozières, P. and de Dominicis, C.T. (1969); Phys. Rev., 178, 1097.

Nozières, P., Gavoret, J. and Roulet, B. (1969); Phys. Rev., 178, 1084.

Roulet, B., Gavoret, J. and Nozières, P. (1969); Phys. Rev., 178, 1072.

Schiff, B. (1954); Proc. Phys. Soc., A67, 2.

Schotte, K.D. and Schotte, U. (1969); Phys. Rev., 182, 479.

Shuey, R.T. (1966); Physik Kondenscierten Materie, 5, 192.

Stocks, G.M., Young, W.H. and Keyer, A. (1968); Phil. Mag., 18, 895.

Stott, M.J. and March, N.H. (1968); in Soft X-Ray Band Spectra, ed. D.J. Fabian, p.283, Academic Press, London.

THEORY OF X-RAY EMISSION BANDS OF TRANSITION METALS

M.A. Blokhin, I.Y. Nikiforov, H. Sommer and I.I. Gegoosin
Rostov State University, Rostov-on-Don, USSR.

1. INTRODUCTION

In spite of the considerable achievements of one-electron energy band calculations, the form of x-ray band spectra of pure metals - not to mention alloys - has not yet been satisfactorily explained. There are only a few theoretical calculations of emission band shapes: for alkali metals (McAlister 1969, McMullen 1970), for aluminium (Rooke 1970), for iron (Nikiforov and Blokhin 1963), and for copper (Nikiforov 1966). These few studies do not change the position in general, which is explained by the fact that theoreticians mostly confine their calculations to electron energy spectra, and pay little attention to the electron wavefunctions. It is impossible without the wavefunction to calculate the transition probability, which influences the form of x-ray emission bands. This situation needs to be changed because it is just those x-ray and electron spectra that are capable of proving the validity of band structures obtained by calculation.

For this reason our programme is to study all the one-electron methods used in band calculations and to calculate the form of x-ray emission bands by these methods. Comparison of the theoretical spectra with measured ones provides an opportunity of finding the disadvantages of the calculation method used, and the limits of its applicability. At the same time, it supplies us with information about those peculiarities of the x-ray excitation process that have not been taken into account, and about the properties of the material studied.

2. CALCULATION PROCEDURE

As a first objective metallic chromium (bcc) was chosen. There have been a number of band-calculations for this metal; but these unfortunately do not meet the requirements - particularly the calculation of wavefunctions and matrix elements - mentioned above. The LCAO method has been used (Asdente and Friedel 1961, Jashi *et al* 1970) and also the APW method (Asano and Yamashita 1961, Snow and Waber 1969).

In the present investigation the band structure of chromium was calculated using the APW, the Green-function, and the cellular methods. The first two have been carried through to give a theoretical form of the emission band of the metal. The only experimental parameter used is the lattice constant. Each of the calculations is characterized by the following steps (figure 1):

(1) finding the crystal potential
(2) computing the radial part of the electron wavefunctions in this potential
(3) evaluating the matrix elements of Hamiltonian and electron eigenvalues (i.e. the dispersion curves)
(4) calculating the electron-state energy distribution
(5) resolving the system of linear equations for the coefficients of the wavefunction expansion into set of basic functions (i.e. finding the wavefunction of an electron in the valence band).

It is then possible, in principle, to find a new crystal potential with the wavefunction obtained and to continue the process until self-consistency is achieved.

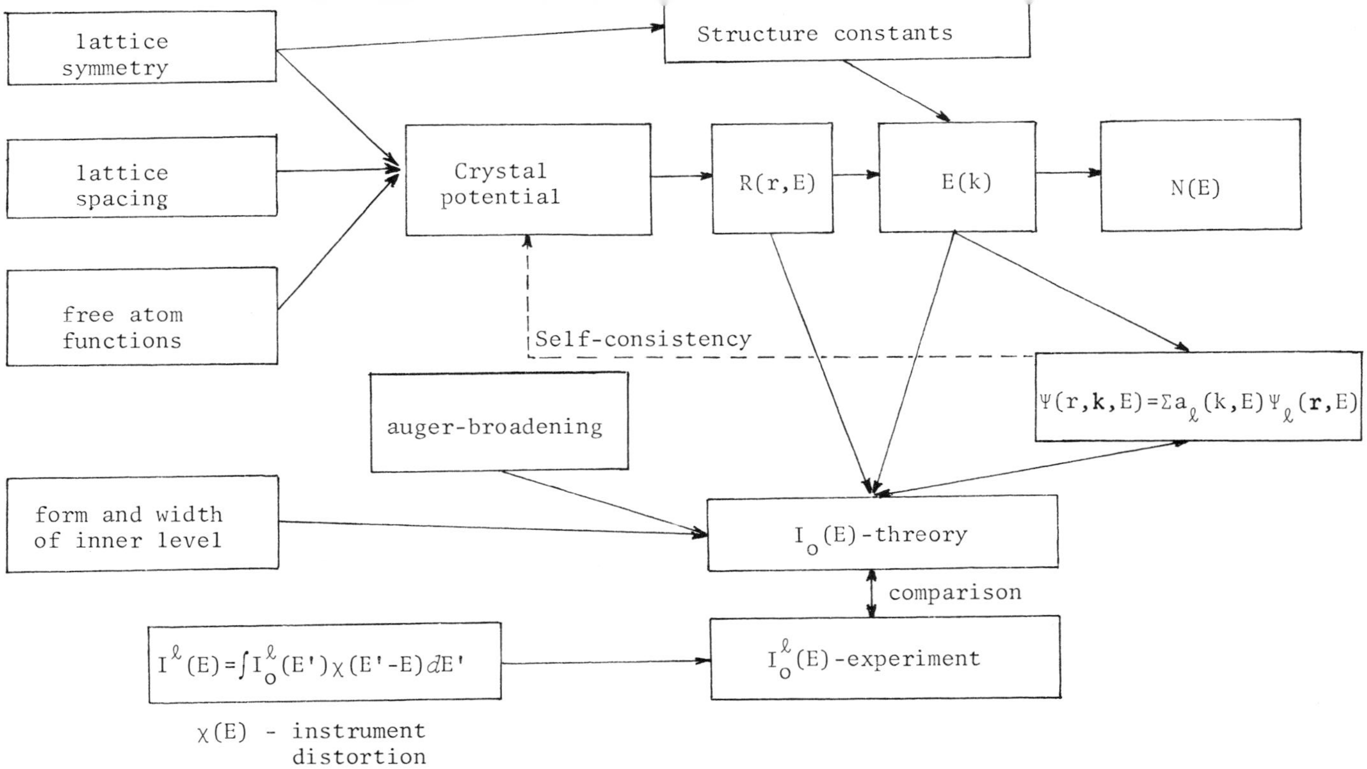

Figure 1 Block diagram of one-electron calculation scheme for obtaining theoretical x-ray emission bands.

The Green-function method differs from other methods in one step, that is the calculation of structure constants, which needs to be done once only for a given type of lattice. Using structure constants already thus obtained the eigenvalue problem is simplified immensely.

The first two steps are common to all the methods, and the most difficult of the two is the crystal potential calculation because of an exchange energy. In the present work the crystal potential field for chromium was calculated taking into account different numbers of neighbouring atoms. The exchange potential was found by the Slater formula and by various modifications of it. A $3d^5 4s$ electron configuration was assumed, and Herman and Skillman (1963) radial functions of a free chromium atom used.

With the APW method electron eigenvalues were found with an accuracy of 0.007 Ryd, achieved using 26 APWs and taking into account all the spherical harmonics up to $\ell=10$. With the cellular method the electron wavefunction was a set of 14 spherical functions (all the s, p and d functions, four f and one g). In the Green-function method all spherical harmonics were taken into account with $\ell=2$. For the three methods programmes were made for calculation of the eigenvalue and wavefunction at a general point in a reciprocal space.

3. RESULTS

The band structure of chromium obtained using the APW and Green-function methods practically coincide; this is easily understood because of the well-known similarity of the methods. In figure 2 the dispersion curves obtained for the [100] direction are shown. These were obtained by the Green-function method, with an exchange potential given by the Slater formula and with a coefficient 3/4 from the exchange term. We note the broadening of the bands obtained in the second case. It was

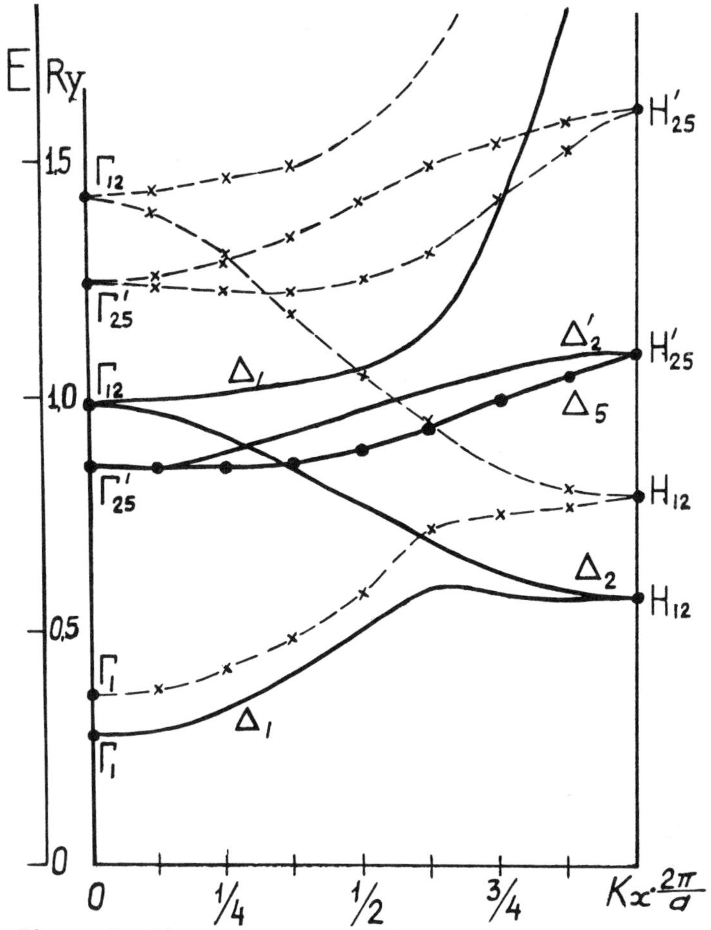

Figure 2 Dispersion curves for metallic chromium (bcc), obtained using the Green-function method.

also established that varying the number of neighbouring atoms, taken into account in calculating the potential, is not as important as the coefficient from the exchange term.

The band structure obtained by cellular method differs considerably from the APW and Green-function results. The energy between the beginning of the s and d bands (that is, the difference in energy between the states Γ_1 and $\Gamma_{25'}$ or Γ_{12}) is less by half than with the APW calculation, and equal to

0.3 Ryd. To find the density-of-states distribution, electron energies were computed for 55 unequivalent reciprocal vectors in a 1/48 part of the Brillouin Zone, corresponding to 1028 points in the whole Brillouin zone. The density of states (figure 3) evaluated using the APW and Green-function methods nearly coincide, and is close to that obtained by Snow and Waber (1969) despite the rather low number of reciprocal vectors used by these investigators. The density of states obtained by Jashi *et al* (1970), using a tight-binding calculation, is similar in form but with 1.5 times narrower bandwidth.

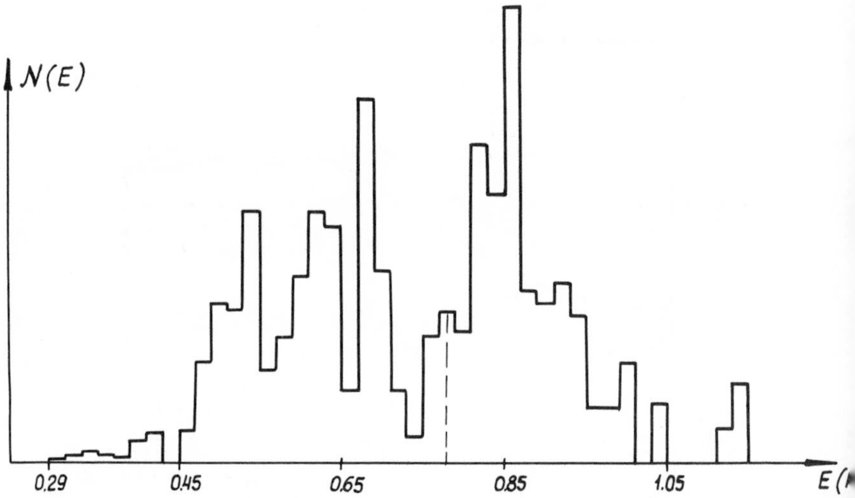

Figure 3 The electron density of states of bcc chromium, evaluated using the APW and Green-function methods.

To obtain the x-ray emission bands the coefficients a_1 of the wavefunction expansion were calculated. The K-band intensity depends on p-state distribution and the $L_{2,3}$ and $M_{2,3}$ bands on the d-state distribution. Our investigation has shown that the contribution of s and d states to the $K\beta_5$-emission band and of s-states to the L and M bands is negligible. The function

$$N_1(E) = \int_S \frac{|a_1(k,E)|^2}{|\text{grad}_k E|} dS$$

was evaluated, multiplied by the radial factor of the transition probability (Nikiforov 1961), and smeared by a Lorentz curve for the inner level. The forms obtained for respectively the K and L bands are shown in figures 4 and 5. They are compared with experimental spectra recorded in our laboratory. There is an encouraging agreement with experiment for K band calculated using the Green-function method. With the APW calculation we observe that using 26 APWs is sufficient for evaluation of the density of states but not enough for the determination of crystal wavefunctions. In particular this affects the K-band because only the distribution of valence p-states is important here. But the relative accuracy of the C_p coefficients is low because of their small values in comparison with the C_d coefficients. This is why $K\beta_5$-band is so sensitive to the crystal wavefunction.

On the other hand, the L_3-emission band depends mainly on the density of d-states which prevail. Thus the difference in crystal wavefunction obtained with the two methods is not important here and the forms of L-emission bands calculated by the two methods are similar.

4. CONCLUSIONS

We conclude that the Green-function method, for the case of chromium investigated, gives more accurate results than the APW-method. It is interesting to note that the time spent on the Green-function calculation was less than with APW-method.

The experimental curves have larger bandwidths than the theoretical ones, which can be attributed to Auger-broadening and instrumental resolution. The width (0.2eV) of the spectrometer 'window' was determined only for the K-spectrum, mea-

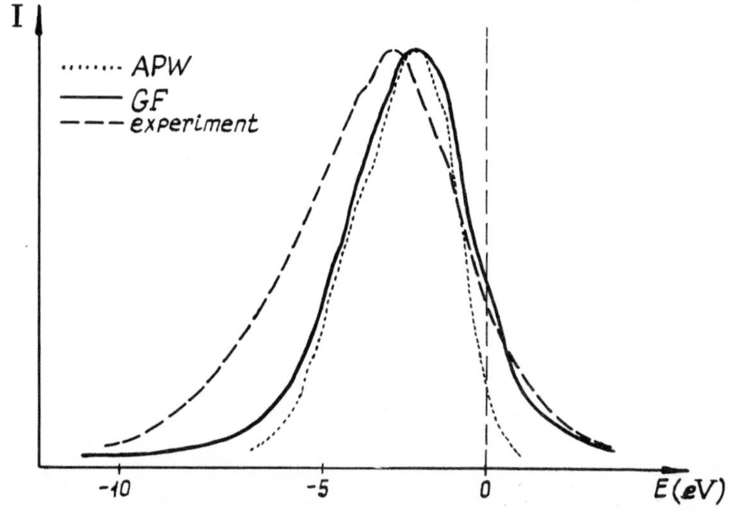

Figure 4 Comparison of the theoretical and experimental forms of the Cr L_3-emission band for metallic chromium.

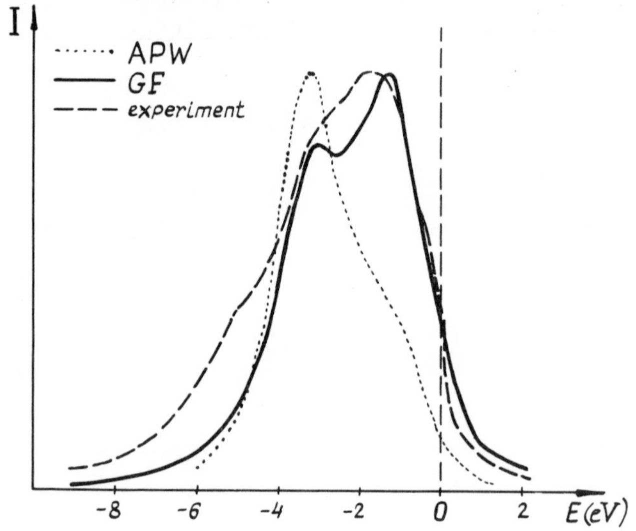

Figure 5 Experimental and theoretical forms of the Cr $K\beta_5$-emission band.

sured with the two-crystal instrument.

REFERENCES

Asano, S. and Yamashita, J. (1961); J. Phys. Soc. Japan, 23, 714.

Asdente, M. and Friedel, J. (1961); Phys. Rev. 124, 384.

Goodings, D.A. and Harris, R. (1969); J. Phys. C. 2, 1808.

Herman, F. and Skillman, S. (1963); Atomic structure calculation, Englewood Cliffs, New Jersey.

Jashi, J., Hayashi, S. and Shumuzu, J. (1970); J. Phys. Soc. Japan, 29, 6.

McAlister, A.J. (1969); Phys. Rev. 186, 595.

McMullen, T. (1970); J. Phys. C. 3, 2178.

Nikiforov, I.Y. (1961); Izv. Akad. Nauk SSSR, Ser. Fiz. 25, 1043.

Nikiforov, I.Y. and Blokhin, M.A. (1963); Izv. Akad. Nauk SSSR, Ser. Fiz. 27, 314.

Nikiforov, I.Y. (1966); "Röntgenspektren und chemische Bindung", Leipzig, 241.

Rooke, G.A. (1970); J. Res. Nat. Bur. Stand. A74, 273.

Snow, E.C. and Waber, J.T., (1969); Acta Met. 17, 623.

MANY-BODY EFFECTS IN X-RAY AND PHOTOELECTRON EMISSION FROM METALS AND ALLOYS

Lars Hedin

Department of Theoretical Physics, Lund, Sweden

1. INTRODUCTION

We shall attempt here to describe the essence of a set of many-body effects, from a unified standpoint, using the concept of 'shake-up'. We shall discuss the different edge and satellite effects, the intensity enhancement in the main band, and phonon effects.

The starting point is the usual golden-rule formula (see, for example, Merzbacher 1970) applied to x-ray spectroscopy

$$W(\omega) = \sum_n |\langle\psi_n|T|\psi_o\rangle|^2 \delta(\omega - E_n + E_o) \qquad (1)$$

Here $W(\omega)$ gives, apart from slowly varying factors, the probability of a transition from the initial state $|\psi_o\rangle$ caused by the transition operator T. The operator T is given by

$$T = \sum_i \vec{n}\cdot\vec{p}_i = \sum_{k\ell} \vec{n}\vec{p}_{k\ell} a_k^\dagger a_\ell \qquad (2)$$

where $\vec{n}$ is the direction of polarization, and $\vec{p}_i$ the momentum operator of electron i. The final expression in equation (2) gives the second-quantized form of T. We see that T is built from operators for electron-hole pairs.

Equation (1) can be interpreted as the energy spectrum of a

fictitious state $|\Phi> = T|\psi_0>$. Expanding $|\Phi>$ as

$$|\Phi> = \sum_n \alpha_n |\psi_n> \qquad (3)$$

we have for the energy spectrum

$$D(E) = \sum_n |\alpha_n|^2 \delta(E-E_n) \qquad (4)$$

and since $\alpha_n = <\psi_n|\Phi> = <\psi_n|T|\psi_0>$

we obtain $W(\omega) = D(\omega+E_0)$ (5)

In the one-electron approximation $|\Phi> = T|\psi_0>$ is an eigenstate differing from $|\psi_0>$ by having an electron-hole excitation, but in general the spectrum D(E) can be far from trivial.

We thus conclude that the x-ray intensity is given by the energy spectrum of the fictitious state $T|\psi_0>$ obtained by operating with the transition operator T (built from electron-hole pairs) on the initial state $|\psi_0>$.

2. SHAKE-UP EFFECTS

To understand edge and satellite effects in metals, it is convenient first to discuss the well understood shake-up effects in noble gases. These effects demonstrate a mechanism for the occurrence of multiple excitations. The initial state here, after primary excitation, is the unrelaxed state in which - using a Hartree-Fock self-consistent-field picture - all electrons except the one ejected retain their old wavefunctions. This unrelaxed state is not an eigenstate and can be resolved into a spectrum of multiple excitations. The validity of this description of shake-up is of course limited to the case when the excitation process occurs very quickly (the 'sudden' approximation) and thus the primary excitation energy is large. With near-threshold excitation the ejected electron leaves so slowly that the remaining electrons have time to readjust.

That no multiple excitation can take place at threshold is also clear from energy conservation.

We can make this discussion more precise by starting from the second-quantized expression for T in equation (2). If the final state contains a high-energy electron (the ejected electron), then this electron can only be produced by the creation operator a_k in T, since the cloud of virtual electrons contained in $|\psi_o\rangle$ has a limited range of energies. The x-ray intensity is therefore given by the energy spectrum of the unrelaxed one-hole states $a_1|\psi_o\rangle$, as we earlier assumed on intuitive grounds.

There is a simple relation between the energy spectrum D(E) in equation (4) and the overlap describing the decay of the fictitious state $|\Phi\rangle$ in equation (3). We have, for the time-evolution of $|\Phi\rangle$,

$$|\Phi(t)\rangle = \sum_n \alpha_n |\psi_n\rangle exp(-iE_n t) \tag{6}$$

and thus for the overlap S, describing the decay,

$$S(t) = \langle\Phi(t)|\Phi(o)\rangle = \sum_n |\alpha_n|^2 exp(-iE_n t) \tag{7}$$

Comparing with equation (4) we find

$$D(E) = \frac{1}{2\pi}\int S(t) exp(iEt) dt \tag{8}$$

Thus we can obtain the energy spectrum, and hence the x-ray intensity, provided we know how the fictitious state $|\Phi\rangle = T|\psi_o\rangle$ decays with time.

3. THE ANDERSON THEOREM AND EDGE SINGULARITIES IN X-RAY PHOTOEMISSION

By using equation (8), and invoking the Anderson (1967) theorem, we obtain a simple argument to demonstrate the edge singularities (Hedin 1971). The Anderson theorem states that the overlap between the ground states of a uniform non-interacting electron gas and an electron gas with a localized potential is

$$N^{-\delta^2/3\pi^2} \qquad (9)$$

where N is the number of electrons and δ is the phase shift for scattering against the potential (assuming only s-wave scattering). The orthogonality can be traced to the large number of electron-hole pairs of low energy created by the localized potential. If we consider x-ray photoemission, where a core electron leaves the solid at high speed, then a_k must describe the photoelectron and a_ℓ the core electron (cf. equation 2); thus the energy-distribution D(E) is given by the original ground state of the valence electrons $|N\rangle$, and the excited states $|N^*,s\rangle$ calculated in the presence of the core hole

$$D(E) = \sum_s |\langle N^*,s|N\rangle|^2 \delta(E-E_s^*) \qquad (10)$$

The ground state contribution to this sum, $|N^*,0\rangle$, vanishes by the Anderson theorem, and therefore the spectrum has no δ-function core-electron peak. We now turn our attention to the decay function S(t). If we assume that at a time t after the photoelectron has been ejected from the ion core, the solid has relaxed to the ground state $|N^*,0\rangle$ within a radius vt (v is of the order of the Fermi velocity), and is unchanged outside that radius, then

$$S(t) \sim (vt)^{-\delta^2/\pi^2} \qquad (11)$$

using equation (9), with $N \sim (vt)^3$.

If, for large t, equation (11) gives the limiting behaviour of S(t), then it follows rigorously, by standard theory of Fourier transforms, that

$$\lim_{E\to 0} D(E) = \left(\frac{1}{E}\right)^{1-\delta^2/\pi^2} \qquad (12)$$

Instead of a δ-function spike at the Fermi edge we find a power-law singularity. However, the coefficient of the exponent is not quite correct (Nozières and de Dominicis 1969); it should have been $1 - 2\delta^2/\pi^2$, but this small defect is not surprising in view of the highly simplified argument employed.

4. EDGE SINGULARITIES FOR X-RAY ABSORPTION AND EMISSION

The situation for x-ray absorption and emission is a little more complicated than for photoemission. The operator a_ℓ, in the transition operator T, still gives the core hole; but, for example in absorption, we now need the energy spectrum of the state

$$\sum_k p_{ck} a_k^\dagger |N\rangle \tag{13}$$

where $|N\rangle$, as before, is the ground state of the valence electrons. The operator of $|N\rangle$ in equation (13) creates a localized state, of the same spin as the core electron, and thus corresponds to a potential giving a phase shift π. In the simplified description of the edge problem so-far considered, only the difference between the potentials enters, and we then have for the singular exponent (Schotte and Schotte 1969, Hopfield 1970)

$$1 - \left(\frac{\delta-\pi}{\pi}\right)^2 - \left(\frac{\delta}{\pi}\right)^2 = 2\left(\frac{\delta}{\pi}\right) - 2\left(\frac{\delta}{\pi}\right)^2 \tag{14}$$

An almost identical argument can be made for emission spectra. While this description produces the correct singularity it must be of very limited validity since clearly emission spectra and photo-electron spectra from core electrons have little similarity except just at the edge.

5. THE MAIN BAND IN X-RAY EMISSION

If we concentrate on the main band in emission instead of

on the edge singularity, we may attempt an approximation of the initial state $|\psi_o\rangle$ as

$$|\psi_o\rangle = (\alpha + \sum_{pq} \alpha_p^q a_q^\dagger a_p)|N\rangle \tag{15}$$

This approximation cannot reproduce the edge correctly since it does not contain products of many electron-hole pairs, but it may be reasonable over most of the main band. From equation (15) it follows that $W(\omega)$ has a one-electron type contribution and a multiple-excitation (ME) term (Hedin 1967)

$$W(\omega) = \sum_k |p_{ck}^{eff}|^2 \delta(\omega - \varepsilon_k) + ME \tag{16}$$

The effective matrix element is

$$p_{ck}^{eff} = \alpha p_{ck} + \sum_q \alpha_k^q p_{cq} \tag{17}$$

admixing states q above the Fermi level. The ME terms are fairly weak over most of the band but give a non-negligible contribution to the Auger tail when compared with the effects of the broadening of the quasi-particle states. The enhancement of the intensity caused by the core hole is much smaller than in the positron-annihilation case, since we have here a pseudo-potential rather than a bare coulomb potential. However, the mixed-in states in p_{ck}^{eff} can easily give a 50% increase in the intensity (but not, say, a factor of five as for positron annihilation), and can thus appreciably distort the predictions from the one-electron picture (Hedin and Sjöström 1971). Since we have summation over q, no sharp Brillouin-zone effects can come from the states above the Fermi surface.

6. PLASMON SATELLITES

The edge singularities come from the cloud of electron-hole excitations present in the state $|\Phi\rangle$. The state $|\Phi\rangle$ also has components from plasmons, giving rise to a plasmon satellite in the spectrum. The competing effect of the potential from the core-hole with that from the jumping valence electron has a large influence on these satellites. Thus in x-ray photoemission, where there are no cancellation effects, the satellite intensity has been estimated as some 30-50% of the main peak (Lundqvist 1969); while in soft x-ray emission the corresponding figure is only 3-5% (Rooke 1963).

In absorption the plasmon effects should rather be comparable with those in x-ray photoemission; however, they are then superimposed and partly obscured by the one-electron features in the spectrum.

A simple estimate of the cancellation effects for the plasmon satellite produced in emission has been given by Ferrell (1965). He simply calculates the probability for plasmon creation when a charged spherical shell, representing the conduction-electron hole, expands with a certain velocity v. The result is then averaged for velocities ranging from zero to the Fermi velocity. The resulting probability for plasmon creation is much smaller than when only the core hole potential is considered, in rough agreement with experiment.

7. PHONON EFFECTS

Overhauser has presented a theory for the smearing effect in x-ray spectra caused by lattice vibrations, described in a review article by Parratt (1959). The picture is simply that the core electron finds itself in different potentials during the motion of the atoms; it changes its energy accordingly, which causes a smearing in the x-ray intensity. The Overhauser picture is the same as that used in the classical Franck-Condon principle. The quantum mechanical version of the Franck-Condon principle (Lax 1952) involves a shake-up problem. We then require the overlap of the initial phonon state with the various excited final-phonon states; these two sets of states are different, due to the change in interatomic potential resulting from the change in occupation of the core state. Overhauser's estimates give a large effect, which appears to be inconsistent with experiment (McAlister 1969). Recent calculations by Bergersen *et al* (1971) and by Hedin (1971) give much smaller values which still correspond to the creation of a phonon cascade (say 5 to 10 phonons), just as for optical transitions in F centres. However, there seems still to be some discrepancy between theory and experiment (Kunz, unpublished communication).

8. CONCLUDING REMARKS

We have tried to describe a set of many-body effects from a unified viewpoint, using the concept 'shake-up', and introducing a fictitious state obtained from the initial state by operating with the transition operator T. Thus, we have studied the energy spectra and decay properties of this fictitious state and related them to the x-ray problem.

Our discussion has only taken-up the fundamental idea, and we have not gone into detail. The details have only been partially explored and we have still a rather uncertain knowledge of, for example, the strengths of the edge singularities and the shapes of the plasmon structures. Progress here is crucial if we wish to extract detailed knowledge of the one-electron properties from the experimental data. The problem of incomplete relaxation is another unsolved question of importance in for example the transition metals.

Thus there is no lack of interesting problems for theoreticians as well as for experimentalists, and we may hope for a very interesting development of x-ray spectroscopy in the current decade.

REFERENCES

Andersson, P.W. (1967); Phys. Rev. Lett. 18, 1049.

Bergersen, B., McMullen, T. and Carbotte, J.P. (1971); Can. J. Phys. 49, 3155.

Ferrell, R.A. (1965); Techn. Rep. 485, Univ. of Maryland.

Hedin, L. (1967); Sol. State Comm. 5, 451.

Hedin, L. (1972); in "X-ray spectroscopy" (edited by L.V. Azaroff) McGraw Hill.

Hedin, L. and Sjöström, R. (1971); in "Electronic Density of States", p269; NBS spec. Publ. 323.

Hopfield, J.J. (1970); Comm. Sol. State Phys. 2, 40.

Lax, M. (1952); J. Chem. Phys. 20, 1752.

Lundqvist, B.I. (1969); Phys. Kondens. Materie 9, 236.

McAlister, A.J. (1969); Phys. Rev. 186, 595.

Merzbacher, E. (1970); "Quantum Mechanics", Wiley.

Nozierès, P. and de Dominicis C. (1969); Phys. Rev. 178, 1097.

Parratt, L.G. (1959); Rev. Mod. Phys. 31, 616.

Rooke, G.A. (1963); Phys. Lett. 3, 234.

Schotte, K.D. and Schotte, U. (1969); Phys. Rev. 182, 479.

INTENSITY OF THE FERMI-EDGE SINGULARITY

P. Longe*

Department of Physics, University of Maryland, USA

1. INTRODUCTION

In the last two or three years an important number of theoretical papers (see, for example, a summary by Hedin 1972) have been devoted to the problem of the singularity appearing at the Fermi edge of the soft x-ray emission and absorption bands of metals. The power expression of this singularity, which was anticipated by several authors (Mahan 1967, Bergersen and Brouers 1969, Nozières *et al* 1969), was finally calculated in a soluble model by Nozières and de Dominicis (1969). They showed that the emission and absorption intensity close to the Fermi edge (with frequency ω_0) is given by

$$I_\ell(\omega) = I_\ell^0(\omega)\ (\xi_0/|\omega-\omega_0|)^{\alpha_\ell} \tag{1}$$

where ξ_0 is the intensity of the singularity, $\omega-\omega_0$ is its frequency measured from the edge, and the power expression is

$$\alpha_\ell = 2\delta_\ell/\pi - 2\sum_{\ell'} (2\ell' + 1)(\delta_{\ell'}/\pi)^2 \tag{2}$$

The δ_ℓ are the electron phase-shifts at the Fermi level, the scatterer being the deep localized hole which is suddenly destroyed or created at the moment of the x-ray emission or ab-

*Present address: Chercheur I.I.S.N., Institut de Physique, Université de Liège, Belgium.

sorption; $I_\ell^0(\omega)$ is the one-electron transition intensity. More precisely, the intensity appears as a sum of two expressions similar to (1), with $\ell = \ell_c \pm 1$ where ℓ_c is the orbital quantum number of the core state. However, for the K emission and absorption bands, only the term $I_1(\omega)$ appears in the expression; while for $L_{2,3}$-spectra the term $I_0(\omega)$ generally dominates, this is particularly well-verified in the case of the alkali metals. The Nozières-de Dominicis model neglects secondary effects, which reduce the sharpness of the spike; these are core-hole finite life-time, temperature, recoil, self-absorption and the many-body character of the electron gas.

Some of these effects have been investigated by other authors, in particular temperature dependence (Ferrell 1969), effects of the many-body correlations (Bergersen *et al* 1971a), and lattice relaxation (Bergersen *et al* 1971b). These effects are generally weak, and following Nozières and de Dominicis we shall neglect them.

The aim of the present paper is to estimate the intensity of the Nozières-de Dominicis singularity; that is, to determine ξ_0. The quantity ξ_0, which is directly related to the potential, is known to be of the order of the Fermi energy. However, no precise value has been proposed until now. The ω-dependence of ξ_0 is also unknown; if it were known, the effect of the singularity beyond the close neighbourhood of the Fermi-edge could be estimated. Calculation of ξ_0 is important because it will show to what extent the one-body theory is satisfactory for interpretation of x-ray band spectra.

We shall be mainly concerned with the case of sodium, which because of its weak band-structure effects is the standard metal for the study of electron interactions. Moreover, the $L_{2,3}$-spectrum of sodium is interesting because it has an edge singularity that depends chiefly on δ_0 (equation 2) which is the largest phase-shift and thus the most sensitive to the structure of the hole potential.

The reason why ξ_0 has not until now been calculated is because all calculations to-date have been performed using the separable-potential method. This approximation is suitable for establishing equation (2) but must be dropped if we wish to determine ξ_0, which requires the use of a realistic potential. We have shown already (Brouers *et al* 1970) that the singularity intensity is very sensitive to the structure of the potential and particularly to its short-range structure.

First we must introduce a non-separable potential, which is done is Section 2, and then we compute the δ_ℓ. We show that, if the open-line and the closed-line portions of the Nozières-de Dominicis problem are computed up to the lowest order in that potential, then we obtain again their solution (equation 1) in the form of a first-order expansion in α_ℓ. This α_ℓ is still given by equation (2) but now the δ_ℓ are expressed in the Born approximation. At first sight, since the Born conditions are not fulfilled, one might regard the values of the δ_ℓ as unacceptable. However, we show in Section 3 that the Born expression for the δ_ℓ is still valid because it allows, paradoxically, account to be taken of the structure of the potential at the core site and incorporates the Friedel sum rule. Moreover we obtain a simple expression for α_ℓ, valid for all the metallic densities and giving numerical results in agreement with those of Ausman and Glick (1969). These authors computed the δ_ℓ and α_ℓ for lithium and for sodium, and show that for $L_{2,3}$-emission the singularity appears as a spike ($\alpha_0 > o$), and that for the other emission bands as a rounding off of the Fermi edge ($\alpha_\ell < o$). However, the method they use is approximate, since they substitute in the Nozières-de Dominicis expressions a non-separable potential which afterwards must be parameterized to satisfy the Friedel sum rule. Our calculations are more self-consistent and we do not encounter these difficulties.

In Section 4 we discuss the validity of the limited expansion of Section 2, chiefly with respect to the intensity of the singularity. The results are presented in Section 5 together with the other effects due to the electron-electron interactions.

2. PERTURBATIVE EXPANSION IN A NON-SEPARABLE POTENTIAL

The intensity of x-ray emission or absorption is given by the Golden Rule

$$I_\ell(\omega) = Re \int_0^\infty ds \; e^{\mp i\omega s} \; F_\ell(s) \tag{3}$$

with

$$F_\ell(s) = \frac{1}{2\ell+1} \sum_{m=-\ell}^{\ell} \sum_{\mathbf{k},\mathbf{k}'} \overline{W}_{\ell m}(\mathbf{k}') \; W_{\ell m}(\mathbf{k}) M_{\mathbf{k}\mathbf{k}'}(s) \tag{4}$$

The $\mp$ signs are related respectively to emission and absorption, and $W_{\ell m}(\mathbf{k})$ is the matrix element of the x-ray transition between the conduction state $\mathbf{k}$ and the core state; ℓ and m are related to the total angular momentum of the core state and the x-ray photon; an average is taken over the $2\ell + 1$ degenerate states.

Dropping the constant factor one can write

$$W_{\ell m}(\mathbf{k}) = k^\ell \; Y_{\ell m}(\hat{\mathbf{k}}) \tag{5}$$

The propagator $M_{\mathbf{k}\mathbf{k}'}(s)$ may be written (Bergersen *et al* 1971b) as

$$M^e_{\mathbf{k}\mathbf{k}'}(s) = \frac{\langle |U'(\infty,s) a_{\mathbf{k}'}^\dagger(s) \; U(s,0) \; a_{\mathbf{k}}(0) \; U'(0, -\infty)|\rangle}{|U'(\infty, -\infty)|} \tag{6a}$$

for emission, and as

$$M^a_{\mathbf{k}\mathbf{k}'}(s) = \frac{\langle |U(\infty,s) \; a_{\mathbf{k}}(s) \; U'(s,0) \; a_{\mathbf{k}'}^\dagger(0) \; U(0, -\infty)|\rangle}{|U \; (\infty, -\infty)|} \tag{6b}$$

for absorption. We use the two-Hamiltonian model which is now standard for this problem (Nozières and de Dominicis 1969, Bergersen *et al* 1971b). When the core state is vacant the evolution of the system is described by the operator U', related to an electron gas interacting with a localized charge +e. When the core state is occupied, the operator U is used and relates to the usual homogeneous electron gas.

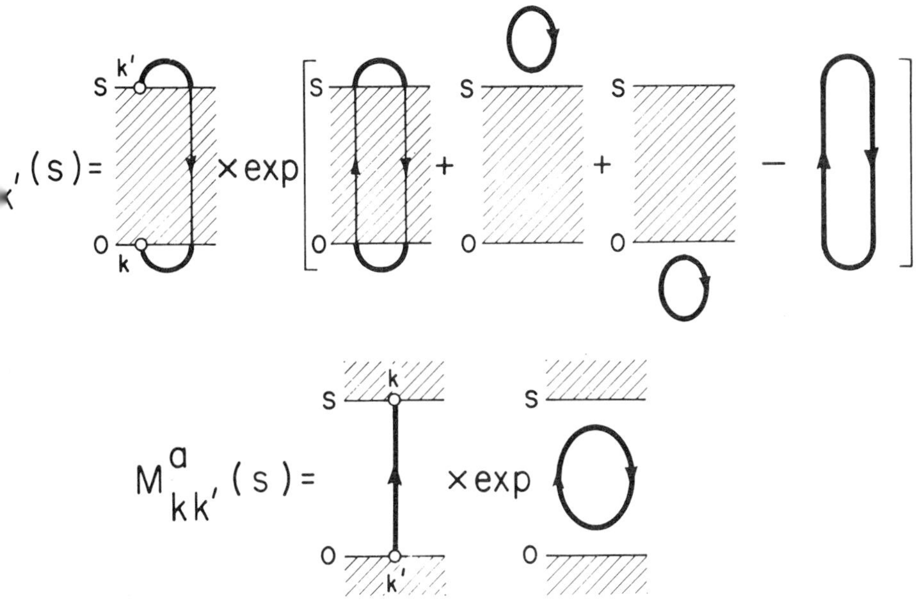

Figure 1 Diagram representation of equations (6): (a) in the emission case, (b) in the absorption case. The core state is occupied during the shaded intervals. The white dots represent the interaction with the radiation field.

The general diagrams describing $M_{kk'}(s)$ are given in figure 1. In the case of absorption, the Nozières-de Dominicis diagrams are easily recognized. The emission case is somewhat more sophisticated because we consider that the scatterer is present in the *initial* state, in contrast to the Nozières-de Dominicis model where it acts in the final state in both emission and absorption. This convention, which we have used before (Bergersen *et al* 1971b) allows us to describe the one-electron states by non-scattered Bloch waves for both cases. However, a relevant vacuum contribution, arising from the renormalizing denominator of equation (6a), must be taken into account. This contribution is represented by the last term between the brackets in figure 1a. The shaded intervals in figure 1 are those during which the scatterer does not operate. The thin and heavy lines are defined as for figure 1 of the Nozières-de Dominicis (1969) model: the thin line denotes the free-electron propagator, the heavy line the renormalized propagator when the scatterer is present.

The Nozières-de Dominicis exponent given by equation (2) contains two terms. The first, $2\delta_\ell/\pi$, comes from the open-line contribution of figure 1 and the second, $-2\sum_{\ell'}(2\ell'+1)(\delta_{\ell'}/\pi)^2$, comes from the closed loop contribution. We shall compute these contributions up to the first order in the potential expansion for the open line and up to the second order for the closed lines. In other words, both contributions are computed up to the lowest significant order in the potential expansion. The diagrammatic expressions of figure 1 are then replaced by the expansion of figure 2. For calculation of the terms of this figure see the Appendix, where we show that the result can be written in the form

$$I_\ell(\omega) = I^0_\ell(\omega)\left[1 + A(\omega) - \alpha_\ell \ ln\ (|\omega-\omega_0|/\omega_0)\right] \qquad (7)$$

where $I^0_\ell(\omega) = \frac{1}{2}\pi\omega^{\ell+\frac{1}{2}}$ is the 'one-body intensity' (the frequencies are measured from the bottom of the emission band). The last term, the only one diverging at the Fermi edge, has a coefficient α_ℓ which is precisely the Nozières-de Dominicis exponent of equation (2), but here the δ_ℓ are given by a Born-type expression as a function of the *non-separable* Coulomb screened potential

$$\delta_\ell = k_0 \int_0^\infty dr\, r^2 \left[R_\ell(r)\right]^2 v(r) \qquad (8)$$

with $k_0 \equiv \omega_0^{\frac{1}{2}}$ (the Fermi momentum).

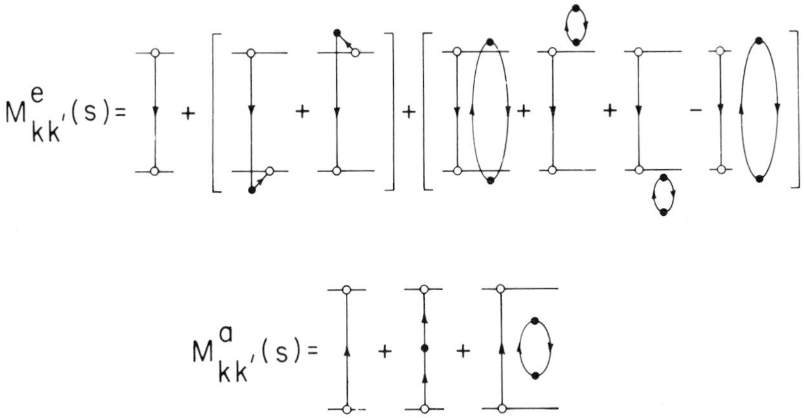

Figure 2 Perturbative expansion of the diagram expressions of figure 1: (a) for emission, (b) for absorption. The black dots represent the interaction with the core hole.

To justify the limited expansion of figure 2, leading to equation (7), we have first to show that the Born δ_ℓ are satisfactory; especially for $\ell = o$. This is done in the following section. Then we must estimate the neglected terms. In Section 4 we show that these terms are negligible over the greater part of the band. However, close to the Fermi-edge, the logarithmic form (equation 7) departs from the exact Nozières-de Dominicis expression (1); but this deviation only occurs close to the edge where the experimental results are also questionable, mainly because of the self-absorption. In spite of this, we can determine the intensity parameter ξ_0 from the non-diverging term of equation (7) by means of the relation

$$A(\omega_0) = \alpha_\ell \, ln \, (\xi_0/\omega_0) \tag{9}$$

3. THE BORN PHASE-SHIFTS

In equation (7), the conduction electrons are supposed to be described by Bloch waves. Since the Bloch structure is important only close to the scattering ion, we can neglect the core effect at the other ion sites, and therefore we replace these waves by spherical waves $R_\ell(r)\, Y_{\ell m}(\hat{\mathbf{r}})$, centred on the hole. In two previous reports (Brouers *et al* 1970, Bergersen *et al* 1971b) we used plane waves; which description is satisfactory as long as we consider the many-body correlations that depend on the dynamic part of the potential, $v_{Cb}(k)\left[\varepsilon^{-1}(k,\omega) - \varepsilon^{-1}(k,o)\right]$. Concerning the effects that depend on the static part

$$v(k) = v_{Cb}(k)/\varepsilon(k,0), \tag{10}$$

which is the Fourier transform of v(r) in equation (8), we suggested as a first approach (Brouers *et al* 1970) that v(k) be replaced by a pseudo-potential, and the plane-wave description retained; i.e. we write

$$R_\ell(r) = j_\ell \, (k_c r) \tag{11}$$

Such a method has the advantage of being simple and it emphasizes the importance of the short-range structure in fixing the intensity of the edge singularity. However, a pseudo-potential is generally not suited to handle the large values of **k** appearing in the intensity term $A(\omega)$ of equation (7) (see equations (A3) and (A4) of the Appendix). This is the reason we introduce orthogonal plane-waves here; more precisely, we orthogonalize $R_\ell(r)$ to the core states - essentially the 1s, 2s and 2p states for sodium, described by means of Slater's rules - and we use the form of $v(k)$ given by equation (10).

We must first check the δ_ℓ values given by equation (8). For this we can use the Friedel sum rule and also the values already computed by Ausman and Glick (1969) for lithium and sodium. Moreover, since the δ_ℓ with $\ell \neq o$ are not very sensitive to the short-range structure, the values computed in the plane-wave approximation (i.e. by means of equation 11) may be considered satisfactory. The Friedel sum rule can then be used to estimate δ_o

$$\delta_o = \frac{\pi}{2} - \sum_{\ell=1}^{\infty} (2\ell + 1)\delta_\ell \tag{12}$$

where the δ_ℓ given by equations (8) and (11) are substituted in the second term.

Here we digress and introduce a useful formula that gives δ_ℓ and α_ℓ. Since the δ_ℓ with $\ell \neq o$ are not very sensitive to short-range structure we could approximate equation (10) by Thomas-Fermi potential

$$v(k) = 4\pi e^2/(k^2 + q_s{}^2) \tag{13}$$

and so integrate equation (8) exactly. This potential has another advantage: if the δ_ℓ with $\ell \neq o$ are computed using equations (11) and (13), and then introduced in the second term of equation (12), then we obtain a δ_o given *also* by the same Born expression, equation (8). This result is due to a property of the Thomas-Fermi potential, equation (13): if the phase shifts of the Fermi-level electrons, scattered by that potential, are computed in the Born approximation, they verify *exactly* the Friedel sum rule. Thus one could compute *all* the δ_ℓ by means of equations (8), (11) and (13), which at first sight means that the short-range structure, neglected in equations (11) and (13), does not play a significant rôle. Such a conclusion looks surprising for δ_o; indeed the short-range structure is, in principle, dominant for $\ell = o$. In fact this structure is taken into account indirectly because the Friedel sum rule imposes, on the potential, a self-consistent short-range structure such that the charge +e of the core hole is neutralized within a finite distance. In other words the Friedel sum rule shows that, by computing δ_o in this way, the Born approximation and the Thomas-Fermi approximation, which taken by themselves are not good for taking account of the short-range structure, cancel each other. The integration of equation (8) gives the known result

$$\delta_\ell = (a_B k_o)^{-1} Q_\ell\left[1 + 2/(\pi a_B k_o)\right] \tag{14}$$

where Q_ℓ is a Legendre function of second kind. Then, using equations (8), (11), (13) and (A2), we obtain $2\sum(2\ell + 1)(\delta_\ell/r)^2 = -\frac{1}{2}(1 + \pi a_B k_o)^{-1}$ and from equations (2) and (14)

$$\alpha_\ell = \frac{1}{2}(x - 1)(x + 1)^{-1}\left[2(x + 1)\, Q_\ell(x) - 1\right] \tag{15}$$

with

$$x = 1 + 2/(\pi a_B k_o) = 1 + 0.33 r_s \tag{16}$$

The expressions (14) and (15), which are simple and depend only on r_s, are plotted in figure 3. In Table I, the numerical results for lithium and sodium are compared with those obtained by Ausman and Glick. The agreement is remarkable,

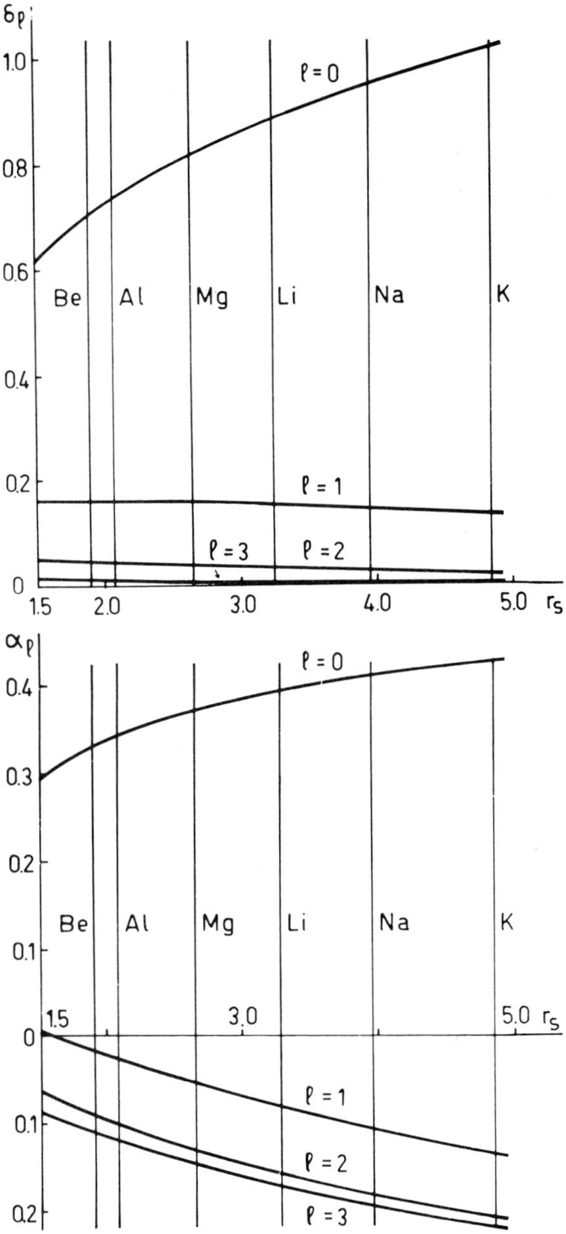

Figure 3 Plots of the phase shifts $\delta_\ell(r_s)$, given by equations (14) and (16), and of the Nozières-de Dominicis exponent $\alpha_\ell(r_s)$, given by equations (15) and (16).

in spite of the different methods and approximations they employ. The method used by Ausman and Glick is not so straightforward. Indeed it consists of substituting a non-separable potential in an expression established by Nozières and de Dominicis (1969) for a separable potential.

Table I: Phase shifts and ND exponents for Li and Na. The numbers in parenthesis are those proposed by Ausman and Glick (1969).

ℓ	δ_ℓ	α_ℓ
	Li(r_s = 3.28);	σ = -0.176 (-0.199)
0	0.891 (0.914)	0.391 (0.409)
1	0.152 (0.149)	-0.079 (-0.104)
2	0.031 (0.025)	-0.156 (-0.183)
3	0.007 (0.005)	-0.172 (-0.196)
4	0.0015 (0.0011)	-0.175 (-0.198)
	Na(r_s = 3.96);	σ = -0.198 (-0.189; -0.232)
0	0.955 (0.921; 1.04)	0.410 (0.398; 0.433)
1	0.145 (0.163; 0.133)	-0.106 (-0.085; -0.148)
2	0.026 (0.024; 0.019)	-0.181 (-0.173; -0.220)
3	0.005 (0.004; 0.003)	-0.195 (-0.186; -0.230)
4	0.0010 (0.0008; 0.0007)	-0.198 (-0.188; -0.232)

Furthermore the expression obtained by these investigators (their equation 43) is nothing other than the solution of an equation established by Kohn (1951) for a very general potential which is solved precisely by the Nozières-de Dominicis method because they use the separable-potential approximation. This introduces in the calculations by Ausman and Glick a number of approximations which are difficult to estimate and which require finally the introduction of a scaling parameter <u>g</u> to satisfy the Friedel sum rule. This parameter is taken to be of the order of 0.6 while it should normally be close to one; moreover it depends, albeit weakly, on the magnetic quantum number of the core hole, which is not to be expected on physical grounds. However, it is interesting to note that the two different methods of computing α_ℓ, the method of Ausman and Glick and the method presented here, give very similar results. Moreover the conclusion reached by Ausman and Glick, according to which α_ℓ is positive only for $\ell = o$ in the cases of lithium and sodium, can now be extended to other metals by means of equation (15). Their conclusion is thus very general. It should be noted that a spike in the K-emission bands ($\alpha_1 > 1$) would require $r_s < 1.6$, which corresponds to a density never found in normal metals.

One must consider whether the values given by equation (14), and also by Ausman and Glick, are not overestimating δ_0. Indeed the values for $\ell \neq 0$ are probably a bit too small, the Thomas-Fermi screening being too strong at small distances. Equation (12) would then indicate that $\delta_0 \sim 0.96$, as given in Table I, is probably overestimated. On the other hand, if we use a pseudo-potential, for instance the Ashcroft pseudo-potential (1968a), we obtain a very small δ_0. Ashcroft proposes for sodium a pseudo-potential having a short-range-structure radius of 0.88 Å. This gives $\delta_0 = 0.54$, which is too small. However, using a comment made by Ashcroft (1968b), his short-range-structure radius can be decreased to take account of L-shell ionization. This allows an increase of δ_0 up to $\sim$0.65 - a more acceptable value, but apparently still too small. If now, we use the exact potential (equation 10) with orthogonal plane-waves we obtain $\delta_0 = 0.70$. This value which is still possibly slightly underestimated, lies between the values of Table I and those given by the pseudo-potential;

we shall consider it as particularly suitable.

This shows that our approach to the Nozières-de Dominicis problem, by the expansion of figure 2 and the use of orthogonal plane-waves, may be considered satisfactory with regard to the value of α_ℓ.

4. THE INTENSITY TERM

We must now estimate the contribution of higher-order diagrams to the intensity term $A(\omega_0)$ of equation (7). The explicit expression of the low-order terms coming from the diagrams of figure 2 is given in the Appendix. The complete calculation of the higher terms appears difficult. Nevertheless the closed loop contribution of figure 2 can be generalized by returning to the exponential form of figure 1, keeping only two vertices per loop. In this way we recover the closed-loop contribution - written by Nozières and de Dominicis; i.e. $(\xi_0/|\omega-\omega_0|)^\sigma$ where σ is the second term of equation (2). Unfortunately such a return to the exponential form is not as simple for the open-line contribution.

Estimation of the higher order terms can be made in the following way. We notice in all the terms that the Fermi-edge divergence comes from factors in the denominator, such as $\sum k^2 - \sum k'^2$. Since the divergence appears when the last k-integration is performed, one finally achieves a denominator of the form $k^2 - k_0^2$. This denominator can be related to a particle-line binding either two hole-potential vertices, or a radiative vertex to a hole-potential vertex, according to the rules given in the Appendix. More precisely, for the first case the divergence appears by a factor

$$\int_0^\infty dk \left[pkD_\ell(p,k)\right] \frac{\eta(k-k_0)}{k^2-k_0{}^2} \left[kqD_\ell(k,q)\right]$$

replacing a factor pqD (p,q) of a lower-order diagram; and for the second case, by a factor*

$$\int_0^\infty dk\ k^{\ell+1}\ \frac{\eta(k-k_0)}{k^2-k_0^2}\left[kqD_\ell(k,q)\right]$$

replacing a factor q^ℓ in the same way.

The step function means that the divergence is due to electrons scattered above the Fermi surface from a quiescent Fermi sea. The substitution for the second case appears only when passing from the zero-order diagram to a first-order diagram, and this is made once and for all in the diagrams of figure 2. It is only the first-case substitution that appears in the successive terms of the expansion and brings about the expansion parameter.

This parameter is thus of the order of

$$\left[k_0{}^2D_\ell(k_0,k_0)\right]^{-1}\int_{k_0}^\infty dk\left[k_0kD_\ell(k_0,k)\right]^2\ (k^2-k_0{}^2)^{-1}$$

Since $k_0kD_\ell(k_0,k)$ decreases with increasing k, this expression is less than

$$k_0{}^2D_\ell(k_0,k_0)\int_{k_0}^\infty dk(k^2-k_0^2)^{-1} = \frac{1}{2}k_0D_\ell(k_0,k_0)\left[\ln 4-\ln\frac{k^2-k_0{}^2}{k_0{}^2}\right]_{k\to k_0}$$

The contribution to $A(\omega_0)$ is given by the convergent part, and we see that the expansion parameter is less than

$$\frac{1}{2}k_0D_\ell(k_0,k_0)\ln 4 = \pi^{-1}\delta_\ell\ \ln 4$$

and may be considered as small. Thus the limited expansion of figure 2 is satisfactory as far as $A(\omega_0)$ is concerned.

*In fact this expression is convergent only for $\ell = 0$ or 1. For other ℓ values, one must use a more accurate expression than (5). As shown in figure 4, this will not significantly modify the result for $\ell = 0$.

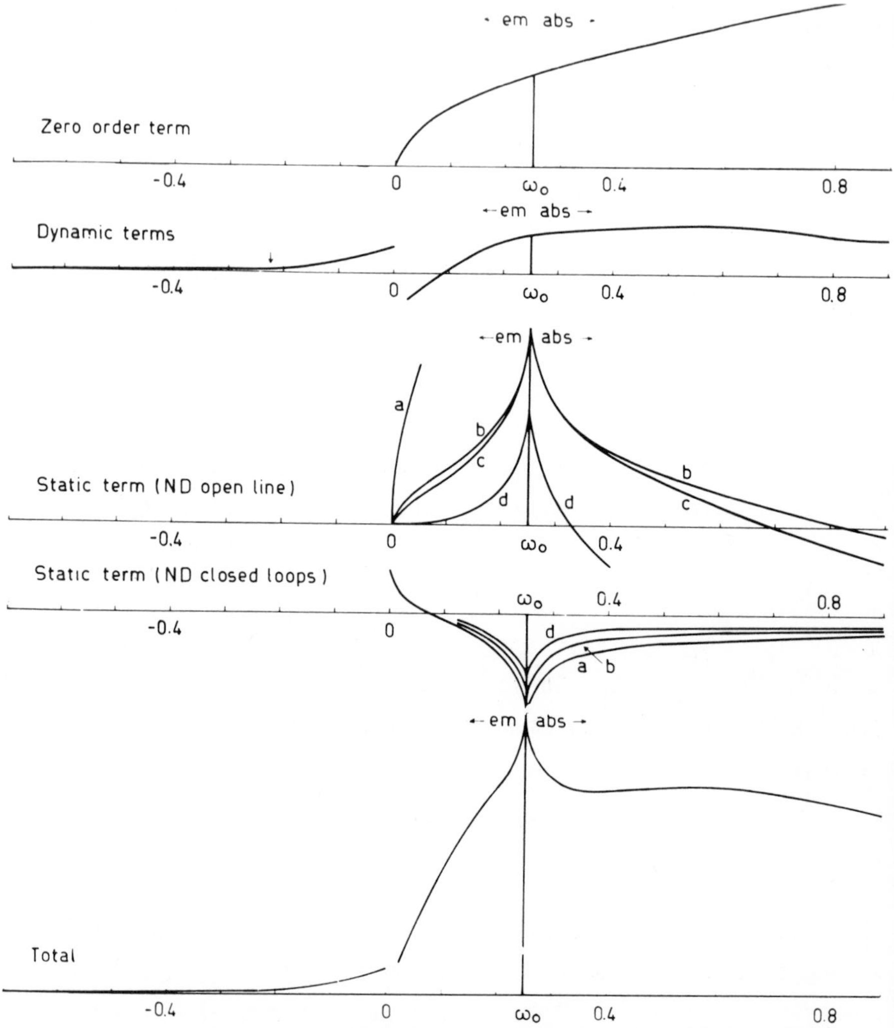

Figure 4 Plots of the various terms contributing to the $L_{2,3}$-emission and absorption bands of sodium. The zero-order term is the 'one-body' term $I_\ell^0(\omega)$. The dynamic terms are those computed by Bergersen *et al* (1971b). The static terms contributing to the singularity, are computed in four approximations : (a) the hole is a point charge with the potential given by equation (10); (b) the hole has a short-range structure and orthogonal plane-waves are used with potential given by equation (10); (c) is the same as (b), but an exact $W_{\ell m}$ is used; (d) the short-range structure is described by means of the Ashcroft pseudo-potential. The last curve gives the total result using curves (b) for the static terms.

5. CONCLUSIONS

The results are given in figure 4. The upper curve represents the zero-order term $I_0(\omega)$ which is quadratic in ω (L-spectra). The next curve represents the dynamic terms which we calculated previously (Bergersen *et al* 1971b), but from which we subtract the terms discussed here. More precisely, the dynamic terms are due to the electron-electron interactions (correlations) *and* to the dynamic part of the electron-hole interaction. These terms are weakly sensitive to the short-range structure of the hole, which can be treated as a point charge; they describe the low-energy features such as tailing and plasmon satellite formation. The third picture of figure 4 shows the Nozières-de Dominicis open-line contribution. The curve (a) is the unacceptable one where the hole is treated as a point charge; i.e. where equations (10) and (11) have been used. In curve (d), the Ashcroft potential is used instead of equation (11), (Brouers *et al* 1970). The middle curves are those discussed here, using orthogonal plane waves. Curve (c) presents a refinement of (b) : we use the exact expression for $W_{\ell m}$ and not equation (5), (see footnote p.355). It is useful to know this effect which is in fact negligible. The fourth picture is analogous to the third but is related to the closed-loop contribution. The last curve is the total result, where all the preceding curves are summed (for the static terms only the curves (b) are considered). This result is very satisfactory if we compare it to the experimental curves observed for example by Rooke (1963) and Haensel *et al* (1969). Finally, the path from $A(\omega_0)$ to ξ_0 is only of formal interest. This we shall discuss in a report now in preparation, where details of the calculations will also be given.

ACKNOWLEDGMENTS

This work was supported in part by the Institut Interuniversitaire des Sciences Nucléaires of Belgium, the National Science Foundation, and the Centre for Theoretical Physics of the University of Maryland. The author is indebted also to the Department of Physics of the University of Maryland for support and hospitality, and to Professor A.J. Glick and Dr. G.A. Ausman for helpful discussions.

APPENDIX

Here we give some details of the calculations of equations (3), (4) and (7). We give also the explicit expression of $A(\omega_0)$ in (7). The complete calculations will be reported in a paper now in preparation.

The following rules are applied to the diagrams of figure 2, to obtain equation (4):

(1) A factor $\left[\eta(k-k_0)\eta(t-t')-\eta(k_0-k)\eta(t'-t)\right]\exp\left[-ik^2(t-t')\right]$ is associated with each particle line ($\eta(x)$ is the usual step function; the mass unit is 2m).

(2) A factor $ikk'D_\ell(k,k')$ is associated with each hole-potential vertex (black dot), with

$$D_\ell(k,k') = \frac{2}{\pi}\int_0^\infty dr\ r^2 j_\ell(kr)j_\ell(k'r)V(r) \tag{A1}$$

Note that, $\delta_\ell = \frac{1}{2}\pi k_0 D_\ell(k_0,k_0)$

(3) A factor $k^{\ell+1}$ is associated with each radiative vertex (white dot).

(4) A general factor ± 1 appears in the emission (-) absorption (+) case (open line) and a factor $-2(2\ell+1)$ for each closed loop.

(5) Integration is performed over $k,k',\ldots$ from 0 to ∞ and over $t,t',\ldots$ in the shaded intervals of figure 1. Summation is performed over ℓ in the closed loops. These rules are connected to the plane-wave rules used in other reports, by the relation

$$\sum_{\ell=0}^{\infty}(2\ell+1)j_\ell(pr)j_\ell(qr)j_\ell(pr')j_\ell(qr') = \frac{1}{2}\int_{-1}^{+1}du\,\frac{\sin kr}{kr}\cdot\frac{\sin kr'}{kr'} \tag{A2}$$

with $\mathbf{k} = \mathbf{p}-\mathbf{q}$ and $u = \mathbf{p}.\mathbf{q}/(pq)$

The intensity $I_\ell(\omega)$ given by equation (3) can then be calculated:

$$I^{e/a}_{\ell,0}(\omega) = \frac{\pi}{2}\omega^{\ell+\frac{1}{2}}\eta(\mp\omega\pm\omega_0)$$

$$I^{e/a}_{\ell,1}(\omega) = 2\int_0^\infty dk'\eta(\pm k_0\mp k')\int_{k_0}^\infty dk$$

$$\times\left[\omega^{\frac{1}{2}(\ell+1)}\right]\;\left[\pi\delta(\omega-k'^2)\right]\;\left[-\omega^{\frac{1}{2}}kD_\ell(\omega^{\frac{1}{2}},k)\right]\;\left[\frac{1}{\omega-k^2}\right]\;\left[k^{\ell+1}\right]$$

$$I^{e/a}_{\ell,2}(\omega) = 2\sum_{\ell'=0}^{\infty}(2\ell'+1)\int_{k_0}^\infty dp\int_0^{k_0}dq\left\{I^{e/a}_{\ell,0}\left[\omega\pm(p^2-q^2)\right]-I^{e/a}_{\ell,0}(\omega)\right\}$$

$$\times\left\{\left[\frac{1}{p^2-q^2}\right]\;\left[-pq\;D_\ell(p,q)\right]\right\}^2$$

The $\mp$ signs and the indices a/b are related respectively to emission and absorption. Indices 0,1 or 2 are related respectively to diagrams or sets of diagrams with 0,1 or 2 hole-potential vertices. The term $I_{\ell,0}$ thus represents the one-body term; and $I_{\ell,1}$ the term we discussed previously (Brouers *et al* 1970).

At the Fermi edge, $(\omega=k_0^2)$, these expressions can be written in the form of equation (7) with

$$A^{e/a}_1(k_0{}^2) = (2/k_0^\ell)\int_{k_0}^\infty dk\,\frac{k^{\ell+2}D_\ell(k_0,k) - k_0^{\ell+2}D_\ell(k_0,k_0)}{k^2 - k_0^2}$$

$$+\frac{2}{\pi}\,\delta_\ell\,ln\,4 \tag{A3}$$

$$A^{e/a}_2(k_0{}^2) = -2\sum_{\ell=0}^{\infty}(2\ell+1)\left\{\int_{k_0}^\infty dpp\int_0^{k_0}dqq\right.$$

$$\left.\times\,\frac{pq\left[D_\ell(p,q)\right]^2 - k_0{}^2\left[D_\ell(k_0,k_0)\right]^2}{(p^2-q^2)^2} + \frac{1}{\pi^2}\,\delta_\ell{}^2\right\} \tag{A4}$$

REFERENCES

Ashcroft, N.W. (1968a); J. Phys. C. (Proc.Phys.Soc.), 1, 232.

Ashcroft, N.W. (1968b); in "Soft X-Ray Band Spectra", ed. D.J. Fabian, p.259; Academic Press, London and New York.

Ausman, G.A. and Glick, A.J. (1969); Phys. Rev. 183, 687.

Bergersen, B. and Brouers, F. (1969); J. Phys. C. (Solid St. Phys.), 2, 651.

Bergersen, B., McMullen, T. and Carbotte, J.P. (1971a); Can. J. Phys. 49, 3155.

Bergersen, B., Brouers, F. and Longe, P. (1971b); J. Phys. F. (Metal Phys.), 1, 945.

Brouers, F., Longe, P. and Bergersen, B. (1970); Solid St. Comm. 8, 1423.

Ferrell, R.A. (1969); Phys. Rev. 186, 399.

Haensel, R., Keitel, G., Schreiber, P., Sonntag, B. and Kunz, C. (1969); Phys. Rev. Letters 23, 528.

Hedin, L. (1972); in "X-Ray Spectroscopy", ed. L.V. Azároff, ch.V; Academic Press, London and New York (to be published).

Kohn, W. (1951); Phys. Rev. 84, 495.

Mahan, G.D. (1967); Phys. Rev. 163, 612.

Nozières, P., Roulet, B. and Gavoret, J. (1969); Phys. Rev. 178, 1072 and 1084.

Nozières, P. and de Dominicis, C.T. (1969); Phys. Rev. 178, 1097.

Rooke, G.A. (1963); Phys. Letters 3, 234.

p-RESONANT STATES AND SOFT X-RAY EMISSION

F.K. Allotey*

UNESCO International Centre for Theoretical Physics, Trieste.

1. INTRODUCTION

In K-emission the parent spectrum arises from an initial state of single ionisation in which the atom has an electronic vacancy in the K shell. Satellite K-emission on the other hand occurs when the inner shell of the atom is in a state of double ionisation; one in which the atom has an electron vacancy in both the K and L shells. However, the K-emission satellite bands of lithium and beryllium occur when both of the two 1s-electrons in the K shell have been removed from the atom.

Unlike most physical properties of a metal, which are determined by the energy of the electron system as a whole (e.g. cohesive energy) or of the electrons at the Fermi surface (cyclotron resonance, electrical conductivity), x-ray emission is capable of giving direct information about the individual energy levels of the conduction or higher bands. Thus x-ray band spectra have been used to define the Fermi energy, E_F, and the effective mass, m*. Emission spectra are particularly valuable in deciding whether the excitation-band picture should be applied to metals as well as to insulators.

---

*Permanent address: University of Science and Technology, Kumasi, Ghana.

Because of these properties, the emission spectra of metals have been well investigated; in particular, the parent and satellite bands of lithium have been measured by several workers. On examination it is found that apart from the overall broadening of the satellite, the shapes of the two bands are very similar (Allotey 1971); see figure 1. At low energies, the intensity follows the $E^{3/2}$-law and there is a premature hump at about 0.5eV below the emission threshold. This similarity (Catterall and Trotter 1958) indicates that the loss of the second 1s-electron from the core does not further change the band.

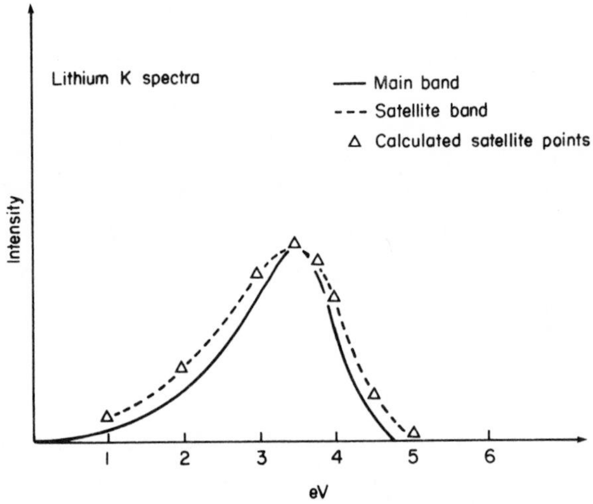

Figure 1 K-emission parent and satellite bands of lithium (Catterall and Trotter 1958).

The parent K-emission band of beryllium has also been widely reported (Aita and Sagawa 1969; Watson, Dimond and Fabian 1968); see figure 2. Following Jones, Mott and Skinner (1934) the observed spectrum has generally been interpreted in terms of band theory only (Altmann and Bradley 1965, and Terrel 1966). Beryllium is a divalent metal and crystallizes in a

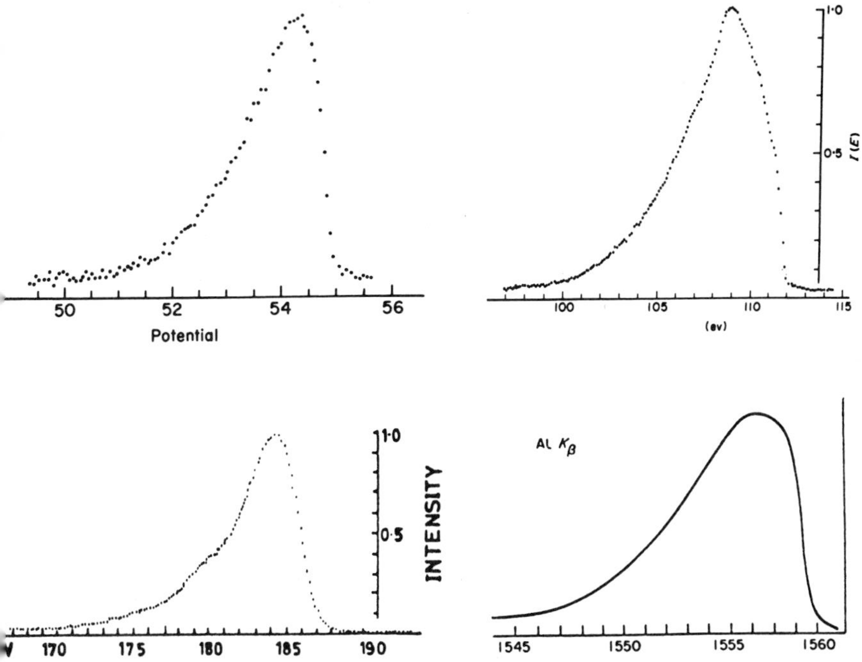

Figure 2 K-emission spectra of lithium, beryllium and crystalline boron (Aita and Sagawa 1968, 1969) and of aluminium (Laüger 1969).

close-packed hexagonal form. The two 2s-electrons lie in the two Brillouin zones, and for interpreting the K-emission it is assumed that the transition probability for the lowest state in the second Brillouin zone is small when the final state has s-symmetry. Thus, from band theory, the x-ray emission spectrum approximately follows the density-of-states curve for beryllium. Calculations are in progress to derive the K-emission spectrum of beryllium using band-structure wavefunctions. However, it should be noted that care is necessary when interpreting soft x-ray emission spectra in terms of density-of-states curves only; recent calculations by Borland and Cooper (1971) on lithium, and by McMullen (1970) on alkali metals, using band-structure wavefunctions have established that band-structure effects are of no importance in the emission spectra of these metals. Measurements of the K-spectrum and L-spectrum of aluminium (see for example Rooke 1968) show that the K-emission band is markedly different from the density-of-states curve while the L band follows the density-of-states curve.

2. CONSTRUCTION OF THE POTENTIAL

To perform a realistic calculation of x-ray emission of a metal, we need to know the potential of its atom with a hole in the inner shell. This potential may be constructed using the Hartree-Fock self-consistent-field method. However, the labour involved in obtaining such a field is considerable and in our calculation we use approximate potentials which produce the observed energy levels of Li^{2+} for the main band of lithium, of Li^{3+} for the satellite band of lithium, and of Be^{3+} for the main band of beryllium.

Since the observed x-ray spectra of lithium and beryllium are continuous, the bound states in the potential that give rise to discrete levels have to be removed from the potential well. This is achieved with the aid of Levinson's theorem, which asserts that for a local potential V(r) satisfying the following conditions

$$\int_0^\infty dr\, r|V(r)| < \infty \tag{1}$$

and

$$\int_0^\infty dr\, r^2|V(r)| < \infty \tag{2}$$

the number of bound states n_ℓ of a fixed angular momentum ℓ is given by

$$\delta_\ell(0) - \delta_\ell(\infty) = \pi n_\ell \tag{3}$$

where $\delta_\ell(0)$ and $\delta_\ell(\infty)$ are the phase shifts at respectively zero and infinite energies. Using this theorem, the strength of the potential is reduced; that is, bound states are removed until, at zero energy, we obtain

$$\delta_\ell(0) = 0 \; . \tag{4}$$

3. THE MODEL

The metal was divided into Wigner-Seitz spheres of radii $r_s = 1.70$ Å and $r_s = 1.25$ Å, for respectively lithium and beryllium. The normal atom in the centre of the sphere was removed and replaced by an atom with the electronic vacancy. Here, we shall refer to an atom with an inner core hole as an impurity atom. With the introduction of the impurity atom inside the centre of the Wigner-Seitz sphere, electrons distribute themselves around it such that the total charge inside the sphere becomes zero. The Hamiltonian, for the electrons within each of the Wigner-Seitz spheres, may be written as

$$H = H_0 + H_h + H_c + H_{ss} + H_{sp} + H_{pp} \tag{5}$$

where H_0 is the kinetic energy of the free-electron system; H_h is the potential energy of the impurity atom; H_c is the lowest energy of the conduction band plus electron-electron interaction outside the cell. In the present calculations H_c replaces the structure of the medium outside the central cell in which the impurity atom is located. It is an optical model potential. If $E_{cohesion}$ is the cohesive energy of the metal, E_i the negative of the ionisation potential and E_m the mean energy of the Fermi gas, H_c is given approximately by

$$E_{cohesion} = (E_m + H_c - E_i) \ . \tag{6}$$

$H_{ss} + H_{sp} + H_{pp}$ represents the electron-electron interaction part of the Hamiltonian. H_{pp} gives the interaction between the p electrons, H_{ss} that between the s-electrons, while H_{sp} gives the interaction between the s and p electrons. In writing equation (5), the d and f electrons have been ignored, since it can be shown that their contribution to the total interaction is negligible. Another argument in favour of this omission is that the K-emission from lithium and beryllium is due only to the p states in the conduction band.

We shall assume that the potential possesses a radial symmetry. Thus the Shrödinger equation for an ℓ electron is

$$\left[\frac{\hbar^2}{2m}\left(-\frac{1}{r^2}\frac{\partial}{\partial r}r^2\frac{\partial}{\partial r}+\frac{\ell(\ell+1)}{r^2}\right)+U_\ell+V(r)\right]R_{k\ell}=E_{k\ell}\,R_{k\ell} \ . \tag{7}$$

$\ell = 0$ for the s electrons, and $\ell = 1$ for the p electrons. $V(r)$ is the potential of the impurity atom which has been obtained semi-empirically. $U_\ell(r_s)$ is the optical potential and it is H_c plus the electron-electron interaction for the ℓ-electrons. If we define what we called the density function f_ℓ by

$$f_\ell = \sum_{\ell=0}^{k_F} \int_0^{r_s} |R_{k\ell}|^2 \, d^3r \ , \tag{8}$$

where $k_F = k_{Fermi}$, we find that for the i-th p-electron

$$U_{pi}(r_s) = H_c(r_s) + J_{pp}\sum_{j=1}^{6} f_{pj} + J_{sp}\sum_{j=1}^{2} f_{sj} - J_{pp}f_{pi} \tag{9}$$

and for the i-th s-electron

$$U_{si}(r_s) = H_c(r_s) + J_{sp} \sum_{j=1}^{6} f_{pj} + J_{ss} \sum_{j=1}^{2} f_{sj} - J_{ss} f_{si} \ . \quad (10)$$

The Coulomb interactions between s-electrons, between s and p electrons, and between the p-electrons are respectively denoted by J_{ss}, J_{sp}, and J_{pp}.

As discussed earlier, the part played by the electrons with angular momentum $\ell \geqslant 2$ is negligible and has been omitted in writing equations (9) and (10). The J's appearing in the above equations are J-effective. As shown previously (Allotey 1967), equations (9) and (10) may be derived from variation principles.

4. X-RAY EMISSION

K emission results when a p-electron in the conduction makes a radiative transition to an empty s state in the core. Optical selection rules indicate that the initial wavefunction should be a p state, while the final wavefunction should be an s state. Ignoring for the moment the many-body effects near the Fermi level, the intensity I is proportional to

$$N(E) |\langle \psi_{core} | \mathbf{P}.\mathbf{A} \ \psi_{k\ell} \rangle|^2 \quad (11)$$

where **P** is the momentum operator and **A** is the vector potential for the x-ray. N(E) is the density of states and is given by

$$N(E) = \frac{\Omega}{8\pi^3} \int \frac{dS_k}{\nabla_k E} \quad (12)$$

Ω is the volume of the crystal and dS_k is a surface element in k-space. ψ_{core} is the solution of equation (7) when V(r) is present, while $\psi_{k\ell}$ is the solution when both V(r) and $U_\ell(r_s)$ are present.

The observed intensity, as given by equation (8), may be affected in two ways: either there is a structure in the density of states N(E), such as Van Hove singularities in the density-of-states curve (points where $\nabla_k E = 0$), or in the matrix elements, such as resonance scattering in the conduction band. The calculations reported here show that there is a p-scattering resonance in the conduction band of lithium and beryllium. In this case, there is a k-dependence in the matrix elements. Van Hove singularities, if they exist, may then be smeared out by various effects due to this k-dependence. Hence the calculations neglect effects due to these singularities. Effects due to instrumental broadening have also been omitted.

Following Nozières and de Dominicis, we now calculate the emission spectrum of lithium and beryllium taking the many-body behaviour of electrons near the threshold into account. Nozières and de Dominicis showed that x-ray emission intensity I near the threshold should be proportional to

$$\sum_{\ell m} |W_{\ell m}|^2 \left(\frac{\xi_0}{E_F - E} \right)^{\alpha_\ell} \tag{13}$$

where

$$\alpha_\ell = \frac{2}{\pi}\delta_\ell - 2 \sum_\ell (2\ell+1) \left(\frac{\delta_\ell}{\pi} \right)^2 \tag{14}$$

E_F is the Fermi energy and $W_{\ell m}$ are the coefficients of the expansion of the x-ray matrix element in spherical harmonics due to transition from the conduction band to the core states. δ_ℓ is the ℓ scattering phase shift ξ_0 is the cut-off energy. In our calculations ξ_0 was taken to be constant and of the order of E_F. This is partly justified by the calculations of Bergersen, Brouers and Longe, who, using Ashcroft's pseudo-potential, obtained $\xi_0 \sim E_F$.

The p-scattered wavefunctions from Li^{2+}, Li^{3+} and Be^{3+} were computed numerically from equation (7) and then used to calculate the emission spectra. U_p (ℓ=1) was taken as a parameter. Calculations were performed for various effective masses from

$m^*/m = 1$ to $m^*/m = 1.8$. The calculated emission spectra, figures 3-5, were not found to be very sensitive to U_p or to the effective mass.

Using our realistic potentials and calculating phase shifts for the s, p and d states, we obtained from equation (14) $\alpha_1 = -1.276$ for the satellite emission of lithium, $\alpha_1 = 1.015$ for the main band of Li, and $\alpha_1 = 1.421$ for beryllium. The negative values are in accordance with the prediction of Nozières and de Dominicis and have the effect of making the calculated intensity go to zero near the threshold, as

$$\left(\frac{\xi_0}{E_F - E}\right)^{\alpha_1} .$$

The d and f phase shifts were found to be very small, while the s phase shift was negative.

The many-body effects at the lower-energy side of the spectra, such as Auger tailing and plasmon satellites, have been ignored in our calculations since we are interested in the main and the threshold part of the spectra.

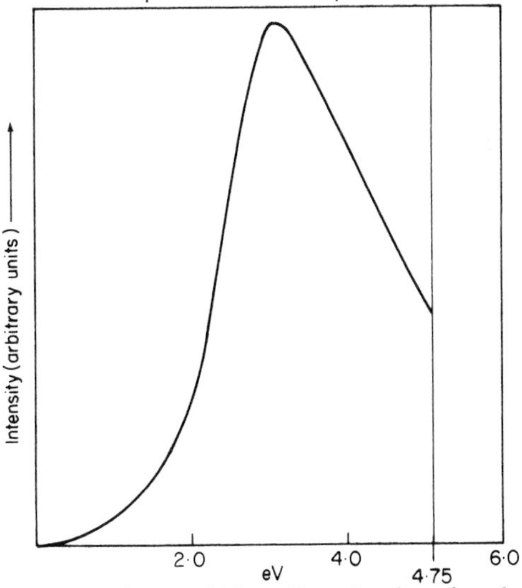

Figure 3 Calculated satellite K-emission band of lithium computed without the threshold correction. Beyond the emission threshold the intensity goes to zero as $[\xi_0/(4.75-E)]^{-1.276}$.

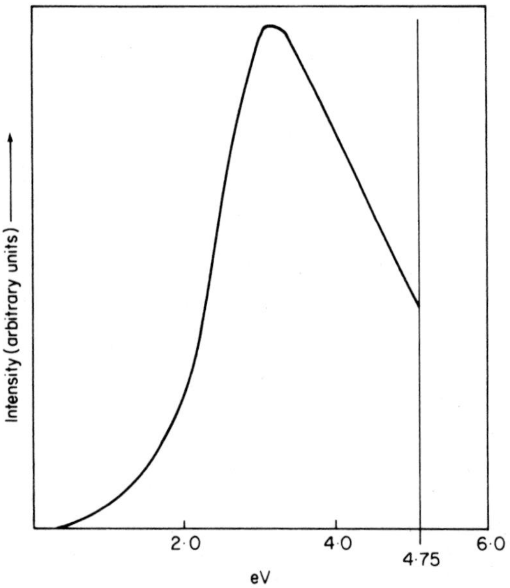

Figure 4 Calculated parent K-emission band of lithium computed without the threshold correction. Beyond the emission threshold the intensity goes to zero as $[\xi_0/(4.75-E)]^{-1.015}$.

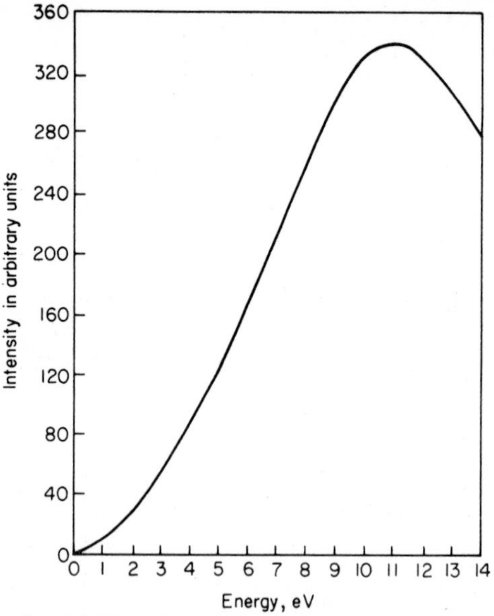

Figure 5 Calculated K-emission spectrum of beryllium computed without the threshold correction. Beyond the emission threshold the intensity goes to zero as $[\xi_0/(14.5-E)]^{-1.421}$.

5. SUMMARY AND CONCLUSIONS

It should be noted that in an accurate treatment of the 'hole' the Friedel sum-rule must be satisfied. Our present treatment does not do this; however, the calculations indicate that if a p-resonance exists in the conduction band of lithium, and if the many-body effects near to the emission threshold introduced by Nozières and de Dominicis are incorporated, the shapes of the main and satellite bands are very similar. The calculation by March (this Volume p297), also shows the similarity. Previous calculations by Goodings (1965) using a rather simplified model gave a similar result. Thus in the double ionisation measurements by Catterall and Trotter (1958) and by Tomboulian (1957), we should not expect to see the effect of the second ionisation. We also show here that as for lithium, part of the peak in the K-emission from beryllium near the threshold could be due to the existence of a p-resonant state below the Fermi level.

The resonance discussed in this paper is caused by a p-electron scattering from the hole in the core. From a pseudo-potential analysis of Harrison (1968) also concluded that such resonance is conceivable. Our previous calculations (Allotey and Hopfield 1966, and Allotey 1967) showed that no resonant states exist in the conduction band of sodium. This may explain why no hump is observed in the L-emission spectrum of this metal.

On observing soft x-ray emission from many substances, such as lithium, beryllium, boron, aluminium and sodium, one usually finds that in those elements which have both L and K bands, the L emission spectra apart from the Nozières-de Dominicis singularities very near the threshold follow the density-of-states curves while the K bands do not. In fact, the K-emission spectra of lithium, beryllium, boron and aluminium (figure 2) all have similar humps. We believe that for the K bands it is the p-resonant states that are important in determining the shapes of the observed spectra, except very near the threshold when Nozières-de Dominicis singularities predominate. If it were only the transient effect of the hole that is important, as discussed by Ausmann and Glick (1969), then we should expect the K spectra also to portray the density-of-states curves except very near the emission edge, but as we have shown previously, this is not the case.

Acknowledgments

The author wishes to thank Professors Abdus Salam and P. Budini, and the International Atomic Energy Agency and UNESCO for hospitality at the International Centre for Theoretical Physics, Trieste. He is also grateful to the Swedish International Development Authority for making available an Associateship at the Centre. Numerical calculations - were performed at the Computer Centre, University of Science and Technology, Kumasi.

REFERENCES

Aita, O. and Sagawa, T. (1968); in "Soft X-Ray Band Spectra and Electronic Structure", ed. Fabian, D.J., p.29; Academic Press, London.

Aita, O. and Sagawa, T. (1969); J. Phys. Soc. Japan 27, 164.

Allotey, F.K. and Hopfield, J.J. (1966); Bull. Am. Phys. Soc. 11, 331.

Allotey, F.K. (1967); Phys. Rev. 157, 467.

Allotey, F.K. (1971); Solid State Commun. 9, 91.

Altmann, S.L. and Bradley, C.J. (1965); Proc. Phys. Soc. (London) 86, 915.

Ausman, G. and Glick, A.J. (1969); Phys. Rev. 183, 687.

Bergersen, B., Brouers, F. and Longe, P. (1972); Solid State Commun., to be published.

Borland, R.E. and Cooper, J.R.A. (1971); J.Phys.F.Met.Phys. 1, 237.

Catterall, J.A. and Trotter, J.A. (1958); Phil. Mag. (8) 3, 1424.

Goodings, D. (1965); Phys. Soc. (London) 86, 75.

Harrison, W. (1968); in "Soft X-Ray Band Spectra and Electronic Structure", ed. Fabian, D.J., p.227; Academic Press, London.

Jones, H., Mott, N.E. and Skinner, H.W.B. (1934); Phys. Rev. 45, 379.

Laüger, K. (1969); Dissertation, University München.

McMullen, T. (1970); J. Phys. C. 3, 2178.

Nozières, P. and de Dominicis, C.T. (1969); Phys. Rev. 178, 1097.

Stott, M.J. (1969); J. Phys. C. 2, 1474.

Terrel, J.H. (1966); Phys. Rev. 149, 526.

Tomboulian, D.H. (1957); Hand Buch der Physik 30, 246.

Watson, L.M., Dimond, R.K. and Fabian, D.J. (1968); in "Soft X-Ray Band Spectra and Electronic Structure", ed. Fabian, D.J., p.45; Academic Press, London.

Wiech, G. (1968); in "Soft X-Ray Band Spectra and Electronic Structure", ed. Fabian, D.J., p.59; Academic Press, London.

THE EFFECT OF ELECTRON-INTERACTION ON SOFT X-RAY AND PHOTOELECTRON SPECTRA OF METALS

V.P. Sachenco, R.V. Vedrinski and J. Richter

Department of Solid State Physics, University of Rostov, Rostov-on-Don, USSR.

1. INTRODUCTION

The singularity in soft x-ray spectra near the Fermi level has been studied theoretically by many authors (Mahan 1967, Langreth 1970, Schotte and Schotte 1969, Friedel 1969). The most complete study was by Nozières and de Dominicis (1969). Their approach was essentially a model one, and they did not take into account the strong conduction-electron interaction, which could modify the x-ray spectra. A special model was used for the interaction-potential of the deep hole with the conduction-electrons. For this reason some authors (Friedel 1969, Rivier and Simanek 1971) question the exactness of the results.

We treat the problem taking fully into account electron-electron interactions, using the realistic Coulomb-potential for the electron-hole interaction. The main approximation of this calculation is the neglect of the periodic lattice potential. Nevertheless, our results relate to the results of Nozières and de Dominicis, and confirms their model.

2. FORMALISM

We use the notation:

H_0 - the conduction-electron kinetic energy (the energy of excitations on the Fermi-surface is chosen as zero)

H_1 - the Hamiltonian of the electron-hole interaction

H_2 - the Hamiltonian of the conduction-electron interactions.

The calculations are performed for x-ray absorption-spectra; the case emission may be treated in a similar way.

Neglecting unimportant constants the intensity of soft x-ray absorption is given by the well-known expression

$$I(\omega) = \omega\sum_f |\langle \Psi_f|H_{er}|\Psi_i\rangle|^2 \, \delta(E_f+\varepsilon_o-E_i-\omega), \tag{1}$$

where ω is the frequency of the photon, ε_o is the binding energy of the deep electron and H_{er} describes the coupling to the photon-field. Ψ_i and Ψ_f are the initial-state and final-state wavefunctions of the conduction electrons and E_i and E_f are the corresponding energies.

One can easily obtain the intensity $I(\omega)$ knowing the function

$$M(x,x';t) = \langle \Psi_i|e^{iH_i t}\,\psi(x)e^{-iH_f t}\,\psi^\dagger(x')|\Psi_i\rangle \tag{2}$$

Here $\psi^\dagger$ and ψ are the creation and annihilation field-operators of a conduction electron,

$$H_i = H_0 + H_2\ ,\ H_f = H_i + H_1$$

By using the standard techniques of many-body perturbation theory and the adiabatic theorem, equation (2) may be replaced (for t>0) by

$$M(x,x';t) = \lim_{\substack{\tau\to t-0\\ \tau'\to +0}} G_1(x,x';\tau,\tau';t)\, G_2(t)\ , \tag{3}$$

$$G_1(x,x';\tau,\tau';t) = -i\frac{\langle 0|T\{\psi(x,\tau)\psi^\dagger(x';\tau')S_1(t,0)S_i(+\infty,-\infty)\}|0\rangle}{\langle 0|T\{S_1(t,0)S_i(+\infty,-\infty)\}|0\rangle}\ , \tag{4}$$

$$G_2(t) = i\,\frac{\langle 0|T\{S_1(t,0)S_i(+\infty,-\infty)\}|0\rangle}{\langle 0|S_i(+\infty,-\infty)|0\rangle}\quad, \tag{5}$$

Where $|0$ is the ground-state vector of the Hamiltonian H_0. T is the usual time-ordering operator and the S-operators are given by

$$S_1(t,0) = e^{iH_0t}\,e^{-i(H_0+H_1)t}\quad,$$

$$S_i(t,t') = e^{iH_0t}\,e^{-i(H_0+H_2)(t-t')}\,e^{-iH_0t'}$$

We shall call the function G_1 the 'one-electron modified Green function'. In the perturbation series for G_1, linked diagrams appear only. We note that the modified one-electron Green function is a function of three arguments τ, τ' and t; thus, performing a Fourier transformation in (4) is not useful.

The perturbation series of the function $G_2(t)$ contains closed diagrams only. Using the linked-cluster theorem one obtains

$$G_2(t) = exp[G_2^c(t)] = exp(\;0|T\{[S_1(t,0)-1]S_i[+\infty,-\infty]\}|0\;_c) \tag{6}$$

Here $G_2^c(t)$ is the sum of all closed loops with at least one interaction line with the deep-hole potential.

Defining the new Hamiltonians $\mathcal{H}_1 = H_1b^\dagger b$ and $H = H_i = H_f = H_o + \mathcal{H}_1 + H_2$, where $b^\dagger$ and b are the creation and annihilation operators for the deep hole, one can easily prove, that $G_2(t)$ is the Green function of deep hole (for t>0):

$$G_2(t) = i\langle\Psi_i|T\{b(t)b^\dagger(0)\}|\Psi_i\rangle = i\langle\Psi_i|e^{iHt}b\,e^{-iHt}b^\dagger|\Psi_i\rangle =$$

$$i\langle\Psi_i|e^{iH_it}\,e^{-iH_ft}|\Psi_i\rangle = i\,\frac{\langle 0|T\{S_1(t,0)S_i(+\infty,-\infty)\}|0\rangle}{\langle 0|S_i(+\infty,-\infty)|0\rangle} \tag{7}$$

The function $G_2(t)$ may be used for determination of the function $w(\varepsilon)$ (Sachenco and Vedrinski 1969). $W(\varepsilon)$ is the probability that the conduction electrons receive the energy ε

above the final ground-state energy by a sudden application of the deep-hole potential

$$w(\varepsilon) = \sum_f |\langle\Psi_f|\Psi_i\rangle|^2 \delta(E_F - E_F^o - \varepsilon) = \frac{1}{\pi} Re\left(-i\int_0^\infty dt\ e^{i(\Delta E+\varepsilon)t} G_2(t)\right), \tag{8}$$

where ΔE is the ground-state energy-shift of the conduction electrons. This function $w(\varepsilon)$ is important in studies of β-spectra of metals.

Now we begin to examine the singularity of the x-ray intensity $I(\omega)$ near the Fermi surface. In order to solve this problem we have to treat the asymptotic behaviour of the matrix-element $M(\mathbf{x},\mathbf{x}',t)$ when $t\to\infty$. Our first task is to analyse the behaviour of the function G_1 alone, when $t\to\infty$ and $|\tau-\tau'|$ is large.

$$G_1(\mathbf{x},\mathbf{x}',\tau,\tau',t) = \sum_\ell (2\ell+1)4\pi\ P_\ell(\cos\theta_{xx'})G_{1\ell}(x,x',\tau,\tau',t)\ ,$$

$$G_{1\ell}(x,x',\tau,\tau',t) = \frac{1}{(2\pi)^3}\int_0^\infty k'dk'\int_0^\infty kdk\ j_\ell(kx)j_\ell(k'x')G_{1\ell}(k,k',\tau,\tau',t$$

$$G_{1\ell}(k,k',\tau,\tau',t) = \overset{o}{G}(k,\tau-\tau')\delta(k-k') +$$

$$\int_0^t dt_1 dt_2\ \overset{o}{G}(k,\tau-t_1)\Sigma_\ell(k,k',t_1,t_2)G(k',t_2-\tau')+\ldots$$

$$= \overset{o}{G} + \overset{o}{G}\Sigma_\ell\overset{o}{G} + \overset{o}{G}\Sigma_\ell\overset{o}{G}\Sigma_\ell\overset{o}{G} + \ldots \tag{9}$$

Here $\overset{o}{G}(k,t)$ is the one-electron Green function and $\Sigma_\ell(k,k\ ,t,t\ )$ represents the mass-operator generated by the Hamiltonian H_1.

Integrating $\overset{o}{G}(k,t_1-t_2)$ over k we obtain a function of time, which decreases as $(t_1-t_2)^{-1}$ when $|t_1-t_2|\to\infty$. The appearance of these terms leads to logarithmic singularities in the time-integrals of equation (9) for large t. To exclude such divergent terms we must separate the slowly decreasing part from $\overset{o}{G}(k,t)$

$$\overset{o}{G}(k,t) = G'(k,t) + G''(k,t)\ , \tag{10}$$

where

$$G'(k,t) = P\,\frac{1}{t}\cdot A\cdot\,\delta(k-k_F)$$

with P denoting the principal part. The constant A we find by integrating equation (10) over a small region of k including the Fermi surface.

$$A = -\,Z(k_F)\left(\frac{d\varepsilon(k)}{dk}\right)^{-1}_{k=k_F}$$

Here Z(k) is the renormalisation constant, and $\varepsilon(k)$ represents the one-particle excitation energy.

Setting equation (10) in (9), we reconstruct the series for $G_{1\ell}$,

$$G_{1\ell} = \overset{o}{G} + \overset{o}{G}\Sigma_\ell\overset{o}{G} + \ldots \tag{11}$$

$$= G'' + \tilde{G}_\ell + G''K_\ell G'' + \tilde{G}_\ell K_\ell G'' + G''K_\ell\tilde{G}_\ell + G''K_\ell\tilde{G}_\ell K_\ell G'',$$

where $K_\ell = \Sigma_\ell + \Sigma_\ell G''K$ and $\tilde{G}_\ell = G' + G'K_\ell\tilde{G}_\ell$. Noting that the most slowly decreasing part in equation (11) is the function $\tilde{G}_1$, we see that the asymptotic behaviour of $G_{1\ell}$ is determined by the asymptotic behaviour of $\tilde{G}_\ell$. Investigating this asymptotic behaviour in the case of large $|\tau-\tau'|$ we may consider the K-operator to be local in time

$$K_\ell(K_F,K_F,t_1,t_2) = K^o_\ell\;\delta(t_1-t_2)\Delta(t_1,t), \tag{12}$$

where $K^o_\ell = -\,\dfrac{\tan\delta_\ell}{\pi A}$

and
$$\Delta(t_1,t) = \begin{Bmatrix}1 \text{ if } t_1\in(0,t)\\ 0 \text{ if } t_1\notin(0,t)\end{Bmatrix}$$

δ_ℓ is the phase shift for scattering of the interacting conduction-electrons at the Fermi surface by the deep-hole potential. Introducing the new function Ψ_ℓ by $\tilde{G}_\ell = A\Psi_\ell\cdot\delta(k-k_F)$ we obtain the integral-equation

$$\Psi_\ell(\tau,\tau',t) = P\,\frac{1}{\tau-\tau'} + \frac{tan\delta_\ell}{\pi}\,P\int_0^t \frac{\Psi_\ell(\tau'',\tau',t)}{\tau-\tau''}\,d\tau'' \quad (13)$$

which corresponds to the equation first obtained by Nozières and de Dominicis in the model case.

The solution of equation (13) is not well defined. Requiring that $\Psi_\ell \to 0$ when $|\tau|\to\infty$ or $|\tau'|\to\infty$, and that the function Ψ_ℓ is continuous for $tan\delta_\ell \to 0$, we find the unique solution

$$\Psi_\ell(\tau,\tau',t) = cos^2\delta_\ell\left(P\frac{1}{\tau-\tau'} - \pi\, tan\delta_\ell \cdot \delta(\tau-\tau')\right)\left(\frac{\tau(t-\tau')}{\tau'(t-\tau)}\right)^{\frac{\delta_\ell}{\pi}} \quad (14)$$

where the phase shifts satisfy the condition $|\delta_\ell|<\pi/2$. Owing to this requirement the solution (14) is a periodic function of the phase shifts with the period π.

The singularities in the function $\Psi_\ell(\tau,\tau',t)$, which appear by letting $\tau\to t$ and $\tau'\to 0$, are spurious. Introducing a cut-off in (14) we obtain for $G_{1\ell}$ the final result

$$G_{1\ell}(t) \simeq \frac{t^{2\delta_\ell/\pi}}{t} \quad (15)$$

This corresponds to the result of Nozières and de Dominicis; we have obtained it for the case of interacting conduction electrons and real deep-hole potential.

Now we wish to study the asymptotic behaviour of the deep-hole Green function $G_2(t)$. Firstly, let us consider the case of noninteracting electrons. Analysing the asymptotic behaviour of $G_2^c(t)$ in this simple case as well as in the general case of interacting conduction-electrons, we are interested in the second term of the asymptotic expansion only, which is proportional to $ln(t)$, (the main term clearly is equal to $-i\Delta Et$).

The perturbation series for $G_2^c(t)$, $(S_i=1)$, is given by

$$G_2^c(t) = -2\sum_{\ell=0}^{\infty}(2\ell+1)\{Sp(V_\ell g) + \frac{1}{2}Sp(V_\ell g V_\ell g) + \dots\} =$$

$$= -2\sum_{\ell}(2\ell+1)\sum_{h=1}^{\infty} Sp(V_\ell g)^h \frac{1}{h} \tag{16}$$

where $V_\ell(k,k')$ is the interaction-potential of the deep hole with conduction electrons in the k,ℓ-representation and g is the free-electron Green function.

$$Sp(AB) = \int_o^\infty dk_1 dk_2 \int_o^t dt_1 dt_2 A(k_1,k_2,t_1,t_2)B(k_2,k_1,t_2,t_1).$$

The factor 2 in equation (16) appears because of spin summation. Separating the slowly decreasing part g' from g in analogy to equation (10) ($g = g' + g''$) and reconstructing (16) we obtain

$$G_2^c(t) = C(t) - 2\sum_{\ell}(2\ell+1)\{Sp(k_\ell g') + \frac{1}{2}Sp(k_\ell g' k_\ell g') + \dots\}$$

$$= C(t) - 2\sum_{\ell}(2\ell+1)\sum_{h=1}^{\infty} Sp(k_\ell g')^h \frac{1}{h} \tag{17}$$

where C(t) does not contain g' and is thus not interesting here, and $k_\ell = V_\ell + V_\ell\, g''\, k_\ell$

To sum up the series $\sum_h \frac{1}{h} Sp(kg')^h$ we multiply every operator k by a factor λ and apply the operator $\lambda\frac{\partial}{\partial\lambda}$ to the series. Defining $k_\lambda = \lambda . k$ we obtain

$$\lambda\frac{\partial}{\partial\lambda} G_{2\lambda}^c(t) = \sum_{h=1}^{\infty} Sp\,(k_\lambda g')^h =$$

$$= Sp(k_\lambda \tilde{g}_\lambda) = A\int_o^t d\tau\, d\tau'\; k_\lambda(\;_F,\;_F,\tau',\tau)\Psi_\lambda(\tau,\tau',t) \tag{18}$$

Here $\tilde{g}_\lambda = g' + g' k_\lambda \tilde{g}_\lambda$ and the quantities A, Ψ are defined by analogy with expressions (10) and (13) for the non-interacting conduction electrons.

We consider now the function $\Psi_\lambda(\tau,\tau')$ in the region of small $|\tau-\tau'|$. At first sight we cannot use equation (12) in this case. However, taking into account that we are interested in only the second term of the asymptotic expansion defined in equation (13) by integration over a large interval of t, we assume that the approximation (12) is possible. According to this we expand equation (14) in powers of $(\tau-\tau')$ and separate the second term

$$\Psi_{\lambda}^{(2)} = \frac{\delta_{\lambda}}{\pi^{2}} \cos^{2}\delta_{\lambda} \left(\frac{1}{\tau} + \frac{1}{t-\tau}\right) , \tag{20}$$

where $\delta_{\lambda}=\arctan(\lambda\tan\delta)$ and δ is the phase shift for free conduction electrons. Putting (20) in (18), taking into account (12) and cutting off the spurious divergent terms, we find

$$\lambda\frac{\partial}{\partial\lambda}(G_{2\lambda}^{c})' = 2\,\frac{\delta_{\lambda}\tan\delta_{\lambda}}{\pi^{2}} \cos^{2}\delta_{\lambda}\, \ln\left(\frac{t}{\Delta}\right) \tag{21}$$

$(G_{2\lambda}^{c})'$ is the second term in the asymptotic expansion of $G_{2\lambda}^{c}$ for $t\to\infty$.

Integrating (21) over λ from 0 to 1 we obtain

$$(G_{2\ell}^{c})' = \frac{\delta_{\ell}^{2}}{\pi^{2}} \ln\left(\frac{t}{\Delta}\right)$$

$$G_{2}(t) \simeq i\, e^{-i\Delta E\cdot t} \left(\frac{\Delta}{t}\right)^{2\sum_{\ell}(2\ell+1)\frac{\delta_{\ell}^{2}}{\pi^{2}}} \tag{22}$$

(Here $|\delta_{\ell}|<\pi/2$, as before).

Considering the general case of interacting conduction electrons, we notice that every diagram in the expansion of $G_{2}^{c}(t)$ may contain, in this case, several linked electron-loops in contrast to (16). In order to find the second term in the asymptotic expansion of G_{2}^{c} we have to separate the second-order contribution of a single electron-loop in the linked diagram. In this manner we mark one electron-loop and sum up all other loops, that gives $\overset{o}{G}$ instead of g, and Σ_{ℓ} instead of V_{ℓ}. The following treatment does not change and so the result (22) is valid for the case of interacting electrons also. Naturally, δ_ℓ is now the phase shift in the system of interacting conduction electrons on the Fermi surface.

3. CONCLUSION

The results obtained above provide the possibility of finding the behaviour for $I(\omega)$ near the Fermi threshold:

$$I(\omega) \simeq (\omega-\omega_F)^{2\sum\limits_\ell (2\ell+\ )\frac{\delta_\ell^2}{\pi} - \frac{2\delta_\ell}{\pi}} \tag{23}$$

Thus we see, that the result of Nozières and de Dominicis is valid in our general case of interacting conduction electrons.

REFERENCES

Friedel, J. (1969); Comm. Sol. State Phys. <u>11</u>, 21.

Langreth, D.C. (1970); Phys. Rev. <u>B1</u>, 417.

Mahan, G.D. (1967); Phys. Rev. <u>163</u>, 612.

Nozières, P. and de Dominicis, C. (1969); Phys. Rev. <u>178</u>, 1097.

Rivier, N. and Simanek, E. (1971); Phys. Rev. Letts. <u>26</u>/8, 435.

Sachenco, V.P. and Vedrinski, R.V. (1969); "X-ray spectra and electronic structure of matter", Vol.II, p.153; Institute of Metal Physics, Academy of Sciences of the Ukr. SSR, Kiev (Russian).

Schotte, K.D. and Schotte, U. (1969); Phys. Rev. <u>185</u>, 509.

A ONE-ELECTRON THEORY OF SOFT X-RAY EMISSION FROM RANDOM ALLOYS

B.L. Gyorffy

HH Wills Physics Laboratory, University of Bristol, Bristol.

and M.J. Stott

Department of Physics, Queens University, Kingston, Ontario.

1. INTRODUCTION

If a band structure calculation is pursued to the point where not only the energy eigenvalues but also the corresponding eigenfunctions are calculated, it is a relatively simple matter to determine the soft x-ray spectrum of the crystal in the one-electron approximation. Recently, a number of such calculations have been performed: by Rooke (1968) for aluminium, by Goodings and Harris (1969) for copper, by McMullen (1970) for the alkali metals, and by Klima (1970) for silicon and germanium. Although they do not take account of electron-electron interaction, it is clear from these investigations that the predictions of the one-electron model can be usefully compared with experiment. However, no such detailed one-electron theory exists for soft x-ray emission from random alloys. Our purpose here is to examine the question of calculating emission intensities from such disordered systems in the light of our present understanding of their electronic structure.

Following our earlier study (Gyorffy and Stott 1971) we

formulate the problem in the language of the Green-function approach to band theory, and derive explicit expressions for the emission intensities. We then discuss a number of approximate ways of evaluating these expressions.

Our aim is twofold. Firstly, we wish to provide a theoretical framework for realistic calculations of the emission spectra from random alloys. This we hope will contribute to a quantitative understanding of much available and reliable experimental data, which to-date has been interpreted in qualitative terms only. Secondly, we wish to bring some of the current ideas in the theory of alloys, such as the coherent potential approximation (Soven 1969), into closer contact with experiment. Until now, almost all theoretical work on the electronic structure of random alloys has been concerned with methods for calculating the averaged density of states. However, this quantity is not directly measurable in any experiment. Although photoelectric emission spectra are often interpreted in terms of the density of states, a quantitative comparison of theory with experiment is greatly hampered by our lack of knowledge of the appropriate matrix elements. By contrast, we argue here that the effects of matrix elements can, in the case of soft x-ray spectra, be handled exactly provided a theory is constructed for the partially averaged local density of states instead of the total average density of states. We shall also show how approximation schemes, for calculation of the latter, such as the virtual-crystal, average t-matrix, and the coherent potential approximations, may be adapted to evaluate local densities of states. Thus we provide a new way of testing the merits and limitations of these approximations by making possible a quantitative comparison of their predictions and experiment.

2. GENERAL FORMULATION

Consider a random binary alloy. In the one-electron approximation the conduction electrons occupy the eigenstates $|\psi_n\rangle$ of the alloy Hamiltonian

$$H = -\nabla^2 + \sum_i v_i(\mathbf{r}-\mathbf{R}_i) \qquad (1)$$

where the $\mathbf{R}_i$ are the positions of the lattice sites, and the potential function v_i associated with an atom at $\mathbf{R}_i$ is $v^A(\mathbf{r}-\mathbf{R}_i)$ if the site is occupied by an A atom but $v^B(\mathbf{r}-\mathbf{R}_i)$ if there is a B atom at $\mathbf{R}_i$. A convenient way of parametrizing $v_i(\mathbf{r}-\mathbf{R}_i)$ is to write

$$v_i(\mathbf{r}-\mathbf{R}_i) = \xi_i\, v^A(\mathbf{r}-\mathbf{R}_i) + (1-\xi_i)v^B(\mathbf{r}-\mathbf{R}_i) \qquad (2)$$

where $\xi_i=1$ if there is an A atom at $\mathbf{R}_i$, but $\xi_i=0$ if the site is occupied by a B atom. Clearly, a particular alloy configuration is defined by a set of numbers $\xi_1, \xi_2, \ldots..\xi_N$, and the corresponding eigenfunctions $\psi_n(r,\xi_1,\xi_2,\xi_3,\ldots\xi_N)$ depend on the particular values of these numbers.

If one removes an electron from the core state $\phi_A^c(\mathbf{r}-\mathbf{R}_n)$ of an A atom located at the site $\mathbf{R}_n$, then any of the conduction electrons may make a radiative transition to fill the hole by emitting an x-ray photon. Because the eigenstates occupied by the conduction electrons are different for each configuration the emitted intensity varies from configuration to configuration. However, in the experiment one does not know the particular arrangement of atoms and only the concentration can be specified, therefore the relevant quantity to calculate is the intensity averaged over an appropriate ensemble of configurations. As we showed previously (Gyorffy and Stott 1971), by using the Fermi Golden Rule and the electric dipole approximation to calculate the rate at which photons are emit-

ted, the emission intensity from an A atom at $\mathbf{R}_n$ may be written as

$$I^A(\omega) \propto \int d^3r \int d^3r' \; \phi_A^c(\mathbf{r}-\mathbf{R}_n)\phi_A^{c*}(\mathbf{r}'-\mathbf{R}_n)\nabla\cdot\nabla' \langle\langle r|\mathit{Im}G^+(\hbar\omega-\varepsilon_c)|r'\rangle\rangle_{\mathbf{R}_n A} \tag{3}$$

where $\langle\,\rangle_{\mathbf{R}_n A}$ denotes the average with respect to the ensemble comprising all configurations that have an A atom at $\mathbf{R}_n$, and

$$\langle \mathbf{r}|G^+(\varepsilon)|\mathbf{r}'\rangle = \sum_n \frac{\psi_n^*(\mathbf{r})\psi_n(\mathbf{r}')}{\varepsilon-\varepsilon_n+i\eta} \tag{4}$$

is the Green function for an electron moving according to the alloy Hamiltonian given in equation (1). Experiments should, of course, be compared with the average of the above expression over all possible ways of choosing the site $\mathbf{R}_n$ at which the radiating A atom is located. However, since for a homogeneous alloy such averaging merely contributes a factor C, where C is the concentration, we shall develop the theory for a particular site $\mathbf{R}_n$ only. We refer to $\langle\,\rangle_{\mathbf{R}_n,A}$ as a partial average in order to distinguish it from the usual average which would be with respect to all possible configurations. In alloy theories one is usually concerned with the fully averaged Green function $\langle G^+(\varepsilon)\rangle$ while here the partially averaged Green function $\langle G^+(\varepsilon)\rangle_{\mathbf{R}_n,A}$ is the function of interest.

As in band theory, we shall now assume that the potential functions $v^A(\mathbf{r}-\mathbf{R}_i)$ and $v^B(\mathbf{r}-\mathbf{R}_i)$ are of the muffin tin kind, each with radius "a" such that they do not overlap. We shall have in mind potential functions constructed according to prescriptions customary in band structure calculations and assume that they are independent of the concentration. This may turn out to be a rather strong assumption in some cases, but the theory of electronic states in random alloys is not yet sufficiently well developed to warrant refinement along these lines.

There are two important points to note about equation (2). The first is that, since $\phi_A^c(\mathbf{r}-\mathbf{R}_n)$ is a very well localized function about the nucleus at $\mathbf{R}_n$, we only need calculate $\langle \mathbf{r}|ImG^+(\varepsilon)|\mathbf{r}'\rangle$ for $|\mathbf{r}-\mathbf{R}_n|\leqslant a$ and $|\mathbf{r}'-\mathbf{R}_n|\leqslant a$. The second point is that the dipole operator ∇ will select out certain well-defined angular momentum components of $\langle \mathbf{r}|ImG^+(\varepsilon)|\mathbf{r}'\rangle$ depending on whether $\phi_A^c(\mathbf{r})$ is an s, p or a d state. Consequently, one should decompose $\langle \mathbf{r}|ImG^+|\mathbf{r}'\rangle$ into angular momentum components about $\mathbf{R}_n$. Both of these features are best kept in sight in the multiple scattering method of solving the Schrodinger equation for a large number of non-overlapping muffin tin potentials; which is the method we use.

We now wish to calculate $\langle \mathbf{r}|ImG^+(\varepsilon)|\mathbf{r}'\rangle$ for a particular configuration $\xi_1, \xi_2, \ldots . \xi_N$. The method we adopt is first to calculate the total scattering operator defined by the operator equation

$$T = V + V\, G_o^+\, T \tag{5}$$

where $V = \sum_i v_i$, and then to use T to find G^+ from

$$G^+ = G_o^+ + G_o^+\, T\, G_o^+ \tag{6}$$

A particularly simple form of the multiple scattering equations follows if we decompose the total scattering operator into contributions from various scattering paths by writing

$$T = \sum_{\ell\ell'} \mathcal{T}^{\ell\ell'} \tag{7}$$

where $\mathcal{T}^{\ell\ell'}$ is the *scattering path operator* which, if operates on an incoming wave to the ℓ'th site, gives the out-going wave from the ℓ th site. By substituting equation (7) into equation (5) we obtain

$$\sum_{\ell\ell'} \mathcal{T}^{\ell\ell'} = \sum_{\ell\ell'} [v_\ell \delta_{\ell\ell'} + \sum_{\ell''} v_\ell\, G_o^+\, \mathcal{T}^{\ell''\ell'}] \tag{8}$$

Hence the equations

$$\mathcal{T}^{\ell\ell'} = v_\ell \ \delta_{\ell\ell'} + \sum_{\ell''} v_\ell \ G_0^+ \ \mathcal{T}^{\ell''\ell'} \tag{9}$$

completely define the scattering path operator $\mathcal{T}^{\ell\ell'}$. Note now that by taking the $\ell''=\ell$ contribution, on the right hand side of equation (9), over to the other side and by dividing both sides by $(1-v_\ell G_0^+)$, this equation may be rewritten as

$$\mathcal{T}^{\ell\ell'} = t^\ell \ \delta_{\ell\ell'} + \sum_{\ell''\neq\ell} t^\ell \ G_0^+ \ \mathcal{T}^{\ell''\ell'} \tag{10}$$

where the scattering operator t^ℓ, which describes the scattering from the ℓ th site when it is alone in free space, is defined by

$$t^\ell = v_\ell + v_\ell \ G_0^+ \ t^\ell \tag{11}$$

Equation (10) is an operator version of the fundamental equations of multiple scattering theory (Lax 1951, Velický *et al* 1968) and we shall take it as the starting point for our calculation of T through equation (7).

For a muffin-tin potential $v_\ell(\mathbf{r}-\mathbf{R}_\ell) = 0$ outside the muffin-tin well, centred on $\mathbf{R}_\ell$. Then from equation (11) it follows that $t^\ell(\mathbf{r}-\mathbf{R}_\ell, \mathbf{r}'-\mathbf{R}_\ell) = 0$ for $|\mathbf{r}-\mathbf{R}_\ell|>a$ or $|\mathbf{r}'-\mathbf{R}_\ell|>a$. Consequently we have from equation (10)

$$\mathcal{T}^{\ell\ell'} (\mathbf{r}-\mathbf{R}_\ell, \mathbf{r}'-\mathbf{R}_{\ell'}) = 0 \text{ for } |\mathbf{r}-\mathbf{R}_\ell|>a \text{ or } |\mathbf{r}'-\mathbf{R}_{\ell'}|>a \tag{12}$$

for non-overlapping muffin tins. This is a major simplification, for it allows us to deal with 'on the energy shell' scattering only. Namely, by analogy with the 'on the energy shell' components of the single scatterer t-matrix (Goldberger and Watson 1964)

$$t_L(\varepsilon) = \int dr_1^3 \int dr_2^3 Y_L(\hat{r}_1) j_L(\sqrt{\varepsilon} r_1) t(\mathbf{r}_1,\mathbf{r}_2;\varepsilon) j_L(\sqrt{\varepsilon} r_2) Y_L(\hat{r}_2) = -(\varepsilon)^{-\frac{1}{2}} \sin \delta_L(\varepsilon) \exp[i\delta_L(\varepsilon)] \tag{13}$$

Where L stands for both the aximuthal and polar quantum numbers, Y_L are the corresponding spherical harmonics, j_L are the spherical Bessel functions, and δ_L are the usual scattering phaseshifts. We may define the 'on the energy shell' components of the scattering path operator $T^{\ell\ell'}$ by writing

$$T^{\ell\ell'}_{LL'}(\varepsilon) = \int dr_1^3 \int dr_2^3 Y_L(\hat{r}_1) j_L(\sqrt{\varepsilon} r_1) T^{\ell\ell'}(\mathbf{r}_1,\mathbf{r}_2;\varepsilon) j_L(\sqrt{\varepsilon} r_2) Y_L(\hat{r}_2) \tag{14}$$

where $\mathbf{r}_1$ is measured from R_ℓ and $\mathbf{r}'$ from $R_{\ell'}$ and therefore L,L' refer to angular momentum about R_ℓ and $R_{\ell'}$ respectively. It is then a relatively straightforward matter using equation (12) to derive from equation (10) an equation for $T^{\ell\ell'}_{LL'}(\varepsilon)$

$$T^{\ell\ell'}_{LL}(\varepsilon) = t^\ell_L(\varepsilon)\delta_{\ell\ell'}\delta_{LL'} + \sum_{\ell''\neq\ell}\sum_{L''} t^\ell_L(\varepsilon) G_{LL''}(R_\ell - R_{\ell''}) T^{\ell''\ell'}_{L''L'}(\varepsilon) \tag{15}$$

where $G_{LL'}(R_\ell - R_{\ell''})$ is the structural Greenian given by

$$G_{LL'}(R_\ell - R_{\ell''}) = \frac{2}{\pi}\int dk^3 \frac{e^{ik(R_\ell - R_{\ell''})}}{\varepsilon - k^2 + i\eta} Y_L(\hat{k}) Y_{L'}(\hat{k}) i^{L-L'} \tag{16}$$

and for our binary alloy system with the alloy potential given by equations (1) and (2)

$$t^\ell_L(\varepsilon) = -\frac{1}{\sqrt{\varepsilon}}\left\{\xi_\ell \sin\delta_\ell(\varepsilon) \exp[i\delta^A_L(\varepsilon)] + (1-\xi_\ell)\sin\delta^B_L(\varepsilon)\exp[i\delta^B_L(\varepsilon)]\right\} \tag{17}$$

Equation (15) is then our fundamental equation of multiple scattering 'on the energy shell'. It gives $T^{\ell\ell'}_{LL'}(\varepsilon)$ in terms of the phase shifts, the lattice and the configuration specified by $\xi_1, \xi_2, \ldots. \xi_N$.

It can be shown that from $T^{\ell\ell'}_{LL'}(\varepsilon)$ one may calculate the total integrated density of states. In fact evaluating the usual definition $N(\varepsilon) = \frac{1}{\pi} Im\int_0^{\varepsilon_F} d\varepsilon\, tr\, G^+(\varepsilon)$, using equation (6) and equation (7), one is led to the formula first derived by Lloyd (1967). However, here we are not only interested in finding

$\langle N(\varepsilon)\rangle$ but wish to calculate $\langle\langle r|\mathit{Im}G^{+}(\varepsilon)|r'\rangle\rangle_{R_n,A}$.

One of our main points is that this can also be done in terms of $T^{\ell\ell'}_{LL'}(\varepsilon)$ provided $\mathbf{r}$ and $\mathbf{r}'$ are both inside one of the muffin-tin wells. To follow this, note that - at least formally - one may rewrite equation (5) as

$$G_0^{+}(\varepsilon) = V^{-1} - T^{-1} \tag{18}$$

and hence, from equation (6), we have

$$G^{+} = \frac{1}{V}\,T\,\frac{1}{V} - \frac{1}{V} \tag{19}$$

Therefore for a real potential

$$\mathit{Im}\ G^{+} = \frac{1}{V}\,\mathit{Im}\ T\,\frac{1}{V} \tag{20}$$

In general it might be very difficult to make sense of equation (20). However, if $\mathbf{r}$ and $\mathbf{r}'$ are within the muffin-tin well centred on $\mathbf{R}_\ell$ and $\mathbf{R}_{\ell'}$ respectively, it is reasonable to interpret equation (20) to mean

$$\langle \mathbf{r}|\mathit{Im}G^{+}(\varepsilon)|\mathbf{r}'\rangle = \frac{1}{v_\ell(\mathbf{r}-\mathbf{R}_\ell)}\mathit{Im}T^{\ell\ell'}(\mathbf{r}-\mathbf{R}_\ell;\mathbf{r}'-\mathbf{R}_{\ell'})\frac{1}{v_{\ell'}(\mathbf{r}-\mathbf{R}_{\ell'})} \tag{21}$$

since in that case $V(\mathbf{r}) = \sum_\ell v_\ell(\mathbf{r}-\mathbf{R}_\ell) = v_\ell$ and $T(\mathbf{r},\mathbf{r}';\varepsilon) = \sum_{\ell\ell'} T^{\ell\ell'}(\mathbf{r}-\mathbf{R}_\ell,\mathbf{r}'-\mathbf{R}_{\ell'}) = T^{\ell\ell'}(\mathbf{r}-\mathbf{R}_\ell,\mathbf{r}'-\mathbf{R}_{\ell'})$. We note in passing that equation (21) is the multiple scattering generalization of a very similar formula for a single scatterer in a paper by Beeby (1967). Its importance to the problem in hand lies in the fact that it may be written in the form

$$\langle \mathbf{r}|\mathit{Im}G^{+}|\mathbf{r}'\rangle = \sum_{LL'} \Delta_L(\mathbf{r})\Delta_L(\mathbf{r}')\ \mathit{Im}\ T^{\ell\ell'}_{LL'}(\varepsilon) \tag{22}$$

where the origin is taken to be at $\mathbf{R}_\ell$ for $\mathbf{r}$ and at $\mathbf{R}_{\ell'}$ for $\mathbf{r}'$, and

$$\Delta_L(\mathbf{r}) = -\frac{\sqrt{\varepsilon}}{sin\delta_L(\varepsilon)} R_L(r_i\varepsilon)Y_L(\hat{r}) \qquad (23)$$

This follows by using

$$T^{\ell\ell} = t^\ell + \sum_{\substack{m\pm\ell \\ m'\pm\ell}} t^\ell G_o^+ T^{m,m'} G_o^+ t^\ell \qquad (24)$$

which arises from equation (10) and from the relation (see, for example, Roman 1965)

$$\langle \mathbf{r}|t(\varepsilon_k)|k\rangle = \sum_L v(r)\left(\frac{2}{\pi}\right)^{\frac{1}{2}} i^L e^{i\delta_L(\varepsilon_k)} R_L(r;\varepsilon_k)Y_L(\hat{r})Y_L(\hat{k}) \qquad (25)$$

where $R_L(r;\varepsilon)$ is the radial solution of the Schrödinger equation for a single scatterer, with the boundary condition $R_L(r;\varepsilon)=cos\delta_L j_L(\sqrt{\varepsilon}r)-sin\delta_L n_L(\sqrt{\varepsilon}r)$, for $|\mathbf{r}|>a$.

Equation (22) is exact, provided $|\mathbf{r}|<a$ and $|\mathbf{r}'|<a$, and is a consequence of the assumption that the alloy potential is of the non-overlapping muffin tin form. Thus, the part of the partially averaged Green function necessary to evaluate the soft x-ray emission intensity from equation (3) may be written

$$\langle\langle \mathbf{r}|ImG^+(\varepsilon)|\mathbf{r}'\rangle\rangle_{\mathbf{R}_n,A} = \sum_{LL'} \Delta_L^A(\mathbf{r})\Delta_{L'}^A(\mathbf{r}')\ Im\ \langle T_{LL'}^{n,n}(\varepsilon)\rangle_{\mathbf{R}_n,A} \qquad (26)$$

where $T_{LL'}^{nn}(\varepsilon)$ is the solution of equation (15), and therefore depends only on the phaseshifts δ_L^A and δ_L^B and on the ensemble of configurations. This is one of the main results of the present paper.

Since $\Delta_L^A(\varepsilon)$ is independent of the environment of the atom at $\mathbf{R}_n$ it may be calculated exactly by a straightforward numerical integration of the Schrödinger equation within the muffin tin well about $\mathbf{R}_n$. The factor $Im\langle T_{LL'}^{nn}\rangle_{\mathbf{R}_n,A}$ takes care that the wavefunction is matched to the wavefunction for the whole system for each arrangement of the atoms included in the ensemble.

Because the potentials on the various sites are non-overlapping, they are manifest through their phase shifts and position only. For an ordered system, namely when the potentials on all sites are the same, the solution of equation (15) for $\mathcal{T}_{LL'}(q)=\frac{1}{N}\sum_{\ell\ell'} e^{iq(R_\ell-R_{\ell'})}\mathcal{T}_{LL'}^{\ell\ell'}(\varepsilon)$ is the inverse of the KKR matrix whose determinant vanishes at the eigenvalues.

For an alloy configuration equation (15) may not be solved so easily, because the lattice symmetry now does not help. Consequently, we cannot calculate $Im\langle\mathcal{T}_{LL}^{nn}(\varepsilon)\rangle_{R_n,A}$ exactly and must resort to the typical approximations of alloy theory. These we shall discuss in the next section. For now we wish only to stress, when the emission intensities are calculated from equation (3) and equation (26), i.e. when we write for the K and $L_{2,3}$ emissions

$$I_K^A(\varepsilon) \propto \frac{1}{3}\left[\int_0^a dr\, r^2 R_{A,o}^c(r)\frac{d}{dr}\,\Delta_1^A(r)\right]^2 \sum_{m=-1}^{m=+1} Im\langle\mathcal{T}_{\ell m,\ell m}^{nn}(\varepsilon)\rangle_{R_n,A}$$

$$I_{L_{2,3}}^A(\varepsilon) \propto \frac{1}{3}\left[\int_0^a dr\, r^2 R_{A,1}^c(r)\frac{d}{dr}\Delta_o^A(r)\right]^2 Im\langle\mathcal{T}_{oo,oo}^{nn}\rangle_{R_n,A}$$

$$+\ \frac{2}{5}\left[\int_0^a dr\, r^2 R_{A,1}^c(r)\frac{d}{dr}\Delta_2^A(r)\right]^2 \sum_{m=-2}^{m=2} Im\langle\mathcal{T}_{2m,\ m}^{nn}\rangle_{R_n,A} \qquad (27)$$

with $R_{A,0}^c(r)$, $R_{A,1}^c(r)$ as the $\ell=0$ and $\ell=1$ radial components of respectively the s and p core states for an A atom, that the matrix elements are independent of the approximation used. Therefore, in so far as the one-electron model and the choice of the muffin tin potential is a realistic one, an experiment directly tests our theory for $Im\langle\mathcal{T}_{LL'}^{nn}(\varepsilon)\rangle_{R_n,A}$. This is most unlike the situation in the theory of photoelectric emission, where one is forced to make some drastic and uncontrollable assumptions about the matrix elements in addition to estimating an approximate averaged density of states. By treating the matrix elements exactly we may expect that the relative magnitudes of various features in the calculated emission spectra, in addition to their positions and widths, will become meaningful. In a density-of-states calculation of the photoelectric emission (Spicer 1967) it is these two features only that are thought to be relevant to experiment. The price we pay for this improvement is that we must now construct a theory for $Im\langle T^{nn}_{LL'}(\varepsilon)\rangle_{R_n,A}$, which is slightly more complicated than for the average density of states.

In order to clarify the meaning of equation (27), we make two further comments. The first is related to the matrix element. For K emission this may be rewritten

$$[\int_o^a dr r^2 R^c_{A,0}(r)\frac{d}{dr}\Delta^A_1(r)]^2 = \frac{[\int_o^a dr r^2 R^c_{A,0}(r)\frac{d}{dr}R^A_1(r)]^2\sqrt{\varepsilon}}{\varepsilon^{-\frac{1}{2}}sin^2\delta^A_1(\varepsilon)} \tag{28}$$

The numerator in equation (29) is clearly the probability per unit time that a scattering electron with energy ε, and angular momentum $\ell=1$, makes a radiative transition to the core hole. The denominator, on the other hand, is the $\ell=1$ component of the total scattering cross section. Thus the ratio is a measure of the relative ease with which the electron scatterer becomes trapped in the core hole. It is intuitively reassuring that just this property of the radiating atom enters into the determination of the emission spectra.

The second comment has to do with the meaning of $Im\langle T^{nn}_{LL}(\varepsilon)\rangle_{R_n,A}$. We argue that it is related to the local density of states, which we shall now define. Since the total density of states may be written

$$n\ (\varepsilon) = -\frac{1}{\pi}\ Im\ \int dr^3 G^+(\mathbf{r},\mathbf{r};\varepsilon) \tag{29}$$

it is reasonable to define the local density of states (per

muffin tin) as the integral

$$n(\varepsilon;\mathbf{R}_n) = -\frac{1}{\pi}\, Im \int_{\Omega_n} d^3r\, G^+(\mathbf{r},\mathbf{r};\varepsilon) \qquad (30)$$

where the integration is over the volume of the muffin-tin well Ω about $\mathbf{R}$. The partially averaged local density of states for an A atom at $\mathbf{R}$ is then

$$n^A(\varepsilon;\mathbf{R}_n) = -\frac{1}{\pi} \sum_{LL'} \int d^3r\, \Delta_L^A(\mathbf{r})\Delta_{L'}(\mathbf{r})\; Im\, \langle T_{LL'}^{nn}(\varepsilon)\rangle \qquad (31)$$

where the angular integration ensures that only diagonal elements of $T_{LL'}^{nn}(\varepsilon)$ need be considered. Hence the ℓ angular momentum component of the partially averaged local density of states is given by

$$n_\ell^A(\varepsilon,\mathbf{R}_n) = -\frac{1}{\pi} \sum_{m=-\ell}^{m=+\ell} \int dr\, r^2 \Delta_\ell^A(r)\Delta_\ell^A(r) \langle Im\, T_{\ell m,\ell m}^{nn}\rangle_{\mathbf{R}_n, A} \qquad (32)$$

We may now rewrite the emission intensities given in equation (27) in the form

$$I_K^A(\varepsilon) \propto \frac{1}{3}\, [\int_0^a dr\, r^2 R_{A,0}^c(r) \frac{d}{dr} \mathcal{R}_1^A(r,\varepsilon)]^2 n_1^A(\varepsilon,\mathbf{R}_n)$$

$$I_{L_{2,3}}^A(\varepsilon) \propto \frac{1}{3}\, [\int_0^a dr\, r^2 R_{A,1}^c(r) \frac{d}{dr} \mathcal{R}_0^A(r,\varepsilon)]^2 n_0^A(\varepsilon,\mathbf{R}_n) \qquad (33)$$

$$+ \frac{2}{5}\, [\int_0^a dr\, r^2 R_{A,1}^c(r) \frac{d}{dr} \mathcal{R}_2^A(r,\varepsilon)]^2 n_2^A(\varepsilon,\mathbf{R}_n)$$

where $\mathcal{R}_\ell^A(r,\varepsilon)$ is proportional to $R_\ell^A(r,\varepsilon)$ but normalized to unity within the muffin tin volume Ω_n. Thus we can say that apart from a now trivial matrix-element factor, which can be calculated exactly, a soft x-ray emission spectrum measures the appropriate angular momentum components of the local density of states. Since these states are filled to the Fermi level, in accordance with the Pauli exclusion principle, we may say also that we are measuring the conduction electron charge in a given energy range inside the muffin-tin well.

In the one-electron approximation equation (33) is exact for muffin-tin potentials, and calculation of the matrix elements is straightforward; therefore it is reasonable to analyse experimental data in terms of local densities of states $n_\ell^A(\varepsilon,\mathbf{R})$ or local charges. This analysis does not lead us back to a representation of soft x-ray emission spectra directly in terms of the density of states because now the energy dependence of the matrix elements has been properly accounted for. On the contrary, once the energy dependence related to the matrix elements has been removed a more coherent pattern of alloying behaviour may emerge. Moreover, the local density of states is more directly related to our qualitative picture of electronic structure in random alloys. For example, if there is to be a common band seen by each constituent of an alloy, as predicted by the rigid-band model, then we would expect the local density of states on an A site, $n_\ell^A(\varepsilon,\mathbf{R})$, to be the same as that on a B site, $n_\ell^B(\varepsilon,\mathbf{R})$. On the other hand, if the virtual bound-state model is applicable, then the local density of states should follow a Friedel sum characteristic of the radiating atom, and therefore should be the same as in the pure metal.

Quantitatively, there are a number of definite calculation methods for estimating the average density of states in a random alloy. In the following section we show how these may be adapted to calculation of the local density of states; equivalent to evaluating equation (27).

3. APPROXIMATE MODELS

In the theory of random alloys there are three widely used approximation methods for calculating the average density of states. In order of increasing complexity these are the virtual crystal, the average t-matrix, and the coherent potential approximations. In all three one has to calculate the *average* Green function by finding the Green function for an equivalent ordered lattice with a medium potential on every site. The methods differ from one another by introducing progressively more complicated prescriptions for calculation of this medium potential, which may be both complex and energy dependent. However, if the real potentials $v^A(r)$ and $v^B(r)$ are of the muffin-tin kind the medium potential $v^M(r)$ will also be zero for $|r|>a$. With such potentials the integrated density of states may be written in terms of the 'on the energy shell' components of the t-matrix $t_L^M(\varepsilon)$ corresponding to $v^M(r)$. Namely we can write

$$\langle N(\varepsilon)\rangle = N^M(\varepsilon) = N^0(\varepsilon) - \frac{1}{\pi}\, Im\, \frac{1}{t} \int d^3q \,||\, t_L^{M-1}\delta_{LL'} - G_{LL'}(\mathbf{q}) \,|| \tag{34}$$

where $\langle N(\varepsilon)\rangle$ is the averaged integrated density of states for the alloy; N^M and N^0 are those for the medium and free space respectively; and $||t_L^{M-1}\delta_{LL'}-G_{LL'}(\mathbf{q})||$ is the KKR determinant for the regular lattice with the scattering described at each site by $t_L^M(\varepsilon)$. Thus, a convenient way of describing the three approximations is to give the corresponding prescription for calculating $t_L^M(\varepsilon)$:

(a) for the *virtual crystal approximation* (Nordheim 1931, Muto 1938, Stern 1967)

$$t_L^M(\varepsilon) = t_L^{\bar{v}}(\varepsilon) = - \, sin\, \delta_L^{\bar{v}}(\varepsilon).(\varepsilon)^{-\frac{1}{2}}.exp[i\delta_L^{\bar{v}}(\varepsilon)] \tag{35}$$

where the phaseshifts $\delta_L^{\bar{v}}$ are to be calculated for the virtual crystal potential

$$\bar{v}(\mathbf{r}) = cv^A(r) + (1-c)v^B(r) \tag{36}$$

(b) for the *average t-matrix approximation* (Korringa 1958, Beeby 1964)

$$t_L^M(\varepsilon) = c\, t_L^A(\varepsilon) + (1-c)t_L^B(\varepsilon) \tag{37}$$

with

$$t_L^A(\varepsilon) = -\,\sin\,\delta_L^A(\varepsilon).(\varepsilon)^{-\frac{1}{2}}.\exp\,i\delta_L^A(\varepsilon)$$

and (38)

$$t_L^B(\varepsilon) = -\,\sin\,\delta_L^B(\varepsilon).(\varepsilon)^{-\frac{1}{2}}.\exp\,i\delta_L^B(\varepsilon)$$

where the phase shifts $\delta_L^A(\varepsilon)$ and $\delta_L^B(\varepsilon)$ are those for $v^A(r)$ and $v^B(r)$.

(c) As argued recently by Soven (1966), Shiba (1971) and also now by Gyorffy (1972), the *coherent potential approximation* (Soven 1969) for muffin-tin model potentials should consist of solving, for $t_L^c(\varepsilon) = t_L^M(\varepsilon)$, the following implicit matrix equation

$$\mathcal{T}_{LL'}^{c,nn}(\varepsilon) = c\,\left[\frac{1}{1-(t^{c-1}-t^{A-1})\mathcal{T}^{c,nn}}\mathcal{T}^{c,nn}\right]_{LL'} + (1-c)\,\left[\frac{1}{1-(t^{c-1}-t^{B-1})\mathcal{T}^{c,nn}}\mathcal{T}^{c,nn}\right]_{LL'} \qquad (39)$$

where $\mathcal{T}^{c,nn}$ is a matrix in the space spanned by the angular momenta L, and its matrix elements are given by the solution of equation (15) with $t_L^\ell(\varepsilon) = t_L^c(\varepsilon)$ as

$$\mathcal{T}_{LL'}^{c\ nn}(\varepsilon) = \frac{1}{t}\int d^3q\ \left[\frac{1}{t^{c-1}-G(q)}\right]_{LL'} \qquad (40)$$

The matrix elements of t^{c-1} and $G(q)$ are $(t_L^c(\varepsilon))^{-1}\delta_{LL'}$ and $G_{LL'}(q) = \frac{1}{N}\sum_{\ell\ell'} \exp[iq.(\mathbf{R}_\ell-\mathbf{R}_{\ell'})\ G_{LL'}\ (\mathbf{R}_\ell-\mathbf{R}_{\ell'})$ respectively.

Our aim is a simple means by which all three of these methods can be adapted to evaluating expression (27) for the emission intensities. We suggest that the appropriate generalization is to represent the partially averaged system by an A impurity at $\mathbf{R}_n$, in an ordered lattice of medium potentials as shown in figure 1, with $t_L^M(\varepsilon)$ determined in the manner discussed above.

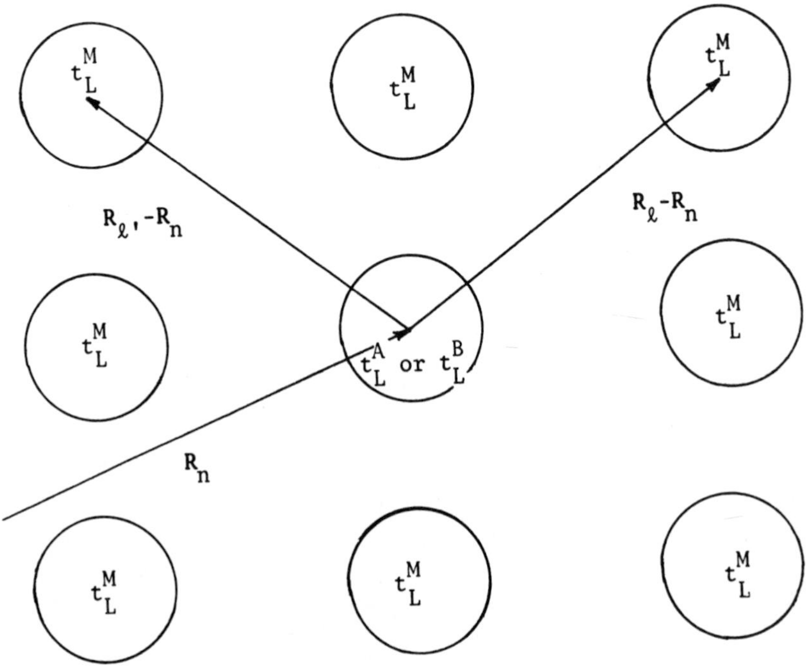

Figure 1 Schematic of the impurity problem in the medium-potential lattice.

In the case of the CPA model this procedure would falsify the original formulation; one should determine $t_L^M(\varepsilon)$ in the presence of the impurity. This would make $t_L^{M,\ell}(\varepsilon)$ depend on the position of the site $\mathbf{R}_\ell$, where the scattering takes place, relative to the position of the impurity, but the added complexity precludes numerical evaluation.

To find $\mathcal{T}_{LL'}^{A,nn}(\varepsilon)$ for an A impurity at $\mathbf{R}_n$ with $t_L^M(\varepsilon)$ on all other sites we solve equation (15) with $t_L^\ell = t_L^c + (t_L^A - t_L^c)\delta_{\ell,n}$. Using lattice Fourier transforms it is relatively straightforward to show that

$$\mathcal{T}_{LL}^{A\ nn}(\varepsilon) = [\frac{1}{1-(t^{M-1}-t^{A-1})\mathcal{T}^{nn}}\,\mathcal{T}^{nn}]_{LL'} \qquad (41)$$

where T^{nn} is the 'diagonal' part of the scattering path operator for the pure medium lattice, and is given by equation (40) with $t_L^c(\varepsilon)$ replaced by $t_L^M(\varepsilon)$. An analogous result was first obtained by Beeby (1967) using very different methods.

Thus, a calculation of the soft x-ray spectra from a random substitutional alloy would consist of

(1) constructing the potentials $v^A(r)$ and $v^B(r)$, and evaluating the matrix elements in equation (27).

(2) evaluating $t_L^M(\varepsilon)$ according to one of the prescriptions discussed above.

(3) evaluation of equation (41) and subsequently the expressions for the emission intensities given in equation (27).

As usual with KKR band-structure calculations, we need only keep the first few angular momentum components of $t_L^M(\varepsilon)$ (at most up to ℓ=2). This simplifies matters, though equation (41) will still be a 9x9 matrix.

For the virtual crystal and averaged t-matrix approximations the above procedure is not only relatively straightforward but also reasonably easy to implement given an average size computer. It would not be significantly more difficult than doing a KKR calculation of a soft x-ray spectrum for a pure metal, such as that reported by Blokhin *et al* (this Volume p.321). However, a CPA calculation will be a much more difficult task, and unfortunately for the most interesting alloys such as NiCu this calculation is the most desirable, because the first two have well-known shortcomings. For example, the virtual crystal approximation works well only when $v^A(r)$ and $v^B(r)$ are similar, while the average t-matrix approach leads to 'unphysical' band gaps (Beeby 1964, Soven 1970). On the other hand, CPA can give realistic densities of states, as shown for NiCu by Stocks, Williams and Faulkner (1971) for example. We hope soon to investigate numerically all three methods.

Finally, we note that once $\langle\langle \mathbf{r}|\mathit{Im}G^{+}|\mathbf{r}\rangle\rangle_{R_{n,A}}$ has been calculated, one can evaluate also the Knight shift contact density P_F^A and the Mössbauer isomer shift from respectively $\langle\langle \mathbf{r}|\mathit{Im}G^{+}|\mathbf{r}\rangle\rangle_{R_{n'A}}\Big|_{\mathbf{r}=R_n}$ and $\int_0^{\varepsilon_F} d\varepsilon \langle\langle \mathbf{r}|\mathit{Im}G^{+}|\mathbf{r}\rangle\rangle_{R_{n'A}}\Big|_{\mathbf{r}=R_n}$, as discussed previously (Gyorffy and Stott 1971).

REFERENCES

Beeby, J.L. (1964); Proc. Roy. Soc. (London) A279, 82.

Beeby, J.L. (1964); Phys. Rev. A135, 130.

Beeby, J.L. (1967); Proc. Roy. Soc. 302, 113.

Goldberger, M.L. and Watson, K.M. (1964); "Collision Theory", pp 286, John Wiley.

Goodings, D.A. and Harris, R.J. (1969); J. Phys. C. 2, 1808.

Gyorffy, B.L. and Stott, M.J. (1971); Solid State Comm. 9, 613.

Gyorffy, B.L. (1972); Phys. Rev. 5, 2382.

Ham, F.S. and Segall, B. (1961); Phys. Rev. 124, 1786.

Klima, J. (1970); J. Phys. 3, 70.

Korringa, J. (1958); J. Phys. Chem. Solids, 1, 252.

Lax, M. (1951); Rev. Mod. Phys. 23, 287.

Lloyd, P. (1967); Proc. Phys. Soc. 90, 207.

McMullen, T. (1970); J. Phys. C., 3, 2178.

Muto, T. (1938); Sic. Papers Inst. Phys. Chem. Res. (Tokyo) 34, 377.

Nordheim, L. (1931); Ann. Physik, 9, 607, 641.

Roman, P. (1965); "Advanced Quantum Theory", Addison-Wesley.

Rooke, G.A. (1968); J. Phys. C. 1, 767.

Shiba, H. (1971); Prog. Theo. Phys. 16, 77.

Soven, P. (1966); Phys. Rev. 151, 539.

Soven, P. (1969); Phys. Rev. 178, 1136.

Soven, P. (1970); Phys. Rev. B2, 4715.

Spicer, W.E. (1967); Phys. Rev. 154, 385.

Stern, E.A. (1967); Phys. Rev. 157, 544.

Stocks, G.M., Williams, R.W. and Faulkner, J.S. (1971); Phys. Rev. B4, 4390.

Velický, B., Kirkpatrick, S. and Ehrenreich, H. (1968); Phys. Rev. 175, 747.

PSEUDOPOTENTIAL THEORY AND IMPURITY STATES IN NOBLE AND TRANSITION METALS

C. Demangeat, F. Gautier, R. Riedinger

Université Louis Pasteur, Laboratoire de Structure Electronique des Solides, Strasbourg, France.

ABSTRACT*

Pseudopotential theory applied to the transition metals provides justification of the calculation from first principles of tight binding 'd' bands hybridized with s-p orthogonal plane-waves (Gautier 1971). The model is applied to the computation of impurity states in noble and transition metals (density of states, anisotropy of relaxation times, etc.). Discussion of the results obtained for Ni-Cr (Demangeat 1971) and Cu-Ni (Riedinger 1971) alloys shows that the virtual bound-state width arises mainly from the host s-d hybridization.

References

Demangeat, C. (1971); International Conference of the E.P.S., Firenze.

Demangeat, C., Gautier, F. and Riedinger, R. (February 1972); to be published in J. Phys., Paris.

Gautier, F. (1971); J. Phys. F. Metal Phys. 1, 382.

Riedinger, R. (1971); J. Phys. F. Metal Phys. 1, 392.

---

*This paper was presented at the Conference on which this Volume is based and has been published in full elsewhere (Demangeat *et al* 1972).

ON THE ELECTRONIC PROPERTIES OF DISORDERED TRANSITION- AND NOBLE-METAL ALLOYS

F. Brouers* and A.V. Vedyayev**

Division of Engineering and Applied Physics, Harvard University, Cambridge, Massachusetts, USA.

1. INTRODUCTION

This paper is concerned with some recent extensions and applications of a two s-d band model relevant to the investigation of electronic properties of disordered transition-metal noble-metal alloys. Such a model Hamiltonian was first considered by Levin and Ehrenreich (1971), referred to here as the LE-model. The density of states of Ag-Au alloys as well as the concentration dependence of the optical absorption edge was calculated by these authors in the framework of the coherent-potential approximation (CPA) introduced and discussed by Soven (1967) and Velický *et al* (1968). The LE-model has been extended recently (Schwartz *et al* 1971, Brouers *et al* 1971) and it has been shown that a non-selfconsistent approximation provides a correct description of the qualitative features of the density of states for a wide variety of scattering strengths and concentrations. It is also possible to calculate the residual resistivity of both s and d electrons and we have discussed and explained in this model the deviations from the Nordheim rule observed in some transition-metal-based alloys (Brouers and Vedyayev 1972a). It is the purpose of the present

*Present address: Laboratoire de Physique des Solides, Orsay-91, France.
**Present address: Dept. of Physics, State University of Moscow, USSR.

paper to review these recent contributions and to present some unpublished extensions of the theory : a self-consistent calculation of the charge transfer (§3B) and a generalization of the formalism to disordered ferromagnetic alloys (§3C).

2. THE COHERENT POTENTIAL APPROXIMATION OF A TWO-BAND MODEL

In this section we discuss a generalized version of the LE two-band model. This model, which contains some of the features of noble-metal transition-metal alloys, emphasizes the effect of s-d hybridization and neglects the structure of the d-bands. The pure metals A and B are assumed to have a broad s-band and a narrow d-band centred at the energies E_d^A and E_d^B. The d-band is supposed to consist of five degenerate but independent sub-bands. In the model discussed by Levin and Ehrenreich the d-band was replaced by a single level of zero width: this assumption leads to the appearance of unphysical hybridization gaps in the density of states. The introduction of a width in the d-band eliminates this difficulty.

In the alloy $A_x B_{1-x}$, the d energy-level at a given site may be E_d^A or E_d^B corresponding to whether the site n is occupied by an A or a B atom respectively with probability x and 1-x. The unhybridized s-bands and the unhybridization parameters are assumed to exhibit virtual-crystal behaviour. By contrast, disorder associated with the d-band is treated in the framework of multiple-scattering theory. For a given configuration of the alloy, the Hamiltonian (cf. Brouers and Vedyayev 1972a) is

$$H = \sum_{k\varepsilon BZ} \varepsilon_s(k)|k_s\rangle\langle k_s| + \sum_{k\varepsilon BZ} \varepsilon_d(k)|k_d\rangle\langle k_d| + \sum_n E_d^{A(B)}|n_d\rangle\langle n_d| + \sum_{k\varepsilon BZ} \bar{\gamma}[|k_s\rangle\langle k_d| + |k_d\rangle\langle k_s|] \qquad (1)$$

The first two terms represent the kinetic energy of the s and d electrons, the third term contains the random d-levels and the last term describes the s-d hybridization. The intersection of the $E_s(k)$ curve with the (spherical) Brillouin zone boundary determines the width $2W_s$ of the unhybridized s-band:

$$E_s(k) = E_0^{A(B)} + \varepsilon_s(k) \quad (2a)$$

with

$$\varepsilon_s(k) = W_s S(k) \text{ and } -1 \leqslant S(k) \leqslant 1 \quad (2b)$$

As indicated above, the s-band and hybridization parameters in the alloy are assumed to satisfy

$$E_0 \to \bar{E} = xE_0^A + (1-x)E_0^B \quad (3a)$$

$$\gamma \to \bar{\gamma} = x\gamma(E_d^A) + (1-x)\gamma(E_d^B) \quad (3b)$$

In addition, it is convenient to suppose that in a given metal the unhybridized s and d bands have the same shape but differ in location and width. If the energy origin is chosen midway between the constituent d-levels we have

$$E_s(k) = \bar{E} + \varepsilon_s(k) \quad (4a)$$

$$\varepsilon_d(k) = \alpha\varepsilon_s(k) \quad (4b)$$

The parameter α, which specifies the relative width of the s and d bands, is a new feature of the model.

The third term of equation (1) is the only part of the Hamiltonian to be treated on the basis of multiple-scattering theory. For a given metal, the parameters $E_0^{A(B)}$, W_s, E_d, $\gamma(\varepsilon_d)$ and α may be chosen approximately to fit existing band calculations. The scattering parameter $\delta = E_d^A - E_d^B$ and the concentration x of atom A are the two parameters characterizing the disorder of the system.

To determine the average properties of the alloy model just described, we shall follow the general scheme used in the one-band model (Velicky *et al* 1968). To calculate the alloy s and d density of states, as well as any macroscopic equilibrium property, we must know the configuration averaged Green function

$$\langle G(\eta)\rangle = \langle\frac{1}{\eta-H}\rangle \tag{5}$$

If we suppose that the electron mean free path is much longer than the interatomic spacing, we can consider an electron as moving in an effective medium characterized by an effective Hamiltonian:

$$\langle\frac{1}{\eta-H}\rangle = \frac{1}{\eta-H_{eff}} \tag{6}$$

We shall restrict ourselves to a simple site solution of (6) i.e. we suppose that the effective Hamiltonian will have the form:

$$H_{eff} = H_{OR} + \Sigma_d \hat{P}_d \tag{7}$$

where H_{OR} is the sum of the first two and the last terms of (1) and

$$\hat{P}_d = \sum_n |n_d\rangle\langle n_d| \tag{8}$$

is the projection operator onto the space of d-states. The self-energy Σ_d has the symmetry of the lattice but will be energy dependent and, in general, complex to simulate the damping of electron states in disordered materials.

In the two-band model, the configuration-averaged Green function is a 2 x 2 matrix which may be written in the $\{|k_s\rangle,|k_d\rangle\}$ representation as

$$\langle G(\eta)\rangle = \begin{bmatrix} \eta-E_s(k) & -\bar{\gamma} \\ -\bar{\gamma} & \eta-\alpha\varepsilon_s(k)-\Sigma_d(\eta) \end{bmatrix}^{-1} \tag{9}$$

The three averaged propagators $\langle G_{ss}\rangle$, $\langle G_{dd}\rangle$, $\langle G_{sd}\rangle = \langle G_{ds}\rangle$ are determined from (9) as:

$$\langle G_{ss}(k,\eta)\rangle = [\eta-E_s(k)-\gamma^2(\eta-\Sigma_d-\alpha\varepsilon_s(k))^{-1}]^{-1} \tag{10a}$$

$$\langle G_{dd}(k,\eta)\rangle = [\eta-\Sigma_d-\alpha\varepsilon_s(k)-\gamma^2(\eta-E_s(k))^{-1}]^{-1} \qquad (10b)$$

$$\langle G_{sd}(k,\eta)\rangle = \gamma[(\eta-E_s(k))(\eta-\Sigma_d-\alpha\varepsilon_s(k)-\gamma^2]^{-1} \qquad (10c)$$

To calculate Σ_d it is convenient to rewrite H at equation (1) as

$$H = H_{OR} + \sigma_d(\eta)\hat{P}_d + \sum_n |n_d\rangle(E_d^{A(B)}-\sigma_d(\eta))\langle n_d| \qquad (11a)$$

$$\equiv \tilde{H} + \sum_n |n_d\rangle V_n \langle n_d| \qquad (11b)$$

with
$$V_n = E_n^{A(B)}-\sigma_d \qquad (12)$$

The potential V_n describes the scattering experienced by an electron at site n occupied by an atom A(B) relative to a reference medium characterized by the 'potential' σ_d.

The 'non-perturbed' Hamiltonian H may be chosen at will. The idea is to start from a Hamiltonian $\tilde{H}$ as close as possible to the unknown H_{eff}. One can derive a formal relation between

$$\langle G\rangle = \frac{1}{\eta-H_{eff}} \text{ and } \tilde{G} = \frac{1}{\eta-\tilde{H}}$$

$$\langle G\rangle = \tilde{G} + \tilde{G}(\Sigma_d-\sigma_d)\ \hat{P}_d\langle G\rangle \qquad (13)$$

The calculation of $\langle G\rangle$ is then carried out most simply in terms of the total scattering operator. For a given configuration of the alloy, the relation between G and the total scattering operator T is simply

$$G = \tilde{G} + \tilde{G}\ T\ \tilde{G} \qquad (14)$$

Averaging (14) over all the possible configurations of the system

$$\langle G\rangle = \tilde{G} + \tilde{G}\langle T\rangle\tilde{G} \qquad (15)$$

Comparison of (14) and (15) yields the relation

$$(\Sigma_d - \sigma_d)\ \hat{P}_d\ (1+\tilde{G}\langle T\rangle) = \langle T\rangle \tag{16}$$

The expression (16) has to be written in 2 x 2 matrix form and it is straightforward to show that, since in the basis $\{|n_s\rangle, |n_d\rangle\}$ the scattering is restricted to the diagonal dd elements, one can write

$$(\Sigma_d - \sigma_d)\ \hat{P}_d = \frac{\langle \hat{T}^d\rangle}{1 + \tilde{G}\langle \hat{T}^d\rangle} \tag{17}$$

Because of the assumed localization of d-states, for any configuration, the T matrix is a sum of single site contributions and can be written

$$\hat{T} = \begin{bmatrix} o & o \\ o & \hat{T}^d \end{bmatrix} \tag{18}$$

$$\hat{T}^d = \sum_n |n_d\rangle T_n^d \langle n_d| \tag{19}$$

The multiple-scattering theory (Velicky *et al* 1968) can be used to express T_n^d in terms of

$$t_n^d = \frac{V_n^d}{1 - \langle n_d|\tilde{G}|n_d\rangle V_n^d} \tag{20}$$

the scattering matrix for the nth atom:

$$T_n^d = t_n^d (1 + \tilde{G} \sum_{m\neq n} |m_d\rangle T_m^d \langle m_d|) \tag{21}$$

The second factor of the right-hand side of equation (21) corresponds to the effective wave incident on site n and composed of contributions from all sites $n \neq m$.

To calculate the self-energy, the T matrix has to be averaged. To obtain a single-site solution consistent with equation (7), we must neglect fluctuations in the effective waves and write:

$$<T_n^d> \;=\; <t_n^d>(1+\tilde{G}\sum_{m\neq n}|m_d><t_m^d><m_d|) \qquad (22)$$

or using definition (19):

$$<T_n^d> = <t_n^d>(1+\tilde{G}<\hat{T}^d>) - <t_n^d>\tilde{G}|n_d><T_n^d><n_d| \qquad (23)$$

which yields:

$$\hat{T}^d = \sum_n <t_n^d>\frac{(1+\tilde{G}<\hat{T}^d>)}{1+<n_d|\tilde{G}|n_d><T_n^d>}|n_d><n_d| \qquad (24)$$

If we substitute (24) into (17) we obtain finally:

$$\Sigma_d = \sigma_d + \frac{<t_n^d>}{1+<n_d|\tilde{G}|n_d><t_n^d>} \qquad (25)$$

which is the single-site solution of (17). Approximation (22) leading to (25) is called single-site approximation (SSA) and has been shown to be equivalent to the neglect of fluctuations in the effective-wave:

$$<(t_n-<t_n>)>\tilde{G}\sum_{m\neq n}(T_m-<T_m>)> = o \qquad (26)$$

Equation (25) can be used in two ways. Either $<t_n^d>$ and $<n_d|\tilde{G}|n_d>$, corresponding to a given choice of σ can be inserted into (25), or the equation

$$<t_n^d> = o \qquad (27)$$

may be used to determine σ. In the second case equation (25) guarantees that $<G> = \tilde{G}$ since $\Sigma = \sigma$.
These two possibilities define two different classes of approximate calculations of Σ within the single-site approximation. In the first case, the second term in (25) describes an effective potential corresponding to the average scattering matrix at the site n. Such a procedure will be referred to as the average t-matrix approximation (ATA). The second approach,

usually referred to as the coherent potential approximation (CPA), requires a self-consistent solution of (25) and is much more difficult to implement numerically than the ATA.

The CPA solution is derived by expliciting condition (27):

$$\frac{X(E^A-\Sigma)}{1-(E^A-\Sigma)\langle n_d|\bar{G}|n_d\rangle} + \frac{(1-X)(E^B-\Sigma)}{1-(E^B-\Sigma)\langle n_d|\bar{G}|n_d\rangle} = 0 \qquad (28)$$

which can be written in a more useful form:

$$\Sigma_d^{CPA}(\eta) = \bar{\varepsilon}_d + \frac{X(1-X)\delta^2 F_{dd}(\eta,\Sigma_d)}{1+(\bar{\varepsilon}_d+\Sigma_d(\eta))F_{dd}(\eta,\Sigma_d)} \qquad (29)$$

with

$$F_{dd}(\eta) \equiv \langle n_d|\bar{G}|n_d\rangle$$
$$\bar{\varepsilon}_d = XE_d^A + (1-X)E_d^B \qquad (30)$$

and

$$\delta = E_d^A - E_d^B$$

The solution of equation (29) has to be determined self-consistently.

The ATA solution, which is non self-consistent, will depend on the choice of the reference Hamiltonian $\tilde{H}$. If we start from the virtual-crystal approximation ($\sigma=\bar{\varepsilon}_d$) and then calculate $\langle t_n^d\rangle$ and Σ from (25) we obtain:

$$\Sigma^{ATA}(\eta) = \bar{\varepsilon}_d + \frac{X(1-X)\delta^2 F_{dd}(\eta,\bar{\varepsilon}_d)}{1 + 2\,\bar{\varepsilon}_d F_{dd}(\eta,\bar{\varepsilon}_d)} \qquad (31)$$

The form $\Sigma^{ATA}(\eta)$ is similar to that of the CPA equation. The ATA may in fact be viewed as the first iteration (i.e. $\Sigma_d \to \bar{\varepsilon}_d$) of equation (29) towards self-consistency.
The function $F_{dd}(\eta,\Sigma_d)$ (30) can be written explicitly

$$F_{dd}(\eta,\Sigma_d) = N^{-1}\, Tr_d\, \bar{G}(\eta)$$

$$= N^{-1}\sum_k \langle k_d|\bar{G}|k_d\rangle = \langle n_d = o|\bar{G}|n_d = o\rangle \tag{32}$$

$$= \Omega_c(2\pi)^3\int_{BZ} d^3k \int_{-\infty}^{+\infty} \frac{\delta[E-\varepsilon_s(k)]}{\eta-\Sigma_d(\eta)-\alpha E-\gamma^2/(\eta-E-\bar{E})} dE$$

where Ω_c is the unit-cell volume.

This expression can be simplified by defining the density of states (per site) in the unhybridized s-band:

$$\rho_{os}(E) = \Omega_c(2\pi)^3\int_{BZ} d^3k\,\delta[E-\varepsilon_s(k)] \tag{33}$$

and its transform

$$F_{os}(\eta) = \int_{-\infty}^{+\infty}\rho_{os}(E)(\eta-E)^{-1}dE \tag{34}$$

Equation (26) can be expressed in terms of $F_{os}(\eta)$

$$F_{dd}(\eta) = [\alpha(E_- - E_+)^{-1}][(\eta-\bar{E}-E_+)F_{os}(E_+)-(\eta-\bar{E}-E_-)F_{os}(E_-)] \tag{35}$$

with

$$E_\pm = (1/2)\left[\eta-\bar{E} + (\eta-\Sigma)/\alpha \pm \surd\{[\eta-E-(\eta-\Sigma)/\alpha]^2 + 4(\gamma^2/\alpha)\}\right] \tag{36}$$

Similarly the functions

$$F_{ss}(\eta,\Sigma_d) = N^{-1}\sum_{k\varepsilon BZ}\langle k_s|\bar{G}(\eta)|k_s\rangle \quad \text{and}$$

$$F_{sd}(\eta,\Sigma_d) = N^{-1}\sum_{k\varepsilon BZ}\langle k_s|\bar{G}(\eta)|k_d\rangle$$

can be shown to be expressed in terms of $F_{os}(\eta)$

$$F_{ss}(\eta) = (E_- - E_+)^{-1}[\{(\eta-\Sigma)/\alpha-E_+\}F_{os}(E_+)-\{(\eta-\Sigma)/\alpha-E_-\}F_{os}(E_-)] \tag{37}$$

and

$$F_{sd}(\eta) = (\gamma/\alpha)(E_- - E_+)^{-1}[F_{os}(E_+)-F_{os}(E_-)] \tag{38}$$

3. APPLICATIONS AND ILLUSTRATION

A. The Density of States

The density of states of s and d electrons is derived from the discontinuity of $F_{ss}(\eta)$ and $F_{dd}(\eta)$ across the real axis. Once equation (29) or (31) has been used to evaluate $\Sigma_d(\eta)$, the density of states for the ten d-electrons is obtained from

$$\rho_d(E) = -(10/\pi)\, Im\, F_{dd}\,(E+io,\Sigma_d) \qquad (39a)$$

and similarly the J-band contribution is given by

$$\rho_s(E) = -(2/\pi)\, Im\, F_{ss}\,(E+io,\Sigma_d) \qquad (39b)$$

The density of states for the model for Au-Ag alloy, discussed by Levin and Ehrenreich and by Schwartz *et al* (1971) has been calculated from equations (39a) and (39b) in the CPA and ATA using respectively equations (29) and (31) for the self-energy Σ_d. Figures 1 (a,b and c) show the result of the numerical calculations for three concentrations x = 0.1, 0.5 and 0.9. The first and last of these figures demonstrate that for dilute alloys the two methods yield essentially the same results. In the equi-concentration case (figure 1b) the ATA and CPA results differ in the energy region between the two sub-bands. The structure in the ATA density of states, near E=0, is spurious and may be traced to the unphysical hybridization gap that exists when α=0. It can be shown that this structure becomes less pronounced if α is increased slightly. These results for AuAg alloys suggest that it is possible to determine a range of parameters for the two-band model within which the ATA and the CPA will agree. Because it is the first iteration of the CPA towards self-consistency, the ATA will be a good approximation if the right-hand side of (29) is relatively insensitive to small variations of Σ_d from the value $\Sigma_d = \bar{\varepsilon}_d$. From this condition it is possible to show that in dilute alloys the ATA is a good approximation over the entire range of

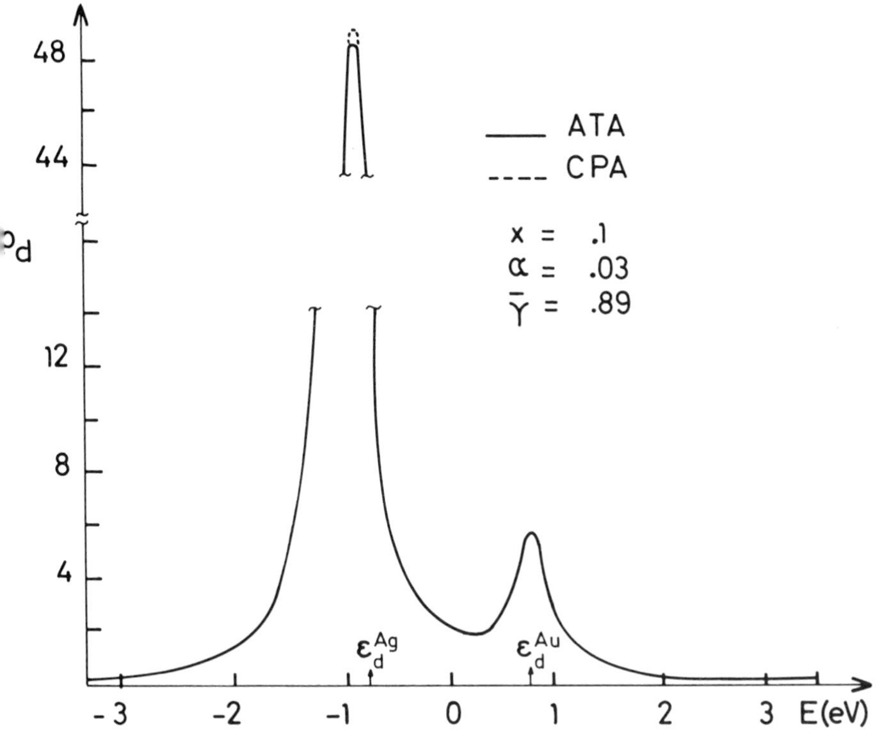

Figure 1a Comparison of the d-band density of states calculated in the CPA and ATA models for the alloy $Au_x\ Ag_{1-x}$, with x = 0.1

energies, and that for high concentrations the ATA fails only in a small region of energies between the two d sub-bands. In the equal concentration case for Au-Ag alloys this region is $|E|<0.3$eV which is roughly the region in figure 1(b) within the ATA exhibits spurious behaviour.

B. The Shift of the Optical Absorption Edge

Levin and Ehrenreich (1971) have emphasized the importance of introducing a concentration dependence of the d resonance levels $E_d^{A,B}$ to obtain the correct shift of the optical absorption edge. This concentration dependence is the direct consequence of a charge transfer occurring from Au to Ag in the LE-model. The charge transfer was supposed to be due essen-

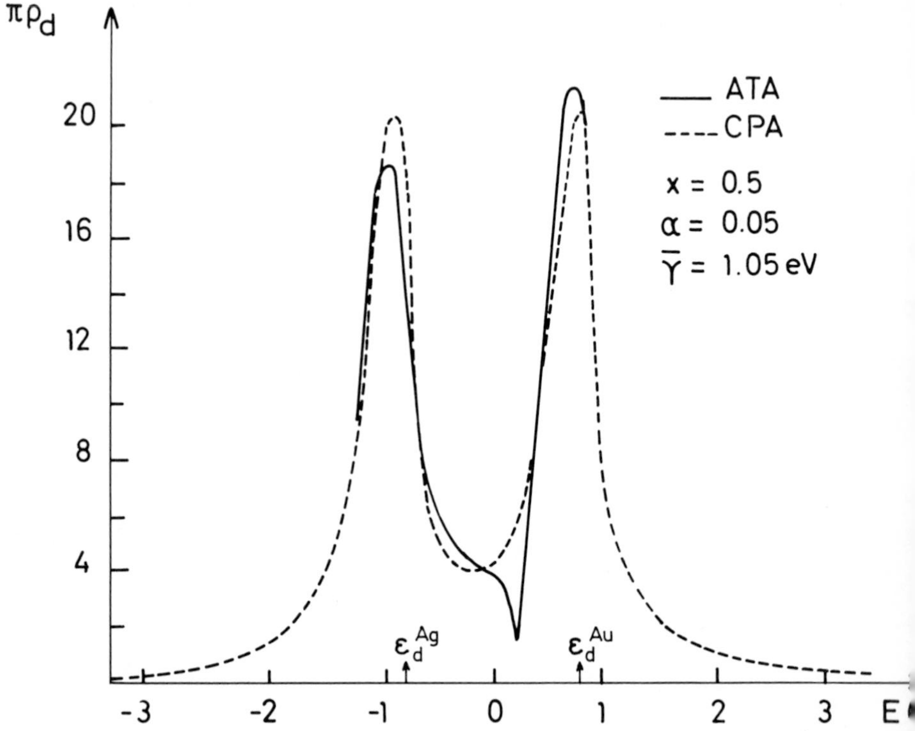

Figure 1b Comparison of the d-band density of states calculated in the CPA and ATA models for the alloy $Au_x\ Ag_{1-x}$, with x = 0.5

tially to conduction electrons and the change in $E_d^{A(B)}$ resulting from the change in s-charge density per atom was calculated in the framework of a renormalized atom model. This description involves cutting off the free-atom s and d wavefunctions at the Wigner-Seitz radius and renormalizing them to a Wigner-Seitz sphere (c.f. Watson *et al* 1970). Neglecting an exchange term, the change in $E_d^{A(B)}$ resulting from a change in s charge density per atom $\Delta Q^{A(B)}$ is written in this picture as:

$$\Delta E_d^{A(B)}(x) = \ell^{-1}\Delta Q_s^{A(B)}(x)[\iint d\mathbf{r}d\mathbf{r}'|\psi_s(\mathbf{r})|^2|\phi_d(\mathbf{r}')|^2|\mathbf{r}-\mathbf{r}'|^{-1}] \quad (40)$$

where $\phi_d(\mathbf{r})$ represents the d atomic wavefunction of A or B and $\psi_s(\mathbf{r})$ is the renormalized atom s wavefunction associated with

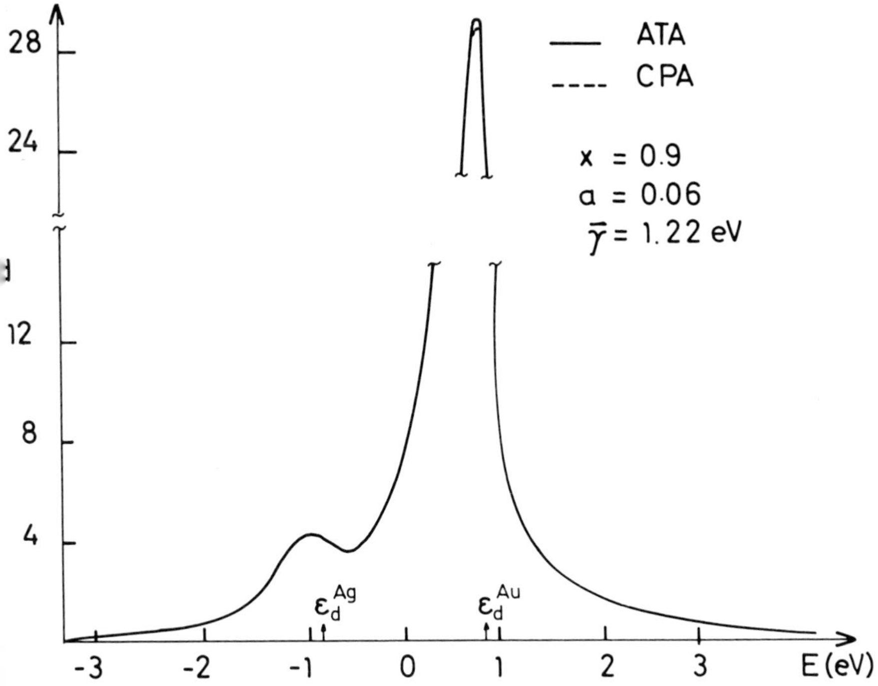

Figure 1c Comparison of the d-band density of states calculated in the CPA and ATA models for the alloy $Au_x\ Ag_{1-x}$, with x = 0.9

A or B before the charge is transferred.

To calculate the parameter $\Delta A^{A(B)}(x)$ Levin and Ehrenreich made the following approximations:

1) the d charge transfer was neglected;

2) for dilute alloys the charge transfer was calculated in the Thomas-Fermi picture;

3) for non-dilute alloys, the charge transfer is interpolated linearly between the dilute-alloy limits:

$$\Delta Q^{Au}(x) = (1-x)\ \Delta Q^{Au}(x \simeq o) \tag{41a}$$

$$\Delta Q^{Ag}(x) = x\ \Delta Q^{Ag}(x \simeq 1) \tag{41b}$$

Moreover, because of the neutrality condition we have the

relation:

$$\Delta Q = \Delta Q^{Ag}(x \simeq 1) = -\Delta Q^{Au}(x \simeq o)$$

The parameter ΔQ was estimated to be 0.3e in order to reproduce the observed optical-absorption edge versus x behaviour in Au-Ag.

In this section we want to show that the CPA can be extended to calculate directly the s and d charge transfers and the concentration dependence of the d resonance level $E_d^{A(B)}(x)$ in a self-consistent way by an iterative procedure. One advantage of the CPA theory is the possibility of decomposing the average density of states into A and B component-densities:

$$\rho_d(E,\Sigma_d) = x\rho_d^{A}(E,\Sigma_d) + (1-x)\rho_d^{B}(E,\Sigma_d) \qquad (42a)$$

and

$$\rho_s(E,\Sigma_d) = x\rho_s^{A}(E,\Sigma_d) + (1-x)\rho_s^{A}(E,\Sigma_d) \qquad (42b)$$

The calculation of $\rho_d^{A(B)}$ is a straightforward generalization of the expressions derived in the one-band model.

$$\rho_d^{A(B)}(E,\Sigma_d) = -(10\pi^{-1})\mathit{Im}\langle n_d = o|\langle(E^{+}-H^{A(B)})^{-1}\rangle|n_d = o\rangle \qquad (43)$$

where $H^{A(B)}$ is the Hamiltonian corresponding to a given alloy configuration with atoms of type A or type B respectively located at site n = 0. Thus equation (43) can be written

$$\rho_d^{A(B)}(E,\Sigma_d) = -(10\pi^{-1})\mathit{Im}\{F_{dd}[1-(E_d^{A(B)}-\Sigma_d)F_{dd}]^{-1}\}_{\eta=E+io} \qquad (44)$$

in an analogous way to equation (4.15) of Velický *et al* (1968). An expression for $\rho_s^{A(B)}(E,\Sigma_d)$ is derived in Appendix 1 and is shown to have the form:

$$\rho_s^{A(B)}(E,\Sigma_d)=-(2\pi^{-1})\mathit{Im}\{F_{ss}+(E_d^{A(B)}-\Sigma_d)F_{sd}F_{ds}[1-F_{dd}(E_d^{A(B)}-\Sigma_d)]^{-1}\} \qquad (45)$$

where F_{ss}, F_{sd} and F_{dd} are given by equations 35, 37 and 38. The expressions (44) and (45) enable us to calculate the charge transfer as a function of concentration. The iteration scheme is the following : The A and B component s and d densities are calculated with $E_d^{A(B)}$ independent of the concentration. The s and d charges at A and B atoms in the alloys are compared with the charges on pure A and B atoms. The shift $E_d^{A(B)}$ with concentration is determined by introducing the calculated s charge transfer into equation (40). The quantities $\rho_{s,d}^{A(B)}$ are recalculated and new charge transfers determined. This procedure is rapidly convergent.

For the LE model of Ag-Au alloys, this method yields the following results

$$-\Delta Q_s^{Ag}(x\sim 1) \simeq + \Delta Q_s^{Au}(x\sim o) \simeq 0.12e \qquad (46a)$$

and

$$\Delta Q_d^{Ag}(x\sim 1) \simeq - \Delta Q_d^{Au}(x\sim o) \simeq 0.05e \qquad (46b)$$

In addition it can be verified that a linear interpolation between the dilute-alloy limits is a good approximation. This result is to be compared with the s charge transfer estimated by LE. A charge transfer of 0.3e was necessary to obtain the observed shift of the optical absorption edge with concentration. However, it has been noticed (Ehrenreich, unpublished communication) that the numerical values for the integrals in equations 3.8 and 3.9 of the LE-paper (Levin and Ehrenreich 1971) are incorrect and have to be multiplied by a factor of approximately 3. If this correction is introduced the s charge transfer of 0.12e gives the correct behaviour for the optical absorption edge *vs* concentration (Rivory 1969) using for E_d^{Ag} the value 2.2eV which is closer to the band structure calculations than the value chosen by Levin and Ehrenreich (2.7eV). The difference in sign of the s and d charge transfers is an indication that the charge density in different regions of the atom, outside or inside the core, may vary differently with concentration; and can explain why different experimental techniques (optical and Mossbauer data, for example) yield different values for the charge transfer.

C. Generalization of the formalism to ferromagnetic alloys

The coherent potential approximation described in §2 can easily be generalized to calculate the electronic and magnetic properties of ferromagnetic alloys. As a first step, we shall take account only of the intra-atomic Coulomb repulsion between two antiparallel d-electrons on the same orbital, at site i, and neglect Coulomb repulsion between d-electrons in different orbitals on the same atom, and also neglect exchange interactions between d orbitals and between d and s orbitals so that

$$H_{cor} = U^{d-d}\sum_i n^i_{d,\sigma} n^i_{d,-\sigma} \tag{47}$$

with

$$U^{d-d} = \int \psi^*_u(\mathbf{r}_1)\psi^*_u(\mathbf{r}_2)\ell^2|\mathbf{r}_1-\mathbf{r}_2|^{-1}\psi_u(\mathbf{r}_2)\psi_u(\mathbf{r}_1)d^3\mathbf{r}_1 d^3\mathbf{r}_2 \tag{48}$$

We assume that the model Hamiltonian (1) includes paramagnetic correlation effects which we consider to be given by

$$H^{ferro}_{cor} - H^{para}_{cor} \tag{49}$$

Adopting the Hartree-Fock approximation to describe the effect of correlations, the model Hamiltonian will have the same form as (1) but with a spin dependence of the d resonance levels.

$$E^{A(B)}_{d,\sigma} = E^{A(B)}_{para} + U_{A(B)}[n^{A(B)}_{d,-\sigma} - n^{A(B)para}_{d,-\sigma}] \tag{50}$$

The number of electrons/atom $n^{A(B)}_{d,\pm\sigma}$ does not depend on the site but depends on the species of atom occupying the site. The coherent potential approximation is easily generalized. The self-energy Σ_d is now spin-dependent:

$$\Sigma_{d,\sigma} = \bar{\varepsilon}_{d,\sigma} - (E^{A(B)}_{d,\sigma} - \Sigma_{d,\sigma}) F_{dd,\sigma} (E^{A(B)}_{d,\sigma} - \Sigma_{d,\sigma}) \quad (51)$$

with

$$\bar{\varepsilon}_{d,\sigma} = x\, E^{A}_{d,\sigma} + (1-x) E^{B}_{d,\sigma} \quad (52)$$

and the s and d densities of states are given by

$$\rho_{s,\sigma}(E) = -\frac{1}{\pi}\, Im\, F_{ss}(\eta, \Sigma_{d,\sigma}) \quad (53)$$

and

$$\rho_{d,\sigma}(E) = -\frac{5}{\pi}\, Im\, F_{dd}(\eta, \Sigma_{d,\sigma}) \quad (54)$$

while the partial d density of states on atoms A and B are respectively

$$\rho^{A(B)}_{d,\sigma} = -\frac{5}{\pi}\, Im\, \frac{F_{dd}(\eta, \Sigma_{d,\sigma})}{1 - (E^{A(B)}_{d,\sigma} - \Sigma_{d,\sigma}) F_{dd}(\eta, \Sigma_{d,\sigma})} \quad (55)$$

Since $E^{A(B)}_{d,\sigma}$ is a function of $n^{A(B)}_{d,\sigma}$, the ferromagnetic solution is obtained by solving the five simultaneous equations with five unknowns E_F, $n^{A(B)}_{d,\pm\sigma}$:

$$n^{A(B)}_{d,\pm\sigma} = \int_{-\infty}^{E_F} \rho^{A(B)}_{d,\pm\sigma}(E)\, d_E = \phi(n^{A}_{d,\pm\sigma}, n^{B}_{d,\pm\sigma}) \quad (56)$$

and

$$x n^{A}_{o} + (1-x) n^{B}_{o} = \int_{-\infty}^{E_F} (\rho_{s,\sigma} + \rho_{s,-\sigma} + \rho_{d,\sigma} + \rho_{d,-\sigma})\, dE \quad (57)$$

where $n^{A(B)}_{o}$ represents the total number of s and d electrons of pure A or pure B metals. When these five equations are solved, the average magnetic moment of the alloy per atom is equal to

$$\bar{\mu} = x\mu_A + (1-x)\mu_B \quad (58)$$

with

$$\mu^{A(B)} = (n^{A(B)}_{d,\sigma} - n^{A(B)}_{d,\sigma})\mu_o \quad (59)$$

where μ_0 is the Bohr magneton.

Equations (58) and (59) are the generalization to the two-band model of equations derived by Kanamori and Hasegawa (1971 for a one-band model. Numerical application of this formalism is described elsewhere (Brouers and Vedyayev 1972b).

D. The Residual Resistivity versus Concentration

Another useful application of the two-band model is the calculation of the residual resistivity in transition-metal alloys. The d bands of these alloys are unfilled and because of the s-d coupling, deviations from the x(1-x) Nordheim law are expected to occur and are actually observed in some transition-metal-based alloys. To calculate the static conductivity for the model described by Hamiltonian (1), we start with a relevant form of the Kubo-Greenwood formula

$$\sigma_{\alpha\beta} = \frac{2\ell^2}{\pi\Omega} \int d\eta \left(-\frac{df}{d\eta}\right) Tr\langle v_\alpha \, Im\, G(\eta+io) v_\beta \, Im\, G(\eta+io)\rangle \qquad (60)$$

The operators v and G are 2 x 2 matrices in the s,d space. The operators v_α and v_β in the model considered here are site independent and therefore we have to calculate essentially the average of a product of two Green-functions <GG>. This average can be written as the produce of two averaged Green functions plus corrections called 'vertex corrections',

$$\langle GG\rangle = \langle G\rangle\langle G\rangle + \text{vertex corrections} \qquad (61)$$

It has been shown (Brouers and Vedyayev 1972a) that in the present model where

(1) the random potentials are short-ranged,

(2) $\varepsilon_{s,d}(k) = \varepsilon_{s,d}(-k)$ and $v_{s,d}(k) = -v_{s,d}(-k)$, and

(3) the coupling constant γ is k-independent, these vertex corrections are zero in the framework of the CPA if we make the assumption that the matrix elements $\langle n_s|R|n_d\rangle$ where $\hat{R}$ is the site position operator are identically zero in accordance

with optical selection rules.

In a cubic lattice, equation (60) can be written as a sum of three contributions

$$\sigma = \sigma_{ss} + 2\sigma_{sd} + \sigma_{dd}$$

represented by the diagrams of figure 2.

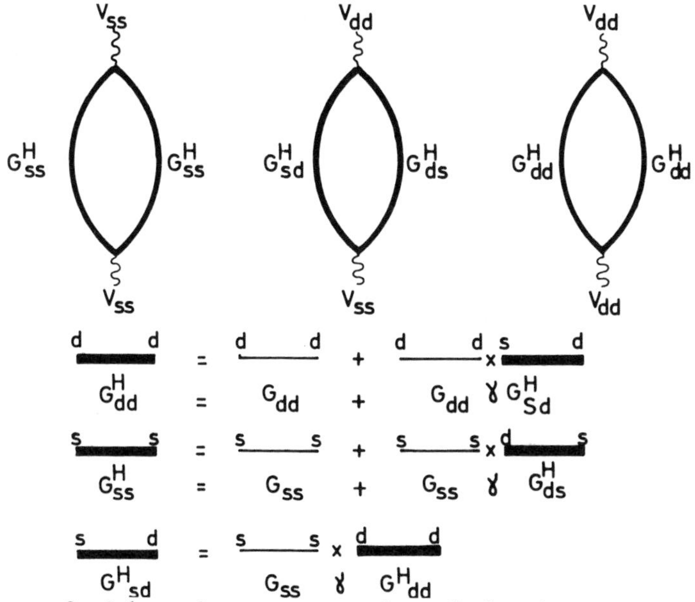

Figure 2 Schematic representation of the three contributions, σ_{ss}, σ_{dd} and σ_{sd}, to the electrical conductivity.

Other diagrams involving $\langle k_s|\Diamond|k_d\rangle$ are zero. This is a consequence of the various approximations mentioned above. The three contributions to the conductivity correspond to the propagation of a pair of s-electrons, a pair of d-electrons and two hybridized s-d electrons. The scattering of s-electrons is viewed therefore as caused indirectly by the randomness of d levels acting through the hybridization interaction s-d. In the dilute-concentration limit, it is possible to reduce this formalism to previous calculations of the impurity induced resistivity by defining, in that limit, an effective s-d scattering potential depending on γ (see Brouers and Vedyayev 1972a).

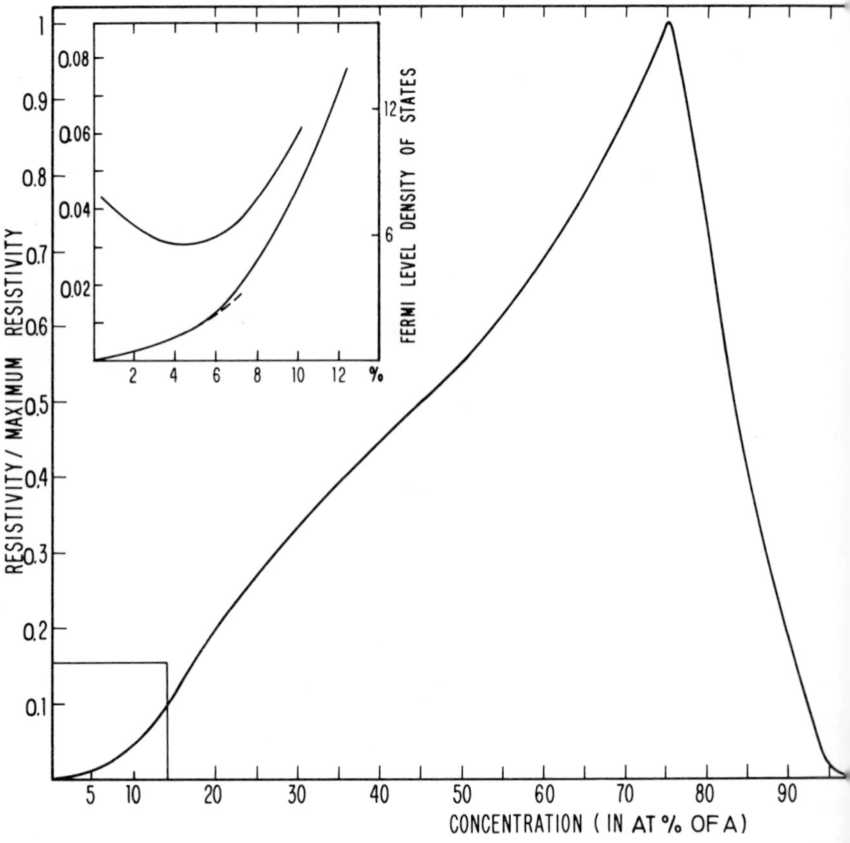

Figure 3 Total residual resistivity *vs* concentration in units of its maximum value. The region of low concentration is magnified in the upper part of the diagram, and shows the change of slope of the resistivity curve correlated to a minimum in the density of states. The asymmetry of the total curve, and the sharp maximum at 75at%, emphasizes the importance of s-d hybridization in this concentration range.

Figure 3 illustrates the result of a numerical application

of the resistivity ($\rho = 1/\sigma$) of the s-d band model. It is not directly related to any particular alloy but the physical parameters are reasonably chosen such that the computed resistivity reproduces qualitatively two types of deviation from the Nordheim behaviour observed in some transition metal based alloys: (a) a change of slope of the resistivity *vs* concentration curve correlated with a minimum of the specific heat observed in Pd-U alloys (Nellis *et al* 1970) and (b) an asymmetry of the resistivity curve with a peak in a range of concentration where the influence of s-d hybridization is important (for instance in palladium noble-metal alloys).

The physical parameters corresponding to figure 3 are $\alpha = 0.05$, $\delta = 0.4$ and $W_s = 7$eV.

The bottom of the s bands Γ_1 is supposed to be the same for A and B. The difference between Γ_1 and E_d^B is chosen to be 3eV. The number of s and d electrons of the two constituents n_B and n_A is taken to be 10 and 6 such that in pure metals, the d-bands are unfilled. More detail is given by Brouers and Vedyayev (1972a).

The main conclusions from the calculations are:

(1) When the Fermi level is far from the d-bands of the alloy, the dominant contribution to the conductivity is σ_{ss}.

(2) When the Fermi level lies in the middle of one of the d-bands, the three contributions ss, sd and dd are of the same order of magnitude due to strong indirect s-d scattering.

(3) The resistivity *vs* concentration law of figure 3 is characterized by a change of slope at about 6at% concentration correlated with a minimum of the density of states and a sharp maximum at 75at%. Such behaviour is related to the motion of the Fermi level. At 6at% the Fermi level is midway between the two d-bands, moving from one band to another (B to A). Between 6at% and 50at% the contribution of s conductivity is dominant and the behaviour is $\sim x(1-x)$. At about 50at% in our

numerical example, the Fermi level lies in the middle of the d-band where the three contributions σ_{ss}, σ_{sd} and σ_{dd} are of the same order of magnitude. This gives rise to a strong increase in resistivity over that predicted by the Nordheim rule, and to asymmetry of the curve. Although we did not introduce a direct s-d scattering potential, our formalism corroborates the physical interpretation described by Mott and Jones (1936) that the asymmetry is caused by a strong s-d scattering.

Acknowledgments

This research is supported in part by a grant (GP-16504) from the National Science Foundation and the Advanced Research Projects Agency. We are grateful also to Professor Ehrenreich for his kind hospitality, for suggesting this investigation, and for numerous stimulating discussions. We wish in addition to acknowledge helpful collaboration with Dr. L. Schwartz, and useful discussions with Dr. K. Levin.

APPENDIX

In this appendix we calculate an expression for the partial s-density of states $\rho_s^{A(B)}(E)$

$$\rho_s^{A(B)}(E) = (2/\pi) Im \left[\langle n_s | \frac{1}{\eta - H_{eff} - |n_d\rangle\langle n_d|(E_d^{A(B)} - \Sigma)} | n_s \rangle \right] \quad (A1)$$

or in the Bloch representation:

$$\rho_s^{A(B)}(E) =$$

$$(2/\pi) Im \left[\frac{1}{N} \sum_{k,k'} \langle k_s | \frac{1}{\eta - H_{eff} - |n_d\rangle\langle n_d|(E_d^{A(B)} - \Sigma)} | n_s \rangle e^{i(k_s - k_s')R_n} \right]$$

(A2)

The right-hand side of A2 can be expanded:

$$\rho_s{}^{A(B)}(E)=(2/\pi)Im\left[\frac{1}{N}\sum_{k,k'}<k_s|\frac{1}{\eta-H_{eff}}\left\{1+|n_d><n_d|\frac{(E_d^{A(B)}-\Sigma)}{\eta-H_{eff}}\right. + \right.$$

$$\left. + |n_d><n_d|\frac{(E_d^{A(B)}-\Sigma)}{\eta-H_{eff}}|n_d><n_d|\frac{E_d^{A(B)}-\Sigma)}{\eta-H_{eff}}|k_s'>e^{i(\mathbf{k}_s-\mathbf{k}_s')\mathbf{R}_n}\right] \quad (A3)$$

We then expand $|n_d>$ in Bloch functions. Since H_{eff} is diagonal in the **k**-space, we can write

$$\rho_s{}^{A(B)}(E) = (2/\pi)Im\left[\frac{1}{N}\sum_k <k_s|\frac{1}{\eta-H_{eff}}|k_s> + \frac{1}{N^2}\sum_{k,k'}<k_s|\frac{1}{\eta-H_{eff}}|k_d> x\right.$$

$$\left. x\ (E_d^{A(B)}-\Sigma)<k'_d|\frac{1}{\eta-H_{eff}}|k'_s> + \ \ldots\right] \quad (A4)$$

or more simply if we introduce F_{ss}, F_{dd} and F_{sd} defined by equations (35), (37) and (38).

$$\rho_s^{A(B)}(E) = (2/\pi)Im\left[F_{ss} + (E_d^{A(B)}-\Sigma)F_{sd}F_{ds} + (E_d^{A(B)}-\Sigma)^2F_{sd}F_{dd}F_{ds}\right.$$

$$\left. + (E_d^{A(B)}-\Sigma)^3\ F_{sd}F_{dd}F_{dd}F_{ds} + \ldots\right] \quad (A5)$$

and finally by summing up the series:

$$\rho_s^{A(B)}(E) = (2/\pi)Im\left[F_{ss} + (E_d^{A(B)}-\Sigma)\ F_{sd}F_{ds}\left[\frac{1}{1-F_{dd}(E_d^{A(B)}-\Sigma)}\right]\right] \quad (A6)$$

REFERENCES

Brouers, F. and Vedyayev, A.V. (1972a); Phys. Rev. B5, 348.

Brouers, F. and Vedyayev, A.V. (1972b); in preparation.

Brouers, F., Ehrenreich, H., Schwartz, L. and Vedyayev, A.V. (1971), Proceedings "Perspective for Computation of Electronic Structures in Ordered and Disordered Solids", Menton, 22-25 September 1971.

Kanamori, J. and Hasegawa, H. (1971); J. de Physique C1-280.

Levin, K. and Ehrenreich, H. (1971); Phys. Rev. B3, 4172.

Mott, N.F. and Jones, H. (1936); The Theory of the Properties of Metals and Alloys, p297: (Dover, New York 1958).

Nellis, W.J., Brodsky, M.B., Montgomery, H. and Pells, G.P. (1970); Phys. Rev. B2, 4590.

Nellis, W.J. and Brodsky, M.B. (1971); J. of Appl. Phys. (in press).

Rivory, J. (1969); Opt. Commun. 1, 53.

Schwartz, L., Brouers, F., Vedyayev, A.V. and Ehrenreich, H. (1971); Phys. Rev. B4, 3383.

Soven, P. (1967); Phys. Rev. 156, 809.

Soven, P. (1969); Phys. Rev. 178, 1138.

Velicky, B., Kirkpatrick, S. and Ehrenreich, H. (1968); Phys. Rev. 175, 747.

THEORY OF X-RAY SPECTRA OF DISORDERED ALLOYS

O.B. Sokolov and Yu. A. Babanov

Institute of Metal Physics, Academy of Sciences of the USSR, Sverdlovsk.

1. INTRODUCTION

X-ray spectroscopy has been used for the study of disordered alloys by many investigators. The accumulated data shows several regularities, but interpretation is difficult due to the absence of a theory of x-ray emission and absorption for disordered substitutional alloys.

The Green-function method has recently been employed for calculation of the x-ray spectrum of a dilute alloy (Babanov and Sokolov 1971). In the present investigation intensity formula for multi-component disordered alloys of arbitrary concentration are derived, taking into account electron-electron interaction in the valence band. Correlation of valence electrons with electrons in deep-state levels has been neglected; an effect which is important in a narrow region of spectrum, close to the Fermi level, investigated in detail by Mahan (1967) and Nozières and de Dominicis (1969).

2. DERIVATION OF FORMULA OF INTENSITY OF X-RAY ABSORPTION AND EMISSION

Let the system of electrons in the field of an electromagnetic wave be described by the Hamiltonian

$$H = H_0 + H_{int} \qquad (1)$$

where H_0 is the Hamiltonian in the absence of the field, and the second term accounts for the interaction of the system with this outer field.

Let us chose a certain orthonormal set of one-electron functions, on which the basis of the secondary quantanization is built. We designate the wavefunctions of inner electrons as $|Rn\rangle$, and valence electrons as $|R\ell\rangle$. The index R refers to the lattice site, ℓ to the valence band, and n to the inner state.

Let us consider the electrons in the deep-lying inner levels as not interacting among themselves nor with the valence electrons. Then the Hamiltonian H_0 may be represented as

$$H_0 = H_0' + \sum_{Rn} \varepsilon_{Rn}\, a^{\dagger}_{Rn}\, a_{Rn} \qquad (2)$$

where $a^{\dagger}$ and a are the operators of creation and annihilation, and ε_{Rn} is the energy of the deep state level. H_0' contains only the operators of creation and annihilation of valence electrons, and includes terms describing their correlation.

The Hamiltonian of interaction of the system with the outer field may be written

$$H_{int} = \hat{H}e^{-i\omega t} + \hat{H}^{\dagger}e^{i\omega t} \qquad (3)$$

where

$$\hat{H} = \sum_{\substack{R_1\ell_1 \\ R_2n_2}} (W_{R_1\ell_1;R_2n_2}\, a^{\dagger}_{R_1\ell_1}\, a_{R_2n_2} + W_{R_2n_2;R_1\ell_1}\, a^{\dagger}_{R_2n_2}\, a_{R_1\ell_1}) \qquad (4)$$

with

$$W = -\frac{e}{mc}\, e^{i\mathbf{qz}} \boldsymbol{A}_0 \rho \qquad (5)$$

in which the electromagnetic field is described by a linear polarized plane wave, with wave vector $\mathbf{q}$ and complex polarization vector $\boldsymbol{A}_0$.

Prior to absorption the system is in the ground state $|0\rangle$; then, for emission, let us consider that the initial state is a certain excited stationary state $|i\rangle$ with a deep-level hole. The electromagnetic field causes the system to transfer from the initial state to the final stationary state. In the case of emission a spontaneous transition is possible, but taking this into account leads only to a change of the coefficient in the intensity formula and does not reflect on the shape of the spectrum. In the linear approximation for the outer field, the transition rate $I(\omega)$, by which is meant the probability of absorption or emission in a time unit of energy quantum ω, is given by

$$I_a(\omega) = 2\pi\sum_m \delta(E_m - E_0 - \omega)\langle 0|\hat{H}^\dagger|m\rangle\langle m|\hat{H}|0\rangle \qquad (6)$$

for absorption

and

$$I_e(\omega) = 2\pi\sum_m \delta(E_m - E_i + \omega)\langle i|\hat{H}|m\rangle\langle m|\hat{H}^\dagger|i\rangle \qquad (7)$$

for emission.

Here E_m and $|m\rangle$ are the eigenvalues of the energy and the eigenstates of the Hamiltonian H_0; that is, $H_0|m\rangle = E_m|m\rangle$.

From the structure of the Hamiltonian H_0 it follows that the state $|m\rangle$ may be characterized by the number and position of electrons in the deep levels. Let N_n^m electrons be in the deep levels of atoms occupying the sites R, noting that some of the states may be empty. Then

$$|m\rangle = \prod_{Rn}^{N_n^m} a_{Rn}^\dagger A_{m'}^\dagger(N_\ell^m)|vac\rangle \equiv \prod_{Rn}^{N_n^m} a_{Rn}^\dagger|m'\rangle \qquad (8)$$

The operator $A_{m'}^\dagger(N_\ell^m)$ acting on the vacuum function creates eigenstates $|m'\rangle$ of the Hamiltonian H_0', corresponding to valence electrons

$$H_0'|m'\rangle = E_{m'}|m'\rangle \tag{9}$$

It is easy to show that the eigenvalue

$$E_m = E_{m'} + \sum_{Rn} \varepsilon_{Rn} \tag{10}$$

of the operator energy H_0, corresponds to the state $|m\rangle$.

Knowing the stationary states between which the transition takes place it is simple to carry out an analysis of the matrix elements of the operator $\hat{H}$ in expressions (6) and (7).

We discuss first the case of absorption. In the ground state for the system there are N electrons, and $N = N_\ell^0 + N_n^0$; N_ℓ^0 is the number of the valence electrons, and N_n^0 the number of electrons in the deep-state levels which in this case are all occupied. Then, according to (8)

$$|0\rangle = \prod_{Rn}^{N_n^0} a_{Rn}^{\dagger} A_{0'}^{\dagger}(N_\ell^0)|vac\rangle \tag{11}$$

We substitute (9) and (11) into the expression for the matrix element $\langle m|\hat{H}|0\rangle$, giving

$$\langle m|\hat{H}|0\rangle = \langle vac|A_{m'}(N_\ell^m)\left(\prod_{Rn}^{N_n^m} a_{Rn}^{\dagger}\right)^{\dagger}\hat{H}\prod_{R_1n_1}^{N_n^0} a_{R_1n_1}^{\dagger} A_{0'}^{\dagger}(N_\ell^0)|vac\rangle \tag{12}$$

The operator $\hat{H}$, equation (4), is a sum of two terms. If we substitute (4) in (12) then the matrix element of the second term in $\hat{H}$ is always equal to zero; this is so because in any term of the sum over R_2n_2 a combination $(a_{R_2n_2}^{\dagger})^2$ arises after a corresponding number of anticommutations, which acting on the vacuum function gives zero. The matrix element of the first term in $\hat{H}$ differs from zero only when there is one hole in the deep level in the $|m\rangle$ state, and $N_\ell^m = N_\ell^0+1$. We characterize the state $|m\rangle$ by an index indicating the place of the hole, $|m\rangle = |m',Rn\rangle$. Then

$$\langle m|\hat{H}|0\rangle = \sum_{R_1\ell_1} W_{R_1\ell_1;Rn}\langle m'|a^{\dagger}_{R_1\ell_1}|0'\rangle \tag{13}$$

paying no attention to sign.
In equation (13) the matrix element $a^{\dagger}_{R_1\ell_1}$ is determined on the states describing the system of valence electrons. Expression (6) now becomes

$$I_a(\omega) = 2\pi\sum_{m'Rn}\delta(E_{m'}-E_{0'}-\omega-\varepsilon_{Rn})|\langle m'|\sum_{R_1\ell_1} W_{R_1\ell_1;Rn}a^{\dagger}_{R_1\ell_1}|0'\rangle|^2$$

Summing over m is substituted in this case by summing over R, n and m'. Using the relation

$$\delta(\omega) = \frac{lim}{\delta\to+0}\frac{1}{\pi}\,Re\int_0^{\infty} dt\; e^{-i\omega t-\delta t}$$

we obtain

$$I_a(\omega) = 2Re\sum_{m'Rn}\sum_{R_1\ell_1}\sum_{R_2\ell_2}\int_0^{\infty} dt\; e^{i(\omega+\varepsilon_{Rn})t-\delta t}$$

$$\times\; W^{\dagger}_{Rn;R_1\ell_1}\langle 0'|a_{R_1\ell_1}(t)|m'\rangle\langle m'|a^{\dagger}_{R_2\ell_2}(0)|0'\rangle W_{R_2\ell_2;Rn}$$

where $a_{R_1\ell_1}(t) = e^{iH_0't}a_{R_1\ell_1}e^{-iH_0't}$

and after summarizing by excited states $|m'\rangle$ we have

$$I_a(\omega)=2Re\,i\sum_{Rn}\sum_{\substack{R_1\ell_1\\R_2\ell_2}}\int_0^{\infty} dt\, e^{i(\omega+\varepsilon_{Rn})t-\delta t}W^{\dagger}_{Rn;R_1\ell_1}G_{R_1\ell_1;R_2\ell_2}(t)W_{R_2\ell_2;Rn} \tag{14}$$

where $G_{R_1\ell_1;R_2\ell_2}(t)$ is the one-particle causal Green function

$$G_{R_1\ell_1;R_2\ell_2}(t) = \frac{1}{i}\langle 0'|T\; a_{R_1\ell_1}(t)a^{\dagger}_{R_2\ell_2}(0)|0'\rangle$$

For convenience we pass to the energetic representation. After expansion of the Green function in Fourier integral, and time integration, we find

$$I_a(\omega) = -Re \sum_{Rn} \sum_{\substack{R_1\ell_1 \\ R_2\ell_2}} W^{\dagger}_{Rn;R_1\ell_1} W_{R_2\ell_2;Rn} \int_{-\infty}^{\infty} \frac{dE}{\pi} \frac{G_{R_1\ell_1;R_2\ell_2}(E)}{\omega+\varepsilon_{Rn}-E+i\delta} \tag{15}$$

It is well known that the causal Green function, being determined on the real axis, coincides at $E<\mu$ with the advanced function G^a, and at $E>\mu$ with the retarded function G^z (μ is the chemical potential of the system). Hence let us split the integral in (15) into two integrals

$$\int_{-\infty}^{\infty} \frac{dE}{\pi} \frac{G_{R_1\ell_1;R_2\ell_2}(E)}{\omega+\varepsilon_{Rn}-E+i\delta} = \int_{-\infty}^{\mu} \frac{dE}{\pi} \frac{G^a_{R_1\ell_1;R_2\ell_2}(E)}{\omega+\varepsilon_{Rn}-E+i\delta} + \int_{\mu}^{\infty} \frac{dE}{\pi} \frac{G^z_{R_1\ell_1;R_2\ell_2}(E)}{\omega+\varepsilon_{Rn}-E+i\delta}$$

$$= \int_{-\infty}^{\infty} \frac{dE}{\pi} \frac{G^a_{R_1\ell_1;R_2\ell_2}(E)}{\omega+\varepsilon_{Rn}-E+i\delta} + \int_{\mu}^{\infty} \frac{dE}{\pi} \frac{G^z_{R_1\ell_1;R_2\ell_2}(E)-G^a_{R_1\ell_1;R_1\ell_1}(E)}{\omega+\varepsilon_{Rn}-E+i\delta}$$

The first of these integrals is equal to zero, since G^a is analytical in the lower semi-plane. Furthermore, using the symbolic identity

$$\frac{1}{\omega+i\delta} = \rho\,\frac{1}{\omega} - i\pi\delta(\omega)$$

and the known relation $G^z_{R_1\ell_1;R_2\ell_2} = (G^a_{R_2\ell_2;R_1\ell_1})^*$ we find

$$I_a(\omega)=2Im\sum_{Rn} \theta(\omega+\varepsilon_{Rn}-\mu) \sum_{\substack{R_1\ell_1 \\ R_2\ell_2}} W^{\dagger}_{Rn;R_1\ell_1} G^a_{R_1\ell_1;R_2\ell_2}(\omega+\varepsilon_{Rn}) W_{R_2\ell_2;Rn} \tag{16}$$

We now discuss the process of emission. As stated earlier, the system prior to emission has a hole in the deep level n at site R, and the valence electrons are unexited. Thus

$$|i\rangle = a_{Rn} \prod_{R_1 n_1}^{N_n^o} a^{\dagger}_{R_1 n_1} A^{\dagger}_{o'}(N_\ell^o)|vac\rangle \tag{17}$$

Following the arguments used in the case of absorption, it can be shown that the matrix element $\langle i|\hat{H}|m\rangle$ has the form

$$\langle i|\hat{H}|m\rangle = \sum_{R_1 \ell_1} W_{R_1\ell_1;Rn} \langle 0'|a^{\dagger}_{R_1\ell_1}|m'\rangle \tag{18}$$

Substituting (18) in the expression for the intensity of emission (7), the same transformations give

$$I_e(\omega) = -2Re\, i \sum_{\substack{R_1\ell_1 \\ R_2\ell_2}} \int_{-\infty}^{o} dt\, e^{i(\omega+\varepsilon_{Rn})t+\delta t} W^{\dagger}_{Rn;R_1\ell_1} G_{R_1\ell_1;R_2\ell_2}(t) W_{R_2\ell_2;Rn}$$

Hence the intensity of x-ray emission can be expressed by a one-particle advanced Green function of valence electrons

$$I_e(\omega) = 2Im\,\theta(\mu-\omega-\varepsilon_{Rn}) \sum_{\substack{R_1\ell_1 \\ R_2\ell_2}} W^{\dagger}_{Rn;R_1\ell_1} G^{a}_{R_1\ell_1;R_2\ell_2}(\omega+\varepsilon_{Rn}) W_{R_2\ell_2;Rn} \tag{19}$$

In deriving this expression we assumed that the initial state for emission was a state with only one hole in the deep level. In the expression for absorption (16), $\sum_{Rn}$ indicates that the electron may be extracted from any state n at any site R.

The observed spectrum of a given element is characterized by a series (K, L etc.). In case of an ideal matrix consisting of atoms of one kind, for the separation of the series expressed in equations (16) and (19) it is necessary to identify the deep-state level involved in the transition. Therefore, if we ignore the different coefficients for emission and absorption, we may write a single expression

$$I^n(\omega-\varepsilon_n) \simeq Im \sum_{\substack{R_1\ell_1\\R_2\ell_2}} W^{\dagger}_{On;R_1\ell_1} G^a_{R_1\ell_1;R_2\ell_2}(\omega) W_{R_2\ell_2;On} \qquad (20)$$

This equation describes the emission spectrum when $\omega<\mu$, and the absorption spectrum when $\omega>\mu$. The site '0' indicates any emission or absorption site, which due to the identity of the sites is located at the origin. ω is the energy renormalized to the value of the deep level energy.

For a multicomponent alloy, in addition to separating the series, it is necessary to produce averaging over all configurations of the environment of the emitting or absorbing atom. If we place this atom at the origin, then the intensity formula for emission and absorption will become

$$I_\alpha(\omega-\varepsilon_{\alpha,n}) \simeq Im \sum_{\substack{R_1\ell_1\\R_2\ell_2}} \overline{W^{\dagger}_{On;R_1\ell_1} G^a_{R_1\ell_1;R_2\ell_2}(\omega) W_{R_2\ell_2;On}} \qquad (21)$$

The bar indicates the averaging over configurations, and the index α denotes the kind of atom at the emitting or absorbing site. Selection rules are included in the matrix elements W of expressions (20) and (21). In all subsequent equations the sign of the Green function a will be omitted.

Thus for calculation of the spectrum intensity, it is necessary to know not only the matrix elements of Green functions, but also the single particle wavefunctions $|R\ell\rangle$ and $|Rn\rangle$. These functions, as already mentioned, form a complete orthonormal set. They can be calculated with the help of atomic functions using the expansion in terms of degree of overlap with any desirable accuracy, as shown by Bogoliubov (1970). Let us consider that the atomic functions of the deep-level electrons at site R do not overlap with atomic functions for the other sites. It is natural that the atomic functions relating to one site are orthonormalized, and that because of this the basic function $|Rn\rangle$ will coincide with the atomic function $|Rn\rangle_a$. The basic function of valence electrons $|R\ell\rangle$ is the linear combination of atomic functions $|R\ell\rangle_a$. For example, in the first order of degree of overlap we have

$$|R\ell\rangle = |R\ell\rangle_a - \frac{1}{2}\sum_{\delta,\ell'} |R+\delta,\ell'\rangle_a \; {}_a\langle R+\delta,\ell'|R\ell\rangle_a \qquad (22)$$

from which we can see that the atomic functions for neighbouring sites, weighted proportional to their overlap, are included in the basic function of valence electrons.

In so far as the basic functions can be found only approximately, a common principle is necessary for the calculation of the Green function G, and the matrix element of transition W. Expansion of the total Hamiltonian H, determined by equation (1), provides this principle. Then the Green function with corresponding Hamiltonian H_o' is sought.

Thus, in this section, expressions for the intensity of emission and absorption have been derived, with the electron-electron correlation in the valence band taken into account.

3. ONE-ELECTRON THEORY OF X-RAY SPECTRA OF DISORDERED BINARY ALLOYS

The theory developed above may be illustrated by an example of a binary alloy, of arbitrary concentration, forming a simple cubic lattice. We assume that each atom has only one valence s-state. The eigenfunction for this state is compact and overlaps only the wavefunctions of neighbouring atoms.

In the one-electron approximation, the Hamiltonian H_o' has the form

$$H'_o = \sum_{R_1 R_2} L_{R_1 \ell; R_2 \ell} a^\dagger_{R_1 \ell} a_{R_2 \ell} \tag{23}$$

with $\hat{L} = \hat{T} + \hat{V}$
where $\hat{T}$ is the electron kinetic energy operator and $\hat{V}$ the lattice potential.

To second order in the overlap of atomic functions $|R\ell\rangle_a$, and using an expansion of the basic function (22), the matrix element of the operator L can be written

$$L_{R_1 \ell; R_2 \ell} = {}_a\langle R_1 \ell | T + V | R_2 \ell\rangle_a - \frac{1}{2}{}_a\langle R_1 \ell | R_2 \ell\rangle_a ({}_a\langle R_1 \ell | T+V | R_1 \ell\rangle_a + {}_a\langle R_2 \ell | T+V | R_2 \ell\rangle_a)(1-\delta_{R_1 R_2}) \tag{24}$$

and separating diagonal and nondiagonal parts

$$L^d_{R_1 \ell; R_2 \ell} = {}_a\langle R_1 \ell | T + V | R_1 \ell\rangle \delta_{R_1 R_2} \equiv E_{R_1} \delta_{R_1 R_2} \tag{25}$$

which takes the value E_A or E_B according to whether site R_1 is occupied by an atom of type A or of type B. The nondiagonal terms differ from zero when R_1 and R_2 are neighbours, then

$$L^{nd}_{R\ell; R+\delta, \ell} = {}_a\langle R\ell | T + V | R + \delta, \ell\rangle_a - \frac{1}{2}(E_R + E_{R+\delta})\, {}_a\langle R\ell | R + \delta, \ell\rangle_a \tag{26}$$

This matrix element does not depend on which atoms occupy the sites R and R + δ.

Thus the Hamiltonian H'_o, with which the Green function should be calculated, is

$$H'_o = \sum_{R_1 R_2} (E_{R_1} \delta_{R_1 R_2} + L \sum_{\delta} \delta_{R_2, R_1 + \delta}) a^\dagger_{R_1 \ell} a_{R_2 \ell} \tag{27}$$

This is the model of an alloy first used by Soven (1969).

For calculating a spectrum using expression (21) it is necessary to know the matrix element W to second order of overlap. Expanding as before the basic function $|R\ell\rangle$, we have

$$W_{On;R\ell} = {}_a\langle On|W|R\ell\rangle_a - \frac{1}{2}\sum_{\delta} {}_a\langle On|W|R+\delta,\ell\rangle_{a\,a}\langle R+\delta,\ell|R\ell\rangle_a \quad (28)$$

The first term of this expression is diagonal due to the fact that the atomic function of the deep level electrons, at the site '0', does not overlap those of other sites. From (28) we see that only the term for which $R+\delta=0$ remains in the summation. Then

$$W_{On;R\ell} = {}_a\langle On|W|0\ell\rangle_a \delta_{RO} - \frac{1}{2}\,{}_a\langle 0\ell|R\ell\rangle_{a\,a}\langle On|W|0\ell\rangle_a(1-\delta_{RO}) \quad (29)$$

We use the designations

$${}_a\langle On|W|0\ell\rangle_a = W \text{ and } -\frac{1}{2}\,{}_a\langle 0\ell|\delta\ell\rangle_a = S \quad (30)$$

Then (29) becomes

$$W_{On;R\ell} = W\delta_{RO} + SW\sum_{\delta}\delta_{R,\delta} \quad (31)$$

Substituting (31) into the intensity expression (21), we have

$$I^n_\alpha(\omega-\varepsilon_{\alpha,n}) \simeq |W|^2 Im\{\overline{G_{oo}(\omega)}+S\sum_{\delta}[\overline{G_{o\delta}(\omega)}+\overline{G_{\delta o}(\omega)}]+S^2\sum_{\delta_1\delta_2}\overline{G_{\delta_1\delta_2}(\omega)}\} \quad (32)$$

noting that site '0' is occupied by an atom of the type α, and that δ, δ_1, δ_2 are the radius-vectors of near neighbours. In the final term of expression (33) we retain only the main contribution $\sum_{\delta}\overline{G_{\delta\delta}}$; i.e. $\delta_1 = \delta_2 = \delta$.

Hence, an expansion of the intensity expression over degrees of overlap of the atomic functions is obtained. In deriving equation (32) we have confined ourselves to the first coordinate sphere. And we have found a regular procedure for such expansion.

Thus the problem of calculating the intensity of x-ray spectra has taken the form of a problem for finding the averaged Green function of valence electrons. We shall calculate these functions in the coherent potential approximation taking account of pair correlation. Namely the Green function of an 'ideal' matrix is given by

$$G_{R_1R_2}(E) = \frac{1}{N}\sum_{\mathbf{K}} \frac{e^{iK(R_1-R_2)}}{E-L\sum_{\delta} e^{iK\delta}-\Sigma(E)} \tag{33}$$

where $\Sigma(E)$ is found from the condition $\overline{t_R} = 0$. Here t_R is the effective scattering potential determined by

$$t_R = \frac{E_R-\Sigma}{1-(E_R-\Sigma)G^o_{oo}} \tag{34}$$

The perturbation introduced by each atom at the site R is equal to $E_R-\Sigma$. Pair correlation is taken account of as a two-impurity problem in finding G_{oo} and $G_{o\delta}$, the impurities being located on neighbouring sites:

$$G_{oo} = G^o_{oo} + G^o_{oo}\, t_o\, G^o_{oo} + \frac{(1+t_o G^o_{oo})^2 G^o_{o\delta} t_\delta G^o_{\delta o}}{1-t_o G^o_{o\delta} t_\delta G^o_{\delta o}} \tag{35}$$

$$G_{o\delta} = G^o_{o\delta}\, \frac{(1+t_o G^o_{o\delta})(1+t_\delta G^o_{\delta o})}{1-t_o G^o_{o\delta} t_\delta G^o_{\delta o}} \tag{35}$$

The Green function $G^o_{o\delta}$, in (35) and (36), is for a simple cubic lattice given by

$$G^o_{o\delta}(E) = \frac{1}{3}\,[1-(E-\Sigma)G^o_{oo}] \tag{37}$$

Using the property $G^o_{o\delta} = G^o_{\delta o}$ it can be shown that $G_{o\delta} = G_{\delta o}$.

In a disordered substitutional alloy each site has the

probabilities C_A and C_B of being occupied by atoms A or B. If the site '0' is occupied by atom A, then, according to expressions (35) and (36), for the averaged Green function, we have

$$\overline{G_{oo}} \equiv g_A = G^o_{oo} + G^o_{oo}\, t_A\, G^o_{oo} + Z(1+t_A G^o_{oo})^2 G^o_{o\delta} G^o_{\delta o}\left(\frac{C_A t_A}{1-t_A G^o_{o\delta} G^o_{\delta o}} + \frac{C_B t_B}{1-t_A t_B G^o_{o\delta} G^o_{\delta o}}\right) \tag{38}$$

$$\overline{G_{o\delta}} \equiv C_A g_{AA} + C_B g_{AB} = C_A G^o_{o\delta}\,\frac{(1+t_A G^o_{oo})^2}{1-t_A^2 G^o_{o\delta} G^o_{\delta o}} + C_B G^o_{o\delta}\,\frac{(1+t_A G^o_{oo})(1+t_B G^o_{oo})}{1-t_A t_B G^o_{o\delta} G^o_{\delta o}} \tag{39}$$

$$\overline{G_{\delta\delta}} \equiv C_A g_A + C_B g_B \tag{40}$$

In equations (38) to (40) Z is the number of nearest neighbours (equal to 6 for a simple cubic lattice), and t_A and t_B are found from the formula (34) with the site R occupied respectively by atom A or B. The function g_B is $\overline{G_{oo}}$ when there is a B atom at site '0', and is determined by formula (38) on replacing A or B. Pair correlation is described by the third term in (38).

Finally, we have for the spectrum intensity for an A atom

$$I^n_A(\omega-\varepsilon_{A,n}) \simeq |W|^2 Im[g_A+2ZS(C_A g_{AA}+C_B g_{AB})+ZS^2(C_A g_A+C_B g_B)] \tag{41}$$

and for a B atom

$$I^n_B(\omega-\varepsilon_{B,n}) \simeq |W|^2 Im[g_B+2ZS(C_B g_{BB}+C_A g_{BA})+ZS^2(C_B g_B+C_A g_A)] \tag{42}$$

The functions g_{BB} and g_{BA}, in equation (42), result from functions g_{AA} and g_{AB} on replacement of A by B; giving $g_{AB} = g_{BA}$.

4. RESULTS AND DISCUSSION

The functions g_A, g_B, g_{AA}, g_{AB} and g_{BB} in expressions (41) and (42) were calculated for a wide range of concentrations of the alloy components, C_B and $C_A = 1-C_B$. The energy separation of the levels E_A and E_B is $E_A-E_B = 2\Delta$. Zero energy is chosen mid-way between these levels, $E_A = \Delta$ and $E_B = -\Delta$. A one sixth part of the width of an ideal matrix was taken for the energy unit of measurement.

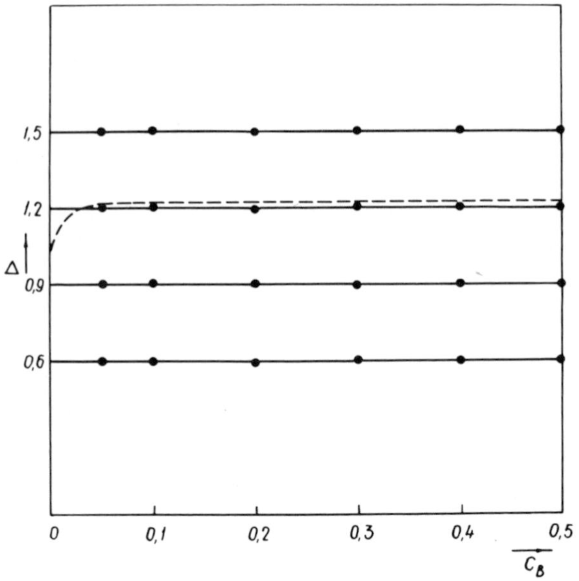

Figure 1 Regimes of the coherent potential approximation for simple cubic lattice. Below the dashed line the band is unsplit. The coordinates of the dots indicate values of parameters Δ and C_B at which the calculation was made.

In figure 1 the coordinates of the points indicate the values of parameters for which the calculation was made. Limitations are imposed upon the parameter Δ because when Δ compares with the half-width of the band of the ideal matrix, the assumption that the transition matrix element of expression

(26) is independent of the kind of atom occupying the appropriate site ceases to be valid. The concentration was always taken such that C_B was within the limits 0-0.5, in accordance with the symmetry of the problem discussed by Velický *et al* (1968). Green functions for concentrations $C_B > 0.5$ are easily found from those for the concentrations $C_B < 0.5$. We have

$$Im g'_A(\omega) = Im g_B(-\omega); \quad Im g'_B(\omega) = Im g_A(-\omega); \qquad (43)$$

and

$$Im g'_{AA}(\omega) = - Im g_{BB}(-\omega); \quad Im g'_{AB}(\omega) = - Im g_{BA}(-\omega). \qquad (44)$$

where the 'prime' indicates the unknown function.

Figure 1 is divided into two regions. Below the dotted line the density of states for the alloy is represented in a single band, and above the line it is splitted into two subbands. This division, into regions of the parameters Δ and C_B, has been discussed by Velický *et al* (1968).

Figure 2 shows the local density curves g_A and g_B calculated with and without pair correlation taken into account, together with the density of states for disordered alloy G^o_{oo} in the coherent-potential approximation. It can be seen from figure 2 that pair correlation changes the shape of the functions g_A and g_B insignificantly when the energy separation the atomic levels is less than one.

The effect of multiple scattering among the neighbouring sites is considerable at $\Delta = 1.2$. It should be noted that this value of Δ is close to critical at which splitting of the band takes place. With decrease of concentration C_B a characteristic peak on the local density curve g_B appears, and sharpens. The peculiarities on the edges of the band (at $\Delta = 1.2$ and $C_B = 0.1$ for example) are the results of inaccurate determination of the band boundaries, which are determined

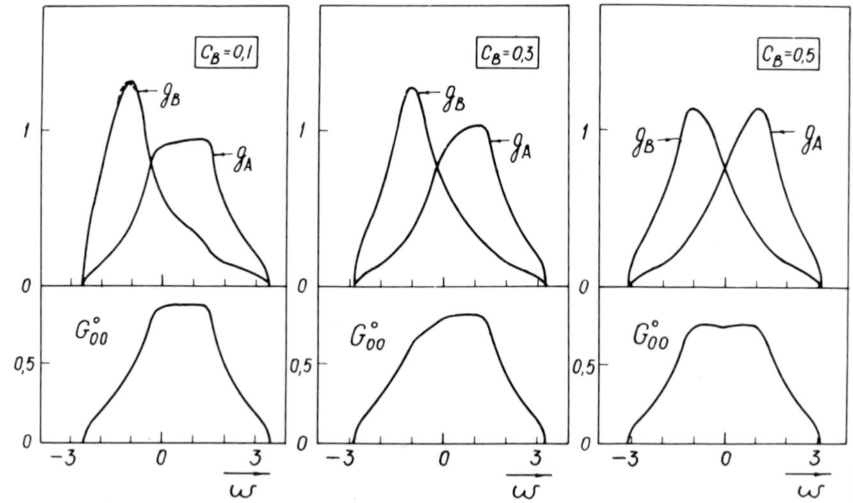

Figure 2 (a)

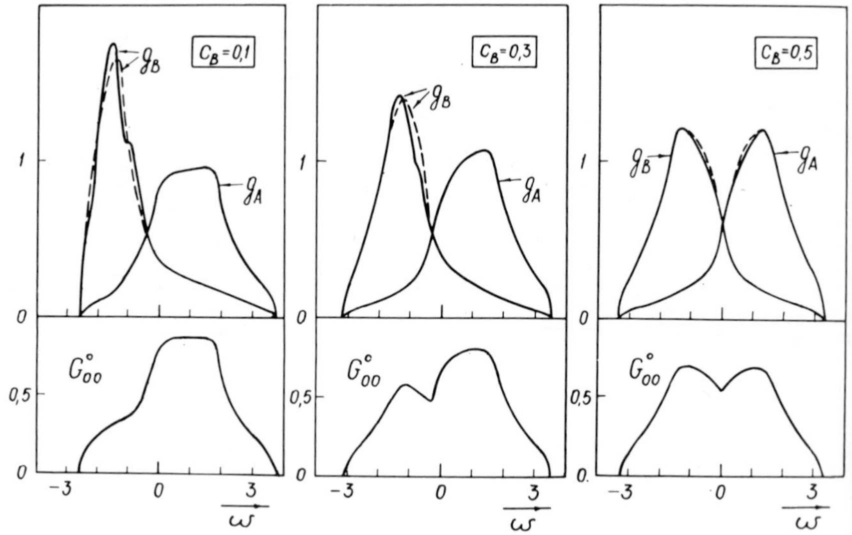

Figure 2 (b)

Figure 2 Local density-of-states curves g_A and g_B, calculated with pair correlation taken into account (solid line) and not taken into account (dashed line), and density of states of the disordered alloy in the coherent potential approximation, for a variety of concentrations C_B, and with: (a) parameter $\Delta = 0.6$; (b) parameter $\Delta = 0.9$; (c) parameter $\Delta = 1.2$.

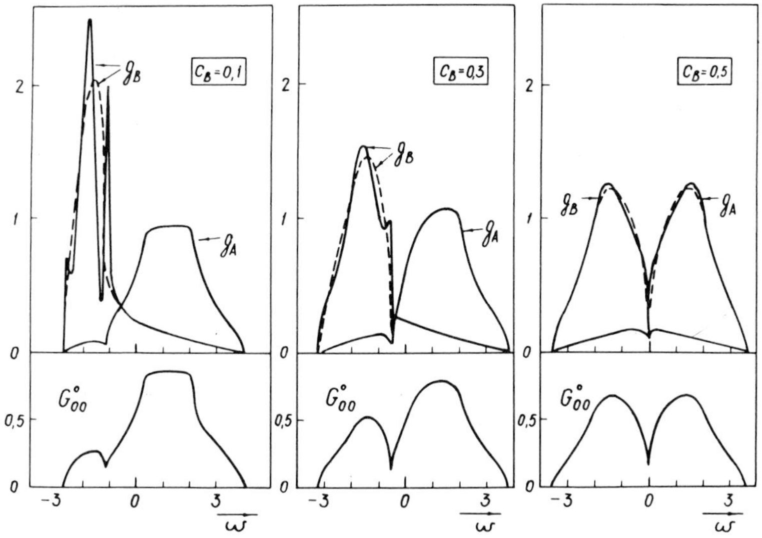

Figure 2 (c)

from the condition $Im\sum = 0$. Therefore, it would be correct to calculate the local density with a Green function G^o_{oo} in which the self-energy $\sum$ is corrected for pair correlation. This was first noted by Lifshits (1964) and Soven (1969). In the case of split bands, the effect of pair correlation is either insignificant (at concentrations ~ 0.5) or leads to a contribution close to the boundaries, where the calculations are then incorrect.

According to expressions (41) and (42), the contribution to the intensity of the x-ray spectrum to first order of overlap of the atomic functions is governed by the functions g_{AA}, g_{AB}, g_{BA}, g_{BB}. These determine the probability of transition of an electron from the site occupied by, say, atom A onto the site occupied by atom B (function g_{AB}). Figures 3 and 4 show the imaginary parts of these functions. The rôle of pair correlation may be seen by comparing them with the Green function of an 'ideal' matrix $G^o_{o\delta}$.

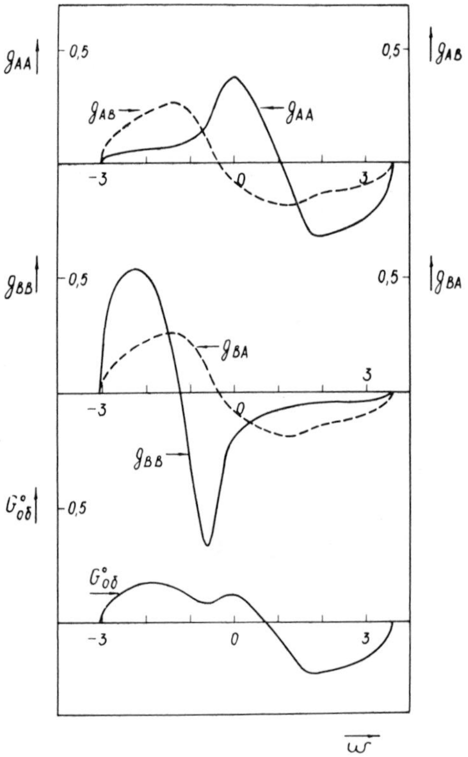

Figure 3 The imaginary parts of the functions g_{AA}, g_{AB}, g_{BA}, g_{BB} and $G^{0}_{0\delta}$, for Δ = 0.9 and C_B = 0.3.

Thus we have discussed the components making up the intensity of the spectrum, whose shape depends on the overlap of the valence-electron wavefunctions. This may be illustrated by examining two extreme cases: S = 0 and S = - 0.1 (see expression 30). At S = 0 the spectrum shape is determined by the local density g_A or g_B shown in figure 2. The case with S = -0.1 may be regarded separately since, for illustration of the theory in section 3, the tight-binding formalism was used.

Figure 5 shows the results for S = - 0.1. First we note the absence of symmetry, which is characteristic of the local density. Asymmetry of the spectrum is caused by the functions g_{AA}, g_{AB}, g_{BA} and g_{BB} giving a contribution in first order of

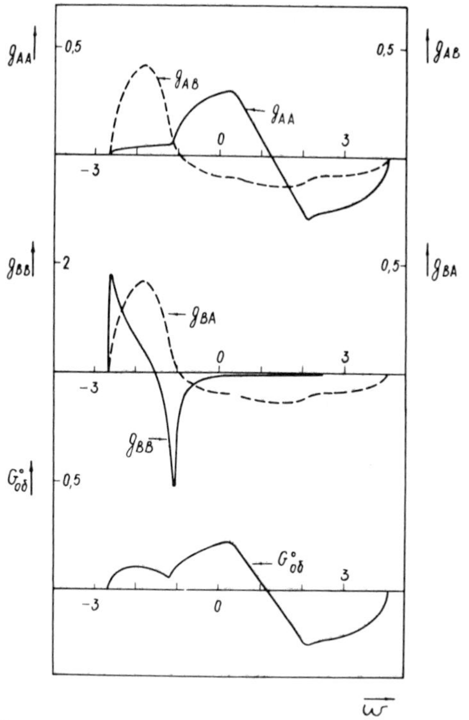

Figure 4 The imaginary parts of the functions g_{AA}, g_{AB}, g_{BA}, g_{BB} and $G^{o}_{o\delta}$ for $\Delta = 1.2$ and $C_B = 0.1$.

overlap, satisfying the correspondence rules (44). The correlation maximum on the intensity curve I_B, for $\Delta = 0.9$ and; $C_B = 0.3$, becomes more pronounced if we compare this curve with that for the local density (figure 2). This, it can easily be seen, is due to the peak of the function g_{BB}, located at the same energy. On going to the alloy, $\Delta = 0.9$ and; $C_B = 0.7$, we note that this peak is not observed in the curve I_A. This is the result of the asymmetry discussed above. In the curves of figure 5b we observe also the characteristic change of this peak. Another interesting peculiarity of the curves under consideration is a 'shoulder' to the right in the intensity curve I_B. This is caused by the second-order term

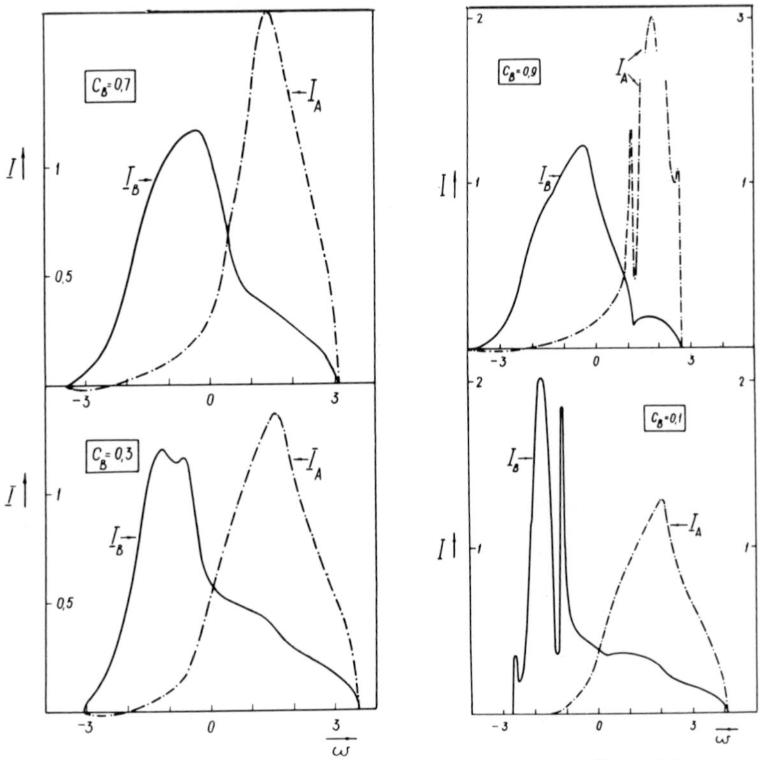

Figure 5 Intensity of the x-ray spectrum for alloys,
(a) : with $\Delta = 0.9$, and overlap constant $S = -0.1$.
(b) : with $\Delta = 1.2$, and overlap constant $S = -0.1$.

of overlap, and reproduces the average density of states of the alloy, as seen from the expressions (41) and (42).

The results obtained allow a qualitative understanding of the features observed in the x-ray spectra of alloys of transition elements. Comparison with experimental data is made elsewhere.

Acknowledgments

The authors wish to thank B.G. Zubko for the calculations, and acknowledge useful discussion with Professor Yu.A. Iziumov and Dr. M.V. Medvedev.

REFERENCES

Babanov, Yu. A. and Sokolov, O.B. (1971); Fiz. Metallov i Metalloved. 31, 675.

Bogoliubov, N.N. (1970); Collected Works, Vol. II, p.287 : Naukova Dumka, Kiev.

Lifshits, I.M. (1964); Uspekhi Fiz. Nauk 83, 617.

Mahan, G.D. (1967); Phys. Rev. 163, 612.

Nozières, P. and de Dominicis, C.T. (1969); Phys. Rev. 178, 1097.

Soven, P. (1969); Phys. Rev. 178, 1136.

Velický, B., Kirkpatrick, S. and Ehrenreich, H. (1968); Phys. Rev. 175, 747.

THE K-EMISSION SPECTRUM FROM TWO PHASES OF LITHIUM METAL

R.S. Crisp

Physics Department, University of Western Australia, Nedlands.

ABSTRACT*

Recent studies of soft x-ray emission spectra, in particular the spectra from binary alloys, have shown that while the chief features of the band structure of the solid may appear in the spectra, the influence of the core states is very strong and dominates the band shape. To explore this further the K-spectrum of lithium metal was recorded for both the normal bcc phase and the close-packed phase to which it transforms spontaneously when cooled to liquid-helium temperatures. The recorded spectrum is the same in both cases. A number of reasons for this are advanced, but we may conclude that the crystal environment and band structure play very little direct part in determining the spectrum, which is dominated by the core states via the transition probability.

REFERENCE

Crisp, R.S. (1972); Phil. Mag. 25, 167.

---

*This paper was presented at the Conference on which this Volume is based and has been published in full elsewhere (Crisp 1972).

Part 4

X-RAY ABSORPTION AND ISOCHROMAT SPECTROSCOPY OF ALLOYS AND DENSITY OF UNOCCUPIED ELECTRON STATES

X-RAY PROBES OF VACANT ENERGY BANDS

D.J. Nagel

U.S. Naval Research Laboratory, Washington, D.C. 20390

1. INTRODUCTION

A. Classification of Band-Spectroscopy Methods

A wide variety of methods is available for experimental study of the basic electronic structure of materials. In order to compare the results of the various techniques with each other and with bonding theory, it is necessary to know how the results of a particular kind of experiment depend on the electronic structure. Methods that depend on the same or similar features of electronic structure can be grouped together and contrasted. In the present paper we outline the various classes of electronic-structure experiment and then discuss empty-band techniques in detail. Emphasis is given to theoretical calculation of spectra. Modulation spectroscopy, which offers relief from the chronic resolution limitations of x-ray techniques, is also discussed.

For conductors, there exists a large class of experimental techniques sensitive to electron states at or very near the Fermi level. A sub-class consists of the various transport methods, including electronic conductivity, thermoelectric power, and the Hall effect. These experiments involve scattering between states near the Fermi level, and generally do not yield simple information on the geometry of the Fermi surface. A second sub-class consists of those techniques that do define the dimensions of the Fermi surface. For example, the de Haas-van Alphen method gives information about the extremeties of the surface and the magnetoacoustic effect information on its linear dimensions. Both ordinary transport and geometrical Fermi-surface methods are relatively well understood. Comparisons of these methods have been available for several years (see for example Ziman 1960, and Kittel 1963).

There is in addition a second large class of experimental electronic structure techniques that probe the energy bands away from the Fermi level. This group also can be subdivided: into those methods that are sensitive only to occupied states, those that link occupied and empty states, and a third set comprising methods that do not involve the occupied bands at all. We shall call the techniques in this second large group broad-band methods, although the terminology is imperfect. Experiments in this class can be used for materials with narrow bands or even, in the case of disordered materials, imperfectly-defined bands. The phrase 'broad band' implies that the methods cover a wide energy range compared to Fermi-level techniques.

The only methods that yield information on the occupied bands alone link them either with core levels or with high-energy bands, which do not leave an imprint on the measured spectra, or else involve the tunnelling of electrons. Valence-band spectra are unique in not involving the empty bands at all. In x-ray valence-band photoelectron and high-energy Auger electron spectroscopy, the excited electron distributions do not reflect the empty-state electronic structure. This situation also prevails for Compton scattering in which electrons are excited from occupied to high-energy empty bands. But a measured Compton profile depends on the electron momentum distribution and does not reflect the density of states. Hence it can be considered to be distinct from band-spectroscopy.

The other occupied-band methods not sensitive to the empty bands involve electron tunnelling and are therefore surface methods. These include field-emission and ion-neutralization spectroscopies. The field emission-method measures the energy distribution of electrons that tunnel through the surface barrier when an electric field is applied (Plummer and Young 1970, Gadzuk 1970). The measured electrons originate in the conduction band and do not occupy normally-empty bands during the process. In ion-neutralization spectroscopy a conduction-band electron tunnels to a vacant state in an ion near the surface of a metal (Hagstrum 1970). The change in energy of the first electron goes into excitation of a second electron, similar to the Auger effect. At present it is not clear whether the second electron tunnels from the conduction band in the surface region or is excited into empty states prior to exit from the sample(Hagstrum, unpublished communication). Both field-emission and ion-neutralization spectroscopies probe only a very limited depth ($\sim$one atom layer) and are very sensitive to surface impurities. In fact spectra from both methods strongly reflect the energy level structure of adsorbed species, and the effect of its interaction with the solid surface. Thus, these methods yield information on the electronic structure of the sample only if the surface is atomically clean and even then, the surface and not the bulk electronic structure is probed. Nonetheless, attention has been given to the relation between the bulk electronic structure and field-emission and ion-neutralization spectra (Gadzuk 1969, Hagstrum 1970).

Many methods involving the occupied bands are also sensitive to the empty bands, into which electrons are excited. It is not clear at what excitation energy range the structure of empty bands becomes important; i.e., at what energy valence-band photoelectron or Auger-electron spectra change from dependence on a single density of states to a joint density of states. There are a number of similar methods, each with a different source of excitation, that involve joint densities of states because electrons are excited from the filled bands to nearby empty bands. If the second electron in ion-neutralization spectroscopy is excited into an empty band by means of the first electron tunnelling to a nearby ion, then the spectra involve a joint density of states. Photon absorption definitely causes electron excitations between the two band-states; i.e., in optical and ultra-violet reflectance, absorption, and photoemission experiments (Nilsson 1970, Connolly 1971). Inelastic scattering of incident electrons also leads to interband electron excitation. Characteristic losses of energy by the incident electrons can be measured in reflection at energies below about 1 KeV (Bauer 1969, 1970) or in transmission at higher energies (Raether 1965). Measurement of the energies of the excited secondary electrons also yields interband energies (Willis *et al* 1971).

Before turning to methods that do not involve occupied bands, we note that two of the techniques linking filled and empty bands have the prospect of being truly band-structure spectroscopic methods. In most broad-band experiments there is only one parameter, energy, but energy bands $E(\mathbf{k})$ consist of two parameters, energy and wave vector. Hence both energy and momentum (i.e., direction or angle) are required experimentally to map out the energy bands of a material. To date, the possibility of measuring actual band structure has been discussed in theoretical detail for ultra-violet photoemission by Mahan (1970), while Gerhardt and Dietz (1971) demonstrated angular variation in the UV-photoelectron intensity from copper. The data obtained by Gerhardt and Dietz were tested against the known energy surfaces of copper, but no band-structure information was derived. Wooten *et al* (1971) have associated a peak observed in the angular distribution of photoelectrons for GaAs with a particular interband transition. Bauer (1969) discussed measurement of the angular dependence of inelastically scattered electrons as a means of obtaining band-structure information (ΔE-Δk pairs). Porteus and Faith (1970) used a combination of inelastic and elastic low-energy electron diffraction results to obtain plasmon dispersion curves (rather than one-electron band structure). Fuller evaluation of the angular sensitivity of both UV-photoemission and inelastic electron scattering as band-structure spectroscopy methods should be forthcoming in the near future.

B. Empty-Band Spectroscopies

There are seven broad-band x-ray methods that do not involve the occupied bands at all, although some involve core levels. The x-ray and electron processes for the seven methods are indicated schematically on an energy scale in figure 1.

The techniques fall into three groups. In the first, core electrons are excited into the normally-empty bands. In a manner similar to that for experiments linking filled and empty bands, the source of excitation energy is different for each of the three methods in this group. Photon absorption and inelastic scattering, respectively, cause core-electron excitation in x-ray absorption and Raman experiments, while in the radiative Auger effect, previous core-level ionization is the source of excitation energy.

In the second group of empty-band methods, Bremsstrahlung radiation emitted by incident electrons provides a probe of the empty states near the Fermi surface. The techniques, short-wavelength limit (SWL) and Bremsstrahlung isochromat (BI) spectroscopy, are experimentally complementary and yield similar information.

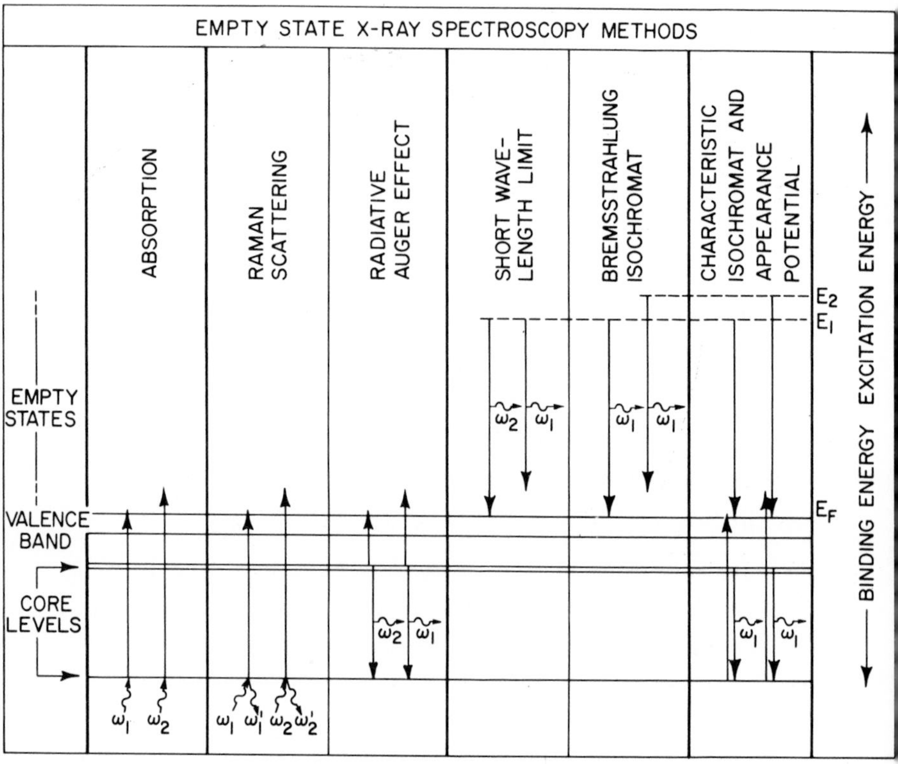

Figure 1 Schematic binding and excitation energy diagram, showing the states probed in 'empty state' x-ray spectroscopic methods. Two values of the experimental variable (x-ray or incident electron energy) are shown for each method, with $\omega_2>\omega_1$ and $E_2>E_1$. Discussion of each method is given in the text.

No core level is involved in this second group, in contrast to the techniques of characteristic-isochromat (CI) and appearance-potential (AP) spectroscopies, which form the final group of methods.

Each of these 'empty band only' x-ray experiments will be discussed in more detail, with emphasis on theoretical calculation of spectra and the similarity of SWL, isochromat (BI and CI) and AP methods. Comparisons of theoretical spectra with experiment, and intercomparisons of respective spectroscopy methods, promote understanding and utilization of spectra. In particular, better interpretation of broad-band spectra in recent years has been due to, and paced by, the results of band theory. For some spectra, only a comparison with the density of states has been made. But now it is widely recognized that transition probabilities prevent direct comparison of the results of band calculations with experiment in most cases (Fabian 1968, Bennett 1972). Hence the second theoretical step from the output of a band calculation to a computed spectrum is most important. This step has already been taken and proved very instructive for several types of broad-band experiments; e.g., optical absorption, UV-photoemission and valence-band x-ray emission. The success of calculations for methods involving occupied bands encourages the application of such calculations to empty-band spectroscopy techniques for which there has been relatively less theory developed.

Spectroscopy calculations are laborious, and require attention to several physical processes involving emission, transport and absorption. However, there are several advantages, beyond a comparison of like quantities. Theoretical spectra aid more detailed interpretation of experimental data, in many cases over a wider energy range. Also, comparisons of spectra calculated from the results of band theory allow separation of (a) one-electron and many-electron effects and (b) band-structure and electron transport or other experimental effects. Success in making such distinctions can show how density-of-states information might be extracted from measured spectra. Finally, comparison of the equations for calculating various spectra places intercomparison of the techniques onto a firmer basis. Sometimes the data from two methods look similar but depend on electronic structure in quite different ways.

2. X-RAY ABSORPTION SPECTRA

Chemical effects on x-ray absorption spectra were first observed in 1920 (Bergengren 1920a, b) and many absorption measurements have been performed since then. Despite this long history, and the recent availability of band calculations, no fully theoretical calculations of absorption spectra have been made until recently, although absorption data have been checked against band energies (see for example Mazalov *et al* 1967). Several other comparisons of empty-band densities of states with absorption spectra have been made (for copper by Burdick 1963, for aluminium by Connolly 1970, and for alkali halides by Brown *et al* 1970) but no x-ray spectra were computed. Irkhin (1961) did a partial absorption calculation for one k-space high-symmetry direction in nickel, in which the density of states was resolved into its p-like component appropriate to a transition from an s-like state.

Very recently McCaffrey *et al* (1973) made a thorough calculation of an x-ray absorption spectrum based on the results of an Augmented Plane Wave (APW) calculation (Dimmock 1971) for nickel. The method used was similar to that described for valence-band emission spectra by Goodings and Harris (1969) and Dobbyn *et al* (1970). The self-consistent potential obtained by Connolly (1967) with a 2/3-Slater exchange, was used for calculation of the energy band structure. The angular part of the transition-probability matrix elements yields the usual dipole selection rules. Hence the density of empty states was resolved according to the angular symmetry of the band states using the ℓ-dependent charge densities that result from the APW method. The resultant partial densities of states for one spin are shown with the corresponding spectra (figures 2 and 3). The radial part of the transition probability is a function of energy only and is shown in figure 4, normalized at the bottom of the conduction band for K and L spectra.

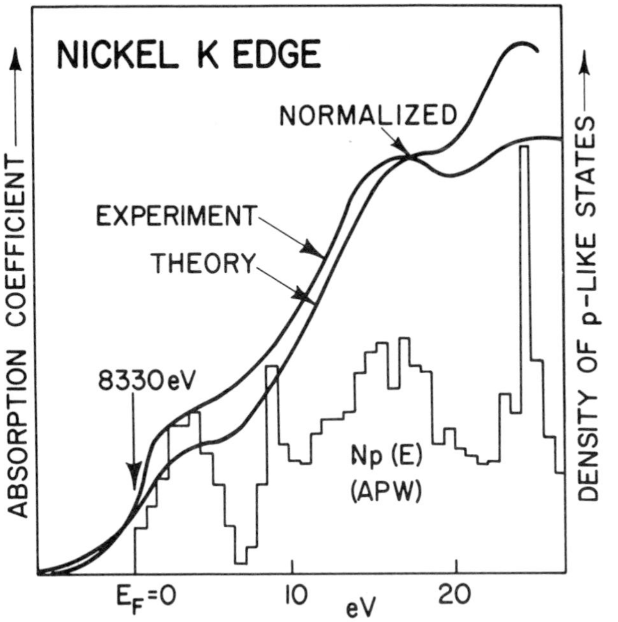

Figure 2 Density of empty p-like states, and the calculated nickel K-absorption edge for the minority spin, compared with the measurements of Beeman and Friedman (1939).

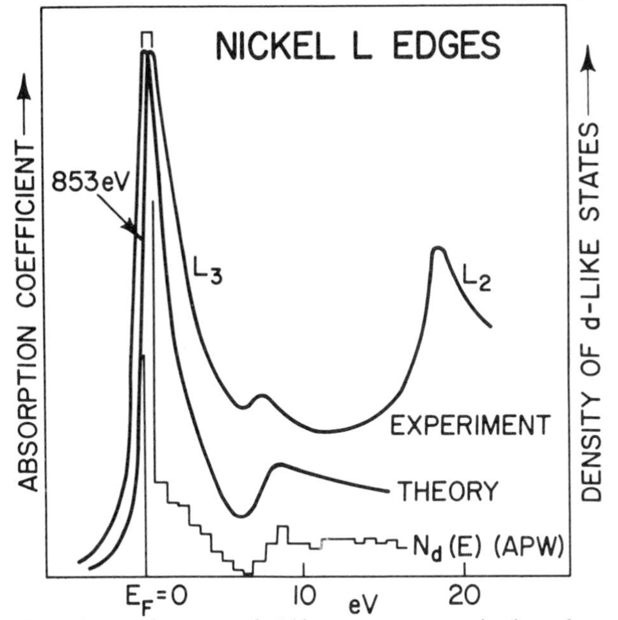

Figure 3 Density of empty d-like states, and the theoretical L_3-absorption edge of nickel for the minority spin, contrasted with the spectrum measured by Bonnelle (1966).

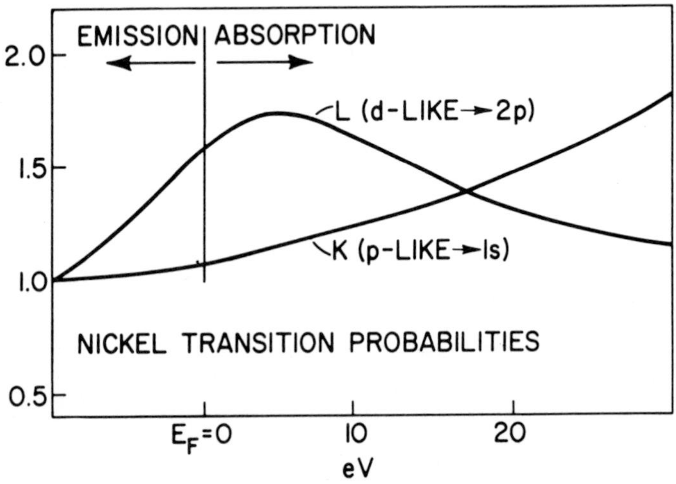

Figure 4 Square of the radial matrix elements for the K and L_3 absorption spectra of nickel, calculated using wavefunctions from the APW and Herman-Skillman programs, normalized at the bottom of the conduction band.

The radial integral is a smooth function of energy so that fine-structure in an absorption spectrum is due to detail in the corresponding partial density-of-states curve. But measured spectra are much smoother than the calculated component density of states because of core-level and final-state ('hot electron') lifetime broadening, and the spectrometer window. These smearing functions are shown for the nickel K and L absorption spectra in figure 5. The spectrometer window is effectively a constant over the energy range involved and, like the core-level lifetime broadening, becomes progressively smaller for softer radiation. The energy-dependent final-state broadening was estimated using

$$\Delta E = \hbar/\Delta t \text{ and } \Delta t = \text{MFP}/(\text{electron velocity})$$

with the velocity simply calculated from the energy.

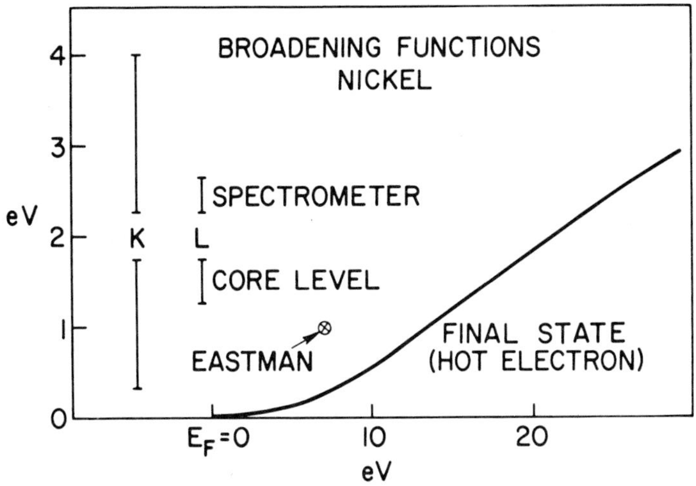

Figure 5 Core-level and excited-state lifetime broadening, plus the spectrometer broadening used to calculate the K and L_3 absorption spectra of nickel.

Only a few, closely-spaced mean-free-path (MFP) values are available for nickel (Eastman 1970), so the available continuous theoretical curve for electron-electron scattering in aluminium calculated by Ritchie (McConnell *et al* 1968) was used. The resulting ΔE-curve is low with respect to the few measurements by Eastman but, at higher energies, the indicated 'hot electron' broadening is large compared to the detail available in ultra-violet photoemission experiments (Spicer this Volume p. 7).

Normalized theoretical and experimental K and L_3 absorption-edge spectra are compared with the component density of states in figures 2 and 3. The calculated spectra agree with experiment on a large scale, but not in detail. Since calculated spectra are sensitive to final-state broadening, detailed comparison must await more precise information on 'hot electron' lifetimes. The energy of the ultra-soft x-ray $M_{2,3}$-edge for nickel (65.8eV) was also calculated, but does not agree well with the experimental spectra which shows multiplet effects (Dehmer *et al* 1971) and delayed oscillator strength (Fano and Cooper 1968), atomic rather than solid-state effects. The complete nickel calculation (both spins) and the Ni $M_{2,3}$-absorption spectrum are discussed in detail elsewhere (McCaffrey *et al* 1973). The general conclusion from calculation of the nickel absorption spectra are: (a) where solid-state effects are dominant, the theory is in fair agreement with experiment and assists in interpretation of the measured spectra; (b) life time broadening at present precludes use of measured spectra to adjust the potential used in band calculations.

3. X-RAY RAMAN SPECTRA

In an x-ray Raman experiment, a core electron is excited by inelastic scattering of an x-ray photon, as shown in figure 1. The energy carried off by the scattered photon depends on the incident energy and on the band into which the core electron is excited. Hence measurement of the scattered x-ray spectrum yields information on the availability of empty electron-states. Experimentally, a Raman spectrum is similar to an absorption spectrum except that it is located on the tail of the ordinary Compton spectrum. Examples of measured x-ray Raman and absorption spectra for lithium and beryllium are shown in figure 6 (Suzuki *et al* 1970). The spectra shown for each element are similar except for greater detail observed in the absorption spectra. Because of this, and the difficulty of measuring Raman spectra, absorption spectroscopy is ordinarily the better of the two methods for empty-band studies.

The similarity of Raman and absorption spectra is not superficial. Mizuno and Ohmura (1967) show that for polycrystals both spectra are proportional to the same function.

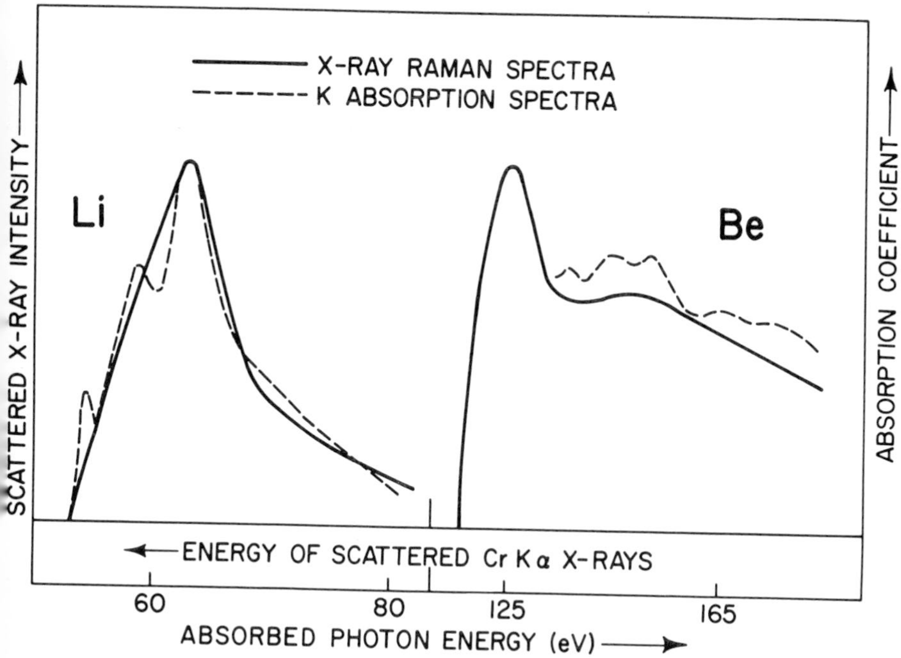

Figure 6 X-ray Raman and K-absorption spectra of lithium and beryllium (Suzuki *et al* 1970).

Kuriyama and Alexandropoulos (1971) pointed out that Raman spectra depend on the electron density correlation function while absorption spectra depend on the electron-current correlation function. Charge conservation relates the electron density and current, and thus provides a link between Raman and absorption spectra.

The scattering angle at which an x-ray Raman spectrum is measured can be varied, but to-date it has usually been chosen or altered to insure separation of the Compton and Raman peaks. In a recent paper Doniach *et al* (1971) show theoretically that the many-body effects (Nozières and de Dominicis 1969) can be altered by changing the scattering angle. Such effects lead to edge-rounding in K-spectra and peaking in L-spectra, for both emission and absorption. But for these experiments, the near-edge many-body effects are not variable, so that separating them experimentally from band-structure effects is impossible. Successful measurement of variable edge effects in an x-ray Raman experiment would help to clear away one obstacle to extracting density-of-states information from x-ray spectra.

4. RADIATIVE AUGER SPECTRA

Usually a core-level vacancy decays by radiative (x-ray) or non-radiative (Auger-electron) emission. However, in some instances shake-off or configuration interaction in the final state leads to simultaneous x-ray and electron emission (Aberg 1971) as indicated in figure 1. For such radiative Auger processes, the emitted photon has an energy lower than usual by the amount of energy given to the excited electron. When the excited electron originates in a core level, the spectrum of emitted x-rays again depends on the availability of unoccupied states, similar to the absorption and Raman processes. For example, the K-radiative Auger spectrum of silicon in SiO_2 is shown in figure 7 (Aberg and Utriainen 1971). The excited electron comes from the L shell in this case, and the L-absorption spectrum of silicon in SiO_2 (Ershov and Lukirskii 1967) is included in the figure. The radiative-Auger and absorption spectra are similar but differ in detail.

There exists a major difference between radiative Auger spectra and both absorption and Raman spectra, despite apparent similarities. The core-level vacancies created in the radiative Auger process interact to produce multiplet splitting in the final state (Aberg and Utriainen 1969). Hence each band in the spectrum has structure beyond that due to the density of empty states. Overlapping bands due to different final-hole configurations, plus exciton structure near the band head for spectra from insulators, further complicate radiative Auger spectra. As indicated in figure 7, radiative Auger spectra are weak relative to the nearby line and the measured spectra presently available do not allow separation of final-state and density-of-states effects. To-date no detailed calculation of a radiative Auger spectra has been made, although Aberg (1971) has calculated the integrated intensity for this process relative to the nearby line.

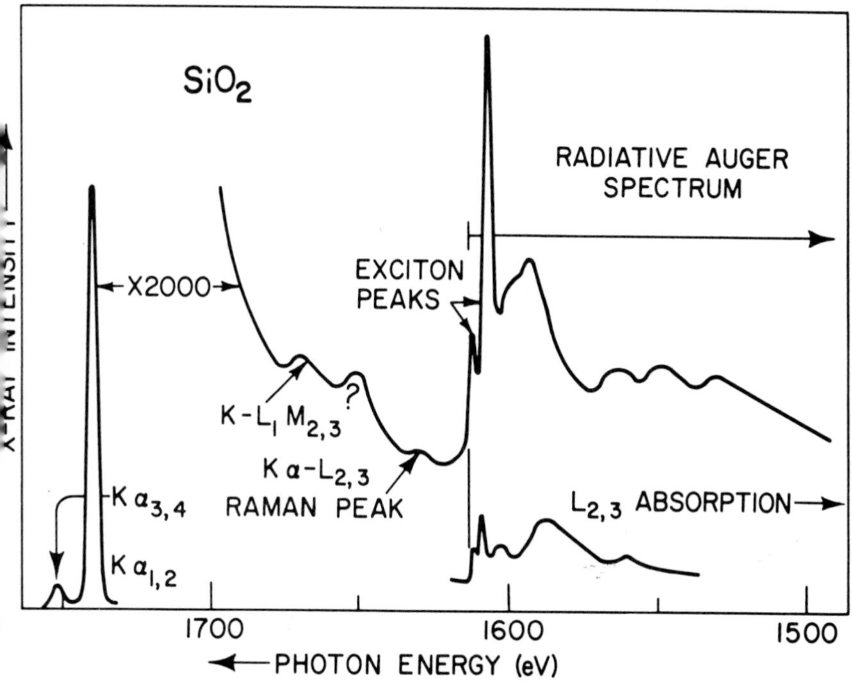

Figure 7 Radiative-Auger spectrum of silicon in SiO_2 (Aberg and Utriainen 1971). The Si $L_{2,3}$-absorption edge of SiO_2 (Ershov and Lukirskii 1967) is aligned with the radiative-Auger spectrum for comparison.

5. GENERAL FEATURES OF SHORT-WAVELENGTH LIMIT, ISOCHROMAT, AND APPEARANCE POTENTIAL SPECTRA

Before discussing each of the remaining four empty-band methods in detail, it is worth examining their differences and similarities. Two of the techniques, SWL and BI spectroscopies, involve Bremsstrahlung emission and, near threshold, provide information on the density of unoccupied states (Ulmer 1969). A constant electron bombarding energy is used for the SWL method and the x-ray Bremsstrahlung is scanned near its high-energy threshold. A BI-spectrum is measured at any single spectrometer setting by varying the incident electron energy across and above the set energy. Figure 8 compares spectra obtained by these methods (Bohm and Ulmer 1971) with the density of unoccupied states of tantalum and tungsten (Mattheiss 1970).

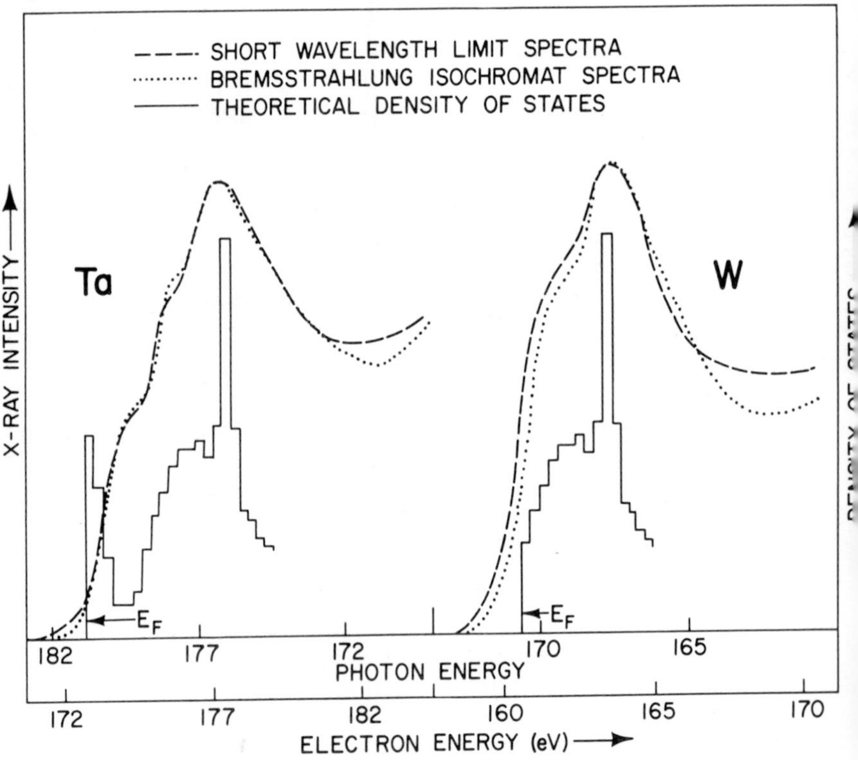

Figure 8 Short-wavelength limit and Bremsstrahlung isochromat spectra of tantalum and tungsten (Bohm and Ulmer 1971) compared to the density of empty states (Mattheiss 1970).

The remaining two techniques, CI and AP spectroscopies, require core-level ionization with emission of characteristic radiation and, near threshold, give information on self-convolution of the density of unoccupied states. For a CI-measurement, the spectrometer is set on a characteristic x-ray line as the electron energy is increased across threshold. Hence, CI spectra are sometimes called excitation curves. AP spectra, which are measured non-disperisvely, are usually obtained by phase-lock techniques yielding a derivative signal. Insofar as this signal is due to characteristic radiation, an AP x-ray spectrum is the derivative of the corresponding CI spectrum. Such a relationship is illustrated in figure 9 which shows examples of CI (Liefeld 1968) and AP (Houston and Park 1972a) spectra for nickel.

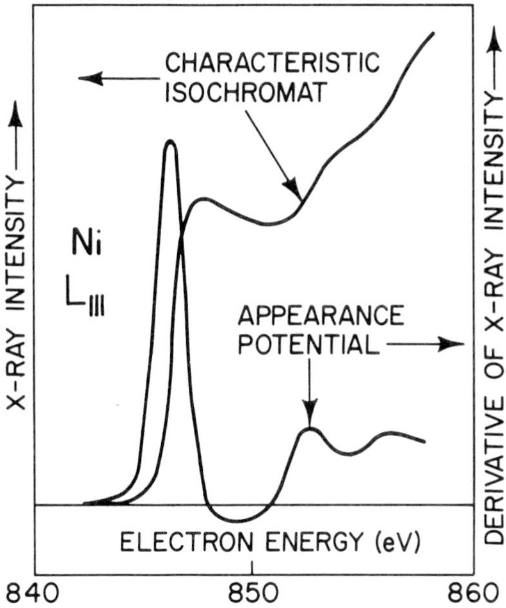

Figure 9 The characteristic isochromat spectrum (excitation curve) of nickel measured by Liefeld (1968) compared with the nickel appearance potential spectrum (Houston and Park 1972a).

All of these four methods have several similarities. They each involve (1) the transport of electrons into the target, (2) an anergy-dependent cross section (for Bremsstrahlung emission or core-level ionization), (3) the density of available unoccupied states or its self-convolution, and (4) transport of x-rays out of the sample. Full calculation of any of the four types of spectra would require consideration of all of these factors. So far this has not been done, but the equations for such calculations are presented below. Success in calculating these spectra and in distinguishing electronic structure from transport effects, might indicate how density-of-states information could be extracted from experimental spectra over a wider energy range than is now possible. At present, the four related empty-band methods yield useful information on the density of states only over a small range of about 10 to 20 volts above the Fermi level. At higher energies, additional electron energy-loss processes occurring during incident electron transport, (over and above the Bremsstrahlung emission or core-level ionization) dominate the spectra and obscure the density of states. The analysis given below differs from the method discussed by Nilsson and Lindau (1972), by which the secondary electron distribution in a UV-photoemission measurement can be made to yield electronic structure information. However, the viewpoint is similar in both approaches, namely consideration of the electron transport problem in an attempt to interpret spectra, and to use parts of spectra that ordinarily do not give information of interest for electronic structure determination.

In order to include absorption effects in the spectrum calculation, it is necessary to know the depth distribution of x-ray production. This can be calculated if the electron energy distribution is known in each layer dx at depth x into the sample. The required electron energy distribution $f(\varepsilon,x)$ is shown schematically in figure 10. ε is the energy loss below the incident energy. The $f(\varepsilon,x)$ distribution can be obtained from solution of the transport equation (Brown *et al* 1969) or by the use of Monte Carlo techniques (Berger 1963).

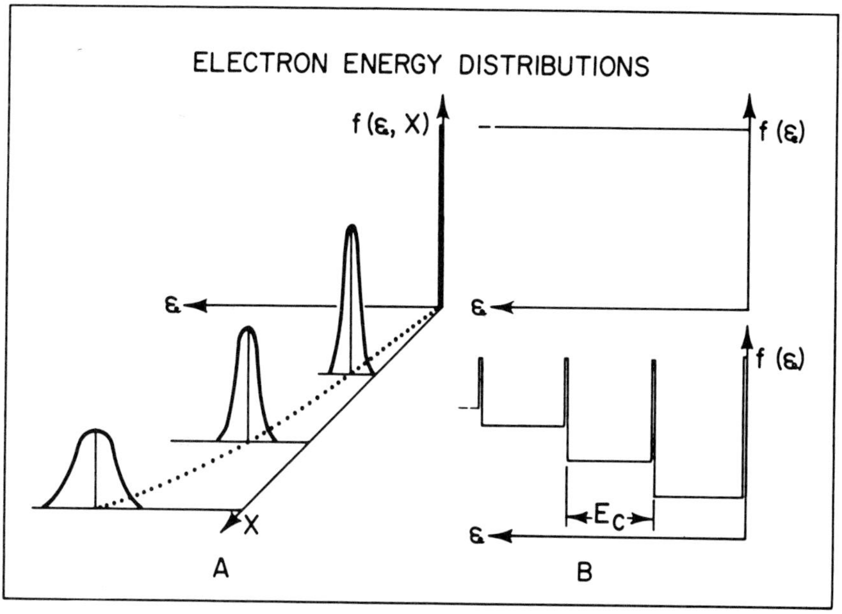

Figure 10 Schematic electron energy distributions as a function of energy loss. In (a) the electron distribution versus depth x into the sample is also given. Distributions in (b) are given with (lower curve) and without (upper curve) discrete energy losses E_c.

As indicated in figure 10a, the distribution moves to lower energies and becomes more diffuse with increasing sample penetration. Calculation of the emitted radiation from each layer would first require integration over ε within the layer. The generated intensity would then be multiplied by the absorption correction for the layer, exp $(-\mu x \sec \psi)$, where μ is the linear x-ray absorption coefficient and ψ is the take-off angle with respect to the surface normal.

The analysis below is complex even without explicit inclusion of the absorption factors. Also, the importance of absorption effects on SWL, isochromat and AP spectra has not yet been demonstrated. This could be accomplished by varying the x-ray take-off angle, as done by Gennai *et al* (1971) to determine the depth distribution of generated radiation for ordinary x-ray excitation. We assume, in effect, that absorption is unimportant so that the x variable is irrelevant, and only $f(\varepsilon) = \int_0^\infty f(\varepsilon,x)\,dx$ is needed.

If it were not for characteristic energy losses, this projection on the $f(\varepsilon)$-plane would be energy-independent since all electrons would slow down approximately smoothly through all energies. Such a flat distribution is shown at the top of figure 10b. At the other extreme of discrete losses only, $f(\varepsilon)$ would consist of a series of spikes. The distribution occurring for both continuous and discrete losses is shown at the bottom of figure 10b. This more realistic distribution will be used in the following analysis. The theory given in the next two sections is an expansion of that given by Dev and Brinkman (1970) who confined their analysis to the near-threshold region and ignored electron transport.

6. SHORT-WAVELENGTH LIMIT AND BREMSSTRAHLUNG ISOCHROMAT SPECTRA

The origins of SWL and BI spectra are discussed in this section using figures 11 and 12. We are interested in calculating the intensity of Bremsstrahlung photons with energy $\hbar\omega$ emitted under bombardment by electrons of energy E_0. As indicated in figure 11, small energy losses ε, up to $\Delta E = E_0 - \hbar\omega$ only can be suffered before it is no longer possible to emit the photon energy $\hbar\omega$. A 'three dimensional' schematic representation, including the density of states N as a function energy ϵ above the Fermi level E_F, is given in figure 12. The electron energy above threshold ΔE forms the third axis. Intensity as a function of ϵ, $dI(E_0,\hbar\omega,\epsilon)$, is shown at two values of ΔE, one below and one above the characteristic energy loss E_c.

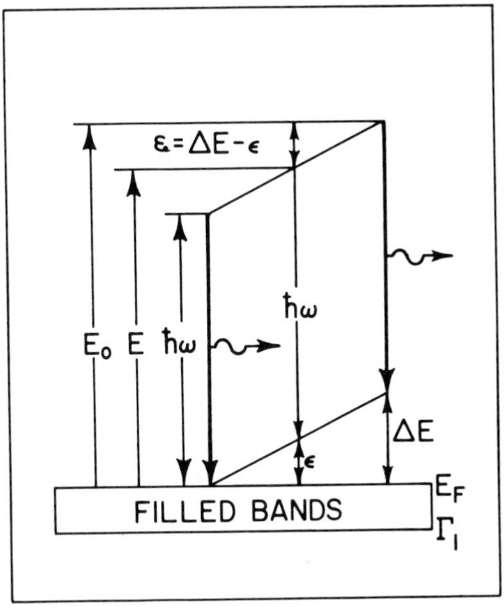

Figure 11 Energy schematic for calculation of short-wavelength limit and Bremsstrahlung isochromat spectra.

To calculate the intensity distribution $dI(E_0, \hbar\omega, \epsilon)$ we refer to figure 11 and consider the number of electrons with energy $E = \hbar\omega + \epsilon$ that are available to emit Bremsstrahlung of energy $\hbar\omega$ and fall into a state with energy ϵ,

$f(\varepsilon)\, d\varepsilon = f(E_0 - E)\, dE = f(E_0 - h\omega - \epsilon)d\epsilon$.

$Q_B(E) = Q_B(\hbar\omega + \epsilon)$ gives the probability for such a transition while $N(\epsilon)$ essentially gives the number of available paths. Hence

$dI(E_0, \hbar\omega, \epsilon) = f(E_0 - \hbar\omega - \epsilon)Q_B(\hbar\omega + \epsilon)N(\epsilon)\, d\epsilon$,

as shown in figure 12. Integrating, we have the intensity from transitions to all accessible states

$$I(E_0, \hbar\omega) = \int_0^{\Delta E} f(E_0 - \hbar\omega - \epsilon)\, Q_B(\hbar\omega + \epsilon)\, N(\epsilon)\, d\epsilon \qquad (1)$$

This is the general equation for both SWL and BI spectra. Q_B may be taken out of the integral if we assume it is a slowly varying function near threshold (Fano *et al* 1958).

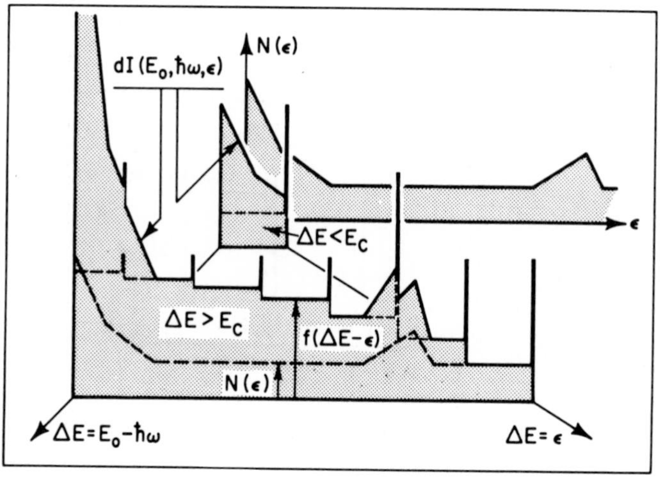

Figure 12 Diagram showing the relation of the distribution of emitted intensity $dI(E_O, \hbar\omega, \epsilon)$ to the density of empty states $N(\epsilon)$ for short-wavelength limit and Bremsstrahlung isochromat spectra.

We note that solid-state effects influence all three of the contributing factors in equation (1). The chemical and structural make-up of the sample determines f. For calculating Q_B, Bloch functions appropriate to the solid should be used, rather than plane waves appropriate to the free atom case. The low-energy final electron state in particular will reflect the crystal potential which also determines N(E).

If discrete losses dominate continuous slowing-down of electrons, then near threshold we can take

$f(\varepsilon) = \delta(0)$, so that $I(E_0, h\omega) \sim N(\Delta E) = N(E_0 - h\omega)$; that is, the measured SWL or BI spectra are proportional to the density of states. Materials with pronounced structure in their density of states, as for tantalum and tungsten in figure 8, show a peak near threshold, while simple metals with their flatter densities of states show only a step. This observed behaviour implies that near threshold continuous slowing-down is relatively unimportant, or at least characteristic losses are appreciable. If continuous slowing-down dominated the electron distribution, $f(\varepsilon)$, then SWL and isochromat spectra would always increase, even for transition metals. Whatever the relative importance of the various loss processes, as the energy ΔE increases (by either increasing E or lowering $h\omega$), the part of f, due to continuous slowing-down, that lies between $\varepsilon = 0$ and $\varepsilon = E_c$ does contribute proportionately more to the spectra. Insofar as they are free of a contribution from continuous slowing-down near threshold, SWL and BI spectra are unique in providing a measure of the total density of (low-lying unoccupied) states essentially free of transition probability effects, as figure 8 indicates.

When ΔE becomes greater than E_c, I suddenly increases due to the step and spike in $f(\varepsilon)$ at $\varepsilon = E_c$. This occurs at the onset of each characteristic loss until the electron distribution becomes essentially flat (still ignoring variation with depth and absorption). The spectra are then dominated by transport rather than density-of-states effects in the intermediate region. At higher energies where $f(\varepsilon)$ becomes energy independent, $dI/d(\Delta E) \sim N(\Delta E)+F(\Delta E)$, where $F(\Delta E)$ depends on both N and f. Hence, the derivative of SWL and BI spectra, well above threshold, could yield information on the density of states if absorption is not too strong, and provided the state density has significant structure and the spectra can be accurately measured. At present, a few band calculations have been made at high energies, but the density of states is given only for aluminium to 45eV above E_F (Connolly 1970). Characteristic loss structure in BI spectra is found at energies greater than this (see, for example, Bergwall and Tyagi 1965).

Before turning to CI and AP spectra, we note that the structure of BI spectra away from threshold is dependent on the electron energy E_0 (Böhm and Ulmer 1969). Such dependence can be ascribed to the energy dependence of $f(\varepsilon)$ in equation (1), provided Q_B is a slowly varying function of energy near threshold for a solid, as it is for free atoms (Fano *et al* 1958). This question merits investigation.

7. CHARACTERISTIC ISOCHROMAT AND APPEARANCE POTENTIAL SPECTRA

Figure 13 is used to analyze the source of radiation in CI and AP spectra. The left-hand part of the figure is simi-

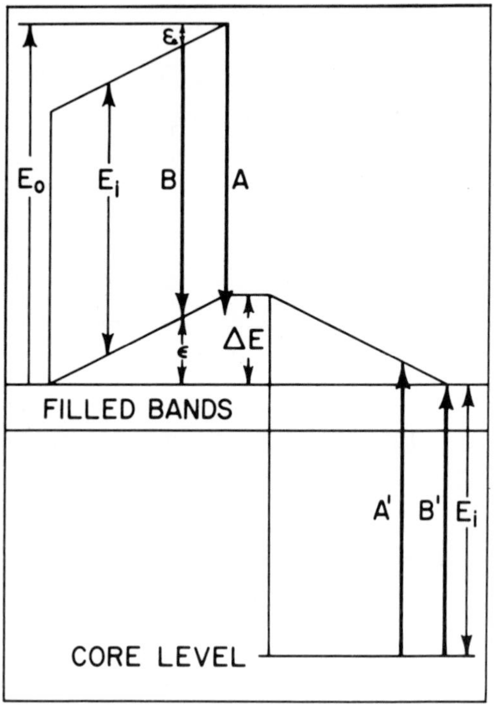

Figure 13 Energy schematic for calculation of characteristic isochromat and appearance potential spectra, without or with consideration of the small energy losses that precede ionization of a core level bound with energy E_i.

lar to figure 11, except that the downward transition of an electron incident with energy E_0 is accompanied by ionization of a core electron, bound with energy E_i relative to E_F, rather than Bremsstrahlung emission. The upward transition of the core electron is shown in the right-hand part of figure 13.

If there are no small (continuous or discrete) energy losses, an incident electron transition to level ϵ, labelled A, will be accompanied by core-electron transition A$'$ to $(E_i + \Delta E - \epsilon) - E_i = \Delta E - \epsilon$. Hence the CI intensity can be written as

$$I(E_0) = \int_0^{\Delta E} f(0)\ Q_i\ (E_0)\ N(\epsilon)\ N(\Delta E - \epsilon)\ d\epsilon \tag{2}$$

where $\Delta E = E_0 - E_i$. Now, $f(0)$ and $Q_i(E_0)$ are both independent of ϵ and therefore equation (2) shows that, ignoring energy losses, the intensity of a CI spectrum at an energy ΔE above threshold varies as the self-convolution of the empty-state density from threshold to ΔE.

An appearance potential spectrum is essentially the derivative of the corresponding characteristic isochromat, as we see in figure 9. This is the case whether characteristic x-rays, Auger electrons, or the Bremsstrahlung emission due to Auger electrons are measured. Each of these is possible (see Houston, unpublished communication) and all signal the occurrence of the two-electron process shown in figure 13. From equation (2), we have for AP spectra (Apostol, 1957)

$$\frac{dI(E_0)}{dE_0} = \frac{dI(E_0)}{d\Delta E} \sim \int_0^{\Delta E} N(\epsilon)\ \frac{d}{d\Delta E}\ N(\Delta E - \epsilon)\ d\epsilon + N(\Delta E)\ N(0) \tag{3}$$

which shows that, even when energy losses are ignored, an AP spectrum is not simply related to the density of states. However, the general shape of the AP spectrum and $N(\epsilon)$ curve may be similar (Houston and Park, 1971).

If we now consider small energy losses, as in the calculation of SWL and BI spectra (still ignoring absorption), the electron distribution function $f(\varepsilon)$ must again be included in the analysis. Transitions A and A' still go together in figure 13. But now additionally a state of energy ϵ can be reached by a downward transition B following energy loss ε. In the case shown in figure 13, no excess energy is given the core electron so that transition B' just reaches E_F. A double integral is necessary, first over the energy loss ε and then over the final-state energy ϵ:

$$I(E_0) = \int_0^{\Delta E} N(\epsilon) \left\{ \int_0^{\Delta E-\epsilon} f(\varepsilon)\, Q_i\, (E_0 - \varepsilon)\, N(\Delta E - \epsilon - \varepsilon)\, d\varepsilon \right\} d\epsilon \quad (4)$$

Now it is important to consider the energy dependence of the ionization cross section Q_i. Unlike the Bremsstrahlung cross-section Q_B, which is slowly varying near the short wavelength limit (at least for free atoms), Q_i is known from Born-approximation calculations to rise rapidly above threshold (again from free-atom calculations). No rigorous calculation of Q_i for a solid, using Bloch functions, is known to the present author. Even with appropriate Q and f available, equation (4) shows that CI spectra bear a relatively complex relation to the density of states. The AP spectrum given by the derivative of equation (4) is even more complex.

As before, we consider the case where discrete losses dominate the continuous slowing-down of electrons so that $f(\varepsilon)$ is approximately $\delta(0)$. Then equation (4) reduces to equation (2). We noted in the discussion of SWL and BI spectra that the assumption $f(\varepsilon) \sim \delta(0)$ gives theoretical spectra that qualitatively agree with experiment. Also, results obtained in the first transition series, by Houston and Park (1972a), are generally consistent with equation (2). Comparison of experimental CI and AP spectra with calculated spectra, using equations (2) and (3), should indicate the importance of Q and f in equation (4). Such calculations are currently being made (Nagel and Criss 1972).

To complete our discussion of SWL, isochromat and AP spectra, it is worth noting additional similarities between Bremsstrahlung and characteristic experiments. Langreth (1971) has shown that for AP spectra, as for ordinary absorption spectra near threshold, the core hole interacts with and essentially screens the ejected electron. Hence for the two methods involving core-level ionization (CI and AP spectroscopy), there is effectively a bare electron present, as for the case of SWL and BI techniques. That is, both these pairs of methods involve a downward transition of a single electron lacking a hole to screen it. The bare electron effectively couples to the plasmon modes, as observed by Houston and Park (1972b), and discussed by Laramore (1972) for AP spectra from simple materials. But Kieser (1971) also observed strong plasmon effects in BI spectra from materials for which the spectra are not dominated by the density of states. The plasmon explanation developed by Laramore to explain the AP data, may be useful also for interpretation of BI (and SWL) spectra of simple materials. We note also a second similarity: SWL, BI, CI, and AP spectra from transition metals are all dominated by the density of states near threshold, while tney exhibit characteristic energy-loss structure at higher energies. Further comparison of measurements made on the same materials, by all four techniques, should be instructive.

8. MODULATION X-RAY SPECTROSCOPY

So far we have discussed two examples of derivative or modulation methods that can be applied to x-ray spectroscopy of empty bands. These are (1) the possibility that derivatives of SWL and BI spectra will yield the density of high-energy unoccupied states and (2) modulation of the exciting-electron energy in AP spectroscopy, to avoid the need for a dispersing element. These examples prompt a more general examination of derivative and modulation x-ray spectroscopy, in which we are concerned with improving the usefulness of x-ray spectroscopy experimentally, in contrast to our earlier emphasis on the theoretical calculation of spectra.

There are three basic elements in any of the spectroscopy experiments: the source of incident radiation, the sample, and the spectrometer. It has been shown in optical spectroscopy (Cardona 1969) that any of these elements may be modulated, and that phase-sensitive detection methods can be used experimentally to improve resolution or to obtain new information about the sample. The same should apply to the x-ray region. The energy of the exciting radiation may be varied, as in AP spectroscopy; or its intensity modulated, as in absorption studies (Kunz, this Volume p503). Motion of the crystal or grating in the spectrometer can also give a derivative spectrum. If sufficient intensity is available, modulation of the source or spectrometer yields data with effectively better resolution.

Modulation of the sample itself by application of stress, by electrical or magnetic fields, or by temperature cycling, also yields more detailed information. Moreover if the applied field destroys symmetries and lifts degeneracies in the energy bands of the sample, then additional electronic structure is obtained such as the location of van Hove singularities. Very little sample-modulation x-ray spectroscopy has been reported. Willens *et al* (1969) have measured the L valence-band emission spectra of nickel and copper, with and without application of a stress. The L-spectrum of copper, and the difference due to application of the stress, are shown in figure 14. The difference, or modulation curve, looks like nothing more than the numerical derivative of the L spectrum itself. The two additional curves in figure 14 are the result of subtracting constant fractions of the L-spectrum derivative from the difference curve. These show that the modulation curve is not merely

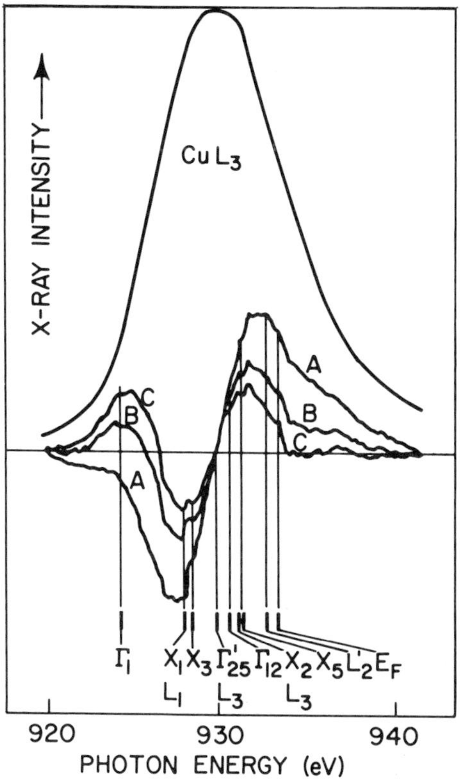

Figure 14 Copper L-emission, and the associated stress modulation x-ray emission spectra obtained by Willens *et al* (1969), compared to energy values calculated by Burdick (1963). (a): Original difference curve, I(load) minus I(unload). (b) and (c): Curve (a) less respectively 0.010 and 0.015 times the derivative of the emission curve.

a simple derivative; i.e., application of stress has induced new structure in the x-ray spectrum. Some of the energy-band points in figure 14 correlate well with structure in the modulation curves, in contrast to the featureless basic L-spectrum. Further work is necessary to prove that sample-modulation x-ray methods are of value for probing empty as well as filled bands; success would make x-ray spectroscopy more competitive, on a resolution basis, with the more common low-energy spectroscopy methods.

9. SUMMARY

In this paper we have developed several themes, first mentioning the wide variety of spectroscopic techniques available to probe occupied and unoccupied bands. We emphasize that empty bands are important in many of these methods, either in conjunction with the filled or partly filled bands or by themselves. Not only are they involved in most broad-band experiments, but also they reflect the potential that determines the occupied bands. Success in using one-electron band theory to calculate spectra involving occupied bands, points toward the value of similar calculations for empty-band methods. Initial theoretical results for absorption spectra are encouraging. Full calculation of SWL, isochromat, and AP spectra, which overall are similar, requires attention to both electron transport and band structure. Modulation techniques offer an experimental means for improving the resolution and the information content of several high-energy spectroscopies. In all, electronic structure can now be probed by many methods, and their usefulness is being improved by both theoretical and experimental advances. Although the present range of techniques appears complete, we can expect new methods of spectroscopy to be developed.

Acknowledgments

The author enjoys pleasant collaboration with D.A. Papaconstantopoulos and J.W. McCaffrey in calculation of band structure and x-ray spectra. Assistance in preparation of this paper was given by D.B. Brown, J.W. Criss, M. Fatemi, and by many authors whose work is referenced. Comments on the manuscript by B.M. Klein and L.S. Birks are gratefully acknowledged

REFERENCES

Aberg, T. (1971); Phys. Rev., 4A, 1735.

Aberg, T. and Utriainen, J. (1969); Phys. Rev. Letters 25, 1346.

Aberg, T. and Utriainen, J. (1971); J. de Phys.

Apostol, T.M. (1957); Mathematical Analysis, Addison-Wesley, New York.

Bauer, E. (1969); Z. Phys. 224, 19.

Bauer, E. (1970); J. Vac. Sci. and Tech. 7, 3.

Beeman, W.W. and Friedman, H. (1939); Phys. Rev. 56, 392.

Bennett, L. (1972); editor, Proc. Third Conf. IMR, "Electronic Density of States", NBS Special Pub. 323, U.S. Government.

Bergengren, J. (1920a); Compt. Rend 171, 624.

Bergengren, J. (1920b); Z. Phys. 3, 247.

Berger, M.J. (1963); in Methods in Computational Physics Vol.1, (Ed. by Alder, B., Fernbach, S. and Rotenberg, M., p.135: Academic Press, New York.

Bergwall, S. and Tyagi, R.K. (1965); Arkiv fur Fysik 29, 439.

Böhm, G. and Ulmer, K. (1969); Z. Phys. 228, 473.

Böhm, G. and Ulmer, K. (1971); J. de Phys.

Bonnelle, C. (1966); Ann. Phys. 1(7-8), 439.

Brown, D.B., Wittry, D.B. and Kyser, D.F. (1969); J. Appl. Phys. 40, 1627.

Brown, F.C., Gänwiller, C., Kunz, A.B. and Lipari, N.O. (1970); Phys. Rev. Ltrs. 25, 927.

Burdick, G.A. (1963); Phys. Rev. 129,138.

Cardona, M. (1969); Modulation Spectroscopy, Solid State Physics Supplement 11, Academic Press, New York.

Connolly, J.W.D. (1967); Phys. Rev. 159, 415.

Connolly, J.W.D. (1970); Intl. J. Quant. Chem. 111S, 807.

Connolly, J.W.D. (1971); Intl. J. Quant. Chem. 4, 419.

Dehmer, J.L., Starace, A.F., Fano, U., Sugar, J. and Cooper, J.W. (1971); Phys. Rev. Ltrs. 26, 1521.

Dev, B. and Brinkman, H. (1970); Nederlands Tijdschrift voor Vacuum techniek 8, 176.

Dimmock, J.O. (1971); in Solid State Physics, Vol. 26, (ed. Ehrenreich, H., Seitz, F. and Turnbull, D., p.104, Academic Press, New York.

Dobbyn, R.C., Williams, M.L., Cuthill, J.R. and McAlister, A.J. (1970); Phys. Rev. 2B, 1563.

Doniach, S., Platzman, P.M. and Yue, J.T. (1971); Phys. Rev. 4B, 3345.

Eastman, D.E. (1970); Sol. St. Comm. 8, 41.

Ershov, O.A. and Lukurskii, A.P. (1967); Sov. Phys. Sol. State 8(7), 1699.

Fabian, D.J. (1968); ed. Soft X-Ray Band Spectra, Academic Press, London.

Fano, U., Koch, H.W. and Motz, J.W. (1958); Phys. Rev. 112, 1679.

Fano, U. and Cooper, J.W. (1968); Rev. Mod. Phys. 40, 441.

Gadzuk, J.W. (1969); Phys. Rev. 182, 416.

Gadzuk, J.W. (1970); Phys. Rev. 1B, 2110.

Gennai, N., Murata, K. and Shimizu, R. (1971); Jap. J. Appl. Phys. 10, 491.

Gerhardt, U. and Dietz, E. (1971); Phys. Rev. Ltrs. 26, 1477.

Goodings, D.A. and Harris, R. (1969); J. Phys. C. (Solid State Phys.) 2, 1808.

Hagstrum, H.D. (1970); J. Res. Nat. Bureau of Stds. 74A, 433.

Houston, J.E. and Park, R.L. (1971); J. Chem. Phys. 55, 4601.

Houston, J.E. and Park, R.L. (1972a); in Electron Spectroscopy (ed. by Shirley, D.A.) p.895, North-Holland, Amsterdam.

Houston, J.E. and Park, R.L. (1972b); Solid State Comm. 10, 91.

Irkhin, Yu. P. (1961); Fiz. Metal, Metalloved 11(1), 10.

Kieser, J. (1971); Z. Phys. 244, 163.

Kittel, C. (1963); Quantum Theory of Solids, Wiley New York.

Kuriyama, M. and Alexandropoulos, N.G. (1971); J. Phys. Soc. Japan, 31, 561.

Langreth, D.C. (1971); Phys. Rev. Ltrs. 26, 1229.

Laramore, G.E. (1972); Solid State Comm. 10, 85.

Liefeld, R.J. (1968); in Soft X-Ray Band Spectra (ed. by Fabian, D.J.) p.153, Academic Press, London.

Mahan, G.D. (1970); Phys. Rev. 2B, 4334.

Mattheiss, L.F. (1970); Phys. Rev. 1B, 373.

Mazalov, L.N., Blokhin, S.M. and Vainshtein, E.E. (1967); Sov. Phys. - Solid State, 8, 1926.

McCaffrey, J.W., Nagel, D.J. and Papaconstantopoulos, D.A. (1973); Phys. Rev. B; to be published.

McConnell, W.J., Birkhoff, R.D., Hamm, R.N. and Ritchie, R.H. (1968); Radiation Res., 33, 216.

Mizumo, Y. and Ohmura, Y. (1967); J. Phys. Soc. Japan, 22, 445.

Nagel, D.J. and Criss, J.W. (1972); Solid State Comm.; to be published.

Nijboer, B.R.A. (1946); Physica 12, 461. An early and excellent treatment of both density-of-states and electron-energy-loss effects on BI spectra.

Nilsson, P.O. (1970); in Optical Properties of Solids (ed. by Haidemenakis, E.O.), p.145, Gordon and Breach, New York.

Nilsson, P.O. and Lindau, I. (1972); this Volume p.55.

Nozières, P. and de Dominicis, C.T. (1969); Phys. Rev. 178, 1097.

Plummer, E.W. and Young, R.P. (1970); Phys. Rev. 1B, 2088.

Porteus, J.O. and Faith, W.N. (1970); Phys. Rev. 2B, 1532.

Raether, H. (1965); in Springer Tracts in Modern Physics, Vol. 38, p.84, Springer-Verlag, Berlin.

Suzuki, T., Kishimoto, T., Kaji, T. and Suzuki, T. (1970); J. Phys. Soc. Japan, 29, 730.

Ulmer, K. (1969); in Proc. of the Int'l. Symp. on X-Ray Spectra and the Electronic Structure of Matter (Kiev 1968), Vol. 2, p.79, Inst. of Metal Physics, Ac. of Sci. Ukr. SSR.

Van Dyke, J.P. (1972); Phys. Rev. B, 15 May. Calculated absorption spectrum of Si including matrix elements.

Willens, R.H. Schreiber, H., Buehler, E. and Brasen, D. (1969); Phys. Rev. Ltrs. 23, 413.

Willens, R.H. and Brasen, D. (1972); Phys. Rev. 5B, 1891. X-ray stress modulation spectrum of nickel.

Wooten, F., Huen, T. and Winsor, H.V. (1971); Phys. Ltrs. 36A, 351.

Ziman, J.M. (1960); Electrons and Phonons, Clarendon Press, Oxford.

ELECTRONIC STRUCTURE OF COPPER-NICKEL-ZINC TERNARY ALLOYS

R.S. Brown† and Leonid V. Azároff

Physics Department and Institute of Materials Science,

Raymond J. Donahue

Metallurgy Department and Institute of Materials Science, University of Connecticut, Storrs, Connecticut, USA.

1. INTRODUCTION

The binary alloys formed by copper, nickel, and zinc have been studied extensively by various experimental methods because of their importance to alloy theory and because they include several important commercial alloys. According to the rigid-band model (Mott and Jones 1936) the binary solid solutions form common 3d and 4s bands whose relative occupation depends on the electron-to-atom ratio of the valence-conduction electrons contributed by the constituent atoms. Thus in Cu-Ni solid solutions, the 3d band should become completely filled when the 'excess' number of 4s electrons supplied by copper atoms just equals the number of 3d holes on the nickel atoms (∿0.6 per atom). This model has been supported by numerous experimental observations that show the expected systematic decline in the nickel 3d-hole density with increasing copper (or zinc) content except when the 'excess' electron-to-atom ratio (e:a) nearly equals or exceeds 0.6. Mott and Jones were aware of this and suggested that some of the nickel 3d

---

†Present address: Department of Mathematics and Physical Sciences, Embrey Riddle Aeronautical University, Florida.

holes remained unfilled because of small amounts of 'undissolved' nickel in Cu-Ni solid solutions containing more than 60at% copper. This view was expanded by Smoluchowski (1951) and Goldman (1952), who suggested the possibility of nickel cluster formation in copper-rich Cu-Ni alloys in order to explain magnetic susceptibility and specific heat data. More recently, Pugh *et al* (1957), Ryan *et al* (1959), and Perrier *et al* (1970) have invoked nickel cluster formation to explain the magnetic properties of Cu-Ni alloys; while Kidron (1969) presented small-angle x-ray scattering data purportedly due to such clusters. However, Moss (1969) has questioned the validity of Kidron's interpretation.

A different approach has been used by Friedel (1956, 1958) who suggested that the nickel 3d states do not form a common band with the copper 3d states in a copper-rich alloy but instead form virtual bound states lying between the copper d states and the Fermi energy. Seib and Spicer (1968) have used this virtual-bound-state model to explain their photoemission and reflectivity data in Cu-Ni solid solutions containing 77 and 90 at% copper.

The x-ray K absorption spectra of the three binary systems formed by copper, nickel and zinc have been studied, most recently by Azároff and Das (1964) in Cu-Ni and by Yeh and Azároff (1967) in Cu-Zn and Ni-Zn. For these alloys, the observed changes in the fine structure of the K absorption edges of the constituent atoms were consistent with the rigid-band model. Therefore, it seemed worthwhile to examine the ternary alloys formed by these atoms in order to obtain a better insight into the rôles played by each of the three constituents.

As shown in the earlier studies, the fine structure appearing near the initial rise in the K absorption edge in all three elements reflects the occupation density of admixed 3d, 4s, 4p states in nickel and 4s, 4p states in copper and zinc. By measuring the areas corresponding to the first maximum, it is possible to deduce the way that these densities change with alloying (Azároff, 1967). Unlike other experiments, which typically are sensitive to the electronic structure of an alloy as a whole, observed x-ray spectra are a superposition of the spectra of individual atoms, so that it is possible to examine separately each kind of atom present. This has frequently been stressed (see, for example, Azároff 1967) and is particularly useful when more than two kinds of atoms are present in an alloy.*

In the absence of other experimental evidence regarding the electronic structure of the alloys under investigation, it was decided to measure their thermoelectric power S. According to the generally accepted model for thermoelectric power (Mott and Jones 1936), the large negative value of S in constantan (58Cu42Ni) is due to the incomplete filling of a nearly full 3d band. (It is actually due to the sharp slope of the N(E) curve near the Fermi energy). When the 3d band is completely filled, the thermoelectric power should go to zero or become positive. Thus a decline in the magnitude of S serves as an indication of the degree of occupation of the 3d band.

Two groups of alloys were selected for the present investigation. In one group the constituents were adjusted to maintain an electron-to-atom ratio close to that of constantan. In the other, it was decided to keep the atomic fraction of

---

*We have in this Volume, (e.g.p.431) calculations indicating that one should use 'local densities of states' rather than alloy densities, the latter being derivable by a suitable linear combination of the constituent atomic densities. However, it should be noted that the results of x-ray absorption studies suggest that this local density of states may change with alloying, so that care must be exercised in forming such combinations.

one constituent (zinc) constant while varying the other two constituents. Since $e/a > 0.6$ in all of these alloys, it was expected that the nickel 3d states would be filled in all specimens and it was hoped that information would be obtained about the rôle of the copper and zinc atoms. However, it turned out that the nickel atoms still retained 3d holes in these alloys and, as shown below, that their density varied inversely with the fractional nickel content.

2. EXPERIMENTAL

The alloy compositions prepared for the present investigation are shown in Table I; the subscripts denote the atomic fractions for each element. The samples were prepared by vacuum melting, from elements of purity better than 99.95%.

Table I
Composition of Cu-Ni-Zn Alloys

Composition	e/a	Composition	e/a
$Cu_{58}Ni_{42}$	0.58	$Cu_{62}Ni_{10}Zn_{28}$	1.17
$Cu_{39}Ni_{52}Zn_{9}$	0.58	$Cu_{56}Ni_{18}Zn_{26}$	1.09
$Cu_{29}Ni_{58}Zn_{13}$	0.55	$Cu_{46}Ni_{29}Zn_{25}$	0.95
$Cu_{22}Ni_{65}Zn_{13}$	0.50	$Cu_{35}Ni_{39}Zn_{26}$	0.88
		$Cu_{25}Ni_{54}Zn_{21}$	0.67

They were subsequently rolled into foils, 10-20μ thickness, with alternate stress-relieving by annealing for periods of one hour at $1000^{o}F$ under argon pressures of 30mm. Subsequently, the foils were checked for homogeneity (single-phase) by metallography and x-ray diffraction and examined under strong light to ensure that no pinholes were present. The foils were finally checked chemically; the compositions are shown in Table I to within 1%.

The metal foils were mounted in a special holder that allows two identical samples to be examined successively; a measurement of the incident beam intensity was taken between each pair of transmitted intensity readings. A two-crystal spectrometer (previously described Azároff, 1965) was equipped with two silicon crystals whose [111] reflections had half-widths of the order of 13 sec. The stability of the x-ray source, scintillation counter, and attendant circuitry was sufficient to ensure an over-all statistical accuracy of better than ±1% for the individual intensity measurements. The measured absorption coefficients were corrected for instrumental effects using a modified computer programme (Porteus 1962).

With a ternary alloy the absorption curve of each constituent can be measured separately in the energy region of its K edge. Subsequently, the absorption of the other constituents should be measured in the same energy region using foils prepared from the pure metals. As previously demonstrated for binary alloys (Azároff and Das 1964) this makes it possible to convert the linear absorption coefficient of the alloy to that of the desired constituent and, concurrently, to derive a quantity that is proportional to the atomic absorption coefficient so that the curves obtained from different alloys will have the same scale and can be compared directly. The chief source of error arises during this procedure, because it requires the determination of the foil thickness by mechanical means. When the alloy is composed of metals that are near-neighbours in the periodic table, another difficulty is encountered because an element of atomic number Z absorbs strongly radiation having energies near the absorption edge of element $Z + 1$ or $Z + 2$. Therefore, this imposes a further limitation on the accuracy whenever small concentrations of the heavier metal are studied.

3. PSEUDO-CONSTANTAN ALLOYS

The first four alloys listed in Table I have $e/a<0.6$ and can be expected to have electronic structures similar to that of constantan, the first-listed binary alloy in the table. In comparing the copper, nickel and zinc K-absorption edges in these alloys, essentially two different procedures can be employed. In one, the edges of the same constituent in different alloys are superimposed to observe any qualitative changes taking place. In the other (Azároff 1967), each absorption curve is divided into three energy regions corresponding to regions containing states of predominantly 3d-4s, 4p, and np character. As previously demonstrated for the binary alloys, this is a relatively straightforward procedure for copper and nickel, whose density-of-states curves are accurately known, but not for zinc. Even for copper and nickel a certain amount of arbitrariness exists in setting the energy limits for these regions although, by using exactly the same limits in all alloys examined, relative changes in the areas within such regions are deemed to be meaningful.

A comparison of the first areas in the Ni K edges due to transitions to empty states having predominantly admixed 3d, 4s, 4p symmetry, showed no variations for the four alloys examined. The inflection point in the initial rise of the absorption curve, presumed to correspond to the position of the Fermi energy, also showed no variation among the alloys studied and it was found to agree with its energy value for pure nickel to within -0.2eV; i.e. all the alloy curves were displaced by that amount from the pure nickel curves. The copper K edges were all shifted by +0.3eV and were quite similar to each other. The increased nickel concentration indicated in Table I causes the first absorption maximum in the Cu curve to rise slightly but the increase in the area of the first region in the Cu K edge (due to 4s-4p holes) was not systematic and is assumed to be changing very slightly. Finally, the Zn K edges are all shifted by +1eV from their energy values for pure zinc, but otherwise also show very little change in the pseudo-constantan alloys examined. This suggests that the electronic structure of these ternary alloys remains similar to that of the binary 58Cu42Ni alloy. This similarity is further borne out by the thermoelectric power S which rises from a value of -41.5 $\mu V/^{o}C$ in constantan to about -37 $\mu V/^{o}C$ in the pseudo-constantan alloys. The observed change is most probably due to scattering effects caused by the presence of zinc (Pollock 1962) rather than to any change in the nickel 3d-hole density.

4. "ELECTRON-RICH" ALLOYS

The second group of ternary alloys in Table I all have $e/a > 0.6$, reaching a maximum of 1.17 in the alloy containing 10at% nickel. Although one would expect from the rigid-band model that the 3d holes in the nickel atoms are completely filled in this case, the x-ray absorption curves do not uphold this expectation. The change in the area of the first region of the nickel K-absorption curve due to transitions to unoccupied admixed d,s,p states, is shown in figure 1. It can be seen that the area declines rather markedly as the nickel concentration increases. (It is doubtful that the marked deviation of the area for the alloy containing 54at% nickel has real significance; it is more likely due to an error in the determination of the area). Concurrently, the corresponding areas in the copper K edge, figure 2, and in the zinc K edge, figure 3, show a systematic increase with nickel concentration. Clearly, this implies a progressive emptying of the admixed 4sp states in these ternary alloys.

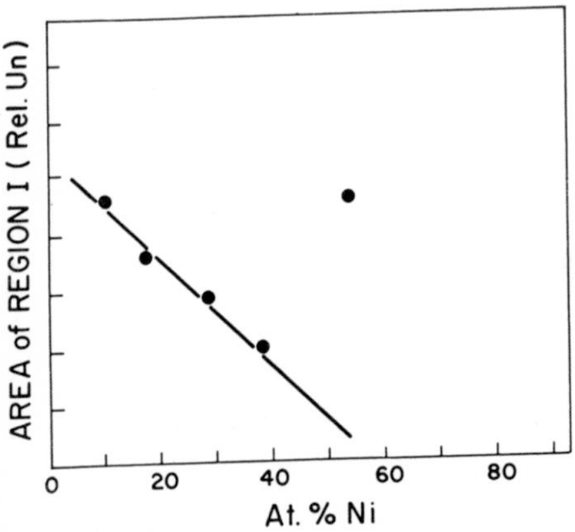

Figure 1 Relative area changes of region I (transitions to empty 3d states) in the nickel K edge.

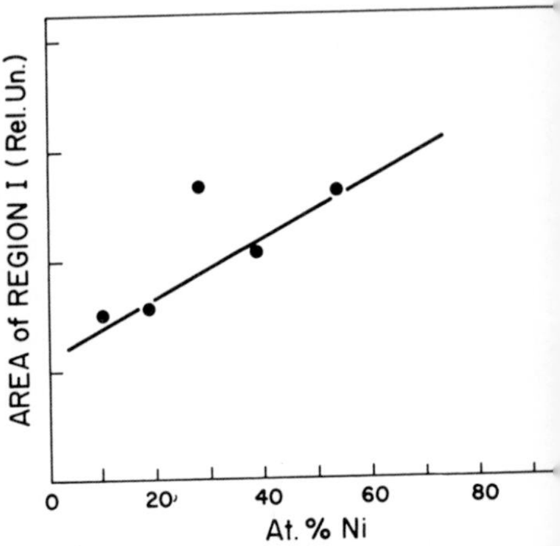

Figure 2 Relative area changes in region I (transitions to empty 4s states) in the copper K edge.

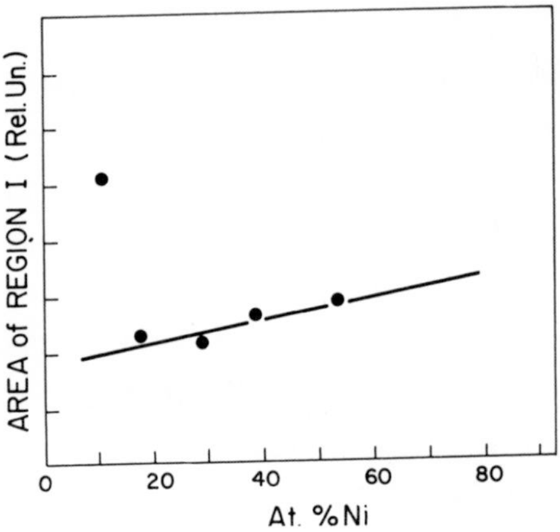

Figure 3 Relative area changes in region I (transitions to empty 4s states) in the zinc K edge.

The three sets of absorption curves all superimpose at the inflection point in their initial rise (Fermi edge) within experimental error. The maxima corresponding to transition to empty 4p states systematically shift to lower energies in the

Cu K edge and to higher energies in the Zn K edge, with increasing nickel content. The successive shifts in the copper 4p peaks go in steps of 0.2-0.3eV whereas the zinc 4p peaks shift by only about 0.1eV. These shifts are probably related to the pronounced change in the lattice constant of the face-centred cubic solid solutions, shown in figure 4, and the

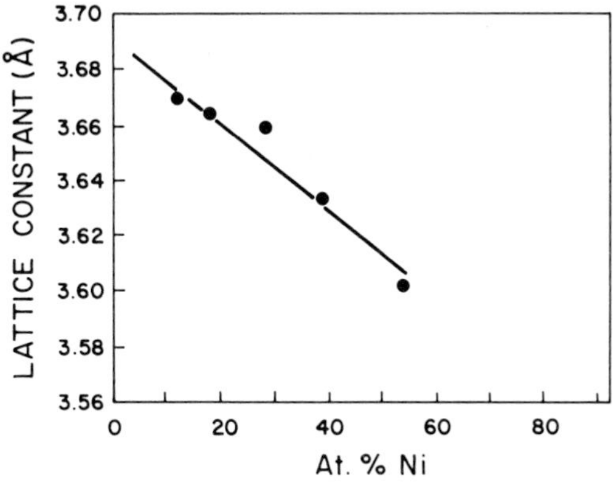

Figure 4 Lattice constants of five ternary Cu-Ni-Zn alloys.

resulting change in the binding energy for 4p electrons. The latter is also affected by the decreased screening due to the 'transfer' of the 4s electrons although, in the absence of exact wave-mechanical calculations, little more can be said on this point.

The thermoelectric power S measured at 30^{0}C in these five ternary alloys is shown in figure 5. It can be seen that S

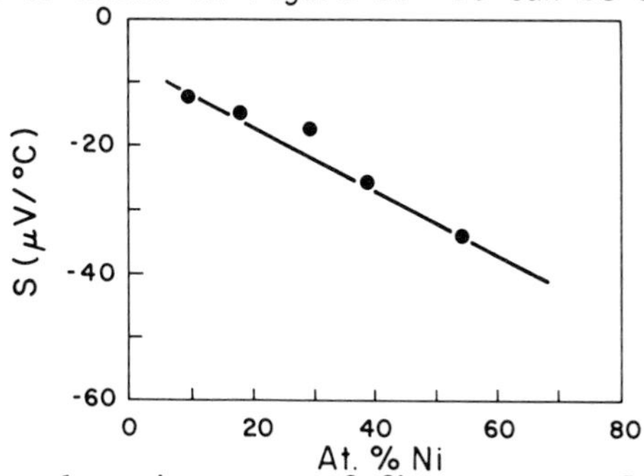

Figure 5 Thermoelectric power of five ternary Cu-Ni-Zn alloys.

declines linearly with increasing nickel content, reaching a maximum negative value of $-34\mu V/^{o}C$ for 54at% Ni. Although the *e/a* value for this alloy has dropped to 0.67, it is still greater than 0.6, the value at which the 3d band should become filled according to the rigid-band model. Nevertheless, the systematic change of S seen in figure 5 corroborates the results of the x-ray study; namely, that the nickel 3d band becomes progressively more occupied as the nickel concentration increases (decline of Area I, as indicated, figure 1). This interpretation of the thermoelectric data is in full accord with the Mott and Jones theory of thermoelectricity, modified by Pollock (1962).

5. DISCUSSION

The pseudo-constantan alloys behave as expected; they show a presence of 3d holes at nickel atoms whose density does not appear to vary significantly with the composition of the ternary alloy. However, the 'electron-rich' alloys clearly do not follow the predictions of the rigid-band model. Noting that the areas of Region I, as shown in figures 1, 2, and 3, are proportional to the density of holes in respectively the 3d-4s band of nickel and the 4s band of copper and zinc, it is clear that increasing the nickel content causes a progressive occupation of the 3d holes in nickel and a progressive emptying of occupied 4s states in copper and zinc. Now, the presence of 3d holes in these alloys is surprising enough, but according to figure 1 the hole density at nickel atoms increases as the nickel concentration decreases; i.e., as the 'excess' electron-to-atom ratio increases.

To understand this phenomenon it is necessary to recall that x-ray absorption is essentially an atomic phenomenon; *viz*, the observed absorption spectrum is a superposition of the absorption 'curves' of individual atoms of one kind in the alloy. Suppose that the nickel atoms tend to form clusters at low concentrations. Then each nickel atom tends to surround itself with other nickel atoms so that it behaves more like an atom in pure nickel; that is, it tends to retain an electronic structure corresponding to an average of 0.6 3d-holes per nickel atom. As the nickel concentration increases, the clusters are gradually broken up causing an increased interaction with unlike atoms so that the probability of 3d hole occupation at nickel atoms increases. Therefore, the formation of nickel clusters would explain the increasing 3d-hole density at nickel atoms evidenced by figures 1 and 5. As already noted, the presence of such clusters has been postulated also to explain magnetic susceptibility and other data. The power of the x-ray spectroscopic results presented here, lies in their capacity to assess independently the rôles played by each of the three constituents of a ternary alloy.

In conclusion it should be noted that the above results do not necessarily contradict the virtual-bound-state model. The spectrometer used was not capable of determining the energy values of the various features of the absorption spectra with sufficient accuracy to test this model. Moreover, exact calculations of the virtual bound states and their energies in the ternary alloys studied are necessary before a meaningful test of the model could be attempted.

Acknowledgments

This research was supported by a grant from the National Science Foundation. The present paper was taken in part from a thesis submitted by one of us, R.S.B., in partial fulfillment of the requirements for the degree of Ph.D. at the University of Connecticut.

REFERENCES

Azároff, L.V. (1965); Advan. X-Ray Anal. 9, 242.

Azároff, L.V. (1967); J. Appl. Phys. 38, 2809.

Azároff, L.V. and Das, B.N. (1964); Phys. Rev. 134, A747.

Friedel, J. (1956); Can. J. Phys. 34, 1190.

Friedel, J. (1958); J. Phys. Radium, 19, 573.

Goldman, J.E. (1952); Phys. Rev. 85, 375.

Kidron, A. (1969); Phys. Rev. Letters 22, 774.

Moss, S.C. (1969); Phys. Rev. Letters 23, 381.

Mott, N.F. and Jones, H. (1936); "The Theory and Properties of Metals and Alloys", Clarendon Press, Oxford (Reprinted 1958, Dover Publ., New York).

Perrier, J.P., Tissier, B. and Tournier, R. (1970); Phys. Rev. Letters 24, 313.

Pollock, D.D. (1962); Trans. AIME 224, 892.

Porteus, J.O. (1962); J. Appl. Phys. 33, 700.

Pugh, E.W., Coles, B.R., Arrott, A. and Goldman, J.E. (1957); Phys. Rev. 105, 814.

Ryan, M., Pugh, E.W. and Smoluchowski, R. (1959); Phys. Rev. 11, 1106.

Seib, D.H. and Spicer, W.E. (1968); Phys. Rev. Letters 20, 1441.

Smoluchowski, R. (1951); Phys. Rev. 84, 511.

Yeh, H.C. and Azároff, L.V. (1967); J. Appl. Phys. 38, 4034.

SOFT X-RAY ABSORPTION SPECTROSCOPY OF METALS AND ALLOYS

C. Kunz

Deutsches Elektronen-Synchrotron DESY, Hamburg, W. Germany

1. INTRODUCTION

Electron synchrotrons have found extensive use as light sources in the soft x-ray and the vacuum ultraviolet range during the past few years. With their use a large number of absorption data on different materials have been accumulated. This report is concerned with two series of metals which have been investigated using the DESY synchrotron during the past few years. These are the transition metals titanium to nickel, near their $M_{2,3}$ edge, the light metals lithium, beryllium, sodium, magnesium and aluminium, in the region of their K or L absorption edges. These investigations have recently led us to an investigation of alloys,which is still in progress and from which the first results are discussed here.

2. EXPERIMENTAL

A. Synchrotron Radiation

Synchrotron radiation (see for example, Haensel and Kunz 1967, Godwin 1969, Gähwiller *et al* 1970) is emitted (in a narrow cone) tangentially to the orbit of the electrons. It is highly polarized and its spectral intensity can be calculated in absolute terms from the known number of orbiting electrons. The spectrum is smooth and extends (down to ∿.05 Å) for a 7.5-GeV accelerator such as DESY. Figure 1 gives, for the DESY synchrotron laboratory, the useful flux of photons into a 2-cm wide collimator at 4¼m from the source. The vertical extensions of the beam depends on the wavelength but is typically ±2cm.

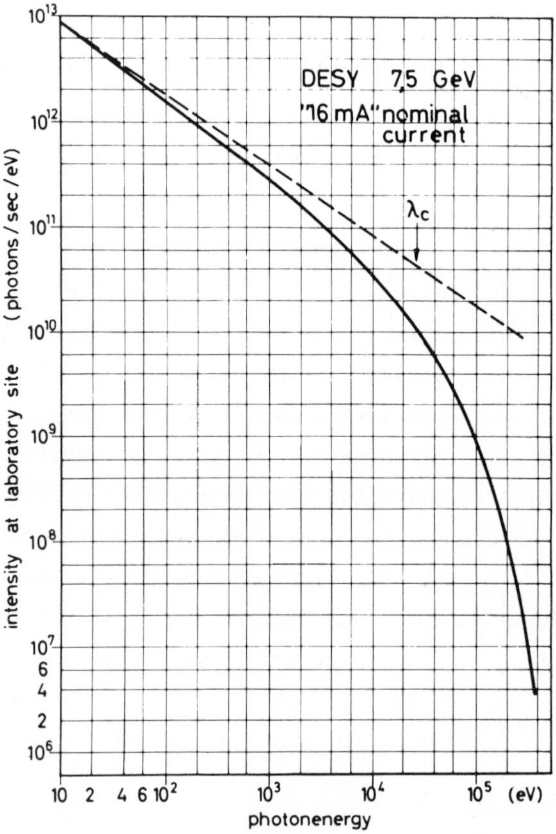

Figure 1 Spectral distribution of photons emitted into a vertical 2cm-wide slit at 40m distance from the electron beam, as a function of photon energy. The dashed line gives the asymptotic behaviour for $\lambda \ll \lambda_c$.

Figure 2 shows the light-beam arrangement at DESY. The experiments need to be directly connected to the synchrotron by a high-vacuum system. The incoming beam is divided, by grazing incidence mirrors, into three separate beams; on each

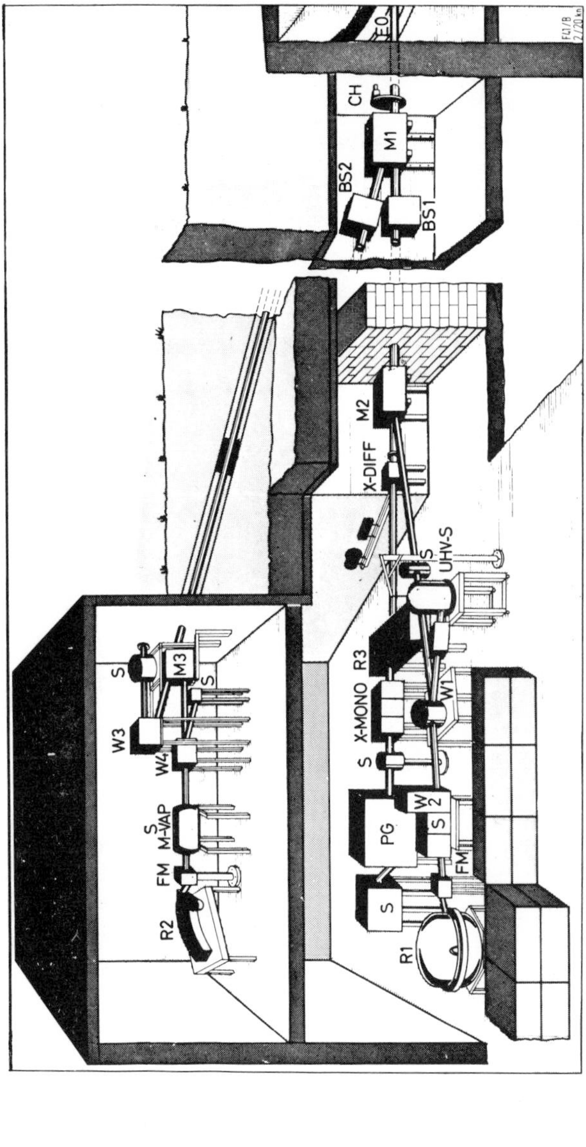

Figure 2 Arrangement of the DESY synchrotron radiation laboratory. The beam is split into several sub-beams by grazing incidence mirrors. At present about ten monochromators and spectrographs are operated at the facility. EO = electron orbit. CH = chopper. M1-M3 = mirrors. B1-B2 = beam shutter. X-DIFF = X-ray diffractometer. UHV S = UHV sample chamber. S = sample. FM = focusing mirror. R1-R3 = Rowland spectrographs. PG = DESY spectrograph. X-MONO = X-ray monochromator. S M-VAP = metal vapours. W1-W4 = Wadsworth monochromators.

several monochromators and spectrographs are arranged. Because of radiation safety requirements all instruments must be operated by remote control. In addition precautions are necessary to protect the electron accelerator from vacuum failure. This makes the experimental arrangement far more complex than with an ordinary x-ray tube.

Nevertheless, such are the advantages that at present more than a dozen synchrotron radiation (SR) laboratories are in operation or are planned throughout the world. Table I gives a survey of these installations.

TABLE I. Accelerators and Storage Rings used as Light Sources

Project Location	E(GeV)	R(m)	I(mA)	λ_c(Å)	Remarks
NBS (Washington)	0.18	0.83	1	800	
INS-SOR (Tokyo)	1.3	4.0	30	10	
Frascati	1.1	3.6		15	
DESY (Hamburg)	7.5	31.7	10-30	0.3	
Wisconsin Storage Ring	0.24	0.54	10	220	exclusively used as a light source
Glasgow	0.33	1.25	0.1	195	
Bonn	2.3	7.65	30	3.5	
Moscow	0.66	∿3		60	
DNPL (Daresbury)	4.0	20.8	40	1.8	SR-Lab near completion
INS-SOR (Tokyo) Storage Ring	0.3	1	100	200	planned for exclusive use as light source starting ∿1973/74
ACO (Orsay) Storage Ring	0.6	1.1	500	30	SR-Lab planned for ∿1973
SLAC (Stanford) Storage Ring	2.5 (3.5)	12.7	250	4.5	SR-Lab planned for ∿1973
DESY (Hamburg) Storage Ring	1.75 3.5	12.12	6000 200	12.7 1.58	SR-Lab planned for 1973/4

Without going into the details of synchrotron radiation theory, a rough estimate of the intensity and the spectral distribution of radiation for a specific accelerator can be obtained from the following two equations

$$\lambda_c(A) = 5.59\ R(m) \times \left[E(GeV)\right]^{-3} \qquad (1)$$

$$I\left(\frac{\text{photons}}{\text{sec.eV.mA.mrad}}\right) = 4.5 \times 10^{12} \times j(mA) \times \left[R(m)\right]^{1/3} \times \left[E_{phot}(eV)\right]^{2/3} \qquad (2)$$

Where R(m) is the radius of the electron orbit in metres, E(GeV) is the electron energy, j(mA) is the beam current and $E_{phot}(eV)$ is the photon energy.

Equation (1) gives the parameter λ_c, commonly referred to the cut-off wavelength. Radiation of wavelengths down to $\lambda \simeq 0.1\lambda_c$ has been used but the intensity is very much decreased at this wavelength (see figure 1).

Equation (2) gives the number of photons emitted into a horizontal angular sector of 1-mrad width and 1-eV energy interval, for the asymptotic region $\lambda >> \lambda_c$ (the dashed curve in figure 1). Note that equation (2) is independent of the electron energy E. Since the dependence on the radius R is weak a good 'rule of thumb' is to compare intensities from different synchrotron accelerators and storage rings by comparing the currents j. For accelerators, the nominal current has to be multiplied by the appropriate duty cycle.

For spectral dispersion of the synchrotron light we use a grazing incidence Rowland monochromator, and also a fixed-exit-slit grating monochromator especially developed for the purpose (Kunz *et al* 1968, Dietrich and Kunz 1971) capable of suppressing the higher-order radiation.

B. Film Preparation

Metal films are prepared by evaporating onto organic

substrates, or onto aluminium. Highly oxidizable metals are prepared *in situ*. The alloy films are evaporated simultaneously from two sources, controlled with an oscillating quartz monitor, onto collodium substrates. In several cases these films are embedded between two carbon films, of thickness 50-100 Å, and the collodium dissolved afterwards.

3. TRANSITION METALS

A. Pure-Metal Spectra

The spectra of the transition metals titanium to nickel were investigated in the region of the 3p-electron transitions (Sonntag 1969, Sonntag *et al* 1969) and are reproduced in figure 3. All the spectra are characterized by prominent structure. Indeed a prominent peak is expected from a one-electron model because of the unfilled 4d states into which the 3p electrons should be preferentially excited. However, the observed structure is much too broad (∿25eV for Cr) to be identified with the density of unfilled d-states which should extend over an energy range of less than 4eV. A similar situation occurs for several transition metals of the third series (Haensel *et al* 1969a), and for the rare earth metals (Haensel *et al* 1970b) where the 4d electrons are preferentially excited into the empty 4f states. In contrast, the $L_{2,3}$-absorption spectra of the transition metals titanium to nickel show a width of only a few electron volts (Bonnelle 1968).

The fact that the magnitude of the structure is considerably reduced in the spectrum of copper where the d-shell is full, shows that this structure is definitely associated with the empty d-states. X-ray photoelectron (XPS) measurements by Fadley and Shirley (1970) have demonstrated a possible mechanism for the spreading of the oscillator strength over a wider energy range. Exchange interaction between the unfill-

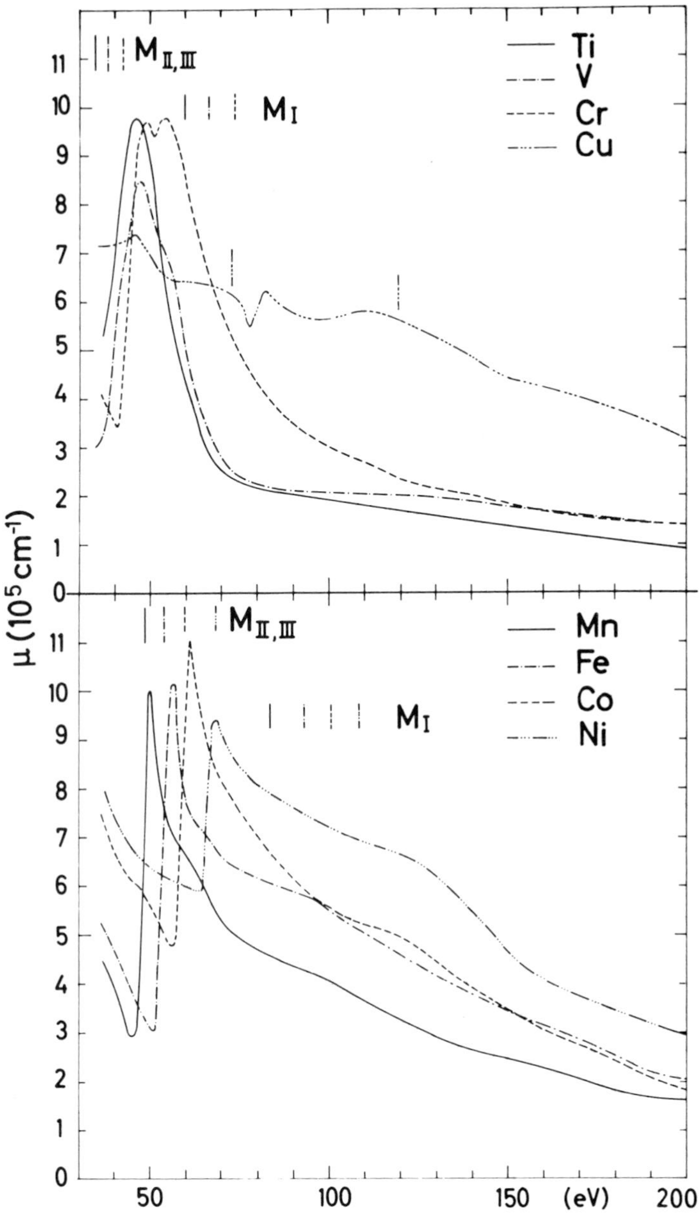

Figure 3 Absorption coefficient for the elements titanium to copper in the $M_{2,3}$ region. (Reproduced with permission from Kunz 1971).

ed d-shell and the 3p hole could split the excited state into several lines. Dehmer *et al* (1970) have placed this on a sound theoretical base; their model calculation shows that exchange splitting could amount to as much as 20eV. To account for the fact that this line structure is not seen in the spectra, one must assume that autoionization interaction with the background continuum broadens and smears the structure (Fano and Cooper 1969). Features at the onset of the manganese and cobalt spectra, and in the platinum spectrum (Haensel *et al* 1969a), resemble the typical depression known for some of the autoionization line shapes (Fano and Cooper 1968). The smaller width of the $L_{2,3}$-absorption spectra (Bonnelle 1968) is understandable because a much weaker exchange interaction exists between states in shells with different principal quantum number than in those with the same principal quantum number.

B. Alloys

If the shape of the spectra is determined mainly by an intra-atomic phenomenon the question arises: how much of the spectrum is determined by band effects? This leads us to investigate the spectra of alloys of transition metals with one another, and to compare the transmission spectrum of an alloy film directly with that of a sandwich film of the separate constituents placed on top of each other.

Figure 4 shows our results for copper-nickel 1:1 (Gudat and Kunz 1971). Within the experimental accuracy no difference can be detected. This indicates that for the alloy the spectra of both atoms are superimposed proportionally. Both types of atom behave independently. Similar results were obtained for the alloys chromium-manganese and iron-manganese. The alloy films were prepared by simultaneous evaporation from two sources without subsequent annealing. Although their metallographic structure is not well defined, this seems to

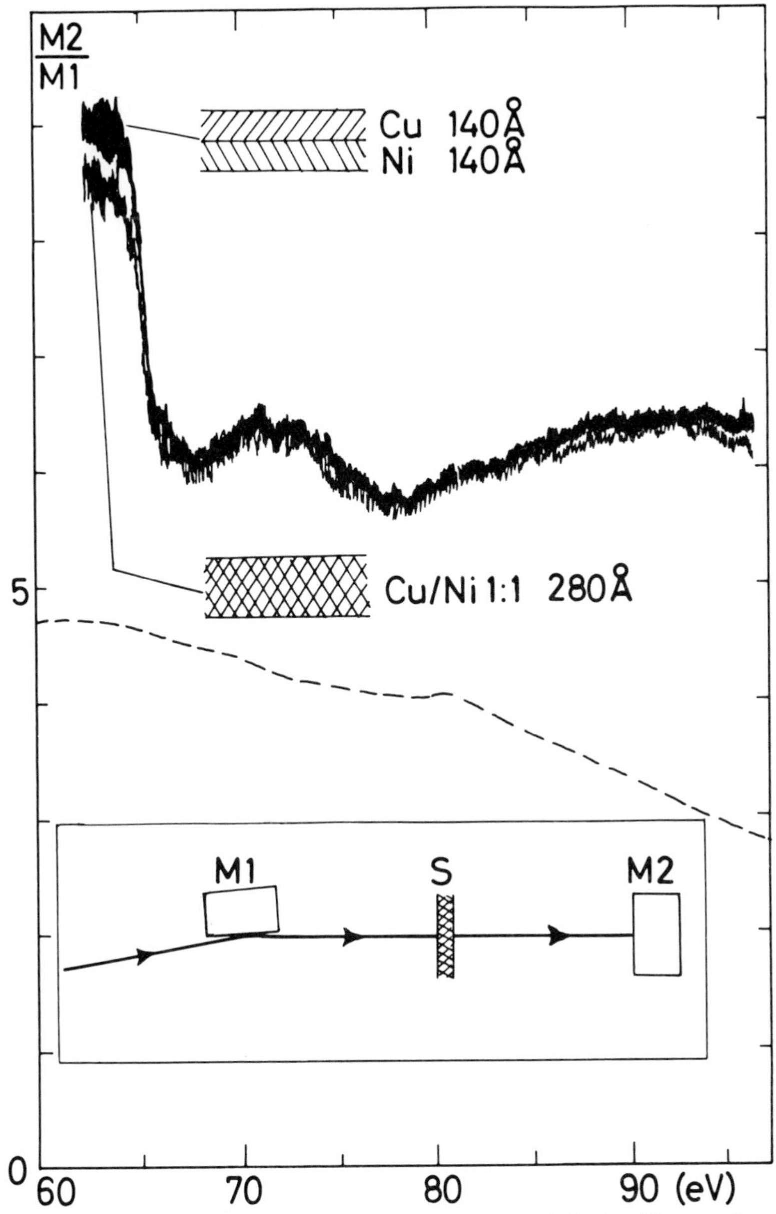

Figure 4 Original spectra of a copper-nickel alloy and a copper and nickel sandwich film. In order to obtain a real transmission spectrum the curves would have to be divided by the spectrum shown as a dashed line. The insert shows the measuring technique. M1 is a combined multiplier and reflector.

be unimportant for the conclusions we reach.

A similar insensitivity of the emission spectra of transition-metal alloys with composition has been reported by Curry *et al* (1968). This indicates that atomic theories give a good first-order approach to the interpretation of these spectra.

4. LIGHT METALS

A. Pure Metal Spectra

Figure 5 shows the $L_{2,3}$ absorption spectrum for aluminium (Haensel *et al* 1970a) which is similar to those of sodium and magnesium. At the edges (which split into L_2 and L_3) a sharp peak is observed similar to more prominent ones for sodium and magnesium. The edge is followed by fine structure, which can be associated with density-of-states effects although to our knowledge has not been identified. The structure culminates in a broad prominent maximum at about 100eV followed by several additional broad maxima. This behaviour for aluminium has already been reported by Fomichev (1967) who found no convincing explanation. The maxima are absent in one-electron atomic calculations (see figure 5) and in recent measurements on sodium vapour (Haensel *et al* 1971).

The structure at the edges is shown in more detail in figure 6 (Haensel *et al* 1969, Kunz *et al* 1970, Gähwiller and Brown 1970, Ejiri *et al* 1970). The L-spectra show a steep increase and a peak at the edge, while the K-absorption edges rise smoothly. This behaviour is closely similar to the edge structure of the equivalent emission spectra and, has been attributed (Mahan 1967, Nozières and Dominicis 1969, Friedel 1969, Hopfield 1969, and others) to the influence of the unshielded part of the hole potential, not only on the ejected electron but also on all the metal electrons. The explanation of this phenomenon goes beyond the one-electron approximation. If the core level is a p-state the result is a singularity in

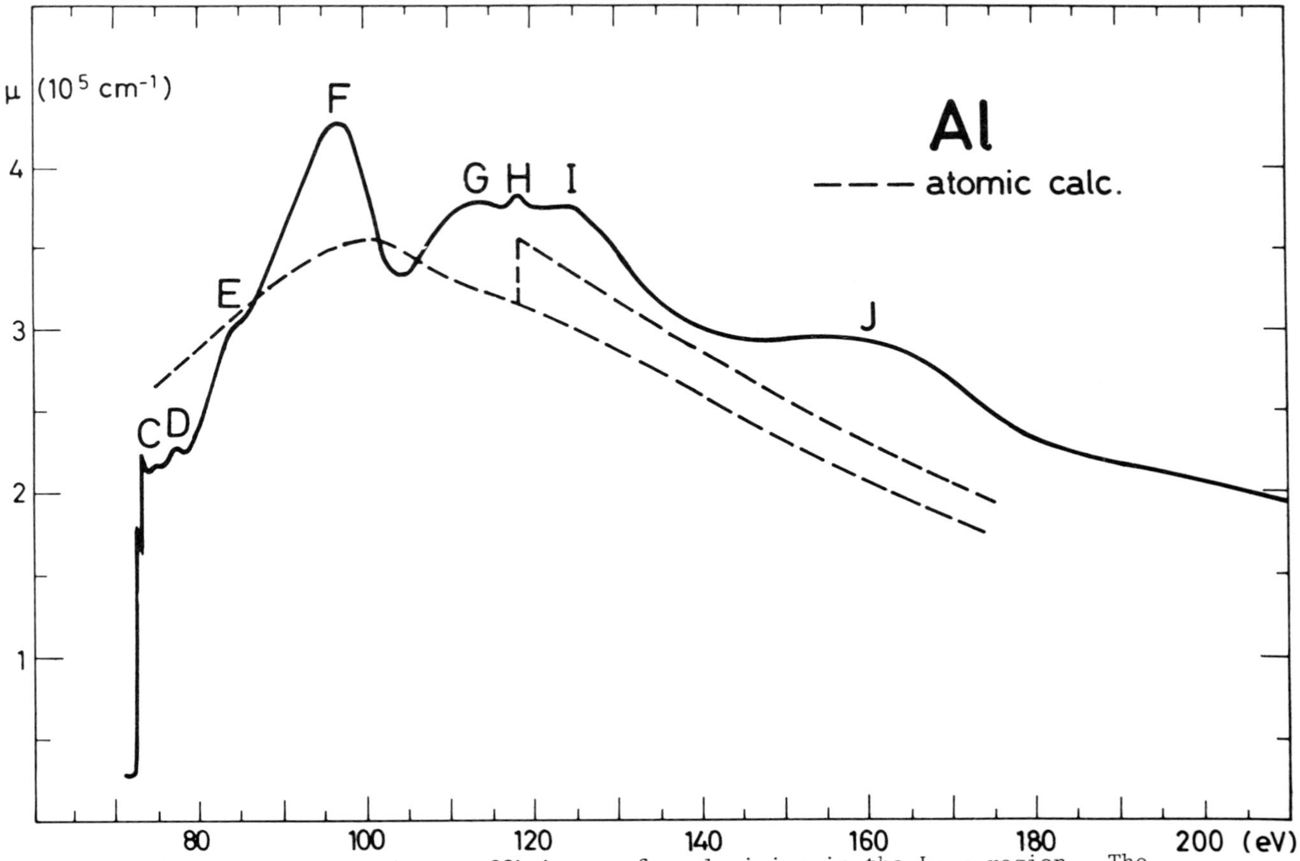

Figure 5 Absorption coefficient μ for aluminium in the $L_{2,3}$ region. The dashed curve gives atomic calculations (Cooper 1967 unpublished communication). (After Haensel *et al* 1970).

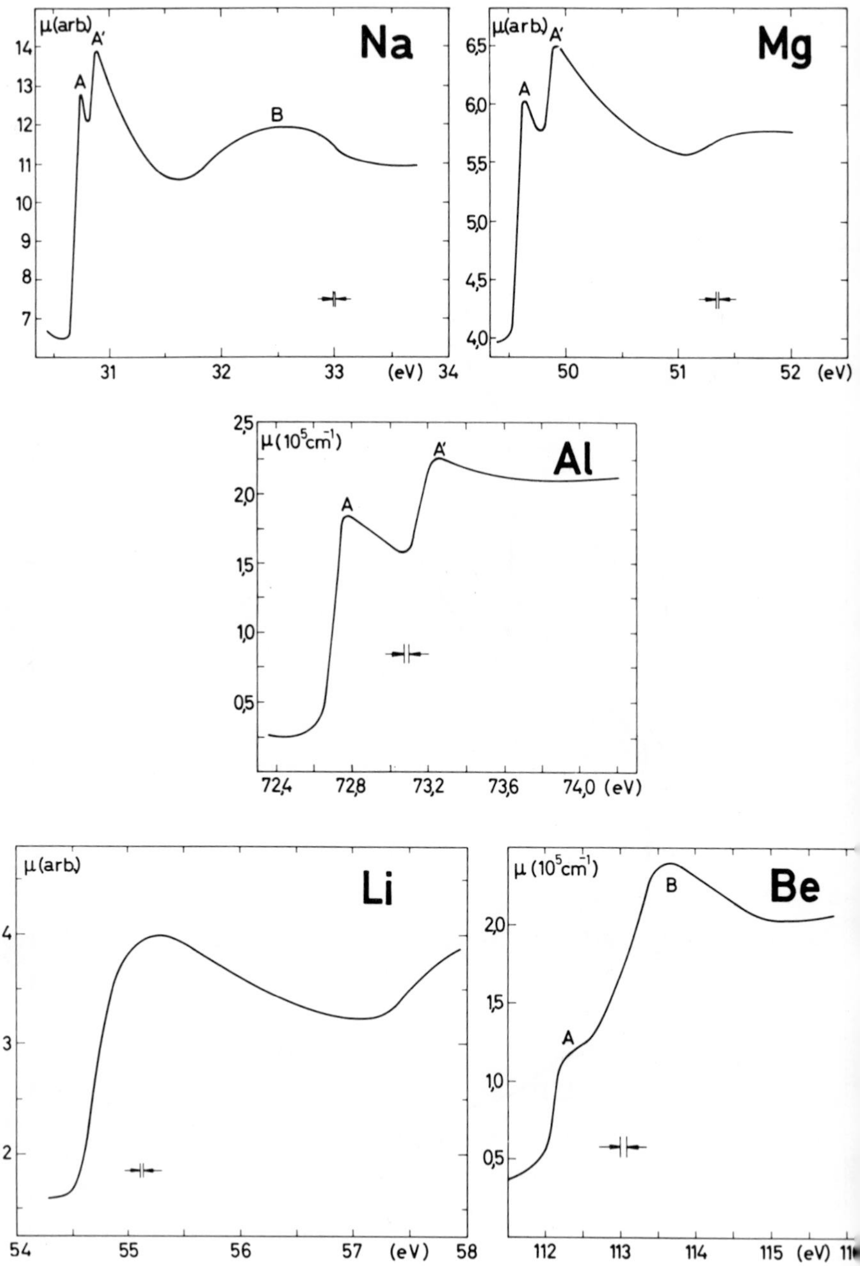

Figure 6 Shape of the $L_{2,3}$-edges for sodium, magnesium and aluminium and of K edges for lithium and beryllium. (Reproduced with permission from Kunz 1971).

the absorption cross-section (μ) at the edge. The shape of the cross-section in the region immediately following the edge, is given by

$$\mu \propto \frac{1}{\Delta E^{\alpha}} \qquad (3)$$

where ΔE is the distance from the edge and α is a positive exponent equal to $\sim$0.5 (Ausman and Glick 1969). (In fact a prominent peak rather than a singularity is expected due to Auger and temperature broadening effects). On the other hand no singularity is expected at the onset of an s-electron transition; for these transitions α is expected to be negative but of small absolute magnitude. The theoretical results are complementary for emission and absorption.

However, it should be mentioned that the peaks in emission are much weaker than in absorption and the absolute magnitude has still to be calculated.

B. Alloys

It is promising to use 'sharp edge' materials such as sodium, magnesium and aluminium, both as a solvent and as a solute for low-concentration alloys. We (and also Yamaguchi *et al* 1971) have commenced measurements on alloys using aluminium as the host. The measurements by Yamaguchi *et al* show clearly the disappearance of the edge-peak and a softening of the edge structure when alloying nickel or manganese with aluminium.

We have investigated aluminium-gold and aluminium-titanium alloys (Gudat *et al* 1971), both prepared by simultaneous evaporation of the two constituents. Figure 7 shows the results. The alloy spectrum is compared to a sandwich film of equivalent composition. The resolution in the present experiments was not sufficient to observe the edge peak. Clearly a broadening and a shift of the edge can be seen even with

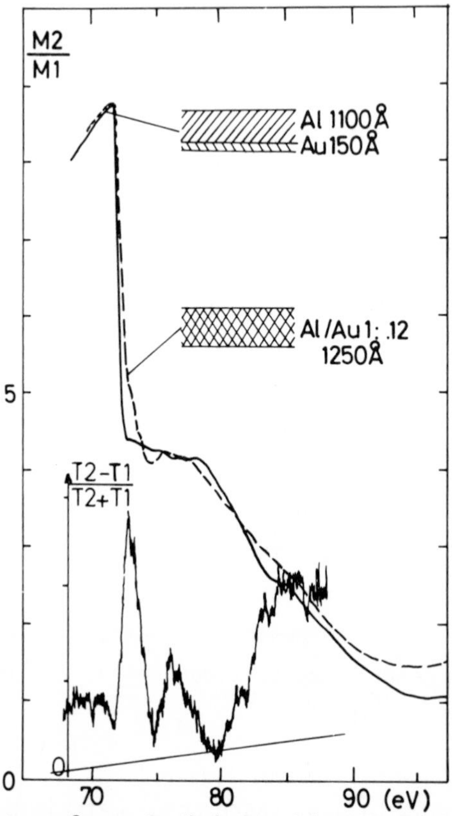

Figure 7 Spectra of an aluminium-gold alloy and an aluminium and gold sandwich film. The insert shows an original spectrum measuring the difference in the transmissivities of the two samples divided by the average. This was done using the beam splitting technique shown in figure 8.

this low resolution. The shoulder at ∿85eV seen also in figure 5 has disappeared and an additional absorption at ∿2.4eV from the edge shows up (∿2eV for titanium). Since the metallographic state of the samples - which were not annealed - was not determined, we do not go into any physical interpretation at present.

The principal interest in these measurements was to test a differential two-beam densitometer for the soft x-ray range (figure 8). A beam splitter, consisting of a mirror rotating synchronously with the pulsed emission from the synchrotron, is mounted behind the exit slit of the monochromator. The transmissivities (T_1 and T_2) of the two films are compared.

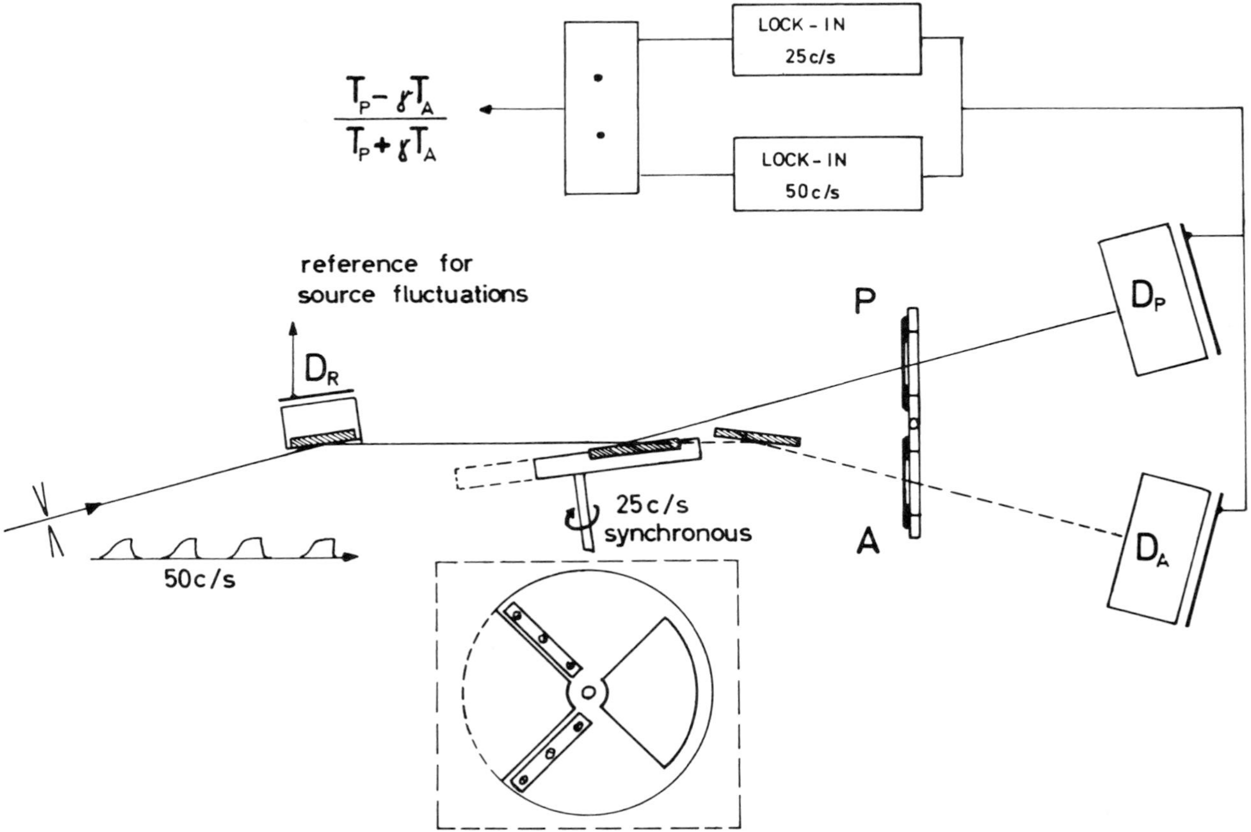

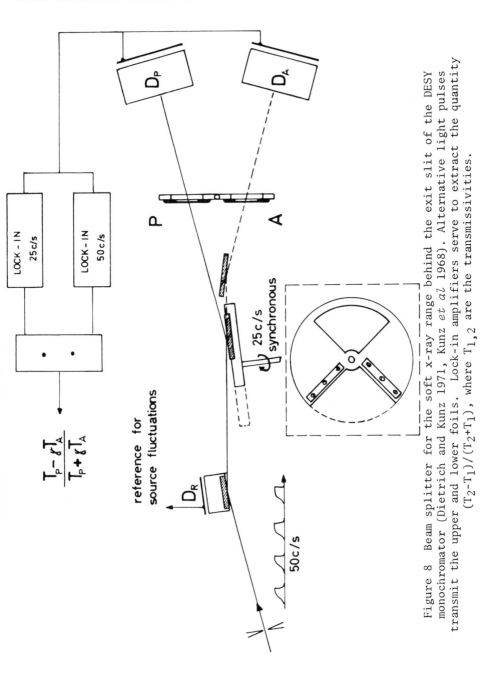

Figure 8 Beam splitter for the soft x-ray range behind the exit slit of the DESY monochromator (Dietrich and Kunz 1971, Kunz *et al* 1968). Alternative light pulses transmit the upper and lower foils. Lock-in amplifiers serve to extract the quantity $(T_2-T_1)/(T_2+T_1)$, where $T_{1,2}$ are the transmissivities.

The quantity $(T_2-T_1)/(T_2+T_1)$ is measured by a lock-in technique. Figure 7 shows also that this spectrum is in reasonable agreement with those obtained directly. We hope to improve the method and to measure this quantity with an accuracy of 10^{-3}. This should enable us to investigate low-concentration alloys where theoretical interpretation is most developed.

5. CONCLUSION

The forgoing examples have shown that the absorption spectra of these metals cannot be considered simply as a reflection of the density of unoccupied states. The vacant hole created during the absorption process causes a non-negligible disturbance to the final states; through exchange interaction in the case of the transition metals, and through the unshielded part of its potential in the light metals in which it causes the edge-peak. The widely held view that such spectra can be regarded as slightly distorted density of states curves appears to be open to critism (see especially Harrison 1968). Nonetheless interesting phenomena occur that deserve attention in their own right.

Acknowledgment

The author wishes to thank W. Gudat for a critical reading of the manuscript.

REFERENCES

Ausman, G.A. and Glick, A.J. (1969); Phys. Rev. 183, 687.

Bonnelle, C. (1968); in "Soft X-ray Band Spectra", ed. by D.J. Fabian, Academic Press, London and New York.

Cooper, J.W. (1967); unpublished communication.

Curry, C. (1968); in "Soft X-ray Band Spectra", ed. by D.J. Fabian, Academic Press, London and New York.

Dehmer, J.L., Starace, A.F., Fano, U., Sugar, J. and Cooper, J.W. (1971); Phys. Rev. Letters 26, 1521.

Dietrich, H. and Kunz, C. (1971); DESY Report SR-71/4 (August 1971) and, Rev. Sci. Instrum. (February 1972), 43, 434.

Ejiri, A., Yamaguchi, S., Saruwatari, M., Yokota, M., Inayoshi, K. and Matsuoka, G., (1970); Optics Commun. 1, 349.

Fadley, C.S., and Shirley, D.A. (1970); Phys. Rev. A2, 1109.

Fano, U. and Cooper, J.W. (1968); Rev. Mod. Phys. 40, 441.

Fano, U. and Cooper, J.W. (1969); Rev. Mod. Phys. 41, 724.

Fomichev, V.A. (1967); Soviet Phys. Solid State 8, 2312.

Friedel, J. (1969); Comments Solid State Phys. 2, 21.

Gähwiller, C. and Brown, F.C. (1970); Phys. Rev. B2, 1918.

Gähwiller, C., Brown, F.C. and Fujita, H. (1970); Rev. Sci. Instr. 41, 1275.

Godwin, R.P. (1969); Springer Tracts in Modern Physics, Vol. 51.

Gudat, W., Karlau, J. and Kunz, C. (1972); to be published.

Gudat, W. and Kunz, C. (1971); reported at the 3rd International Conference on VUV Radiation Physics (Tokyo) and to be published.

Haensel, R. and Kunz, C. (1967); Z. Angew Phys. 23, 276.

Haensel, R., Radler, K., Sonntag, B. and Kunz, C. (1969a); Solid State Commun. 7, 1495.

Haensel, R., Keitel, G., Schreiber, P., Sonntag, B. and Kunz, C. (1969b); Phys. Rev. Letters 23, 528.

Haensel, R., Keitel, G., Sonntag, B., Kunz, C. and Schreiber, P. (1970a); Phys. Stat. Sol. 2, 85.

Haensel, R., Rabe, P. and Sonntag, B. (1970b); Sol. State Commun. 8, 1845.

Haensel, R., Radler, K., Sonntag, B. and Wolff, H.W. (1971); Reported at the 3rd International Conference on VUV Radiation Physics (Tokyo), to be published.

Harrison, W.A. (1968); in "Soft X-Ray Band Spectra", ed. by D.J. Fabian, Academic Press, London and New York.

Hopfield, J.J. (1969); Comments on Solid State Physics 2, 40.

Kunz, C., Haensel, R. and Sonntag, B. (1968); J. Opt. Soc. Am. 58, 1415.

Kunz, C., Haensel, R., Keitel, G., Schreiber, P. and Sonntag, B. (1970); Proc. 3rd IMR Symposium, Electronic Density of States, Nat. Bur. Stand. (US), Spec. Publ. 323.

Kunz, C. (1971); Journal de Physique, 32, Suppl. 10, C4-180.

Mahan, G.D. (1967); Phys. Rev. 163, 612.

Nozières, P. and de Dominicis, C.T. (1969); Phys. Rev. 178, 1097.

Sonntag, B., Haensel, R. and Kunz, C. (1969); Solid State Commun. 7, 597.

Sonntag, B. (1969); Dissertation.

Yamaguchi, S., Sato, S., Ishiguro, E., Aita, O., Hanyu, T. and Koike, H. (1971); reported at the 3rd International Conference on VUV Radiation Physics (Tokyo).

ISOCHROMAT SPECTROSCOPY OF ALLOYS

K. Ulmer

Physikalisches Institut, Universität Karlsruhe,
Karlsruhe, W. Germany.

1. SPECTROSCOPIC METHODS

There are many different methods for studying the electronic structure of solids. In this report the term "electronic structure of solids" is taken to mean exclusively the density Z(E) of electron states as a function of energy E. For studying these electronic structures one must apply spectroscopic methods.

Figure 1 shows schematically a classification of the most prominent x-ray spectroscopic methods, in terms of well known one-electron energy levels (for a more detailed scheme see Nagel, this volume p457). In the present report only one-electron energy-level schemes will be used. Figure 1 may be applied to the conduction band of metals, to which we shall confine our considerations. The well known emission spectroscopy for the occupied states and absorption spectroscopy for the unoccupied states are represented on the left-hand side of figure 1. We should mention that we do not concern ourselves here with transition probabilities; there are only two cases where different transition probabilities appear to be of essential importance to the subject and we draw attention to these in due course.

An inherent feature of these classical methods is the

necessary participation of core-level holes or core-level electrons in the respective elementary processes. The methods shown schematically on the right-hand side of figure 1 - summarized under 'isochromat spectroscopy' - are free from this restriction. The pertaining transitions take place to or from high-lying normally unoccupied levels. Moreover the rôles are exchanged: absorption for the occupied states and emission for the unoccupied states. The inclusion of photoemission supplements our classification and makes it more symmetrical. We shall return to this point later. The main interest of this contribution is related to the last column - termed bremsstrahlung - and is discussed more closely shortly. However, before proceeding, two important points should be mentioned:

(1) Investigation of the density of states in the conduction band by classical x-ray spectroscopy (see for example Ulmer 1969, Fabian 1971) requires correction for the finite width of the participating core levels. This width may be large, especially for targets composed of elements with high atomic number. Isochromat spectroscopy is free from this feature. Additionally, the free choice of one of the high (normally unoccupied) levels as the final-state level represents a new free parameter of the experiment; for the experimentalist, a new degree of freedom. Nonetheless the two spectroscopic methods supplement each other in a promising manner, last not least because of different transition probabilities. This, then, is our first point - that the two spectroscopies supplement one another - and we note that transition probabilities are of essential importance.

(2) The ionization states for the four columns in figure 1 are different. In the first we have a missing electron, and in the last there is an additional electron. We shall not be concerned with these differences.

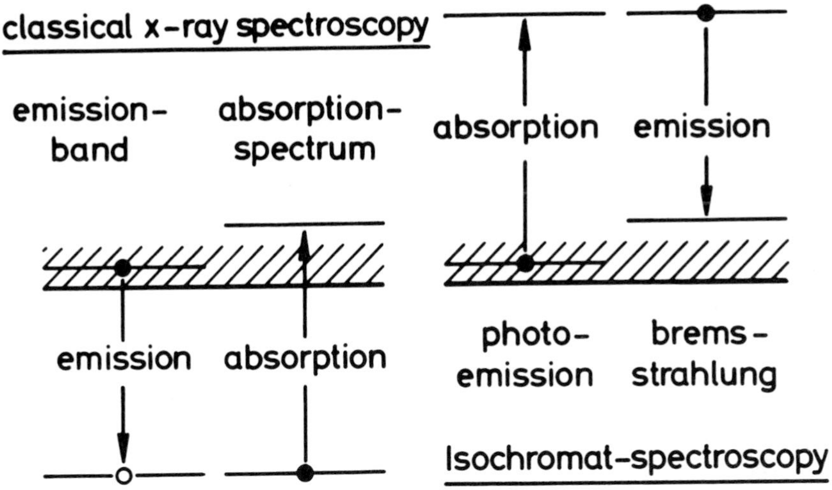

Figure 1 Classification schemes for classical and isochromat spectroscopies.

Figure 2a illustrates the two possible variants of bremsstrahlung-isochromat spectroscopy, i.e. the last column of figure 1. The transition diagram to the left illustrates the case of the classical bremsstrahlung experiment. E' is the energy of the incoming primary electron; this energy is fixed, and quanta of different energies, $\hbar\omega$, are generated. The diagram on the right refers to an experimentally more convenient variant: E' is varied (by varying the voltage of the x-ray tube) and only quanta of *one* fixed energy $\hbar\omega$ are recorded using an x-ray monochromator. This, strictly speaking, is the case of *true* isochromat spectroscopy in the meaning of this report.

In figure 2b we show experimental results for such measurements. The plotted points and the full curve provide an experimental comparison of the two variants, of which the second is experimentally the much simpler (Eggs and Ulmer 1968). The full line is an example of a so-called 'bremsstrahlung-isochromat', for rhodium. All measured curves that follow are such

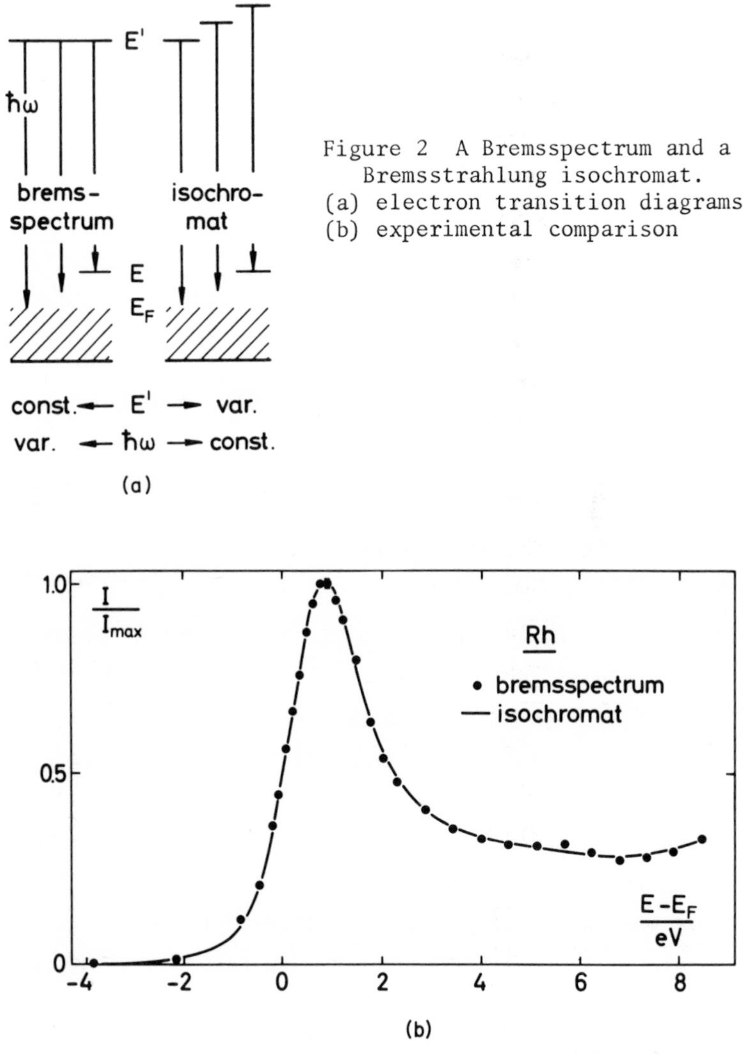

Figure 2 A Bremsspectrum and a Bremsstrahlung isochromat.
(a) electron transition diagrams
(b) experimental comparison

bremsstrahlung isochromats, or in short, 'isochromats'. We take the rhodium isochromat to be a direct image of the distribution of the electronic density of states in the unoccupied part of the conduction band of rhodium. The steep rise with increasing excitation energy E' represents the threshold between occupied and unoccupied levels (the Fermi-edge), and immediately beyond, at the peak, we see a high density of unoccupied 4d-states. For photoemission we would have all arrows reversed in direction, i.e. beginning in the *occupied* part of the conduction band. Then we would again interpret the arrows as representing the electron transitions and would then recognize in the photoemission also two different variants. To-date the experiment corresponding to the resulting right-hand scheme, which necessitates the use of an electron spectrometer, seems to be the favoured method in photoemission experiments. However, in principle it should be possible to vary the illuminating photon energy and record the intensity of emission of electrons with a particular energy.

2. SPECTROSCOPY OF ALLOYS

We have reviewed the essential spectroscopic methods for studying densities of states and we turn now to the problem of alloys. Our aim is to contribute some simple physical pictures or models that appear to hold - at least qualitatively - when applied to the measured isochromats of some alloys.

We formulate now the special problem, which will be discussed exclusively below. The question is: how is the density of states for alloys related to the density of states of the constituent elements? We must assume of course that the term 'density of states' is sensible not only for the elements but also for alloys.

There arises at once a well-known difficulty: in the idealized regular lattice of an element, all lattice points are geometrically alike and physically indistinguishable, and *one* such lattice point together with its proper unit cell and suitable boundary conditions suffices for the determination of the homogenous conduction band. But in an alloy the lattice is disturbed by compositional disorder at least. (We exclude from consideration amorphous substances with their additional structural disorder, and also stoichiometric compounds with their additional interaction energies). Now, in small volumes of alloy - in the extreme case containing one lattice atom only - we have large fluctuations of composition; whereas in large volumes we have negligible fluctuations of the composition. We further restrict our consideration to binary systems exhibiting a continuous substitutional solid solution, for all compositions, without change in structure.

To what degree fluctuations of composition, over small volumes, are 'seen' by a particular experiment is a matter of the weighting and averaging properties of that experiment. It appears, that this is a salient point in the alloy problem; a point which has not always been fully realized. A trivial example of an experiment that averages the weighting over macroscopic dimensions is the measurement of buoyancy, of a macroscopic piece of matter, in water. An example of an extremely non-localized elementary process is Bragg reflection. On the other hand, an elementary process with extreme localization is the transition of an electron from the conduction band to a free core level; that is, to a hole at a fixed atom. Such localization takes place not only with alloys but also with elements. But then we have to assume, that emission spectroscopy - for example as based on this localized elementary process - *always* measures some sort of a localized density of states. It is interesting, that only measurements on alloys were to reveal this localization and to bring it into perspective, since, as we can see for *elements* localized and averaged densities are obviously the same.

Consequently we cannot expect to measure averaged densities of states with any highly localizing experiment. A well-known example is the $L_{2,3}$ emission spectra of the magnesium-aluminium alloys (Neddermeyer, this volume p153); not even with ordered alloys can we expect to measure averaged densities of states, since even the smallest unit cell contains more than one lattice point, and emission spectroscopy explores essentially only one lattice point and its immediate surroundings.

To prepare the ground for isochromat spectroscopy we dwell for a moment on emission spectroscopy and take a look at the general situation, illustrated by figure 3. Here we have

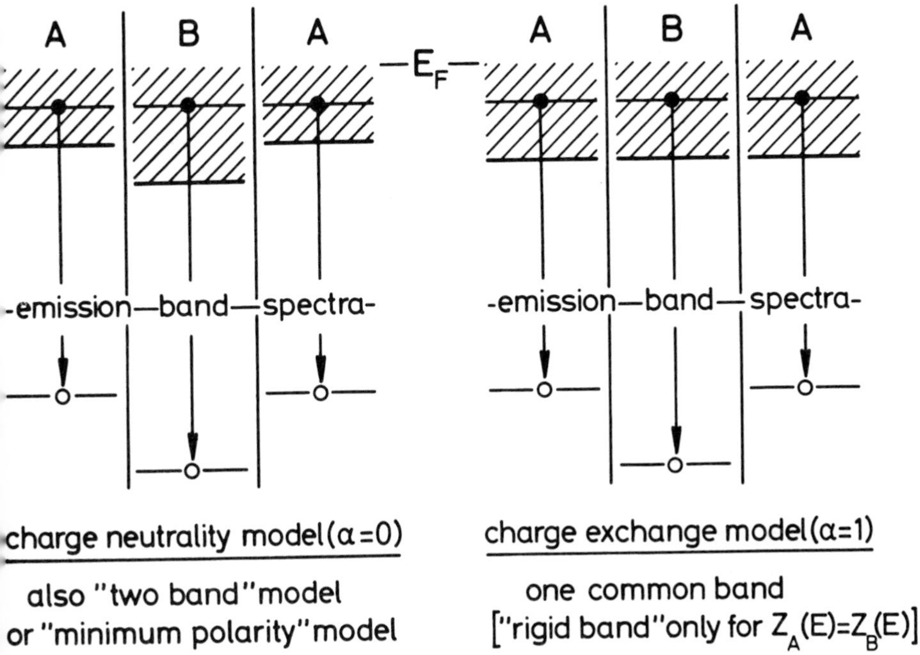

Figure 3 Valence band models.

energy-level schemes for an alloy composed of two elements A and B. For simplicity we consider only samples where the second element has one more valence electron than the first; i.e. for which

$$n_B = n_A + 1 \tag{1}$$

The indicated transitions to core levels (we indicate adjacent lattice atoms) may suggest the localization of emission spectroscopy. The right-hand scheme, suggesting an averaged common band, is ruled out by former considerations, at least for emission spectroscopy. Ours is a simplified discussion (c.f. for example, Fabian 1971). Indeed there are some experimental results that show a tendency towards this model in special cases; for example measurements on aluminium-palladium alloys (Watson *et al* 1971) with strongly different electron configurations of the two constituent elements.

The term "charge neutrality" in figure 3 refers to the fact that for this model each lattice point including its surroundings is neutral; whereas "charge exchange" indicates that some charge transfer among adjacent lattice points is necessary. Such charge exchange has a polarizing effect. Accordingly, the notation "minimum polarity" specifies the case with no charge exchange at all among the two kinds of atom. The parameter α indicates the relative amounts of charge exchange, if this should take place.

Later we shall recognize that the cases $\alpha=0$ and $\alpha=1$ can be considered as the two limiting cases of a generalized "partial charge exchange" model. By Z_A and Z_B we designate the densities of states, in electrons per electronvolt per atom, of the constituent elements A and B, irrespective of the occupancy of these states. The reference to "rigid band" will be discussed later.

We are now at the point where we can turn finally to isochromat spectroscopy, and we first concern ourselves with the question: do we observe localization with isochromat spectroscopy? This is indeed a salient point, which we shall return to at the end of the report.

At first sight we cannot see any localizing effect. Both participating electron states can be expected, approximately to extend over the lattice. Hence it can be assumed that, in both elements and alloys, they extend over volumes that are sufficient for theoretical construction of common energy bands to be possible. Then there arises the possibility with isochromat spectroscopy - in contrast to classical x-ray spectroscopy - that we could observe something like a common band both for elements and for alloys. To anticipate the result reached later: measured isochromats suggest that a common band applies in alloys, for suitable weighting and averaging experiments, but they cannot prove this because isochromat spectroscopy - similar in this respect to classical x-ray spectroscopy - is itself a method with selective weighting and averaging.

3. THE SUPERPOSITION MODEL

To examine this model we try the construction of a density of states curve $Z_{AB}(E)$ resulting from the two densities of states Z_A and Z_B of the constituent elements, which are assumed to be known. It turns out that our model will bear some resemblance to the two-band model of Varley (1954), and in a sense it is a generalization of the Varley model.

Our procedure is simply a linear interpolation - or superposition - of Z_A and Z_B. The corresponding formula

$$Z_{A_xB_y}(E - E_F) = \frac{1}{x + y}\left[x.Z_A(E - E_F) + y.Z_B(E - E_F)\right] \quad (2)$$

is the main equation for the model. Here x and y are the concentrations of the components A and B of the alloy. We do not distinguish between dilute and concentrated alloys, but apply our formula only to concentrated binary solid solutions. A crucial point is that the formula demands counting all energies relative to the Fermi-edge E_F for the alloy; but this does not necessarily coincide with the Fermi-edges for the component metals within their individual conduction bands. This means - before applying formula (2) - that we are faced with the problem of finding the position of the alloy-Fermi-edge in the two conduction bands of the components.

To elucidate on this question, and on the concepts behind our formula, let us imagine macroscopic slabs of the two metals A and B brought into contact with each other. Then the two Fermi-edges balance at a common level, by developing a dipole-layer at their interface. The direction of electron transfer at this interface is determined by the sign of the electronegativity difference between the two metals. Electron-transfer takes place without disturbing the electronic condition in the bulk of the individual metals, because the volume of the dipole-layer is negligible compared with the volumes of the two metals. This means also that the levelling-up of the two Fermi-edges does not influence the position of these Fermi-edges relative to their individual energy bands.*

Now let us imagine our two metals dissolved into one another until an atomically dispersed mixture is reached which then represents the alloy. The volume of the interface layers between the atoms is then no longer negligible, but is comparable to the atomic volumes. We recognize, from the pertaining electron transfer that there may result a considerable shifting of the individual Fermi-edges relative to their individual density-of-states curves. The partial densities of states are fitted together at the *new* positions of their Fermi-limits separating now the occupied from the unoccupied states. The fitting level is then the Fermi edge E_F of the alloy. Finally the mutually displaced partial densities of states can be superimposed according to formula (2).

Thus the concepts behind our interpolation formula assume that the partial densities of states-before superposition- are unaffected by alloying, except for an eventual displacement to-

*Features of a common energy band - even for the case of macroscopic contact - should be found if we could succeed in finding an experiment that averages over macroscopic volumes. Such an experiment could, for example, be the measurement of the mean inner potential with electron interferences, developed for some time by Moellenstedt (see, for example, Gaukler and Schwarzer 1971).

wards each other, and that merely the individual Fermi-edges are eventually shifted.

From these features we expect our model to be applicable chiefly to alloys of transition metals. Indeed our experimental examples below refer to transition metals only.

A definition of the parameter α, determining the shifting of the individual Fermi edges, may now be given:

$$\alpha = \Delta N_A + \Delta N_B = \left[\begin{matrix}\text{number of electrons} \\ \text{taken up per A-atom}\end{matrix} + \begin{matrix}\text{number of electrons} \\ \text{supplied per B-atom}\end{matrix}\right] \quad (3)$$

With $\alpha=0$, we have charge neutrality; with $\alpha=1$, we have charge exchange; and the case $0<\alpha<1$ represents partial charge exchange. Note that generally the two terms on the right-hand side of equation 3 are not equal. To obtain equality we must multiply each term by the concentration of the respective atoms in the alloy, according to:

$$x.\Delta N_A = y.\Delta N_B \quad (4)$$

All our formula so far do not refer to any special spectroscopy. They merely give the procedure used for obtaining an averaged density of states curve from the individual densities of states Z_A and Z_B. To examine how to apply the interpolation formula (2) to the unoccupied part of the conduction band of an alloy, we illustrate the simplest case of two identical density-of-states curves, and equal concentrations of A and B, in the schematic example of figure 4. On the left we have the case of charge neutrality. As for the case of macroscopic contact, we must average the two original densities of states without shifting the individual Fermi-edges within their individual energy bands. The resulting averaged density of states Z_{AB} differs from the primary density of states $Z_A = Z_B$ for all concentrations, even for this simplest case of identical original densities of states. On the right we have the case of charge

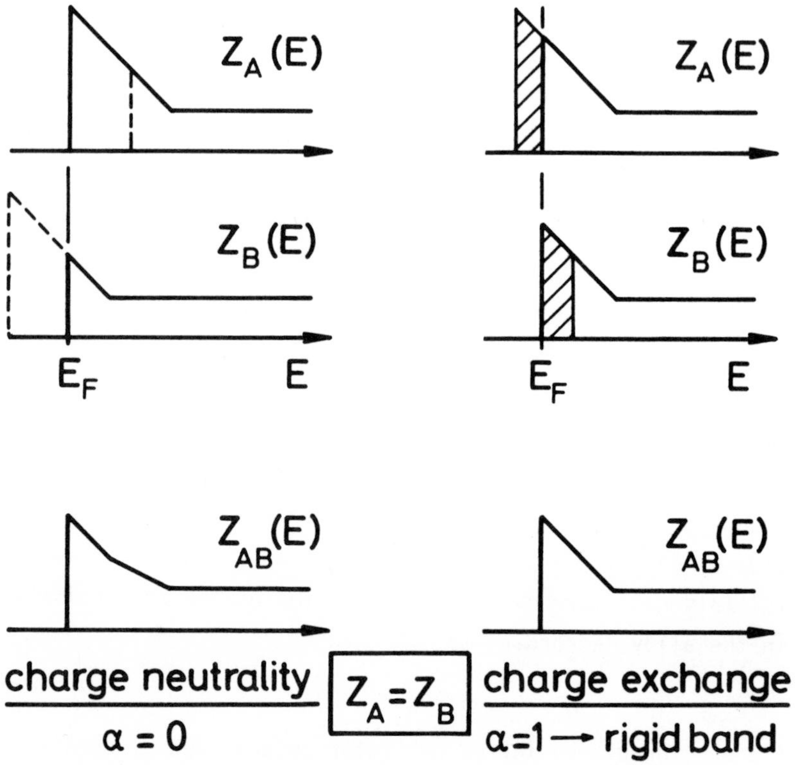

Figure 4 Density-of-states superposition scheme for the case $Z_A(E)=Z_B(E)$, with A and B in equal concentrations.

exchange, with $\alpha=1$, again for the otherwise similar elements. Before averaging, the B atoms supply half of an electron per B atom, and the A atoms pick up half an electron per A atom. Only after shifting the Fermi-edges of the A and B component densities of states to their new individual positions, can the superposition take place. This yields the same shape density-of-states curve as the original, and it can easily be shown that this remains valid for all concentrations of the two components; giving us the case of the well-known 'rigid band', which we have now classified.

$Z_A = Z_B$, with $\alpha=1$, is also the *only* case for which a rigid band can come into existence. Whether or not such a rigid band

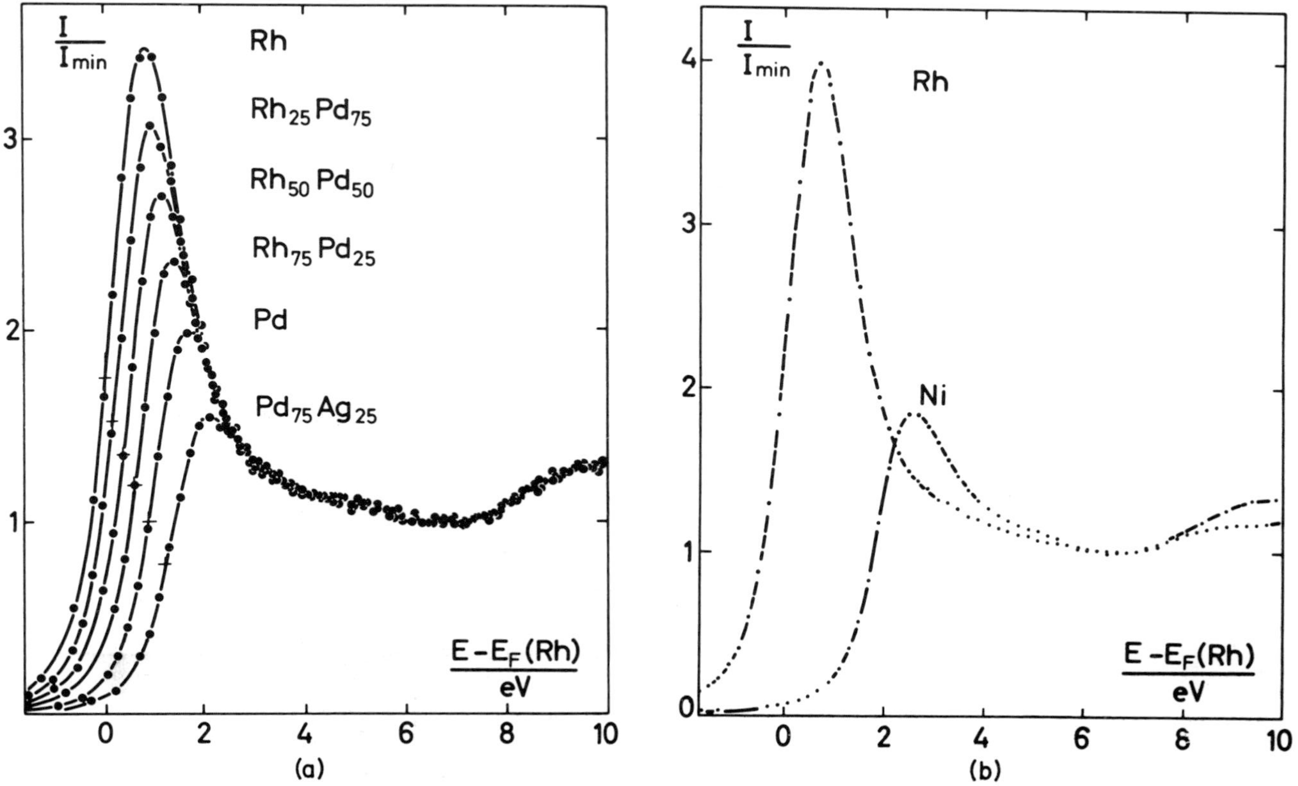

Figure 5 Measured isochromats for (a) rhodium-palladium, and palladium-silver alloys; (b) pure rhodium and pure nickel.

is realized in nature, *or* manifests itself in a special experiment, is a question determined by the particular alloy-system *and* the form of experiment. We must bear in mind that our interpolation formula (2) refers only to averaged densities of states without any reference to a special experiment.

4. EXAMPLES OF ISOCHROMAT MEASUREMENTS

On the left of figure 5 we show measured isochromats for the rhodium-palladium system; these can indeed be described by a rigid-band model (Eggs and Ulmer 1968). The normalization factor for these curves fluctuates only within the experimental reproducibility of the measurements, ∿5%. It is interesting that we can derive from our measurements absolute values for the densities of states (Eggs and Ulmer 1968). The isochromats of rhodium and palladium give the densities of states for the pure elements, which we need for our formalism.

That is, we can assume weighting factors P (for our element-isochromats) that are independent of energy, according to

$$I_A(E) = P_A . Z_A(E) \tag{5}$$

and

$$I_B(E) = P_B . Z_B(E) \tag{6}$$

For the rhodium-palladium system these weighting factors are even equal to each other: $P_A = P_B = P$. Then our superposition model (2) suggests a relationship of the form of equations (5) and (6) for also the alloy isochromats. The measured isochromats show that these conditions are indeed fulfilled for the rhodium-palladium system. Thus all isochromats of this alloy system are proportional to the corresponding averaged densities of states with the same weighting factor P.

We note that equations (5) and (6) and their consequences represent the second case (we cited the first in §1.) in which transition probabilities - included in our weighting factors -

are of essential importance. Further measurements (Fuchs unpublished), not illustrated here, show that a rigid band holds also for the alloy system cobalt-nickel, but with a different band and a different weighting factor. Now we can expect that in, for example, the rhodium-nickel system a rigid band should be impossible, because $Z_A \neq Z_B$. This is confirmed by the measured isochromats of rhodium and nickel, shown to the right in figure 5.

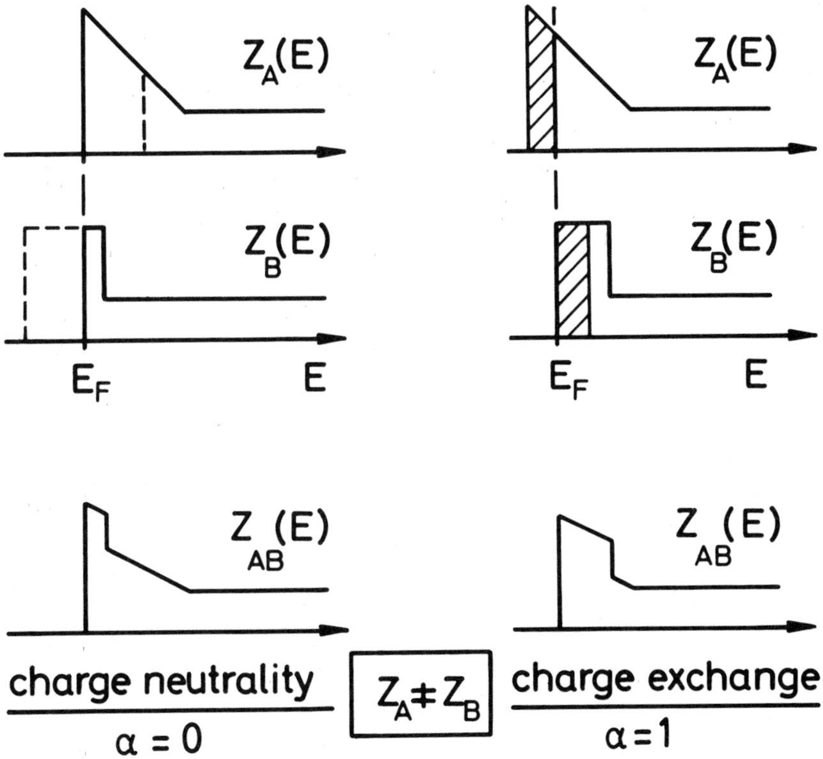

Figure 6 Density-of-states superposition scheme for the case $Z_A(E) \neq Z_B(E)$.

But our model is not confined to the rigid band situation. In figure 6 we see how it operates for different densities of states Z_A and Z_B. For simplicity we again consider, in this schematic illustration, equal concentrations; that is x=y. As before, we have on the left an example of charge neutrality, and on the right the other limiting case, charge exchange. All the densities of states shown are different from one another. Again, we must bear in mind that we are considering densities of states, whereas our measurements give us isochromats. However, we can anticipate what the isochromats should look like, and this we do. The opposite would of course also be possible: to confirm the averaged densities of states from our measured isochromats. For the rhodium-nickel system (our second example) the experimentally determined weighting factors are different; i.e. $P_A \neq P_B$. Then our model predicts that the proportionality of the alloy isochromats to the averaged densities is destroyed. However, it turns out - within our experimental accuracy - that we can replace the proportionality by the following relationships:

$$I_{A_xB_y}(E-E_F)=\begin{cases} \frac{1}{x+y}\left[x.I_A(E-E_F) + y.I_B(E-E_F)\right] \\ \frac{1}{x+y}\left[x.P_A.Z_A(E-E_F)+y.P_B.Z_B(E-E_F)\right] \end{cases} \tag{7}$$

The first line reproduces our experimental findings and the second gives the interpretation within our model. Fitting our measured isochromats to these expressions yields the amount of charge exchange; that is, the value of the parameter α.*

Figure 7 shows the results obtained for a particular alloy of the rhodium-nickel system. The full curve is the measured isochromat. The dashed curves result from superimposition according to the lower expression of equation (7), for the cases of charge neutrality (C N, α=0) and charge exchange (C E, α=1). We see that charge exchange gives the best fit,

*α is the only free parameter in our model; if α can be quantitatively related to electronegativity then we have no free parameter at all.

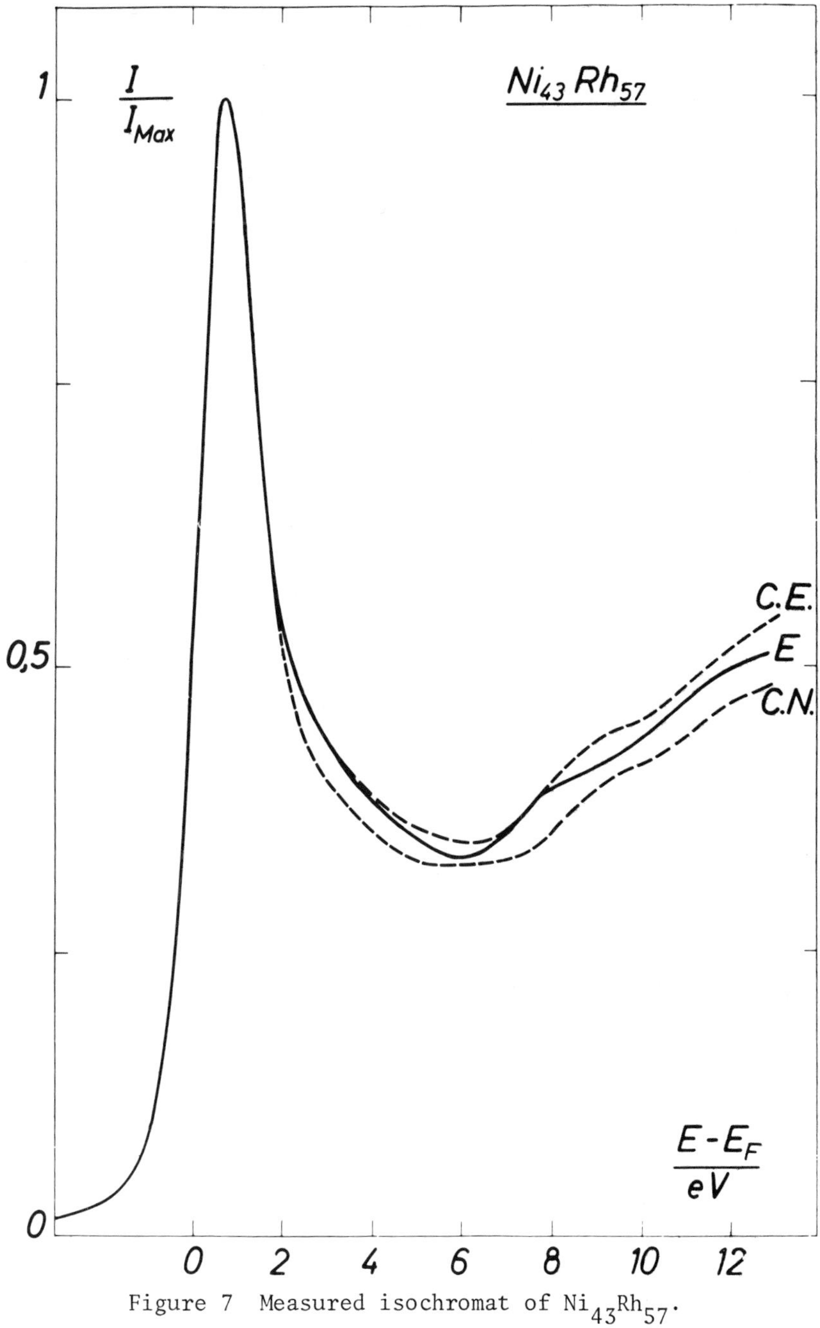

Figure 7 Measured isochromat of $Ni_{43}Rh_{57}$.

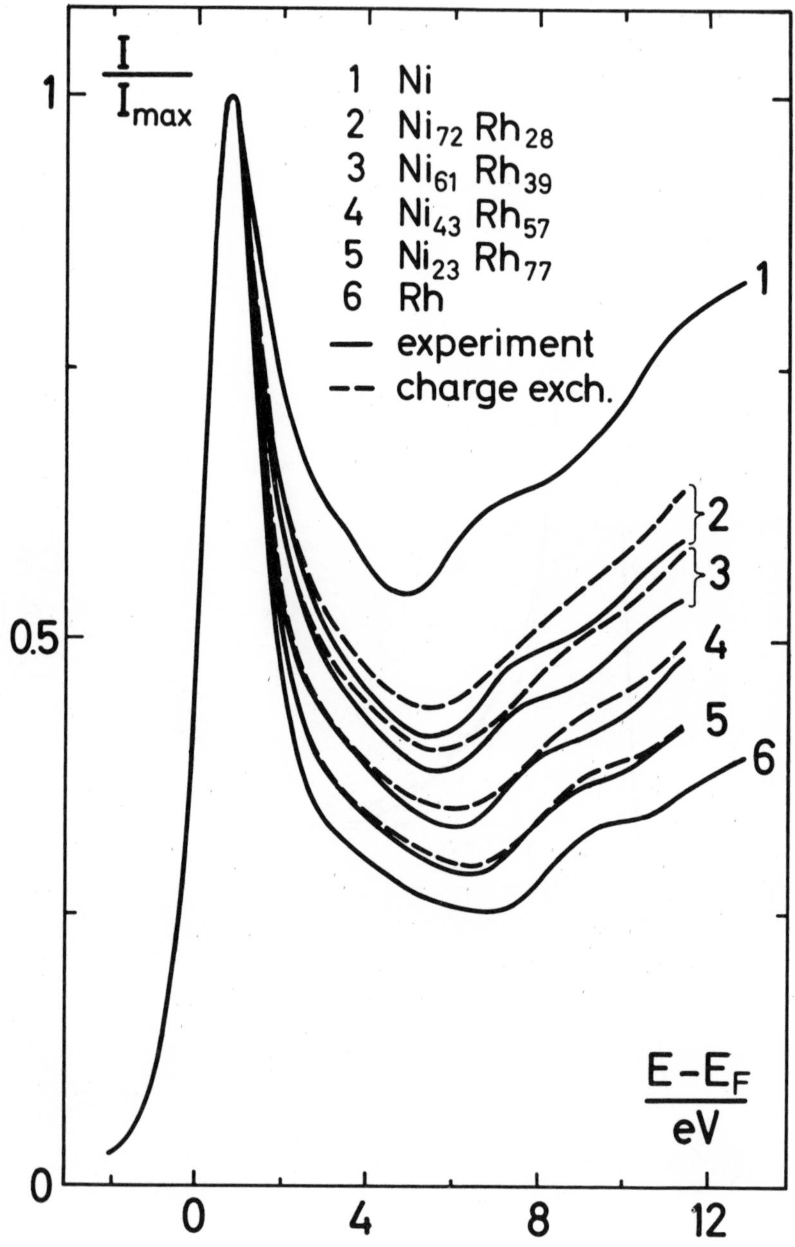

Figure 8 Measured isochromats for nickel-rhodium alloys.

especially if we take into account that the shape at low energies is the most important due to disturbing secondary effects at higher energies (those that are further from E_F). In figure 8 the results for the whole rhodium-nickel system are summarized (Kleber, unpublished). We recognize that the charge-exchange model is a good description for this system, for which the rigid-band model does not hold. More detailed measurements, using higher precision, are in progress on this system of alloys.

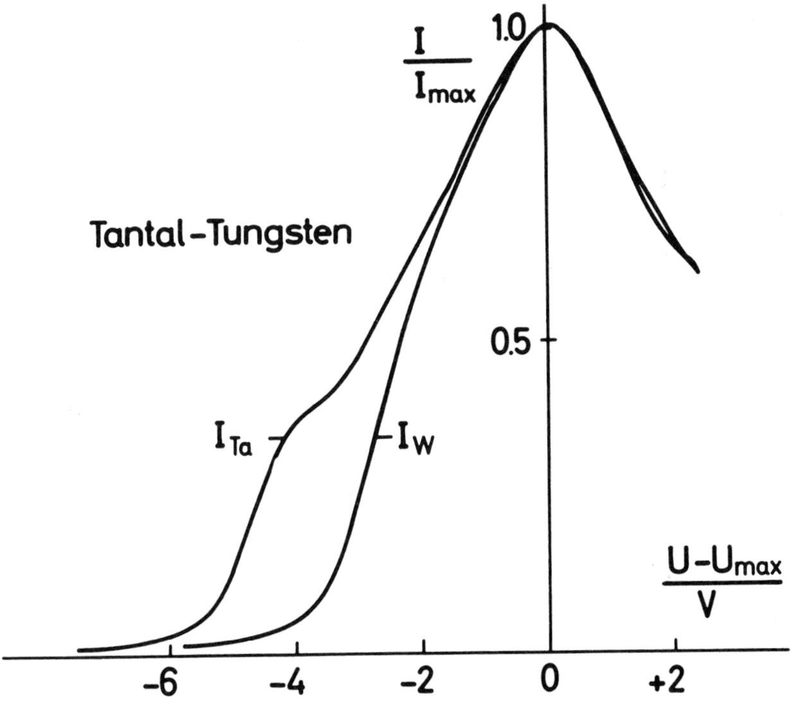

Figure 9 Measured isochromats of pure tantalum and pure tungsten.

Our final example concerns the system tantalum-tungsten. In figure 9 we show the isochromats of the two elements (Weimann, unpublished). The absolute intensities at their maxima correspond to within the experimental accuracy of about 1%. For some time this system has been our standard example of a rigid-band alloy series. However, recently at Karlsruhe we have made a more precise investigation, the results of which are illustrated in figure 10. Here we show the measured isochromat for the particular alloy $Ta_{60}W_{40}$, and examine how our model fits with it for various α. The different curves - based on the two-element isochromats of figure 9 - differ only in the assumption made concerning the charge exchange parameter α.

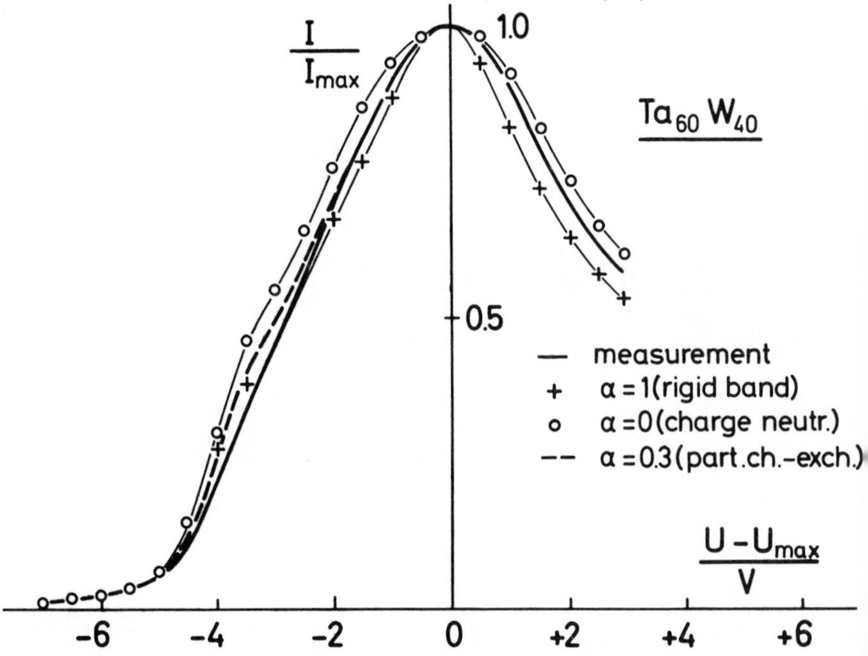

Figure 10 Measured isochromat of $Ta_{60}W_{40}$.

The important and surprising result is that the parameter α is predicted to deviate significantly from one. The value α=0.3 gives the best fit and rules out the rigid-band model. The reasons for this deviation may be more or less trivial.

We can give a short but probably important discussion of this. The *more* trivial reason attributes this deviation to macroscopic clusters in the alloy. When we recall our earlier model involving macroscopic contact of two metallic slabs, we easily recognize that such clusters must diminish the value of α locally, at the positions of these clusters, and thereby cause an averaged smaller α to apply. At present we cannot decide whether or not this is the full explanation of our result. The *less* trivial reason originates from the same phenomenon: for statistical reasons some clustering on an atomic scale must also be expected for substitutional alloys because of the binomial distribution of A and B atoms in the neighbourhood of every considered atom. Among other factors this problem should be considered carefully; before, for example, comparing experimental results with the coherent potential model (e.g. Shiba 1971).

5. THE PROBLEM OF LOCALIZATION

Finally in this report, we must again directly attack the question: can localization be manifest in isochromat spectroscopy? Let us begin with the relations (5) and (6), and (7). According to these expressions the weighting factors P_A and P_B are generally different from one another, and apparently retain their values in substitutional (and disordered?) alloys. We can interpret this as indicating that the different kinds of lattice atom cause different efficiencies of our primary-electron radiation transitions. We may have some localization embodied in this point. Let us speculate a little, without referring to our last example of tantalum-tungsten alloys in which we saw strong hints of localization.

Localization may originate from the necessity for balancing not only energy but also momentum in our elementary process. The momentum of the generated photon is smaller than the momentum of the primary electron. Therefore at least one phonon must be generated, because of the short duration of the elementary process. This would be some sort of a localized phonon, or phonon wave packet, not yet coupled to the thermal phonon system*.

In this respect the lattice atoms should behave like nearly free atoms. Indeed the measured ratio of the weighting factors $P_{Rh}/P_{Ni} \sim 2.9$, compares favourably to the value of about 2.6 for free atoms. This discussion of localization - if applicable at all - should apply not only to isochromat spectroscopy but also to x-ray spectroscopy of the valence band in general.

Acknowledgments

This research has been performed within "Sonderforschungsbereich 66: Electronic properties of solids" of the Deutsche Forschungsgemeinschaft.

I thank my coworkers Dr. J. Eggs, Chr. Fuchs, H. Geiger, R. Kleber and K. Weimann for performing the experiments and for making the measurements that have been used in this report.

REFERENCES

Eggs, J. and Ulmer, K. (1968); Z. Physik, 213, 293.

Fabian, D. (1971); CRC Rev. Sol. State Sci., 1, 255.

Gaukler, K.H. and Schwarzer, R. (1971); Optik, 33, 215.

Shiba, H. (1971); Progr. Theor. Phys., 46, 77.

Ulmer, K. (1969); in "X-ray spectra and electronic structure of matter" II, Kiev Institute of Physics of Metals, p.79.

Varley, J.H.O. (1954); Phil. Mag., 45, 887.

Watson, L.M., Kapoor, Q. and Nemoshkalenko, V.V. (1971); "Proc. de Colloque Int. du CNRS no. 196" Paris, Jnl. Physique, 32, C4-325.

*With Δx= the 'local extension' of the phonon wave packet, Δt= duration of the elementary process, and v_{ph}= group velocity of the phonon wave packet, we can write;
$\Delta x \simeq v_{ph} \times t \simeq 3 \times 10^5$ cm sec$^{-1} \times 10^{-16}$sec $= 3.10^{-11}$cm
This is fulfilled for the radiation transition taking place at one lattice atom only. (The other extreme would be the Mossbauer-effect).

ON THE TEMPERATURE DEPENDENCE OF BREMSSTRAHLUNG ISOCHROMATS

H. Merz
Physikalisches Institut, Universität Karlsruhe, West Germany.

1. INTRODUCTION

High-resolution bremsstrahlung isochromats give information on the electronic structure of solids (Ulmer 1969): isochromat spectroscopy which is in many respects complementary to photo-emission spectroscopy. For transition metals the first part of the isochromat spectrum - up to the plasmon echo - can be interpreted using a simple one-electron model, and gives a picture of the density of states in the unoccupied part of the conduction band (above the Fermi energy). An indirect transition process (Kieser, this volume p557), which predominates for the non-transition elements with low atomic number, is not important for the initial part of bremsstrahlung isochromats of transition metals. In the present report only transition metals will be treated. Therefore, in the following discussion the isochromat maximum - some few eV above the starting energy of the isochromat - can be identified with the unoccupied part of the d-band, and its high density of states.

2. TEMPERATURE-DEPENDENT EFFECTS

We use here a very simple model (Ulmer 1969) showing all the features essential for the discussion of the experimental results. The theoretical intensity $I'(E)$ is expressed in terms of the transition probability P(E) and the density of unoccupied states Z(E);

$$I'(E) = P(E) \cdot Z(E) \cdot \{1-f(E,T)\} \tag{1}$$

$$f(E,T) = \frac{1}{1+exp(E-E_F)/kT} \tag{2}$$

Because of the finite resolution of the apparatus the measured intensity I results from the folding of I' with the instrumental window function A

$$I = I' \star A \tag{3}$$

The folding with the window function introduces no new temperature dependence in equation (3); so it is sufficient to discuss the different temperature effects that appear in equation (1). Only the term in brackets depends on the temperature explicitly through the Fermi function; the influence of this factor on the measured isochromats will be discussed in §3. However, there are other temperature effects on the isochromats originating from changes in the transition probability or the density of states.

Temperature effects concerning the transition probability have been treated only the case of soft x-ray spectroscopy (Gyorffy, this volume p641) but not as yet for isochromat spectroscopy. The temperature-dependent changes in the density of states can be divided into two groups:

(1) Direct effects, dependent on the temperature directly (for example, caused by thermal variation of the mean lattice parameter, or thermal fluctuation of the local lattice parameter); here the size of the temperature effect is directly related to the value of the temperature change.

(2) Indirect effects, dependent on phase transformations at a fixed temperature; here the effect on the density of states is

independent of the size of the temperature change, and it is only necessary for the temperature concerned to be either below or above the transition temperature.

These two density-of-states effects are treated in §4.

3. INFLUENCE OF THE FERMI DISTRIBUTION

A. Experimental Results

The breadth of the Fermi distribution f(E,T), of the order of 4 kT is proportional to the temperature. Thus, after the folding procedure, the steepness of the initial part of the isochromat should depend on the temperature of the probe. This temperature dependence - solely the influence of the Fermi distribution - can be seen most clearly in the case of palladium (Mechelke and Ulmer 1972). The anode (a ribbon 25 x 1 x 0.01 mm) is heated to the lower temperature T_2 by the power dissipation of the emission current; an additional current, through the anode, raises the temperature to the higher value T_1. This additional heating current is intermittently switched on and off with a heating phase of 12 seconds and a pause of equal duration; simultaneously the pulses from the detector are switched into one or the other of two counting channels. With this arrangement it is possible to diminish the influence of thermal and long-term fluctuations of the apparatus; this 'modulation' technique permits the measurement of isochromats while at the same time maintaining nearly the same cleanliness of the probe surface at both temperatures. A special technique ensures that no voltage drop across the anode spoils the resolution during the actual measuring time.

As expected, the measured low-temperature isochromat of palladium (figure 1) has a steeper rise to the maximum; the intensity of the maximum is slightly higher than that of the high-temperature curve. The crossing point of the two curves lies significantly above the midpoint of the rise. The

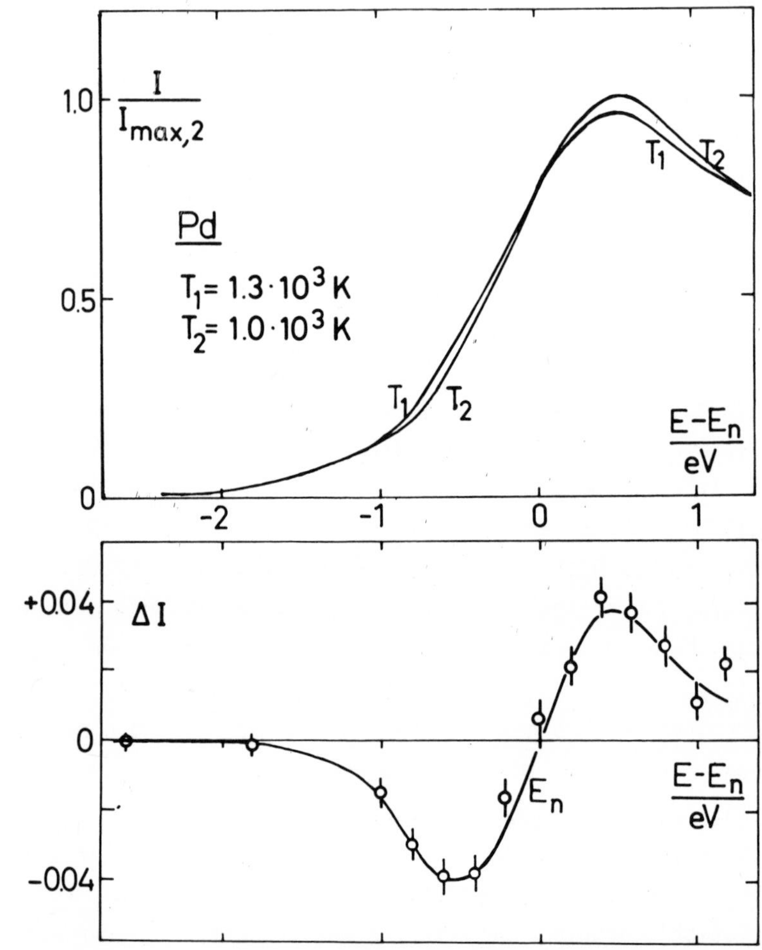

Figure 1 Bremsstrahlung isochromats (with $\hbar\omega_0$=1250eV) for palladium, measured at two temperatures. Abscissa zero is at the intersection point of the two isochromats. The lower diagram shows the quantity $\Delta I=[I(T_2)-I(T_1)]/I_{max,2}$.

difference between the two isochromats is small, yet the lower part of figure 1 shows that this difference is statistically significant. The difference curve is nearly an odd function with respect to the zero position E_n. The following discussion of this difference curve offers a new answer to the well-known question; where does the Fermi level lie on the measured curve?

B. The Position of the Fermi Level

In general the Fermi energy E_F depends weakly on the temperature. Calculation shows that this temperature shift can be neglected when compared with the accuracy presently achieved; equation (1) shows that it is at the Fermi level that the curves for the two different temperatures intersect.

This is no longer valid when the finite resolution of the apparatus is taken into account. The effect of the folding expressed by equation (3) has been investigated numerically (Mechelke and Ulmer 1972) and the results are briefly summarized in figure 2. The shape of the difference curve depends on the shape of the density of states within a region of a few kT around the Fermi energy, where the Fermi distribution f(E,T) varies strongly. We have two different models, expressed graphically:

(1) The density of states independent of energy, i.e. Z(E) = const, on the left in figure 2; the upper diagram gives the density of unoccupied states for the two temperatures of the palladium measurements of figure 1. The difference curve (middle diagram) is an odd function with respect to the Fermi energy E_F (zero point of the abscissa). After folding with the apparatus window A (Gaussian, with halfwidths-1.3eV), the folded curve is still an odd function, and the zero position remains at the Fermi energy.

(2) Linear approximation of the density of states, within the critical region around the Fermi energy (to the right in figure 2) $Z(E) = Z(E_F) + Z'(E_F) \times (E - E_F)$; in this case the folded curve exhibits asymmetry, and the zero position E_n is shifted away from the Fermi energy E_F (abscissa zero). If one introduces an asymmetry index

$$\mu = \frac{|\mu_1| - |\mu_2|}{\mathrm{Max}(|\mu_1|,|\mu_2|)}$$

then calculation shows that a unique relation exists between

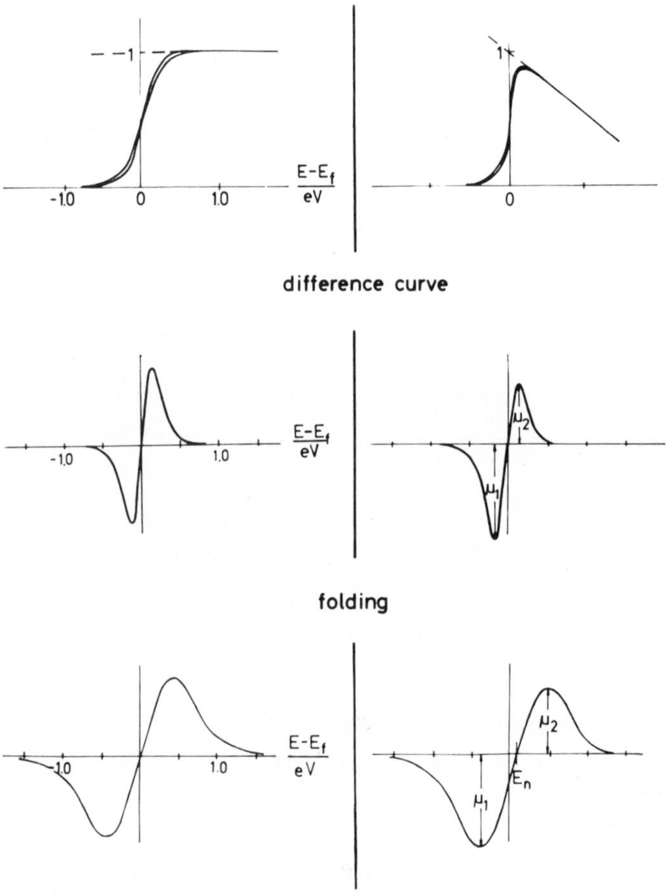

Figure 2 Model calculations for two different densities of states. Left: $Z(E)$ = const; right: linear approximation near the Fermi energy $Z(E) = Z(E_F) + Z'(E_F)(E - E\ )$. Upper diagrams: number of unoccupied states for the temperatures $T_1 = 1300^oK$, $T_2 = 1000^oK$. Centre diagrams: difference of the two upper curves. Lower diagrams: difference curve folded with the instrumental window function A.

the index μ and the shift $E_n - E_F$; this relation is nearly independent of the slope of the density of states $Z'(E_F)$, and of the choice of the window function (Gaussian or Lorentzian). Therefore, in this linear model, the value of the asymmetry index - taken from an experimental difference curve ΔI - yields the calculated shift E_n-E_F and, together with the measured zero position E_n of the difference curve ΔI, the exact position of the Fermi level E_F on the isochromat structure.

For the palladium isochromat (figure 1) the asymmetry index $\mu_{Pd} = 0.06 \pm 0.15$ gives a shift E_n-E_F = 10meV, which is negligible within the present experimental accuracy. Accordingly in figure 1 the Fermi level lies at the point where the isochromats for the two different temperatures intersect; i.e. at 77±5% of the full height of the low-temperature isochromat. This result agrees well with earlier findings using a different method for positioning the Fermi level (Merz 1970a).

Accordingly, a measurement of isochromats at different temperatures offers a possible method for determining the position of the Fermi level. Of course this method is applicable only when the Fermi distribution is the sole factor in equation (1) that depends on temperature. This condition seems to hold in the case of palladium.

4. TEMPERATURE-DEPENDENT CHANGES OF THE DENSITY OF STATES

A. Direct Effects

Direct temperature-dependent effects are most clearly found in the case of tantalum (figure 3); niobium, tungsten and molybdenum show analogous behaviour (Merz and Ulmer 1966; Mechelke and Ulmer 1972). Experimentally the changes in the isochromats vary linearly with the temperature differences during the measurements. The initial part of the tantalum isochromat shows the influence of the Fermi distribution but above 1241eV an additional temperature effect is observed.

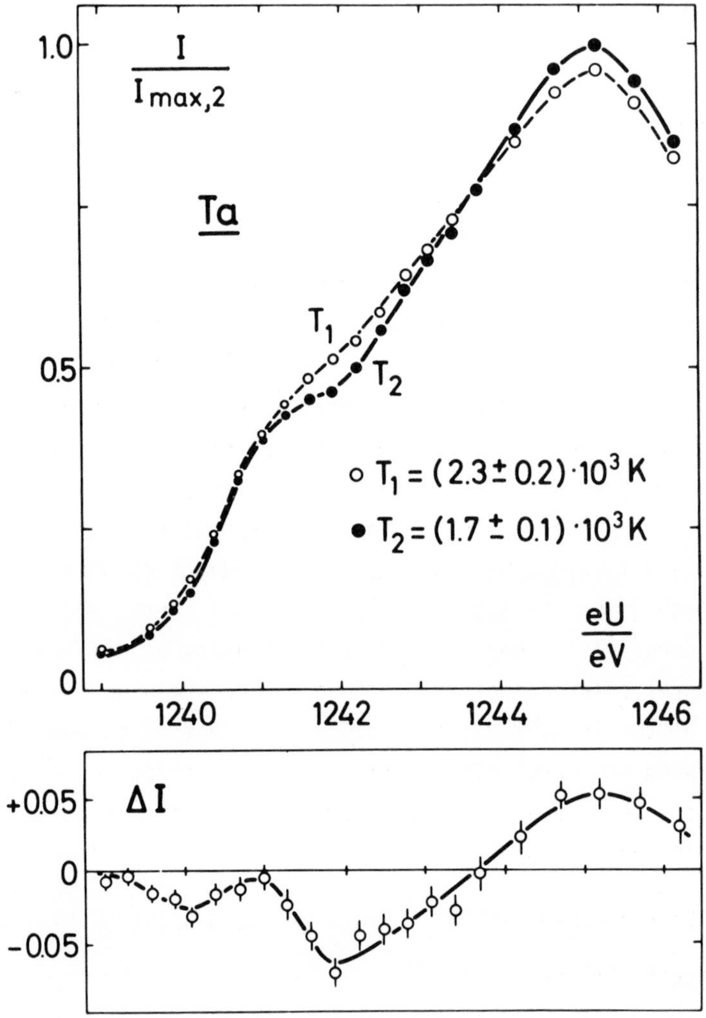

Figure 3 Bremsstrahlung isochromats of tantalum measured at two temperatures. The diameter of the circles expresses the statistical uncertainty of the measured points. Lower diagram: difference of the two isochromats.

With rising temperature the intensity of the maximum decreases and the structures are broadened and smeared out. Similar 'melting' effects have been reported from x-ray absorption measurements (Ronani and Sharkin 1963); this effect cannot be caused by the influence of the Fermi distribution. When the temperature rises the lattice parameter changes and this may lead to a variation of the density of states. Also with rising temperature the thermal smearing of the lattice increases and this results in a growing smearing or 'melting' of structures in the density-of-states curve. This thermal smearing might be the essential effect, for if one relates the decreases of the intensity of the maximum to high-temperature values of the Debye-temperature θ (Gschneidner 1964), then the relative decrease

$$\alpha = \frac{\Delta I/I}{\Delta T/\theta}$$

is practically the same for both the 4d and 5d metals:

$$\alpha_{4d} = -(2.7 \pm 0.3) \times 10^{-2}$$

$$\alpha_{5d} = -(1.8 \pm 0.2) \times 10^{-2}$$

The lower value for the 5d-metals is easily explained because the bands are broader (Merz and Ulmer 1968), and therefore the same smearing brings about a smaller decrease of the peak intensity.

One conclusion that can be drawn from the observed decrease with rising temperature is that classical phonon-assisted transitions are not important in the isochromat process.

B. Temperature-induced Phase Transitions

Density of states changes in the case of phase transitions induced by temperature change are indirect temperature effects; when the temperature is altered primarily the phase changes; and only if the two phases involved have different density of states will the shape of the two corresponding isochromats be different. In order to exclude direct effects the isochromats have been measured as closely above and below the temperature of the phase transition as possible. Two types of phase transformations have been investigated: (1) transition from ferromagnetism to paramagnetism, and (2) allotropic changes of the symmetry of the lattice. The temperatures chosen are normally about 70°K above and below the transition temperature; in the ferromagnetic phase the measuring temperature was chosen so that at least 60% of the saturation magnetization at 0°K should be present.

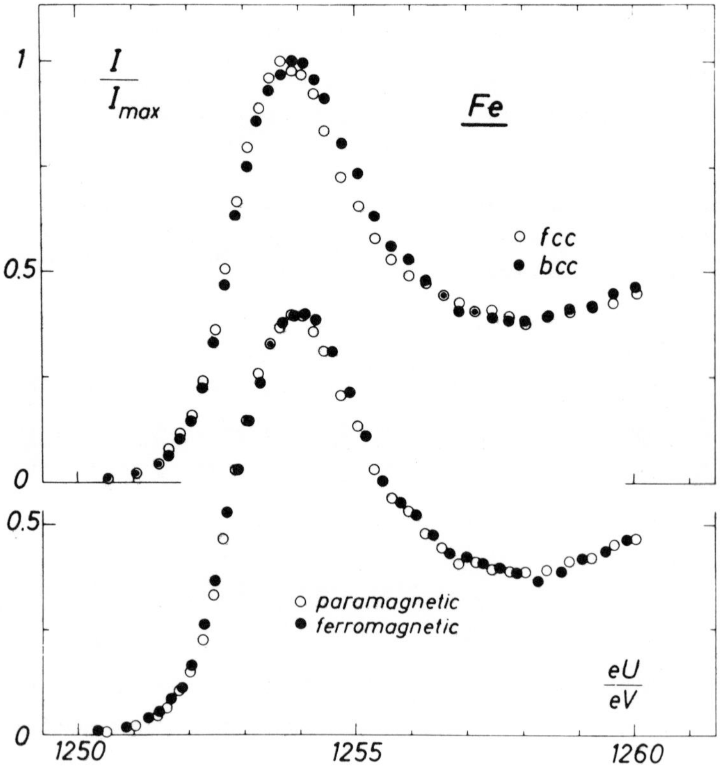

Figure 4 Bremsstrahlung isochromats of iron measured in different phases. Upper curves: fcc and bcc iron (both paramagnetic); lower curves: paramagnetic and ferromagnetic iron (both bcc). The diameter of the circles expresses the statistical uncertainty of the measured points.

In the case of iron (figure 4) both types of transitions could be investigated on the same probe. Figure 4 shows that only a negligible effect is found at the Curie temperature while for the high-temperature fcc phase the unoccupied part of the d-band is narrower than in the bcc phase.

For Co and $Co_{60}Fe_{40}$ the same result was found: no significant change of the isochromats at the Curie temperature, and yet a significant change when a phase transformation of the lattice occurs. Likewise only small effects at the Curie temperature have been observed in photoemission experiments (Fadley and Shirley 1970); an explanation of this is given using an atomic model of ferromagnetism (for example, Harrison 1970).

Figure 5 shows a further example of allotropic phase transformations: the transition from hexagonal low-temperature phase to the bcc high-temperature phase of titanium, zirconium, and hafnium (Merz 1970b). These metals show a similar behaviour: the intensity in a region just above the Fermi energy (1253-1254eV) decreases when going to the bcc phase. This effect confirms a theoretical statement (for example, Friedel 1969) predicting a minimum in the density of states in the midpart of the d-band for bcc transition metals. This minimum should not - or not so distinctly - occur for other lattice types. For titanium, zirconium and hafnium the interesting central region of the total d-band is unoccupied, and hence can be investigated by the isochromat technique. The observed change of the intensity agrees well with the theoretical expectation.

5. CONCLUDING REMARKS

Different types of temperature-dependent effects appear in isochromat measurements of transition metals, all of which can be understood within the normal description of the isochromat

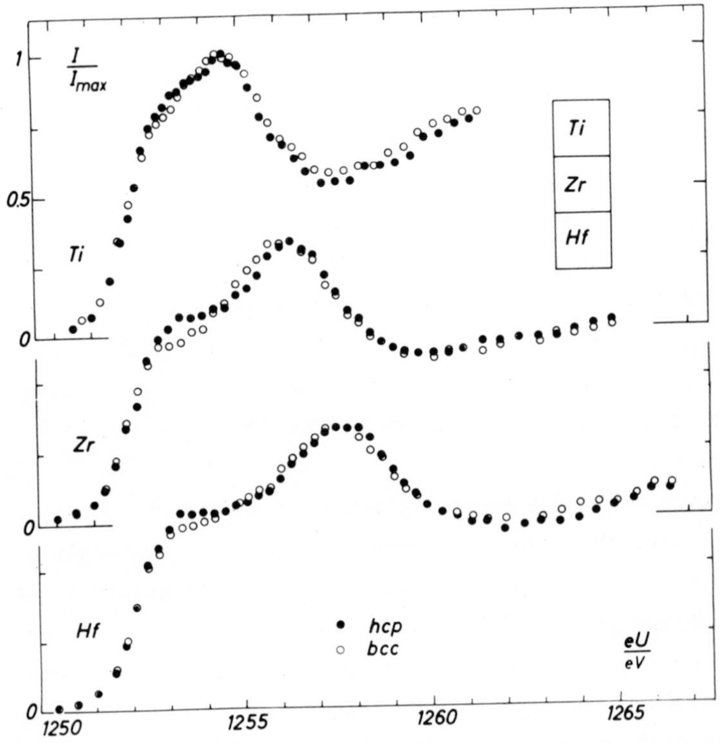

Figure 5 Bremsstrahlung isochromats of titanium, zirconium, and hafnium, for the hcp low-temperature phase and the bcc high-temperature phase. The diameter of the circles expresses the statistical uncertainty of the measured points.

process. One common feature is that all of these temperature effects cause only small changes in the observed isochromats and are not nearly so distinct as the differences between the isochromats of two different elements. Therefore, it is very useful to apply a difference-measuring technique to obtain significant results.

Acknowledgments

This investigation was performed within the Sonderforschungsbereich 66 "Elektronische Eigenschaften fester Körper" of the Deutsche Forschungsgemeinschaft.

It is a pleasure to thank Professor K. Ulmer for his support and for numerous stimulating discussions.

REFERENCES

Fadley, C.S. and Shirley, D.A. (1970); J. Res. NBS, 74A, 543.

Friedel, J. (1969); in The Physics of Metals Vol. I, (ed. Ziman, J.M.), p.354 : University Press, Cambridge.

Gschneidner, K.A. (1964); in Solid State Physics Vol. 16, (ed. Seitz, F. and Turnbull, D.), p.276 : Pergamon Press, New York, London.

Harrison, W.A. (1970); "Solid State Theory", p.485 : McGraw Hill, New York.

Mechelke, G. and Ulmer, K. (1972); Phys. Stat. Sol. *in press*.

Merz, H. (1970a); Phys. Stat. Sol. (a), 1, 707.

Merz, H. (1970b); Phys. Lett. 33A, 53.

Merz, H. and Ulmer, K. (1966); Z. Phys. 197, 409.

Merz, H. and Ulmer, K. (1968); Z. Phys. 210, 92.

Ronani, G.N. and Sharkin, O.P. (1963); Trans. Bull. Acad. Sci. USSR, Phys. Ser. 27, 824.

Ulmer, K. (1969); in X-ray Spectra and Electronic Structure of Matter Vol. II, p.79 : Academy of Sciences, Kiev.

A NEW TRANSITION PROCESS CONCERNING THE GENERATION OF BREMSSTRAHLUNG IN SOLIDS

Jörg Kieser

Physikalisches Institut, Universität Karlsruhe, Karlsruhe, W. Germany.

1. INTRODUCTION

The isochromats of several non-transition elements with low atomic number have been measured. The results obtained for graphite and silicon are reported here (figures 1 and 2), and those for additional non-transition elements elsewhere (Kieser 1971a, b).

In contrast to transition metals, where corresponding measurements represent the electronic density of states above the Fermi level, new effects occur in the case of the investigated elements with low atomic number and with small or vanishing density of states at the Fermi level. The experimental results demand the introduction of a new transition process which seems to play a dominant rôle. The measured isochromats are explained in terms of this 'non-direct transition' process, which is strongly affected by characteristic energy losses as well as by direct interband transitions.

2. EXPERIMENTAL

Samples were used in both the monocrystalline and polycrystalline modifications and were of purity better than 99.99%. To obtain clean surfaces during the measurements, a steady disintegration of the target surface was achieved by electron bombardment (for further experimental details see Edelmann 1967).

3. RESULTS

A. Graphite

Figure 1 shows the isochromat of graphite. The diameter of

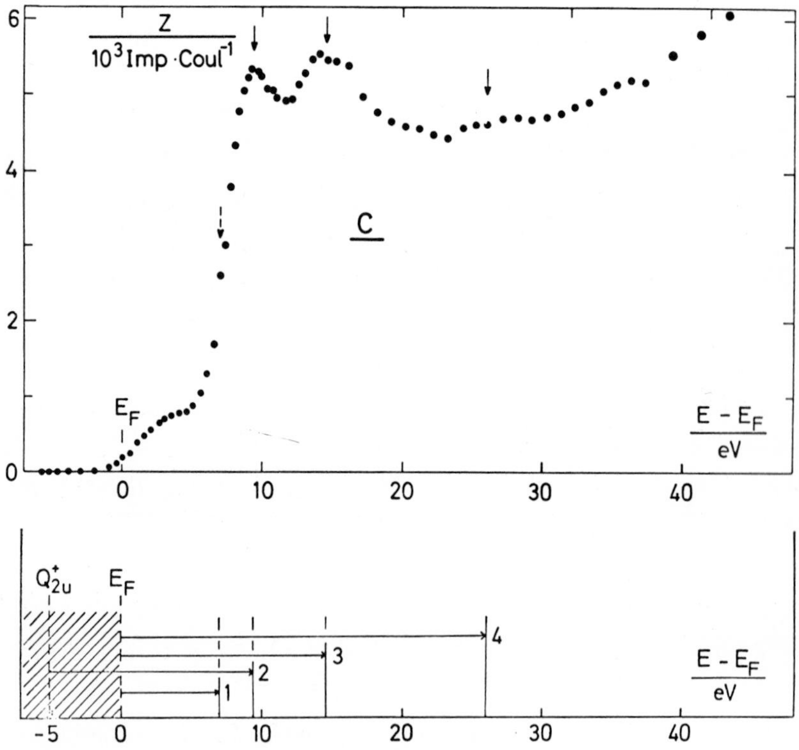

Figure 1 Isochromat spectrum of graphite measured at a threshold voltage of∿1keV. The Fermi level has been determined to an accuracy of ±0.2eV. Below the isochromat we show the positions for the isochromat maxima calculated from CEL data. These positions are indicated by arrows above the isochromat. The numerals refer to the characteristic energy losses (CEL) : (1) ΔE_{CEL} = 7eV; (2) ΔE_{CEL} = 14.5eV; (3) ΔE_{CEL} = 14.5eV; (4) ΔE_{CEL} = 26eV.

the points is smaller than the statistic error. The Fermi level has been determined experimentally (Merz 1970). Comparison of the isochromat structure with density of states calculations by Linderberg and Mäkilä (1967) gives qualitative agreement in a range of approximately 5eV above the Fermi level. We discuss the proposed transitions, shown schematically below the spectrum, in §4.

B. Silicon

Figure 2 shows isochromats of mono- and polycrystalline silicon; the statistical error is smaller than the point diameter. As for the case of graphite, the Fermi level has been determined experimentally. Most remarkable is the great difference between the shapes of these two isochromats; this precludes any explanation of the structure in terms of the density of states.

4. DISCUSSION

A. The 'nondirect transition process

Any explanation of the results on the basis of the simple one-electron transition process, valid for the transition metals, fails. Obviously the measured isochromats consist of two parts:

(1) A picture of the density of states, generated by the well known isochromat process and dominating the isochromat structure of the transition metals. This process is weak in the case of all measured non-transition elements.

(2) A to-date unknown process that produces peaks in a region of several electron volts above the Fermi level. From this we conclude that characteristic energy losses (CEL) play an important rôle in the process generating these isochromat maxima. If we consider the energy position of these maxima relative to the Fermi level we find agreement with CEL values for

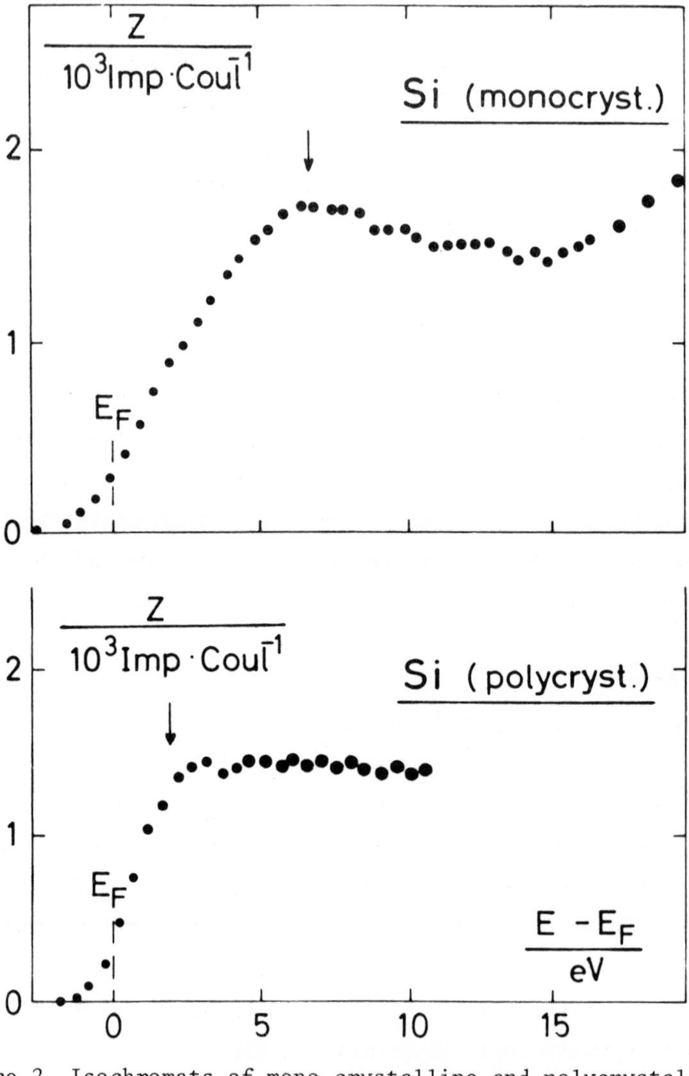

Figure 2 Isochromats of mono-crystalline and polycrystalline silicon measured at threshold voltages of ∿1keV. The Fermi level has been determined to an accuracy of ±0.2eV. The calculated positions for the isochromat maxima are indicated by arrows.

the particular element considered. A closer look gives excellent agreement in the case of graphite, and rather poor agreement in the case of silicon. A further difficulty is introduced by the following consideration: if a maximum at a distance from the Fermi level corresponding to a CEL value is caused by a plasmon alone, it can only be the picture or 'plasmon echo' of a density of states maximum at the Fermi level. But such a maximum does not exist in our case.

It is possible to overcome the above difficulties by assuming that the final level for the measured isochromat transitions is fixed. Such a mechanism yields a picture of the energy *vs* intensity distribution of the primary electrons, provided we take the isochromat measuring principle into consideration.

The above considerations establish a model for a new elementary process that seems to play a dominant rôle in the isochromats of all the investigated non-transition elements:

(a) A CEL 'plasmon', which is generated by a primary electron, excites a direct interband transition.

(b) The primary electron then makes a radiative transition into the resulting empty state.

Figure 3 shows such a process schematically. The energy distribution of the primary electrons is caused by a CEL. Such a CEL removes an electron from the initial level of a direct interband transition at an energy ΔE_A below the Fermi level. The primary electron then makes a radiative transition into the resulting hole. A stepwise increase of E' therefore yields a picture of the intensity *vs* energy distribution of the CEL involved in this process.

If we consider this process quantitatively the energy conservation for a quantum yields, according to figure 3:

$$\hbar\omega_o = (E' - E_F) - \delta E_{CEL} + \Delta E_A \qquad (1)$$

Application of the transformation, $E' - E = h\omega_o$, then gives

$$(E - E_F) = \delta E_{CEL} - \Delta E_A \qquad (2)$$

and the position of the maximum relative to the Fermi level is therefore given by:

$$(E_{max} - E_F) = \Delta E_{CEL} - \Delta E_A \qquad (3)$$

ΔE indicates the energy where the CEL intensity distribution has its greatest value.

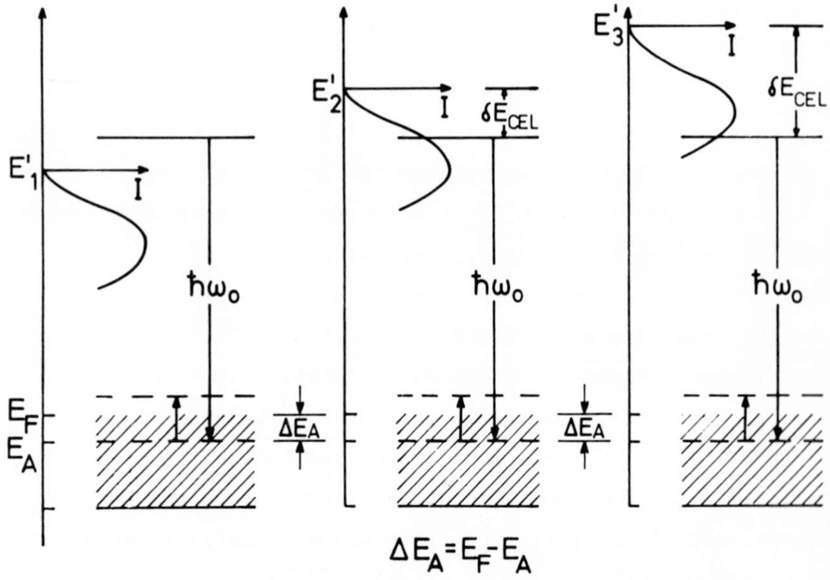

Figure 3 Schematic representation of the non-direct transition process. E' means the variable primary energy of the electrons. The fixed monochromator energy is indicated by $\hbar\omega_o$. The initial state of the interband transition is located at ΔE_A below the Fermi level. The curve just below E' gives the energy *vs* intensity distribution of the primary electrons caused by a CEL. This distribution is scanned by stepwise increase of E'.

B. Application of the non-direct transition process

(1) Graphite: Marton and Leder (1954) investigated the well known plasmons of 7eV and 25eV. Recently Zeppenfeld (1968) detected a third with an energy of 14.5eV.

The optical properties of graphite have been investigated by Taft and Philipp (1965). They found the onset of interband transitions to come at very low energies. If we combine these small transitions with the known plasmons, according to equation (3), we obtain positions of the resulting maxima that are in good agreement with experiment, as shown in the lower part of figure 1.

Graphite also exhibits a very strong interband transition at 14.5eV; this was observed by Taft and Philipp (1965) and Zeppenfeld (1968). Its position in the E(**k**)-scheme has been discussed by Bassani and Pastori (1967), by Greenaway *et al* (1969) and by Tosatti and Bassani (1970). Accordingly, its initial state lies at E_A=-5.2eV (E_F=0eV). A maximum in the isochromat structure caused by this interband transition according to the non-direct transition process should appear at a position, also indicated in the lower part of figure 1. Agreement with experiment is very good, hence our measurements confirm the calculated value for the initial state of this interband transition.

(2) Silicon: Careful investigations of CEL values have been performed by Creuzburg (1963), Raether (1965) and Zeppenfeld and Raether (1966). According to these authors monocrystalline silicon exhibits a surface plasmon of 10eV, while the polycrystalline modification possesses a lower surface plasmon of about 5eV.

Silicon exhibits a direct interband transition $X_4 \rightarrow X_1$, according to Philipp and Ehrenreich (1963), with an initial state at E_A=-3.2eV (Kane 1966, Bauer 1969). A connection of the different surface plasmons with this interband transition yields two different positions for the maxima as indicated in figure 4. Agreement with the experiments shown in figure 2 is good.

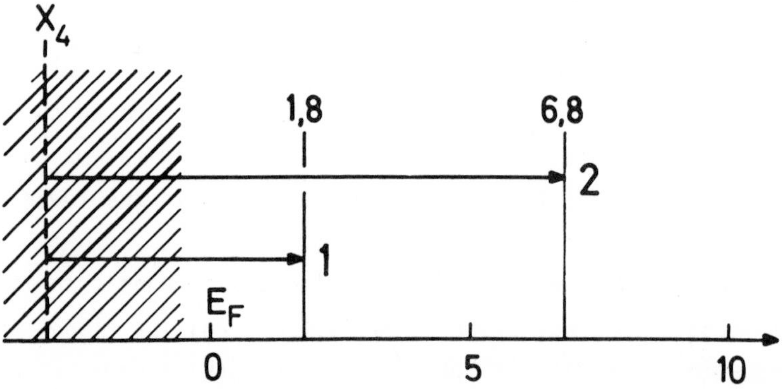

Figure 4 Schematic diagram showing the calculated positions for the isochromat maxima in silicon. The numerals refer to the CEL in the poly-crystalline and mono-crystalline cases: (1) ΔE_{CEL} = 5eV; (2) ΔE_{CEL} = 10eV.

5. CONCLUSIONS

The isochromat structures of graphite and silicon have been investigated and discussed in terms of a newly introduced 'non-direct transition' process. In the case of the investigated non-transition elements this process prevails strongly over the one-electron transition process which is valid in the isochromats of the transition metals. Its mechanism seems to be influenced strongly by characteristic energy losses as well as direct interband transitions. The connection of the non-direct transition process with plasmons is most clearly demonstrated by the completely different shape of the isochromats of monocrystalline and polycrystalline silicon in agreement with the different surface plasmon values of both modifications. The rôle played by interband transitions needs additional confirmation and this will be subject to further investigation.

Acknowledgment

This work has been performed within Sonderforschungsbereich 66 "Electronic Properties of Solids" of the Deutsche Forschungsgemeinschaft.

REFERENCES

Bassani, F. and Pastori Parravicini, G. (1967); Nuovo Cimento B50, 95.

Bauer, E. (1969); Z. Physik 224, 19.

Creuzburg, M. (1963); Z. Physik 174, 511.

Edelmann, F. (1967); Z. Physik 205, 476.

Greenaway, D.L., Harbeke, G., Bassani, F. and Tosatti, E. (1969); Phys. Rev. 178, 1340.

Kane, E.O. (1966); Phys. Rev. 146, 558.

Kieser, J. (1971a); Z. Phys. 244, 163.

Kieser, J. (1971b); Z. Phys. 244, 171.

Linderberg, J. and Mäkilä, K.V. (1967); Sol. St. Comm. 5, 353.

Marton, L. and Leder, L.B. (1954); Phys. Rev. 94, 203.

Merz, H. (1970); Phys. stat. sol. (a) 1, 707.

Philipp, H.R. and Ehrenreich, H. (1963); Phys. Rev. 129, 1550.

Raether, H. (1965); Springer Tracts in Modern Physics 38, 84.

Taft, E.A. and Philipp, H.R. (1965); Phys. Rev. 138, 197.

Tosatti, E. and Bassani, F. (1970); Nuovo Cimento B65, 161.

Zeppenfeld, K. and Raether, H. (1966); Z. Physik 193, 471.

Zeppenfeld, K. (1968); Z. Physik 211, 391.

NOTE: ON ION IMPACT-EXCITATION OF X-RAY SPECTRA

Is ion excitation useful for electronic structure studies?

D.J. Nagel

US Naval Research Laboratory, Washington D.C., USA.

Photons in the x-ray region of the electromagnetic spectrum may be generated by several processes, three of which are useful for studying the electronic structure of materials. Table I summarizes the spectroscopic methods that make use of these processes. X-ray fluorescence or electron impact excitation is usually used for all the listed x-ray methods. However, high-energy ion impact also produces core-level ionization. Ion excitation has already been put to use for elemental x-ray chemical analysis of thin surface layers on bulk materials, such as aluminium metal (Hart *et al* 1968), and of microsamples, for example air pollution particles collected on thin substrates (Johansson *et al* 1970). The present note is addressed to the question asked in the title. Recent measurements of ion-excited x-ray spectra are briefly described in considering the question.

Following earlier low-resolution observations made with a Si(Li) detector, which indicated that ion-impact produces broadened and shifted x-ray spectra (Richard *et al* 1969), a flat-crystal Bragg spectrometer was constructed for the measurement of shifts at high resolution (Nagel, unpublished). The spectrometer is controlled, and the data recorded and reduced, using programmes written for the SEL-840A computer at the 5-MV Van de Graaff accelerator at the Naval Research Laboratory (Knudson, unpublished). Spectra are stripped into components for peak location and intensity determinations using other routines initially designed for non-dispersive x-ray analysis (Burkhalter 1971). Ions, with energies in the range 0.1-10 MeV per atomic mass unit, excite x-rays with energies below about 1-5 keV with sufficient intensity for convenient measurement using a high-resolution flat-crystal spectrometer.

Table I. Excitation processes in x-ray spectroscopy.

X-Ray Generation Process	Spectroscopic Technique
Emission, following core-level ionization	Valence-band emission Radiative Auger Characteristic isochromat Appearance potential
Electronic Bremsstrahlung	X-ray absorption Continuum isochromat Short-wavelength-limit
Magnetic Bremsstrahlung (Synchrotron Radiation)	X-ray absorption

In the case of aluminium bombarded by 5-MeV N^+ ions, the broadened and shifted K spectrum was found to consist of six components, the parent $K\alpha_{1,2}$ line and five satellites corresponding to one through five L-shell vacancies (Knudson *et al* 1971). Similar observations of x-ray satellites due to multiple ionization produced by heavy-ion impact were made by Cunningham *et al* (1970), Mokler (1971), Burch *et al* (1971), Datz *et al* (1971), and Der *et al* (1971). Subsequently, for aluminium, it was found that even light ions produce enhanced satellite intensity due to direct multiple ionization (Knudson *et al* 1972). The degree of satellite enhancement, relative to the Al $K\alpha_{1,2}$ line, was also discovered to be strongly dependent on the ion energy for both light and heavy ions (Knudson *et al* 1973).

The ability to produce relatively intense satellite spectra using ion-excitation suggests a new approach to the old problem of identifying satellite structure present in a measured x-ray valence-band spectrum. In order to obtain a valence-band spectrum relatively free of satellite structure, threshold excitation (with electrons) is used. However, such measurements are difficult to perform due to the low intensity of the emission. Rather than attempting to suppress satellite emission, we could use ion excitation to enhance the relative intensity of satellite structure and facilitate its identification.

Figure 1 shows the $L\alpha$ region for germanium, the only available example where both threshold (Blokhin *et al* 1969) and near threshold (Deslattes 1968) electron excitation, as well as light-ion excitation (Burkhalter *et al* 1973a), have been used. As illustrated in figure 1, the enhancement and energy dependence of ion-generated satellite spectra can be used to distinguish satellites from diagram spectra. The Ge $L\alpha_1$ line is, of course, not a valence band spectrum, such as would commonly be used in an electronic structure study, and furthermore its satellites are quite separate from the line itself. However, the $L\alpha_1$ line becomes a valence-band spectrum just two elements below germanium, and here there is difficulty in distinguishing satellite structure from the main band, which is found of use. Hence, figure 1, or comparable data taken directly for the transition elements, can be used for identifying satellites in the 3d-metal L spectra, such as those for example observed by Fischer (1970).

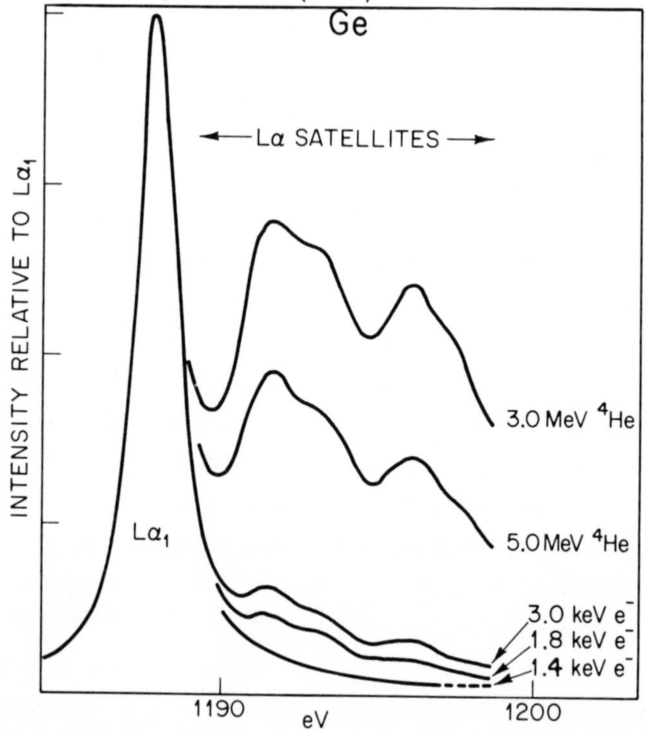

Figure 1 Germanium $L\alpha_1$ line and satellites excited by electrons of 1.4 keV (Blokhin *et al* 1969), and 1.8 and 3.0 keV (Deslattes 1968), and by $^4He^+$ ions (Burkhalter *et al* 1973a). Ion excitation provides easy identification of satellite structure, especially in cases where satellites are not as well resolved from the diagram spectrum as they are in this case.

X-ray satellites are known to be sensitive to chemical effects, especially the $K\alpha$ satellite of the third-row elements. When these satellite spectra become directly useful for studying bonding in materials, then ion excitation will provide an advantage for electronic structure determination; at present, the ability to extract useful new information from satellite spectra has not been demonstrated. In some cases chemical effects are understood but have not been used: for example, the change in $K\alpha_3/\alpha_4$ intensity-ratio from the metal to the oxide for magnesium, aluminium and silicon is due to the loss of M-shell electrons from the metal during oxide formation and the attendant decrease in probability of Coster-Kronig transitions which robs the α_4 component of intensity in the metal (Demekhin and Sachenko 1967a). Also, the shifts of satellite peaks to higher energies for the oxides, which are similar for light ion and electron excitation (Burkhalter *et al* 1973b), are due to the loss of core-level screening that accompanies oxidation (Demekhin and Sachenko 1967b). Effective charges of atoms in compounds can be obtained from satellite shifts, but they are more simply obtained from diagram line-shifts. In many cases, changes in satellite spectra with bonding are not understood and can therefore not yield electronic structure information. One example is the variation of the $K\alpha_3/\alpha_4$ intensity-ratio with composition in aluminium alloys (Fischer and Baun 1967).

If ion bombardment is used to increase satellite intensities for electronic structure studies, then as light an ion and as low an energy as possible should be used, consistent with yielding adequate overall intensity and the desired satellite enhancement. Projectile ions with masses approaching that of the target atom cause significant target-atom recoil. This can lead to two undesirable effects, as discussed by Burkhalter *et al* (1973b): (1) doppler broadening of spectra from recoiling ions moving with an appreciable fraction of the speed of light, and (2) spectral changes due to crystal-field effects on displaced atoms in compounds. Also, bombardment by heavy ions, with their associated electron clouds, produces such copious multiple ionization that in many cases only a broad indistinct satellite structure is obtained (especially for L and M spectra). For example, figure 2 shows the germanium L spectrum obtained by Ne^+ ion impact (Burkhalter *et al* 1973a). The spectrum, which spans over 100eV, is not as

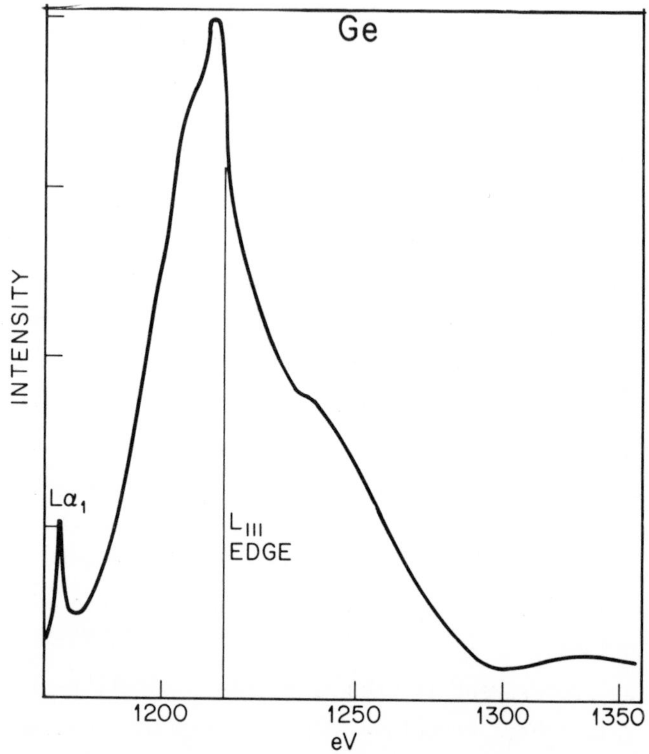

Figure 2 $L\alpha_1$ line and the dominant satellite structure due to copious multiple ionization, produced by the impact of $^{20}Ne^+$ ions on germanium at 5 MeV (Burkhalter *et al* 1973**a**).

useful for identifying satellites associated with particular vacancy configurations as the spectra shown in figure 1.

The abrupt drop in intensity just to the right of the peak in figure 2 is due to intervention of the L_3 edge of germanium. The measured spectrum is strongly affected by self absorption. This observation and the large bandwidth, such as that in figure 2, suggest that satellite spectra produced by heavy-ion impact can be used as sources for absorption spectroscopy. Such spectra occur on the high-energy side of x-ray lines and bands, and span regions of over 200eV in some cases (Knudson *et al* 1971, Der *et al* 1971). Hence targets can be found to cover most desired soft x-ray and UV energy ranges. Although the intensity variations in broad-band ion-excited spectra might be a nuisance for absorption measurements, such spectra do appear to be continuous.

Ion-excited satellite spectra must be compared with other sources for soft x-ray and UV absorption measurements. Comparisons are very dependent on experimental parameters, such as beam energy and current and monochromator efficiency. Table II contrasts the intensities available, after monochromation, from three types of sources for a photon energy near 1500eV.

Table II

Source	Energy Current	Monochromator Efficiency	X-rays/ sec-eV	Reference
Ion Impact ($Ne^+ \rightarrow Al$)	10 meV 0.1 μa	Crystal 10^{-4} rad.	10^4	Knudson *et al* 1973
X-ray tube (Rh target, 5 mil Be window)	45 keV 20 ma	Crystal 10^{-4}	10^3	Gilfrich 1971
Synchrotron (DESY)	6 GeV 10 ma	Grating (Unknown)	10^7	Godwin 1969

The intensities cited in the table are subject to qualification. For example, increasing the ion energy by a factor of 2 can increase the intensity by about one order of magnitude - depending on the shape of the ionization cross-section curve. Also, using a windowless Rh-target tube would improve the intensity by a factor of about 50. These uncertainties, and notwithstanding the lack of a common monochromator, the figures imply a substantial intensity superiority for synchrotron radiation. This advantage should improve with decreasing photon energy because of decreasing values of fluorescent yields for ion and electron-impact excitation. But, low-energy ion-generated satellite spectra and electron-Bremsstrahlung spectra will probably still be useful as x-ray and UV light sources. Ion accelerators, and of course x-ray tubes, are more commonly available than synchrotrons and storage rings.

In summary, and in answer to the question posed, ion-excitation of x-rays appears to have some auxiliary use for electronic structure studies. Its major value is in identification of satellite structure, although ion-excitation may also find some use as a source of x-rays for other experiments.

Acknowledgments

Continual collaboration with A.R. Knudson and P.G. Burkhalter, and their comments on this note, are acknowledged with pleasure.

REFERENCES

Blokhin, M.A., Zommer, G., Volkov, V.F. and Monastyrskii, L.M. (1969); Sov. Phys. Solid State, 11(1), 12.

Burch, D., Richard, P. and Blake, R.L. (1971); Phys. Rev. Ltrs. 26, 1355.

Burkhalter, P.G. (1971); in "Multi-element X-Ray Analysis of Ores Using Semiconductor Detectors and Radioisotopic Excitation", (ed. Ziegler, C.A., p.147; Gordon and Breach, New York.

Burkhalter, P.G., Nagel, D.J. and Knudson, A.R. (1973a); J. Physics A, *in press*.

Burkhalter, P.G., Knudson, A.R., Nagel, D.J. and Dunning, K.L. (1973b); Phys. Rev. A, *in press*.

Cunningham, M.E., Der, R.C., Fortner, R.J., Kavanagh, T.M., Khan, J.M., Layne, C.B., Zaharis, E.J. and Garcia, J.P. (1970); Phys. Rev. Ltrs. 24, 931.

Datz, S., Moak, C.D., Appleton, B.R. and Carlson, T.A. (1971); Phys. Rev. Ltrs. 27, 363.

Demekhin, V.F. and Sachenko, V.P. (1967a); Bull. Ac. Sci. USSR 31, 921.

Demekhin, V.F. and Sachenko, V.P. (1967b); Bull. Ac. Sci. USSR 31, 913.

Der, R.C., Fortner, R.J., Kavanagh, T.M. and Khan, J.M. (1971); Phys. Ltrs. 36A, 239.

Deslattes, R.D. (1968); Phys. Rev. 172, 625.

Fischer, D.W. and Baun, W.L. (1967); J. Appl. Phys. 38, 229.

Gilfrich, J.V. (1971); unpublished communication. See also Gilfrich, J.V., Burkhalter, P.G., Whitlock, R.R., Warden, E.S. and Birks, L.S. (1971); Anal. Chem. 43, 933 and Brown D.B. and Gilfrich, J.V. (1971); J. Appl. Phys. 42, 4044.

Godwin, R.P. (1969); in "Springer Tracts in Modern Physics", Vol. 51, (ed. Höhler, G.), p.1; Springer-Verlag, New York.

Hart, R.R., Olson, N.T., Smith, H.P. and Khan, J.M. (1968); J. Appl. Phys. 39, 5538.

Johansson, T.B., Akselsson, R. and Johansson, S.A.E. (1970); Nucl. Inst. & Methods 84, 141.

Knudson, A.R., Nagel, D.J., Burkhalter, P.G. and Dunning, K.L. (1971); Phys. Rev. Ltrs., 26, 1149.

Knudson, A.R., Burkhalter, P.G. and Nagel, D.J. (1972); "Proceedings of Conference on Inner Shell Ionization Phenomena", to be published.

Knudson, A.R., Nagel, D.J. and Burkhalter, P.G. (1973); Physical Review A, *in press*.

Mokler, P.H. (1971); Phys. Rev. Ltrs. 26, 811.

Richard, P., Morgan, I.L., Furuta, T. and Burch, D. (1969); Phys. Rev. Ltrs. 23, 1009.

Note added in proof:
Overlap of x-ray absorption and emission spectra allows relative absorption spectra to be obtained from the ratio of emission taken under different conditions of target self-aborption (Fischer, D.W. and Baun, W.L., 1967, J.Appl. Phys. 38, 4830; Liefeld, R.J., 1968, in Soft X-ray Band Spectra, Ed. D.J. Fabian, p.153, Academic Press, London). Since the extent of emission spectra can be greatly increased with heavy-ion excitation, it should be possible to obtain absorption replicas over a wider energy range by taking ratios of ion-excited spectra measured with different take-off angles.

Part 5

LIQUID AND AMORPHOUS METALS AND ALLOYS

COMPOUNDS AND CHEMICAL BONDING

ON THE DENSITY OF STATES OF LIQUID METALS AND ALLOYS

G. Busch and H.J. Güntherodt

Laboratorium für Festkörperphysik ETH, Zürich, Switzerland

Sections of this review, concerned with liquid transition and rare-earth metals, are based on theses and discussions of:

H.U. Kunzi, H.A. Meier, L. Schlapbach, A.ten Bosch and A. Zimmermann

1. WHAT DO WE WISH TO KNOW FROM BAND-STRUCTURE SPECTROSCOPY?

Over the past few years there has been an increasing interest in the properties of non-crystalline conductors such as liquid metals and amorphous semi-conductors. The strict periodicity or long-range order in the atomic arrangement, the basic assumption in solid-state theory, is lost in these materials. Liquid metals are the best suited to the investigation of the non-crystalline state because of their well-defined atomic distribution and because the results show good reproducibility; although many experimental problems such as those associated with high temperatures and the effects of corrosion are encountered in their study. Furthermore, liquid metals have the advantage of ideal miscibility. This allows us to investigate the properties of alloys over the whole concentration range and to vary their properties continuously. In the solid state the ideal miscibility is confined to dilute alloys or to some special examples that form solid solutions over the whole concentration range such as copper-nickel, and some palladium and rare-earth alloys.

Until recently the main interest in liquid metals and alloys was focused on their magnetic and electronic transport properties. Such investigations suggest certain behaviours for the density of states in the liquid state. Liquid normal metals (sodium, mercury, aluminium, tin or copper), with s and p conduction electrons, have a free-electron density of states. Liquid transition metals show nearly the same density of states as the solid state. For example, in liquid transition metals such as iron, cobalt and nickel one expects partially filled 3d-bands and in liquid rare-earths metals such as cerium, praseodymium and neodymium one expects narrow 4f-bands of localised states below the Fermi energy.

Liquid alloys should make it possible to investigate the change of the density of states on alloying as a function of concentration and therefore to test the different models for the density of states in alloys originally proposed for the solid state: the rigid-band model, the virtual-crystal model, the virtual-bound-state model and the coherent potential approximation; also for the liquid state: the pseudopotential approach and the new Cluster model. All of these models for pure liquid metals and alloys must be tested by experimental band-structure spectroscopy.

Little experimental data exists on the band-structure spectroscopy of liquid metals and alloys. The data for aluminium, indium, mercury, and gold, will be described in the section on liquid normal metals.

In the following sections we give a brief review of the available experimental data on liquid normal and transition metals and their alloys. Then we shall show how their properties are related to the density of states, or in what manner these properties are influenced by the 3d or 4f electrons.

Finally, in connection with our results on electronic and magnetic properties, we shall suggest experiments for UV-photoemission spectroscopy (UPS), x-ray photoelectron spectroscopy (XPS) and soft x-ray spectroscopy (SXS), which will certainly be valuable for a better understanding of liquid and solid metals and alloys.

2. LIQUID NORMAL METALS AND THEIR ALLOYS

A. A simple model of liquid normal metals

Electronic and magnetic measurements suggest that in liquid normal metals the valence electrons are completely separated from the positive ions and behave like conduction electrons. The simple free-electron model seems to be valid for these conduction electrons. This can be tested by measuring the Hall coefficient and the Pauli-susceptibility. The distribution of the ions can be described by classical kinetic theories, and can be experimentally determined by x-ray or neutron scattering experiments.

Many physical properties such as the electrical resistivity can only be explained by interaction of the ions with the conduction electrons. Only the Pauli-susceptibility is directly related to the density of states at the Fermi energy, but it is helpful also to consider the Hall coefficient and the electrical resistivity.

B. The Hall coefficient

The experimentally determined Hall coefficient of almost all liquid normal metals is equal to the value predicted by free-electron theory:

$$R_H = -\frac{1}{ne} = -\frac{A}{eLn_A D} < 0 \tag{1}$$

where n is the concentration of conduction electrons per unit volume, A is the atomic weight, L is Avogadro's number and D the density. The number of conduction electrons per atom n_A is equal to the group number in the periodic table. Sodium and copper have 1, mercury and zinc 2, gallium and indium 3, tin and germanium 4 and antimony 5 conduction electrons per atom in the liquid state. The free-electron model for the Hall coefficient is also valid for liquid zinc and liquid cadmium, which show a positive sign for their Hall coefficients in the solid state, and for liquid germanium, which is a typical semiconductor in the solid state. Deviations from the free-electron values are found only for the liquid metals thallium, lead and bismuth. A possible explanation could be the effect of spin-orbit coupling (ten Bosch and Baltensperger 1971).

If the free-electron model is valid, the Fermi surface is a sphere with radius

$$k_F = (3\pi^2 n)^{1/3} \qquad (2)$$

The value of k_F can be determined directly from the measured Hall coefficient without prior knowledge of the volume density.

The Hall coefficient for liquid alloys of normal metals corresponds also to the free-electron value. In a copper-tin alloy, copper provides one and tin four conduction electrons per atom. Therefore, the wave number $2k_F$ can be shifted simply by varying the concentration.

C. The Pauli-susceptibility

It is noted that both the Pauli-susceptibility

$$\chi_P = 2\,\frac{A}{D}\,\mu_B^2\,N(E_F) \qquad (3)$$

and the Pauli-Landau susceptibility

$$\chi_{PL} = \frac{2}{3}\,\chi_P \qquad (4)$$

for free non-interacting electrons, are explicitly related to the density of states at the Fermi energy. Theoretical calculations and experimental measurements of the density of states must be compared with information on $N(E_F)$ deduced from χ_P.

There are three different experimental methods for determining χ_P or χ_{PL}: conduction-electron spin resonance (CESR), NMR Knight shifts, and total susceptibility. All three have disadvantages.

The CESR measurements are feasible for only sodium, potassium, and lithium, due to the large spin-orbit coupling exhibited by the other simple metals. Knight shifts measure the product of the contact density $|\psi(o)|^2$ with χ_P from which it is seldom possible to separate an accurate value of χ_P.

Total susceptibility is easily measured for liquid metals and almost all the data used for χ_{el}, the electronic susceptibility, for comparison with χ_{PL}, are taken from total susceptibility measurements. To obtain χ_{el} from the total susceptibility,

$$\chi = \chi_{Ion} + \chi_{el} \tag{5}$$

we require the value of the ionic susceptibility χ_{Ion} which is uncertain and usually of the same order of magnitude as χ_{el}. Busch and Yuan (1963) have shown that the value of χ_{el} for liquid normal metals is in agreement with the value of χ_{PL} from the free-electron model. (Exceptions are Hg, Zn and Cd). Therefore, the density of states at E_F is also in agreement with the free-electron model.

Dupree and Seymour (1970) do not reach the same conclusion. These investigators use another χ_{Ion} value, and modify χ_P and χ_{PL} by taking into account the electron-electron interaction, resulting in an enhanced susceptibility

$$\chi_{P_{e-e}} = \chi_P \frac{1 + A_1}{1 + B_o} \tag{6}$$

where A_1 and B_o are coefficients of an expansion in Legendre polynomials of the electron-electron interaction in the Landau theory of a Fermi liquid. Thus, interpretation of the susceptibility data of pure liquid normal metals must await accurate density-of-states measurements. These should help to decide which χ_{Ion} values are the best, and how large is the contribution of the electron-electron correlation to the susceptibility.

The same problems arise for liquid alloys of normal metals.

It is not easy to compare the density of states deduced from the susceptibility data with the directly measured density of states. A large diamagnetic contribition is often observed in alloys. Examples are Cu-Sn and Ag-Sn, near 20at%Sn (where the Fermi wavenumber $2k_F$ is equal to the nearest neighbour distance). This diamagnetic contribution (see also Friedel 1958) is to-date unexplained and can be due to the ionic or to the electronic part of the susceptibility. Accurate density-of-states measurements for alloys should enable us to answer this question.

D. Electrical Resistivity

We include a short discussion of the electrical resistivity because the pseudo-potential approach has been quite successful in explaining the electrical resistivity of liquid alloys. From this we hope to learn more about the density of states, or of related quantities for alloys, by using the same approach.

The electrical resistivity of liquid normal metals is given by the well known Ziman formula

$$\rho = C_F(1/k_F) \int_0^{2k_F} |U(K)|^2 K^3 dK \tag{7}$$

with

$$|U(K)|^2 = a(K)|V(K)|^2 \tag{8}$$

V(K) is the pseudopotential, a(K) the Fourier transform of the pair correlation function and k_F the Fermi wavenumber. Assuming that V(K) and a(K) are nearly the same for all liquid normal metals, then different values of the electrical resistivity mainly arise from different values of $2k_F$. The Ziman formula explains the following experimental results: increase of the electrical resistivity with increasing number of the conduction electrons, or with increasing values of $2k_F$; linear dependence of the electrical resistivity on temperature for monovalent, trivalent and quadruvalent metals (e.g. Na, Al, and Sn or Sb);

and lastly negative temperature coefficients, which arise because $2k_F$ is equal to K_p - the wavenumber of the first peak of a(K). With a temperature dependence of $|U(K)|^2 a(K)$ at K_p, the temperature derivative of equation (7) becomes negative.

The behaviour of liquid alloys can be explained by an extension of the Ziman formula (7) with

$$|U(K)|^2 = C_A C_B |V_A - V_B|^2 + C_A^2 |V_A|^2 a_{AA} + C_B^2 |V_B|^2 a_{BB} + 2C_A C_B V_A V_B a_{AB} \quad (9)$$

where C_A, C_B and V_A, V_B are respectively the concentration and the pseudopotentials of the pure elements A and B, and a_{AA}, a_{BB} and a_{AB} are the partial correlation functions of the alloy. This Faber-Ziman formula includes the assumption that the pseudopotentials of the pure components alone are sufficient to explain the resistivity on alloying.

The Faber-Ziman formula divides the alloys of liquid normal metals into three groups according to respective values of $2k_F$ and K_p: the Ag-Au group with $2k_F < K_p$, the Ag-In group with $2k_F = K_p$, and the In-Sn group with $2k_F > K_p$. Liquid mercury alloys form a fourth group. The detailed behaviour of these four groups has been described by Busch and Güntherodt (1971).

E. The density of states of liquid normal metals

Experimental data The only available experimental results for the density of states of liquid normal metals or their alloys are the soft x-ray $L_{2,3}$-emission spectrum of liquid aluminium (Catterall and Trotter 1963; and Fabian *et al*, to be published), and the photoemission data for liquid indium (Koyama and Spicer 1971) and liquid gold (Eastman 1971). Measurements on liquid mercury by Cotti *et al* (unpublished) are not conclusive and further measurements by these investigators, using higher resolution, are in progress. The data for indium and gold indicate that a free-electron density of states is valid for at least these two liquid normal metals. Also, the measured density of states at the Fermi energy agrees with the value predicted by the free-electron model. The data for liquid and solid gold indicate no drastic change for the filled 3d-band on melting. To-date no data exist for the density of states of liquid alloys.

Theoretical calculations Many calculations have been reported for the density of states of pure liquid normal metals. In most cases the results are in fairly good agreement with a free-electron behaviour. The density of states for liquid zinc has been calculated by Watabe and Tamaka (1964); for liquid aluminium, zinc and beryllium by Ballentine (1966); for liquid sodium, potassium, aluminium and lead, by Schneider and Stoll (1967); for liquid lithium, cadmium and indium by Shaw and Smith (1969); for liquid beryllium by Rousseau *et al* (1970); for liquid bismuth by Chan and Ballentine (1971a); for liquid copper by Schwartz and Ehrenreich (1971); and additionally for liquid aluminium by Pant *et al* (1971).

In the case of liquid mercury there is a difference of only ∿10% between the calculated density of states and that predicted by the free-electron model (Chan and Ballentine 1971b). These calculations for mercury were performed using a large pseudopotential, which also gives good agreement with the experimental resistivity value (Evans 1970).

Calculations of the density of states of liquid alloys (Itami and Shimoji 1972) have been made using an extension of the self-energy expression originally used by Ballantine (1966) for pure liquid metals. The self-energy expression contains the product of the pseudopotential, the correlation function and a logarithmic term. The extension to liquid alloys is carried out in the same way as in the Faber-Ziman formula for the electrical resistivity. Calculated densities of states for the liquid alloys Na-K, Pb-Sn, Cu-Sn, Hg-Bi and Hg-In come close to the free-electron model.

What do we wish to know about the density of states? A better interpretation of the available data requires the answers of many questions on the density of states of liquid normal metals and their alloys.

Is the simple free-electron model really valid? How good is the agreement between the directly measured density of states and the values deduced from the Pauli or Pauli-Landau susceptibility? What can we conclude about the contribution of the electron-electron correlation to the susceptibility?

Of particular interest is the density of states for liquid mercury. Is it a free-electron density of states, or is there a minimum in the density of states at the Fermi energy - as suggested by Mott (1966)? We can then hope to understand the large difference between χ_{el} and χ_{PL} for liquid mercury and also the behaviour of mercury on alloying with indium.

How does the density of states change at the melting point? Particularly, what is the effect at the melting point for the divalent metals zinc and cadmium, which have minima in their densities of states at the Fermi level for the solid state, and for germanium which is a semiconductor in the solid state and a metal in the liquid state?

How does the density of states of two liquid normal metals change on alloying? And what is the origin of the large diamagnetic contribution to the susceptibility for liquid alloys of the Ag-In group?

Finally, what happens to the K-conservation selection rule at the melting point; is there a change from direct to non-direct transitions?

3. LIQUID TRANSITION AND RARE EARTH METALS

A. Influence of the 3d and 4f electrons

The experimental data on liquid normal metals and their alloys can be interpreted in terms of the free-electron model.

This model may not be valid for the liquid transition and rare-earth metals, and their alloys due to the presence of the 3d and 4f states. For the liquid state new effects are observed in the Hall coefficient, in the electrical resistivity and in the magnetic susceptibility, for metals and alloys with 3d and 4f electrons.

The influence of the 3d and 4f electrons appears to depend on the position of these electron states relative to the Fermi energy, and on the magnitude of the density of states at the Fermi level.

Alloying of transition metals with liquid normal metals may shift the Fermi energy, making it possible to vary the relative position of the 3d or 4f states and the magnitude of the density of states at the Fermi energy. For a review of solid transition metals see Mott (1964).

<u>The Hall coefficient</u> Many of the transition metals and rare-earth metals show positive Hall coefficients in the liquid state. Therefore, an interpretation using the simple free-electron model is not appropriate, and the advantage of measurements of the Hall coefficient of liquid normal metals and their alloys - namely the determination of the wave number k_F - is lost.

A relation has been observed between the values of the experimental Hall coefficients and the position of the elements in the periodic table. But to-date the quantitative influence of the 3d and 4f electrons on the Hall coefficient is not known.

<u>Number of the conduction electrons</u> It is difficult to define the number of s-like conduction electrons in a liquid transition metal due to the presence of the 3d states. We would expect from our measurements of the susceptibility and electrical resistivity, small numbers of s-like conduction electrons for the pure liquid transition metals iron, cobalt and nickel. This we discuss in more detail later. From mag-

netic measurements on pure liquid transition metals it is possible to determine the number of d electrons. Subtracting this from the total number of s and d electrons we find, as a consequence of s-d hybridisation, that the number of s-like conduction electrons is ~ 1. The effect of alloying on the resistivity and susceptibility verifies that the number of s-like conduction electrons is small. Therefore equation (2) leads to small values of k_F, and to small values of the energy calculated from

$$E_F = \frac{\hbar^2}{2m} k_F^2 \tag{10}$$

where E_F is the Fermi energy measured from the bottom of the conduction band. However, we must adjust the bottom of the band, as in band-structure spectroscopy, and this Fermi level is determined by

$$E_F = E_{const} + \frac{\hbar^2}{2m} k_F^2 \tag{11}$$

The k_F-values for pure liquid transition metals are of about the same magnitude as for monovalent noble metals, and considerably smaller than the k_F-values for polyvalent normal metals.

The Fermi energy for liquid alloys of transition metals in monovalent noble or polyvalent normal metals, can be taken as varying continuously between the two components. This agrees with recently observed data on the change of the number of 3d electrons in solid alloys of nickel with copper and of aluminium with respectively the transition metals manganese, iron, cobalt and nickel (Wenger *et al* 1971). Measurements using soft x-ray spectroscopy of the change in the number of 3d electrons on a Ni atom in Ni-Cu alloys suggest strongly that the number of 3d electrons does not change on alloying. On the other hand, an increase is observed in the number of 3d electrons on the atoms of manganese, iron, cobalt and nickel in binary alloys of these metals with aluminium. The conduction electrons from the aluminium fill not only the s band of the transition metals but also the partially empty 3d band, which results in a rapid decrease of the conduction electrons of aluminium.

Electrical resistivity The 3d band and 4f electrons also influence the electrical resistivity. The resistivity is sensitive to the position of the resonance energies E_{3d} or E_{4f} relative to the Fermi energy E_F.

Evans *et al* (1971) propose an expression for the electrical resistivity of pure liquid transition metals. The t-matrix of the muffin-tin potential replaces the pseudopotential in equation (7) and

$$|U(K)|^2 = a(K)\ |t(K)|^2 \tag{12}$$

with

$$t(K) = -\frac{2\pi\hbar^3}{m(2mE)^{1/2}} \times \frac{1}{\Omega_o} \sum_\ell (2\ +1) sin\ \eta_\ell(E) e^{i\eta_\ell(E)} P_\ell cos\ \theta \tag{13}$$

The electrical resistivity can be written in terms of the d-phase-shift η_2 of the same potential.

$$\rho \simeq c'(1/k_F)\ a(2k_F) sin^2 \eta_2(E_F) \tag{14}$$

or

$$\rho \simeq c'(1/k_F)\ a(2k_F) \frac{\Gamma^2}{(E_{res}-E_F)^2+\Gamma^2} \tag{15}$$

where Γ is the width of the resonance, and E_{res} the resonance energy, at approximately the centre of the 3d band.

Equation (15) explains qualitiatively the experimental resistivity data for pure liquid transition metals (see figure 1).

The electrical resistivities of the liquid metals manganese, iron, cobalt, nickel and copper decrease with increasing number of 3d electrons, due to the increasing difference between the resonance energies of the 3d electrons and the Fermi energy for these metals. E_{res} and E_F are close together for manganese, with its approximately half-filled 3d band. As the number of

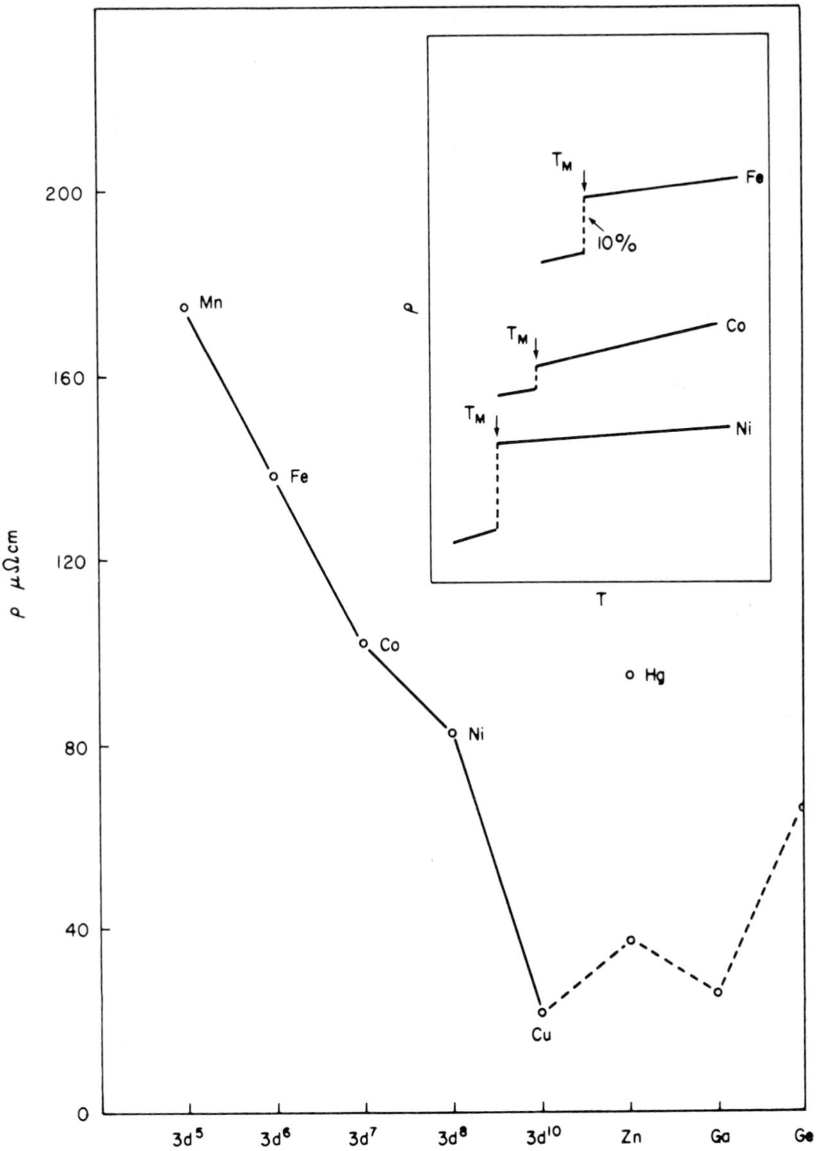

Figure 1 Electrical resistivity of liquid transition metals at their melting points.

3d electrons increases the difference ($E_{res}-E_F$) increases and the resonance scattering, which is the main contribution to the electrical resistivity of liquid manganese, iron, cobalt and nickel decreases.

A smaller contribution arises from the pair correlation function, which is nearly the same for all of these liquid transition metals.

Most of our experimental data on liquid alloys of transition metals with normal metals suggest an extension of the Faber-Ziman formula (equation 9) to these alloys (Evans *et al* 1972), with

$$|U(K)|^2 = C_A|t_A|^2(1-C_A+C_A\ a_{AA}) + C_B|t_B|^2(1-C_B+C_B\ a_{BB}) + C_AC_B(t_At_B^*+t_Bt_A^*)\ (a_{AB}-1) \qquad (16)$$

where C_A, C_B and t_A, t_B are respectively the concentrations and the t-matrices for the components A and B, and a_{AA}, a_{BB} and a_{AB} the partial correlation functions of the alloy. In alloys of transition metals with normal metals the t-matrix of the normal metal has no d-resonance, as for a normal pseudopotential

Equation (16) offers a qualitative explanation for the fact that the liquid alloys of the transition metals nickel, cobalt, iron, and manganese can be divided into two groups: namely the Au-Ni group (transition metals with monovalent noble metals) and the Co-Ge group (transition metals with polyvalent normal metals).

The Au-Ni group shows almost similar behaviour to the Au-Ag group; the Nordheim rule is applicable and positive temperature coefficients of the electrical resistivity are observed. On the other hand, the Co-Ge group behaves like the Ag-In group. A large maximum in the electrical resistivity is observed as a function of concentration, and temperature coefficients of the electrical resistivity are negative. These negative temperature coefficients occur over a large concentration range, which - for alloys with a given polyvalent normal metal - increases from copper to iron. The maximum electrical resistivity and the minimum temperature coefficient do not come at the same alloy composition.

The important difference between the two groups is the position of $2k_F$-value for the pure monovalent noble metal, or the polyvalent metal, compared with the K_p-value of the first peak in a(K) of the pure liquid transition metal. The K_p-value for the transition metals is assumed to be representative of the K_p-value of the partial correlation function a_{AB} for the alloy. Then we have the relation: $2k_F$ (Au) $< K_p <$ $2k_F$ (Ge). For the Au-Ni group the $2k_F$-values are always smaller than the K_p-values for the transition metals.

In the Co-Ge group the small $2k_F$-value for the transition metals increases and, on alloying with polyvalent normal metals, moves through the K_p-value for the pure transition metals. The contribution of the correlation function a(K) increases, and negative temperature coefficients occur.

Results of numerical calculations on the concentration dependence of the electrical resistivity for the liquid alloys Ni-Au, Fe-Au and Ni-Sn (Dreiach *et al* 1972) are in reasonable agreement with experiment. A difference between calculation and experiment occurs for pure liquid iron and iron-rich Fe-Ge alloys. This difference suggests a contribution to the conductivity from d-electrons, as proposed by Mott (1971). The conductivity of the d-electrons should be proportional to the density of states at the Fermi energy.

The electrical resistivity of pure liquid rare-earth metals can probably be similarly understood with resonance scattering of the 4f states. Particularly, we can explain the fact that cerium and lanthanum have nearly the same electrical resistivity.

B. Magnetic susceptibility

In the following sections we first describe the magnetic behaviour of pure liquid 3d-metals and alloys; we then discuss briefly the magnitude of the susceptibility and its temperature dependence; and finally we report on magnetic measurements for some rare-earth metals and their alloys.

Pure liquid transition metals The reciprocal susceptibilities of the liquid metals nickel, manganese, iron and cobalt (Kohlhaas 1965 and 1969, Urbain and Uebelacker 1967) are shown as a function of temperature in figure 2.

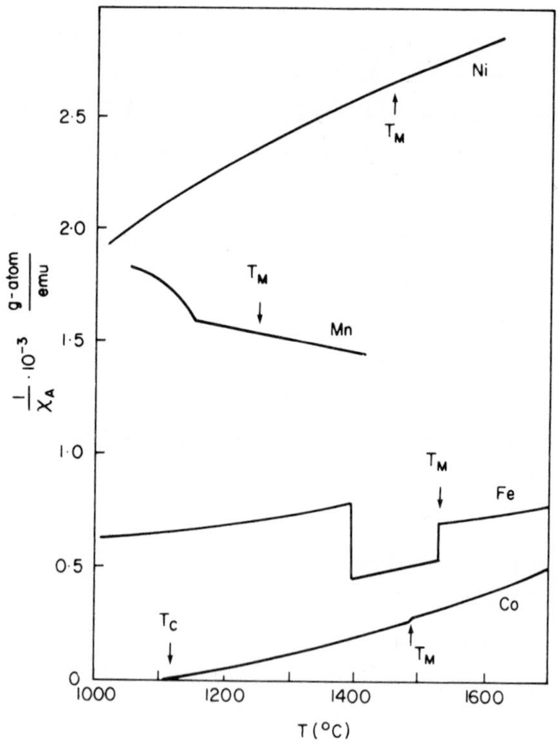

Figure 2 Reciprocal magnetic susceptibility for liquid transition metals.

The variation of χ_A^{-1} expressed as a function of temperature, for example for pure liquids iron, cobalt or nickel, can be fitted at high temperatures to a variety of formula that result from the models of paramagnetism: the itinerant electron model, the Kondo model, or the model of localised magnetic moments.

The last gives the well known Curie-Weiss law

$$\chi = \frac{C}{T-\Theta} \tag{17}$$

where Θ denotes the Curie temperature. C, the Curie constant, is given by

$$C = \frac{N_L p^2 \mu_B^2}{3k_B} \tag{18}$$

with N_L the Avogadro number, p the effective number of Bohr magnetons μ_B per atom, and k_B the Boltzmann constant.

From the slope of the Curie-Weiss relation the effective magnetic moment

$$\mu = p\mu_B \tag{19}$$

per atom can be deduced. Mostly it has been common to interpret the susceptibility data for pure liquids iron and cobalt in terms of the Curie-Weiss relation. The values found for μ/μ_B are larger than those obtained from saturation magnetization measurements in the solid state, but comparable with the values for the free ion.

Nickel shows no change at the melting point in its susceptibility, or in the temperature coefficient. Although the susceptibility decreases smoothly with increasing temperature, the dependence does not conform exactly to a Curie-Weiss relation. However, it is possible for the purpose of general discussion to examine the results above the melting point in terms of a Curie-Weiss relation. In contrast to the three liquid metals iron, cobalt and nickel, pure liquid manganese shows an increase in susceptibility with increasing temperature above the melting point.

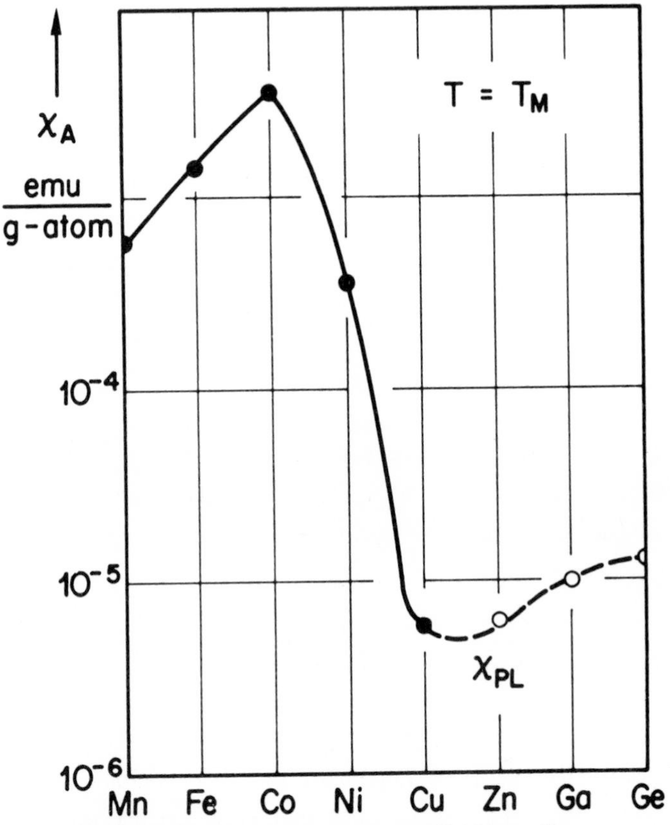

Figure 3 Magnetic susceptibility for first-series liquid transition metals.

Figure 3 shows the magnitude of the susceptibility to increase from manganese to iron, reach a maximum at cobalt and then decrease through nickel to copper. The open circles are the experimental data for the alloys of the transition metals (Nakagawa 1956). For comparison the Pauli-Landau susceptibilities for some liquid normal metals are also shown (closed circles).

<u>Liquid alloys of transition metals</u> The change of the tem-

perature coefficient of the reciprocal susceptibility, and the magnitude of the susceptibility, are the quantities to be discussed in the case of liquid alloys of transition metals with normal metals.

At first liquid concentrated alloys were classified, according to the temperature coefficient of their susceptibilities, into two groups which can be represented by cobalt-germanium and gold-nickel. The same classification is used for dilute alloys in the solid state.

In the Co-Ge group, the Curie-Weiss law is obeyed by the pure liquid transition-metal-rich alloys. For certain large concentrations of normal metals the Curie-Weiss relationship is lost and the temperature coefficient changes sign, which in terms of the Curie-Weiss law means that the magnetic moment disappears. We find a transition from a magnetic to a non-magnetic state. In the Au-Ni group the Curie-Weiss behaviour is observed for almost all concentrations.

Classification into these two groups is no longer necessary. Extensive experimental data suggest that the change of sign of the temperature coefficient of the susceptibility depends on the relative number of conduction electrons for the normal metal and 3d-electrons for the transition metal.

The shape of the susceptibility curve, as a function of concentration, clearly distinguishes the cobalt and nickel alloys on the one hand and the iron alloys on the other. On alloying, the susceptibility of cobalt and nickel alloys decreases rapidly as a function of concentration of the normal metal. The susceptibility of iron and manganese alloys, as a function of concentration of the normal metal, shows a maximum and then decreases to the susceptibility of the pure normal metal.

Discussion concerning the susceptibility We shall discuss now the magnitude and the temperature coefficient of the susceptibility of liquid alloys with small concentrations of the transition metal component. The experimental data to-date have been analysed in two ways:

(1) The magnetic case. The susceptibility is explained in terms of the ionic model. The magnetic moments are deduced from the slope of the Curie-Weiss relation. No correlation is made between the magnitude of the susceptibility and the density of states at E_F. As an example, we refer to dilute alloys of manganese in indium, tin, antimony and bismuth (Tamaki 1967, Collings 1970).

(2) The non-magnetic case. The susceptibility is analysed in terms of the models developed by Friedel (1958) and Anderson (1961), based on localised impurity states and their effect on the density of states.

The susceptibility is given by

$$\chi = \mu_B^2 \frac{N(E_F)}{1-(U+4J)N(E_F)/10} \tag{20}$$

where U and J are respectively the usual Coulomb and exchange correlation energies.

Because of lack of information on the electronic specific heat in the liquid state, a numerical analysis in terms of the Friedel-Anderson model is not as direct as in the solid state (Klein and Heeger 1966). However, discussion along these lines has been presented by Tamaki (1968a and b) for alloys of iron, cobalt and nickel with bismuth, antimony and tin.

The magnetic to non-magnetic transition has been investigated by Collings (1971) for manganese in Al-Sn alloys, and by Gruber and Gardner (1971) for manganese and iron in Cu-Al alloys. Collings discusses the transition in terms of the Anderson model. Gruber and Gardner (1971) describe the impurity susceptibility of the copper-rich alloys in terms of Kondo perturbation theory above the Kondo temperature, as for the solid state. For the aluminium-rich alloys the Kondo bound-state model, and the weak-coupling spin fluctuation model, both fail to describe the rapid increase of susceptibility with temperature.

The results for more concentrated alloys (Wachtel 1967, Nakagawa 1956) are interpreted only in so far as the Curie-Weiss law holds for an ionic model. No explanation is given of the value of the susceptibility as a function of concentration.

Our measurements suggest explanations for two behaviour characteristics of concentrated alloys. One is the magnetic to non-magnetic transition; the other is the value of the susceptibility.

(1) Transition from magnetic to non-magnetic state. A transition from the magnetic to the non-magnetic state is also observed in more concentrated alloys (Busch *et al* 1969, 1971) and we expect that the explanation of the Friedel-Anderson model can be extended to these alloys. The transition from the non-magnetic state then depends on the Coulomb and exchange correlation, the bandwidth and shape of the density of states, the value of the density of states at the Fermi energy, the values of the resonance and of the Fermi energy.

(2) Value of the susceptibility. The continuous change of the experimentally determined susceptibility at the magnetic non-magnetic transition suggests an expression for both the magnetic and non-magnetic states of the form of the Friedel-Anderson model for the non-magnetic state

$$\chi_A = \frac{N_L \mu_B^2 N(E_F)}{1-\bar{U}_{eff} N(E_F)} \qquad (21)$$

where $\bar{U}_{eff}$ is the average exchange energy per pair of electrons per atom.

For a simplified discussion we assume that the variation of the susceptibility can be explained by the dependence of the density of states on concentration. The exchange energy $\bar{U}_{eff}$ should change little on alloying.

As already mentioned the shape of the susceptibility as a function of concentration differs for the iron alloys from the cobalt alloys. This is further evidence for a correlation of the susceptibility with the density of states at the Fermi energy, even for the magnetic case.

The similarity of the susceptibility curves, as functions of concentration, for alloys of the transition metals with first monovalent and second polyvalent normal metals could be ascribed to a similar shape of the density of states for both categories of alloy. The susceptibility values are larger for the liquid alloys of transition metals with monovalent noble metals than for the alloys with polyvalent normal metals. This could be due to a larger density of states at the Fermi energy in the case with noble metals.

Pure liquid rare-earth metals The magnetic susceptibilities of liquids cerium, praseodymium, and neodymium follow the Curie-Weiss law up to room temperature. The effective moment corresponds to one localised 4f-electron for cerium, two for praseodymium and three for neodymium. At higher temperatures in the liquid state a deviation to larger values of the susceptibility occurs, due to the occupation of adjacent states above the ground state. Lanthanum, which has no 4f-electrons, shows the usual temperature independent Pauli paramagnetism enhanced by a factor of 10 compared with the χ_{PL} for liquid normal metals.

Liquid rare-earth alloys The magnetic susceptibility of the investigated liquid alloys of cerium with copper, lanthanum and praseodymium varies nearly linearly with concentration. Cerium retains its local magnetic moment in those alloys.

This picture does not hold for liquid Co-Ce alloys (Doriot

et al 1971). For the mid-concentration range of these alloys the magnetic susceptibility shows a minimum and is temperature independent; that is, the local magnetic moment of cerium disappears.

The susceptibility of the rare-earths and their alloys. The solid light rare-earth metals can be divided into two groups. First, the normal rare earths (Pr, Nd) which, independent of pressure and temperature, have three conduction electrons and an integral number of 4f-electrons per atom. These are described by the ionic model and by classical s-f exchange. Second, the anomalous rare-earths (Ce, La) for which the number of conduction electrons and the number of 4f-electrons can vary with pressure and temperature. In these metals the presence of a narrow 4f-level close to the Fermi level produces a large resonant scattering effect.

The Anderson model has been extended to describe the magnetic behaviour (Coqblin 1971) of the anomalous rare-earth metals, and the variation with pressure, temperature and, for alloys, the nature of the host metal.

A similar duality to the pure metals is found for alloys with rare-earths. Alloys with normal rare-earths as solute correspond to the ionic model, while the resonant scattering model is valid for alloys with cerium.

We shall discuss cerium and its alloys as representative of the anomalous rare earths. In the β and γ phases solid cerium is characterised by the presence of a narrow 4f-level close to the Fermi energy, with $E_{4f} < E_F$, and in these phases cerium exhibits a local magnetic moment. Solid cerium shows no local magnetic moment under pressure in the α-phase (McPherson *et al* 1971), and in the α'-phase it even becomes superconducting (Wittig 1968).

The disappearance of the local magnetic moment at the γ-α

transition has been explained by means of the extended Anderson model, assuming that the 4f states move upwards with respect to the Fermi level and are emptied as the pressure increases. An additional simple theory for the α-γ transition is given by Ramirez and Falicov (1971).

For liquid Co-Ce alloys, because $E_F(Co) < E_F(Ce)$, we expect the Fermi energy to move downwards with respect to the energy of the local 4f-level. Therefore, the localised 4f-electron becomes a conduction electron, which together with the (approximately three) normal conduction electrons of cerium leads also to a disappearance of the magnetic moment of cobalt. Magnitude and temperature independence of the susceptibility suggest that these Co-Ce alloys are Pauli-paramagnets like α-cerium and lanthanum.

To better our understanding of the behaviour of cobalt-cerium we have investigated alloys of cobalt with praseodymium and lanthanum. In liquid cobalt-lanthanum the magnetic moment of cobalt disappears, as is characteristic for trivalent normal metals alloyed with cobalt. In Co-Pr alloys the magnetic moment of praseodymium does not disappear.

C. Density of states for transition and rare-earth metals

The density of states of transition and rare-earth metals has not been determined experimentally in the liquid state. We do not expect a drastic change at the melting point and the data for the solid state should provide a fairly good approximation (Baer *et al* 1970, Eastman 1970, Hedén *et al* 1971).

There are some theoretical investigations of the density of states of liquid transition metals. A general account without detailed calculations is given by Cyrot-Lackmann (1967, 1968). Anderson and Mcmillan (1967) and Keller and Jones (1971) have calculated the density of states for liquid iron. Both calculations give a broad 3d-band with two peaks that form in the region of the resonance energy. The Fermi energy lies to the low-energy side of the second peak.

The cluster model (Lloyd 1967) used by Keller and Jones (1971) offers the simplest description of the density of states of liquid transition metals, and possibly also of their alloys. The basis of this model is to evaluate the density of states for a few atoms situated in an atomic environment appropriate to a cluster in the liquid. The atoms of the transition metal are represented by their muffin-tin potentials with 3d-electron resonance states.

No theoretical or experimental study of the density of states of liquid alloys of transition metals has yet been reported. We expect theoretical treatments using the coherent potential approximation (CPA) and the cluster model to be adequate, and to agree with the virtual-bound-state model of Friedel and Anderson for the dilute alloys.

For a final interpretation of the susceptibility of alloys with transition metals we require the density of states of some characteristic liquid alloys such as gold-nickel, cobalt-germanium, iron-gold, iron-germanium, cobalt-cerium. We can predict a probable behaviour from the extensively investigated solid copper-nickel alloys. This solid solution should be representative of the paramagnetic state of the liquid gold-nickel system.

The CPA-calculations (Stocks *et al* 1971) of a Cu-Ni density states are consistent with experimental photoemission measurements (Seib and Spicer 1970) and with soft x-ray emission results (Clift *et al* 1963, Azaroff and Das 1964), in as far as comparison with data from these experiments is possible.

For solid copper-nickel alloys the copper d-states are located in the energy region associated with d states of pure copper, and similarly the nickel d-states are located in the energy region associated with d states in pure nickel. Therefore, the resonance energy for nickel is the same for pure nickel as for the virtual bound state of nickel in copper.

A consequence of these experimental and theoretical results is the failure of the rigid-band model to describe the density of states or the magnetic properties of copper-nickel alloys. We expect that the resonance energy of the transition metal varies on alloying with a polyvalent normal metal, between the value for the virtual bound state and the value for the transition-metal-rich alloy.

What do we wish to know about the density of states of liquid transition metals and their alloys? A better interpretation of the available data requires answers to many questions on the density of states in liquid transition and rare-earth metals and their alloys.

What are the features of the density of states in liquids nickel, cobalt, iron and manganese and the liquid light rare-earth metals? The occupied as well as the empty states are of interest.

What are the values of the bandwidth, the resonance energy, and the density of states at the Fermi energy? How does the density of states change at the melting point? How does it change in the liquid state as a function of temperature?

How does the density of states of transition metals change on alloying in the virtual-bound-state region, and in the more concentrated alloys? Is the E_{res} of transition metals on alloying with monovalent noble metals really unchanged? How does the E_{res} of the transition metals, and the difference $(E_{res}-E_F)$ change in alloys with polyvalent normal metals?

How does the number of 3d and 4f electrons change on alloying with monovalent and polyvalent metals? What is the density of states in liquid cobalt-cerium alloys; particularly for the dilute alloys of cerium in cobalt and vice versa?

4. CONCLUSIONS

Discussion of the experimental data in this review shows the advantage of metal physics in the liquid state. This should also be true of experimental band-structure spectroscopy, with a further advantage arising from break-down of the k-conservation selection rule.

The interpretation of the properties of liquid transition and rare-earth metals and their alloys is based on extensive experimental data not given in detail here. The aim was to give a more general discussion of the electronic and magnetic properties in relation to the density of states.

Clearly density-of-states measurements could prove or disprove our hypotheses, could answer a lot of unsolved questions, and perhaps alter some of our conclusions.

Acknowledgments

The authors wish to thank P.W. Anderson, N.W. Ashcroft, W. Baltensperger, B.R. Coles, R. Evans, D.J. Fabian, J. Keller, B.T. Matthias, A.R. Miedema, N.F. Mott, T.M. Rice and J.M. Ziman for helpful discussions. D.T. Pierce assisted with reading the manuscript.

Partial financial support for this investigation was received from the "Schweizerische Nationalfonds zur Förderung der wissenschaftlichen Forschung" and from the Research Centre of Alusuisse.

REFERENCES

Anderson, P.W. (1961); Phys. Rev. 124, 41.

Anderson, P.W. and McMillan, W.L. (1967); in Proc. Int. School of Physics, Enrico Fermi, (edited by W. Marshall), p.50, Academic Press, New York.

Azároff, L.V. and Das, B.N. (1964); Phys. Rev. 134, A747.

Baer, Y., Hedén, P.F., Hedman, J., Klasson, M., Nordling, C. and Siegbahn, K. (1970); Phys. Scripta 1, 55.

Ballentine, L.E. (1966); Can. J. Phys. 44, 2533.

Busch, G. and Yuan, S. (1963); Phys. kondens. Mat. 1, 37.

Busch, G., Güntherodt, H.-J. and Meiser, H.A. (1969); Phys. Letters 30A, 111.

Busch, G. and Güntherodt, H.-J. (1971); in Phase Transitions, Innerscience Publishers.

Busch, G., Güntherodt, H.-J., Künzi, H.U. and Meier, H.A. (1971); J. de Phys., suppl. 32, C1-338.

Busch, G., Güntherodt, H.-J. and Künzi, H.U. (1971); First EPS Conference on the Physics of Condensed Matter, Florence.

Catterall, J.A. and Trotter, J. (1963); Phil. Mag. 8, 897.

Chan, T. and Ballentine, L.E. (1971a); Phys. Chem. Liquids 2, 165.

Chan, T. and Ballentine, L.E. (1971b); Phys. Letts. 35A, 385.

Clift, J., Curry, C. and Thompson, B.J. (1963); Phil. Mag. 8, 593.

Collings, E.W. (1970); Sol. State Com. 8, 381.

Collings, E.W. (1971); J. de Phys. suppl. 32, C1-516.

Coqblin, B. (1971); J. de Phys. suppl. 32, C1-599.

Cyrot-Lackmann, F. (1967); Adv. Phys. 16, 393.

Cyrot-Lackmann, F. (1968); Thesis, University of Paris.

Doriot, P.A., Güntherodt, H.-J. and Schlapbach, L. (1971); Phys. Letts. 37A, 213.

Dreirach, O., Evans, R., Güntherodt, H.-J. and Künzi, H.U. (1972); J. of Phys. F. to be published.

Dupree, R. and Seymour, E.F.W. (1970); Phys. kondens. Materie 12, 97.

Eastman, D.E. (1970); Techn. of Metals Research VI, to be published.

Eastman, D.E. (1971); Phys. Rev. Letts. 26, 1108.

Evans, R. (1970); J. of Phys. C3, 5137.

Evans, R., Greenwood, D.A. and Lloyd, P. (1971); Phys. Letts. 35A, 57.

Evans, R., Güntherodt, H.-J., Künzi, H.U. and Zimmermann, A. (1972); Phys. Letts. 38A, 151.

Friedel, J. (1958); Nuovo Cimento 2, 287.

Gruber, O.F. and Gardner, J.A. (1971); Phys. Rev. B4, 3994.

Heden, P.O., Löfgren, H. and Hagström, S.B.M. (1971); Phys. Rev. Letts. 26, 432.

Itami, T. and Shimoji, M. (1972); Phil. Mag. 25, 229.

Keller, J. and Jones, R. (1971); J. of Phys. F1, L33.

Klein, A.P. and Heeger, A.J. (1966); Phys. Rev. 144, 458.

Kohlhaas, R. (1965); Archiv f. Eisenhüttenwesen 36, 437.

Kohlhaas, R. and Weiss, W.D. (1969); Z. Naturforschung 24a, 287.

Koyama, R.Y. and Spicer, W.E. (1971); Phys. Rev. B4, 4318.

Lloyd, P. (1967); Proc. Phys. Soc. 90, 207.

Mott, N.F. (1964); Adv. Phys. 13, 325.

Mott, N.F. (1966); Phil. Mag. 13, 989.

Mott, N.F. (1971); private communication.

MacPherson, M.R., Everett, G.E., Wohlleben, D. and Maple, M.B. (1971); Phys. Rev. Letts. 26, 20.

Nakagawa, Y. (1956); J. Phys. Soc. Jap. 11, 855.

Pant, M.M., Das, M.P. and Joshi, S.K. (1971); Phys. Rev. B4, 4363.

Ramirez, R. and Falicov, L.M. (1971); Phys. Rev. B3, 2425.

Rousseau, J., Stoddart, J.C. and March, N.H. (1970); Proc. Roy. Soc. A317, 211.

Schneider, T. and Stoll, E. (1967); Adv. Phys. 16, 731.

Schwartz, L. and Ehrenreich, H. (1971); Ann. Phys. 64, 100.

Seib, D.H. and Spicer, W.E. (1970); Phys. Rev. B2, 1676.

Shaw, Jr., R.W. and Smith, N.V. (1969); Phys. Rev. 178, 985.

Stocks, G.M., Williams, R.W. and Faulkner, J.S. (1971); Phys. Rev. B4, 4390.

Tamaki, S. and Takeuchi, S. (1967); J. Phys. Soc. Jap. 22, 1042.

Tamaki, S. (1967); J. Phys. Soc. Jap. 22, 865.

Tamaki, S. (1968a); J. Phys. Soc. Jap. 25, 379.

Tamaki, S. (1968b); J. Phys. Soc. Jap. 25, 1602.

ten Bosch, A. and Baltensperger, W. (1971); EPS Conference on the Physics of Condensed Matter, FLORENCE.

Urbain, G. and Übelacker, E. (1967); Adv. Phys. 16, 429.

Wachtel, E. and Maier, J. (1967); Z. Metallkde. 58, 885.

Watabe, M. and Tanaka, M. (1964); Progr. Theor. Phys. 31, 529.

Wenger, A., Bürri, G. and Steinemann, S. (1971); Phys. Letts. 34A, 195.

Wenger, A., Bürri, G. and Steinemann, S. (1971); Sol. State Com. 9, 1125.

Wittig, J. (1968); Phys. Rev. Letts. 21, 1250.

Wilson, J.R. (1965); Metall. Rev. 10, 381.

ELECTRONIC STRUCTURE OF LIQUID SEMICONDUCTORS

J.E. Enderby

Department of Physics, University of Leicester, England

1. INTRODUCTION

A. The Electrical Behaviour of Liquid Semiconductors

We shall focus attention on binary liquid alloys for which the electron transport parameters are outside the range characteristic of the metallic state for at least some compositions.* We can refer to those alloys as "true" liquid semiconductors because they exhibit most of the properties that characterise conventional solid semiconductors.

It was originally thought (Joffe and Regel 1960) that a wide variety of materials could be classed as liquid semiconductors. However, the experimental evidence now available (see for example, Enderby and Simmons 1969, Allgaier 1969) rules out this classification for systems like $AuTe_2$, Bi_2Te_3, Sb_2Te_3, etc. in spite of the fact that $d\sigma/dT$ is positive for these liquids. There are very cogent reasons (Enderby 1972) for believing that the electron transport in these liquids can be handled theoretically by the methods first discussed by Faber and Ziman (1965).

Other systems behave in a qualitatively different way. These include liquid Cu-Te, Ag-Te, In-Te, Mg-Bi and all liquid

*Generally accepted values for the transport parameters in the metallic state are: conductivity (σ) in excess of $\sim 3,000\ \Omega^{-1}\ cm^{-1}$; thermopower (S) less than $\sim \pm 50 \mu V deg^{-1}$; Hall coefficient (R) approximately equal to R_o, the free electron value.

Table I Liquid Semiconductors

Liquid Alloy	Critical Composition	Conductivity Ω^{-1} cm^{-1}
S - Ag	Ag_2S	200
S - Pb	PbS	110
S - Cu	Cu_2S	50
S - Sn	SnS	24
S - Ge	GeS	1.35
S - Tl	Tl_2S_3	1.7×10^{-2}
	Tl_4S_3	6.5×10^{-3}
S - Sb	Sb_2S_3	1.5×10^{-2}
Te - Cu	Cu_2Te	200
Te - Ag	Ag_2Te	150
Te - Fe	$FeTe_2$	400
Te - Tl	Tl_2Te	70
Te - Cd	CdTe	40
Te - Zn	ZnTe	40
Te - In	In_2Te_3	25
Te - Ga	Ga_2Te_3	10
Bi - Mg	Mg_3Bi_2	<45
Bi - Li	Li_3Bi	?

References

Allgaier (1969)

Enderby and Collings (1970)

alloys involving selenium as one component. Table I gives a list of some of the alloy systems that fall into this category, together with the value of their electrical conductivities at the composition of particular interest. To be specific we shall focus attention on two groups of liquid alloys that

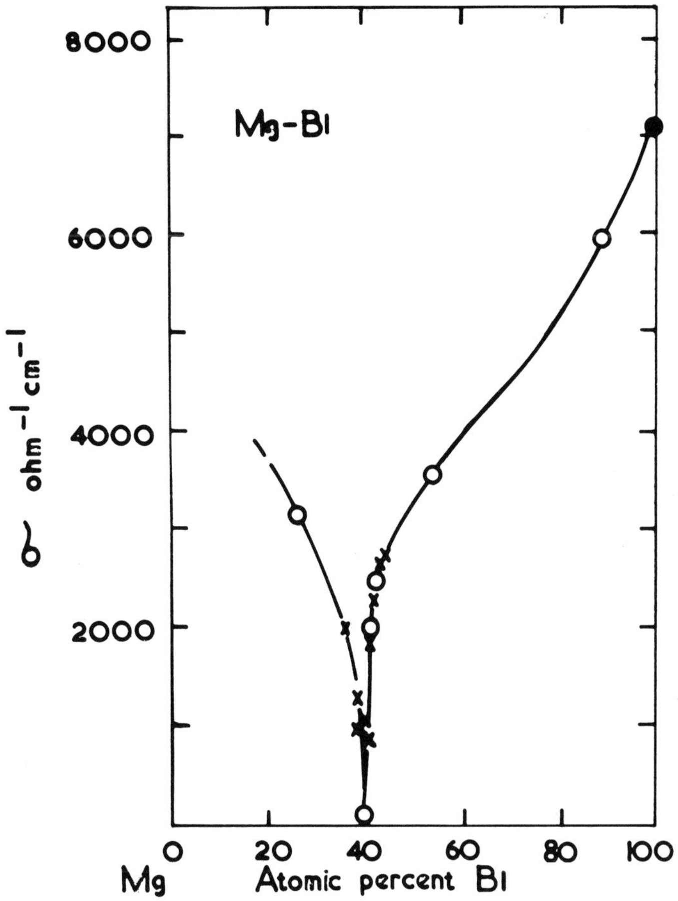

Figure 1 The conductivity of liquid Mg-Bi
x - Ilschner and Wagner (1953). o - Enderby and Collings (1970).

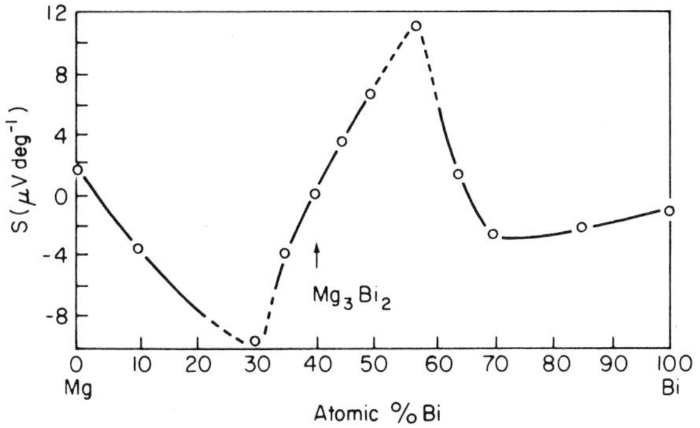

Figure 2 The thermoelectric power of liquid Mg-Bi as a function of composition at temperatures 100°C above the liquidus.

constitute systems of this type. Let *M* and *S* refer respectively to pure liquids that have metallic and semi-metallic electrical properties.

The first group comprises *M-M* systems which are of particular interest because it is possible continually to follow the transition from metallic behaviour to semiconducting behaviour. Liquid Mg-Bi, Mg-Sb and Li-Bi, represent some of the alloys known to fall in this group (Ioannides and Enderby unpublished. Experimental results for conductivity σ, and for thermoelectric power S, have been reported for liquid Mg-Bi by Ilschner and Wagner (1953) and by Enderby and Collings (1970); a selection of the data is given in figures 1 and 2. At the composition Mg_3Bi_2, the S changes sign and σ falls to a minimum value. It is not yet possible to say exactly what this value of σ is, although it is known that $d\sigma/dT$ is positive for several compositions around the stoichiometric liquid Mg_3Bi_2. Phase diagrams for Mg-Bi, Mg-Sb and Li-Bi are available (Hanson 1958) and information concerning the thermodynamic behaviour of Mg-Bi is to be found in the compilation given by Hultgreen *et al* (1963). It is clear from this evidence that a major change in the bonding characteristics takes place as we proceed from pure liquid magnesium. Presumably the same situation obtains for Li-Bi. There is also some evidence that substantial electromigration occurs in liquid Mg_3Bi_2, with the Mg drifting towards the cathode and the Bi drifting towards the anode.

The second group comprises *M-S* systems which include Ag-Te, Cu-Te (but *not* Au-Te), Ga-Te and Tl-Te; they represent the most widely studied liquid semiconductors. Phase diagrams for Cu-Te and Tl-Te are available (Hanson 1958), and a selection of the experimental data due to Dancy (1965), Cutler and Mallon (1966), and Enderby and Simmons (1969) is given at figures 3 and 4. The conductivity falls to a minimum value for Tl_2Te, at which composition elementary valence considerations are

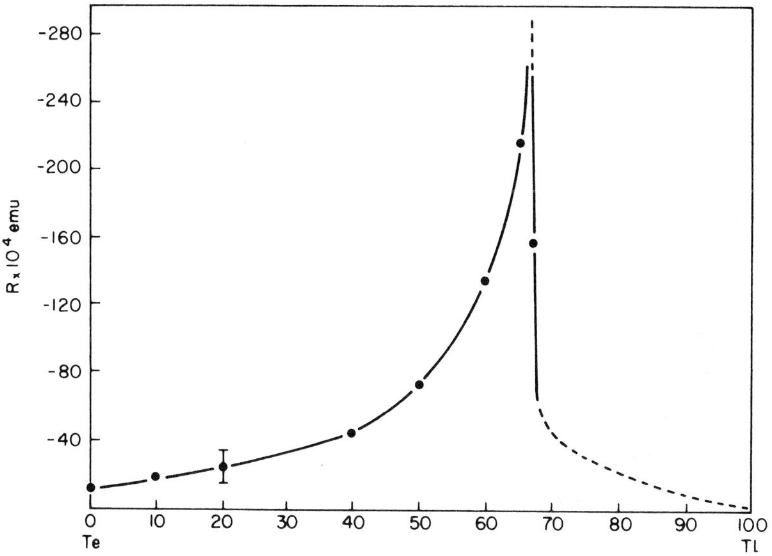

Figure 3 The Hall coefficient of liquid Te-Tl as a function of composition at 585°C. The dotted portion refers to the two-phase region.

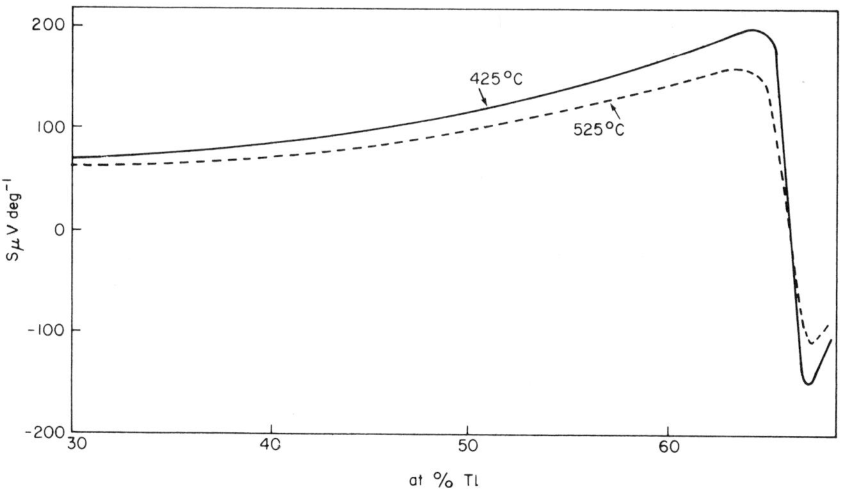

Figure 4 Thermoelectric power for liquid Te-Tl (After Cutler and Mallon 1966).

satisfied. Although the system Tl-Te has received special attention, the following remarks appear to be valid to all alloys within this second group:

(1) the alloys possess a two-phase liquid region, (liquid imiscibility) often in the range $70 \leq X_M \leq 100$, where X_M is the atomic percentage of the metallic component.

(2) R/R_0 is significantly different from unity; R itself is negative at all compositions, and apparently achieves a maximum absolute value at the composition corresponding to minimum conductivity.

(3) S varies rapidly close to, but not necessarily precisely at the stoichiometric composition; in some cases a change in sign is observed, although this change is not reflected in the Hall coefficient.

Thermodynamically Te-Tl closely resembles Mg-Bi and Li-Bi (Nakamura and Shimoji 1971). In particular, the thermodynamic properties show a similarity at the stoichiometric composition Tl_2Te.

B. Outstanding Problems

We shall consider four questions which we regard as outstanding problems:

(1) What is the most useful way to characterise the structure of liquid semiconductors?

(2) How can the distribution of binary liquid semiconductors (within the periodic table) be understood?

(3) What will be the form of the density of states as a function of composition and temperature?

(4) What is the mode of electron transport, particularly around the stoichiometric composition?

2. THE STRUCTURE OF LIQUID SEMICONDUCTORS

A. Possible Models

Close to the stoichiometric composition two models have

been suggested. The first proposed by Cutler and Field (1968) and Enderby and Simmons (1969), involves the existence of well-defined molecular groups such as Tl_2Te or Mg_3Bi_2. However, an equally plausible description, particularly in view of the electromigration data for liquid Mg_3Bi_2, is the second model in which the liquid alloy is assumed to be essentially *ionic* in character. The 'covalent' and the 'ionic' models represent the two extremes; while a mixed bond type (Nakamura and Shimoji 1971) cannot, at this stage, be ruled out. Nevertheless the substantial change in bonding as we proceed from pure metallic to either covalent or ionic is entirely consistent with the thermodynamic evidence (Hultgreen *et al* 1963).

At compositions other than stoichiometric, several possibilities can be envisaged. Cutler (1971) has argued that the atomic arrangement in liquid alloys with excess tellurium is chain-like, with basic units consisting of $\{Tl\text{-}(Te)_n\text{-}Tl\}$. Other models include the concept of a gradual change from ionic to metallic bonding as the composition changes from (say) Mg_3Bi_2 to Mg.

Detailed structural experiments can, in principle, resolve these different models, and we now describe some recent work (Enderby and Hawker 1972) that attempts to do this.

B. Structural Studies Involving Neutron Diffraction

We introduce generalised *interference functions*, $a_{\alpha\beta}$, defined by

$$a_{\alpha\beta}(q) = 1 + \frac{4\pi n}{q} \int dr\ [g_{\alpha\beta}(r) - 1]\ r\sin qr \qquad (1)$$

where $g_{\alpha\beta}$ represents the distribution of α-atoms observed from a β-atom at the origin, n is the number density of atoms irrespective of type, q is the conventional momentum transfer variable, and α and β are dummy suffices which take on the values 1 and 2 for a binary alloy. In terms of the differential scattering cross-section, which for a given type of radiation is a quantity that can be determined experimentally, it can be readily shown that

$$\frac{d\sigma}{d\Omega} = c_1 f_1^2 + c_2 f_2^2 + F(q). \tag{2}$$

Here c_1 and c_2 are the atomic concentration of two constituents, f_1 and f_2 are the appropriate atomic (or nuclear) scattering factors, and F(q) is given by

$$F(q) = c_1^2 f_1^2 (a_{11} - 1) + c_2^2 f_2^2 (a_{22} - 1) + 2c_1 c_2 f_1 f_2 (a_{12} - 1). \tag{3}$$

Thus, to discuss liquid semiconductors from a fundamental standpoint, three partial interference functions a_{11}, a_{12} and a_{22} are required. To obtain these, three separate determinations of $d\sigma/d\Omega$ for the same alloy must be made, by changing f_1 or f_2. It is now clear that the only feasible means of carrying out such experiments is to combine neutron diffraction with the use of isotopically enriched samples (Enderby *et al* 1966).

As a first step towards a fuller understanding of liquid semiconductors, experiments have been carried out on the liquid alloy CuTe which forms part of a system in which semiconducting behaviour has been observed. Three samples of CuTe were prepared using natural tellurium with: (1) natural copper, (2) copper enriched to 99% with ^{63}Cu, and (3) copper enriched to 99% with ^{65}Cu. The coherent scattering cross-sections for ^{nat}Cu, ^{63}Cu, ^{65}Cu, and Te are in the proportion 1:0·73:1·96:0·51. We conclude that:

(1) Te-Te correlations will be most evident in the diffraction pattern for liquid $^{63}CuTe$;

(2) Cu-Cu correlations will dominate in the diffraction pattern for $^{65}CuTe$.

The experimental F(q) for each specimen are shown in figure 5. It is instructive to compare the position of the first few extrema in the alloy with those of pure liquid tellurium and

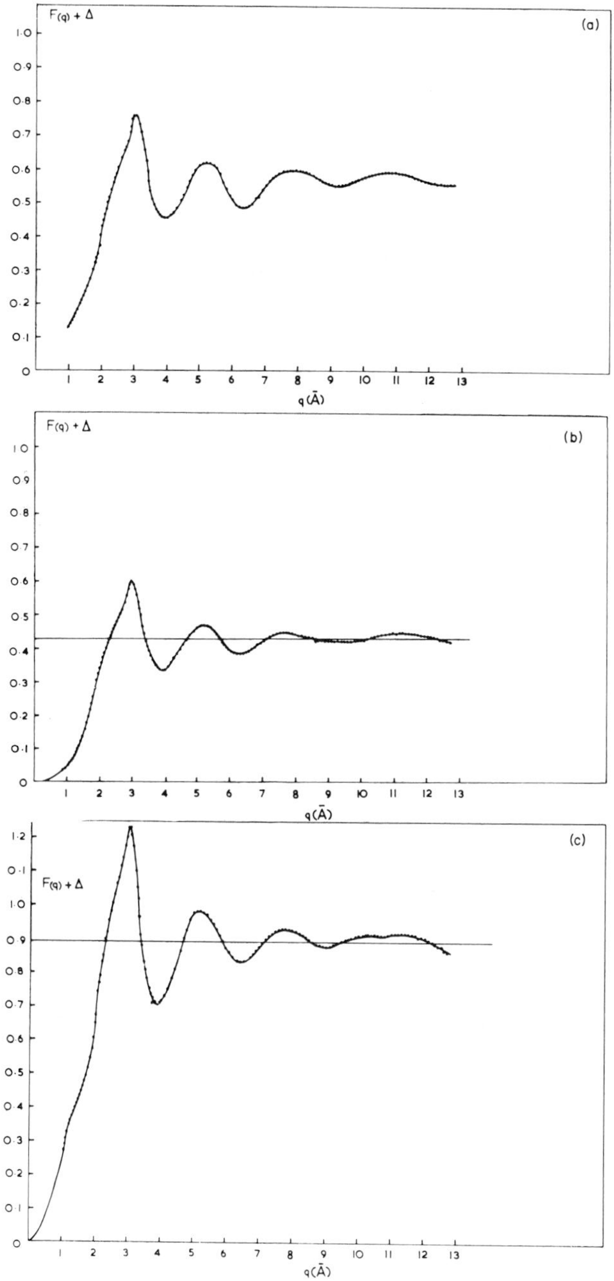

Figure 5 $F(q) + c_1 f_1^2 + c_2 f_2^2$ as a function of q for liquid: (a) ^{nat}CuTe, (b) ^{63}CuTe, (c) ^{65}CuTe.

pure liquid copper (Table II). The surprising fact is that features characteristic of pure copper and pure tellurium *persist* in the alloy.

Table II Position of Extrema

Extrema (A^{-1})	Liquid CuTe	Liquid Cu(a)	Liquid Te(b)
	(2.2)(c)		
1st Max	3.0	3.0	2.0
2nd Max	5.3	5.4	4.3
1st Min	3.9	3.9	2.8
2nd Min	6.5	7.0	5.3

(a) See Wagner *et al* (1965).
(b) Unpublished data due to Hawker, Howe, Howells and Enderby.
(c) The feature at 2.2Å^{-1} is most noticeable in ^{63}CuTe, but is just visible in ^{65}CuTe.

On the basis of these results and without further detailed analysis, Enderby and Hawker were able to conclude that:

(1) Models in which copper atoms are located either substitionally within the tellurium chains (Te-Cu-Te-Cu-Te-Cu) or at the ends of the tellurium chains (Cu-Te-Te-Cu) contradict the experimental facts*. The former model implies $a_{11} = a_{22} = a_{12}$ and leads to an F(q) with first maximum at ∿2Å^{-1} The second model will involve a peak in the Cu-Te partial interference function at ∿2Å, so that the low-angle shoulder (which we attribute to the Te-Te partial interference function) should *increase* as we change from ^{63}CuTe to ^{65}CuTe. Neither of these models is sustained experimentally.

---

*It cannot be over-emphasised that this conclusion applies only to liquid Cu-Te. It would be dangerous to extend it to the Te-Tl system where the chain model proposed by Cutler (1971 explains many of the observed electrical properties.

(2) A model based on hard spheres is also inappropriate. This structure form would give rise to a broad maximum in F(q) centred at 2.5Å^{-1}.

However, the data were consistent with the 'ionic' model of liquid semiconductors. Enderby and Hawker suppose that some degree of electron transfer takes place as copper is added to tellurium and that liquid CuTe is weakly ionic, though still with a substantial electron density (more complete ionic behaviour occurs in Cu_2Te, at which composition the electrical conductivity is a minimum). This electron rearrangement leaves the Cu-Cu and the Te-Te first-neighbour distances relatively unchanged compared with those for the pure materials. It follows that a maximum in the scattering pattern will appear at 3.0Å^{-1}, due to the partial interference function a_{11} (i.e. a_{Cu-Cu}), and will be most pronounced in the alloy ^{65}CuTe. Similarly, a peak will occur at 2Å^{-1} corresponding to the maximum in a_{Te-Te}, although this feature will be masked in the alloy pattern by the relatively strong scattering characteristic of copper. The peak will however be most evident in ^{63}CuTe, and this was indeed found to be the case.

3. OCCURRENCE OF SEMICONDUCTING LIQUID ALLOYS

The existence of a fairly well-defined molecular group like Tl_2Te or Mg_3Bi_2 or of the ionic assembly Tl^+, Te^{--} or Mg^{--}, Bi^{+++} cannot be approached from a free-electron theory because it involves the formation of bound or localized states. Stern (1966) for example has shown that such states can arise if the atomic potentials of the two species are very different. In the dilute alloy problem, the method of Koster and Slater (1954) indicates how the formation of real or virtual bound states depends crucially on the strength of the impurity potential.

A rough measure of the difference in potentials between,

say, Mg and Bi or Te and Tl, can be obtained from published tables of electronegativities or stability ratios (Sanderson 1960), or the well-depths that characterise the Heine-Abarenkov-Animalu model potential (see for example Animalu and Heine 1965). The connection between pseudopotentials and electronegativity implied by this remark has recently been justified by Heine and Weaire (1970). The broad distribution of liquid alloys that possess semiconducting properties is entirely consistent with this approach. However, to make really quantitative predictions will involve calculations like those used in the coherent potential approximation (CPA) method (Velicky *et al* 1968) together with a careful and self-consistent treatment of the electron screening (Ziman 1967).

4. THE ELECTRONIC STRUCTURE OF LIQUID SEMICONDUCTORS

Consider an *M*-*M* system like Mg-Bi. In pure liquid lithium the density of states n(E) will be essentially free-electron-like. Suppose we now add a few atoms of a strongly electronegative element like bismuth. A modification to n(E), entirely consistent with the ionic model postulated in §2 is predicted by the Slater-Koster theory. If the impurity potential is strong enough, a bound state appears below the bottom of the band; alternatively a virtual bound state within the band enhances n(E) around the resonant energy. We assert, without proof, that this process continues as the impurity content increases in the schematic manner illustrated in figure 6.

A possible experimental test of these ideas is to investigate the conduction-electron spin resonance (CESR) for a variety of liquid alloys in which lithium or sodium form one component. Such experiments have been reported by Enderby and Nguyen (1972) and we now briefly review these:

A spin-flip cross-section is defined by

$$\sigma = \frac{1}{N\nu_F c T_1} \tag{4}$$

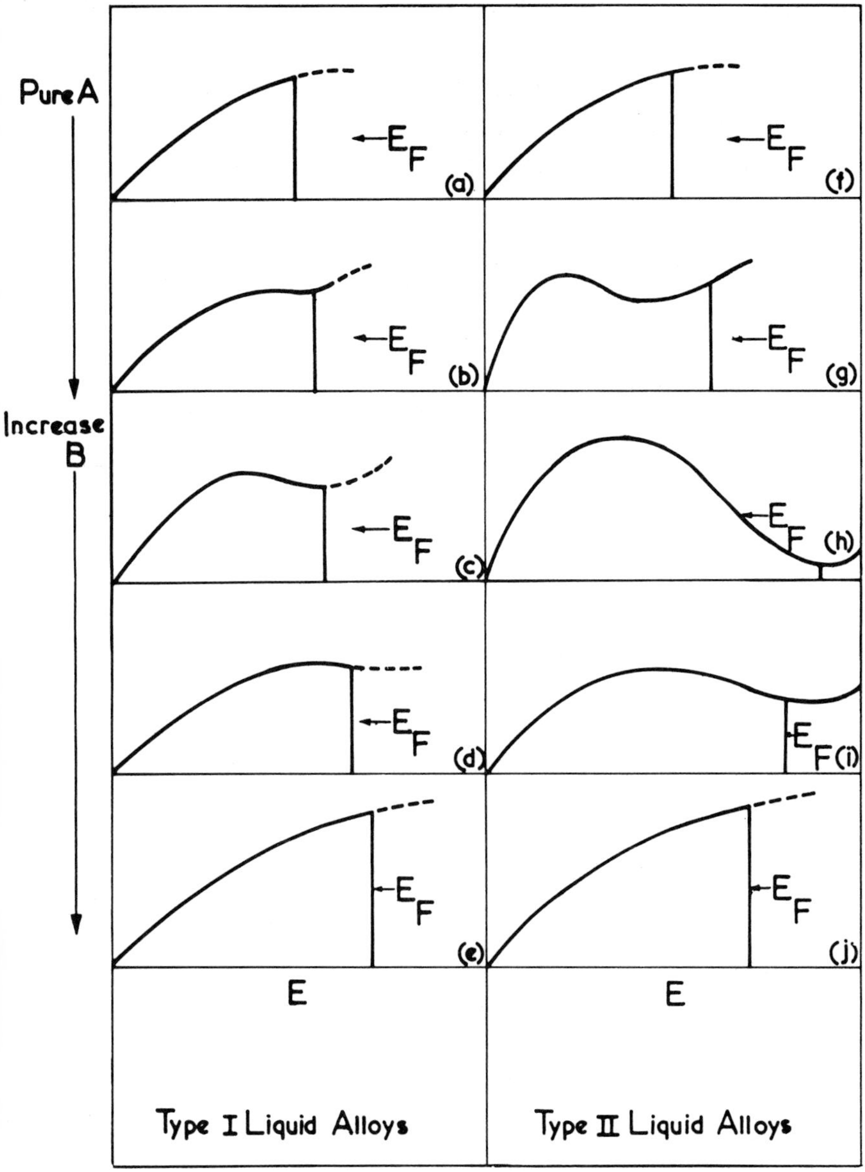

Figure 6 Modifications to n(E) in liquid semiconducting alloys: Type I alloys include Au-Te, Na-Bi, Cu-Sn, Type II alloys include Cu-Te, Mg-Bi, Li-Bi (Enderby and Collings 1970). The stoichiometric compositions (Au_2Te, Na_3Bi, Cu_6Sn_5 Cu_2Te, Mg_3Bi, Li_3Bi) correspond to (c) and (h).

where c is the atomic concentration of the solute, N is the number density of solvent atoms, ν_F is the Fermi velocity, and T_1 is the spin-lattice relaxation time. T_1 was determined by Enderby and Nguyen using CESR techniques; a summary of the results obtained, together with other data, is given in Table III.

Table III Spin-Flip Scattering Cross-Sections for Various Solutes in Liquid sodium and Liquid lithium

Solvent	Na		Alloy Type		Li		Alloy Type	
Solute	$\sigma(10^{-17}cm^2)$		Ref.(e)	Ref.(f)	$\sigma(10^{-17}cm^2)$		Ref.(e)	Ref.(f)
Au	*2.9*	(a)	I	A	*6.8*	(a)	I	A
Hg	*9.6*	(a)	I	A	*11.0*	(a)	I	A
Tl	*27*	(a)	I	A	*9.7*	(c)	I	B*
Pb	*20*	(a)	I	B	1.6	(c)	II*	C*
Bi	0.73	(b)	I	B	$4.1x10^{-3}$	(d)	II	C
Te			II	C	$7.5x10^{-4}$	(d)	II	C

References

(a) Asik *et al* (1969). Note: Cross-sections in italics refer to data derived from solid alloys. The classification into alloy type is based on the electrical conductivity (Freeman and Robertson 1961) and Ioannides and Enderby (unpublished).

(b) Cornell and Slitcher (1969).

(c) Enderby, Heath and Titman (unpublished).

(d) Enderby and Nguyen (1972).

(e) Enderby and Collings (1970).

(f) Allgaier (1969).

*Data at present incomplete.

The probability of a spin flip in lithium or sodium based alloys is dominated by the presence of the solute atoms. Wignall *et al* (1965) and Asik *et al* (1969) suggested that the magnitude of σ can be understood in terms of the relaxation mechanism first suggested by Elliott (1954), in which the importance of spin-orbit coupling was emphasised. A trend to increased σ-values as the solute atoms become heavier is to be expected on this model.

It is quite clear that breakdown of this simple rule is particularly severe for liquid Li-Bi. Any model based on spin-orbit effects will predict substantial values of σ for bismuth in lithium. Indeed the change in conductivity as bismuth is added to liquid lithium is substantial (Ioannides and Enderby unpl.), but there is no corresponding change in the rate of spin flips. Enderby and Nguyen point out that a phase-shift analysis indicates that the reduction in σ, over that expected from the spin-orbit model, requires substantial p-wave screening even for solutes of valency 4.

This behaviour is entirely consistent with the ionic model illustrated in figure 6. It explains, for example, the differences between σ for bismuth alloyed with lithium (type II alloy) and bismuth alloyed with sodium (type I alloy). The rôle of localization in the Mott sense of the term will become increasingly important as the critical composition Li_3Bi is approached; further experiments aimed at elucidating the CESR behaviour in this composition region are at present underway.

5. ELECTRON TRANSPORT IN LIQUID SEMICONDUCTORS

A. Introduction

According to Mott (1971) the magnitude of the *d.c.* conductivity is of crucial significance because it enables the magnitude of Λ, the electronic mean free path, to be estimated. If d represents the average interatomic spacing then the three cases of interest are:

(a) $\Lambda > d$: This is the so-called weak scattering limit and applies for $\Lambda \gtrsim 3\text{Å}$ which corresponds to conductivities in excess of $\sim 3000\ \Omega^{-1}\text{cm}^{-1}$. Many liquid metals and alloys fall into this category and a prescription that enables Λ to be evaluated in terms of the pseudopotential has been given by Ziman (1961).

(b) $\Lambda \sim d$: The theory of electronic conduction in this regime is not yet fully developed and most of the arguments used are due to Mott (1966, 1967, 1970, 1971).

Following the theoretical work of Anderson (1958) Mott concludes that a lower limit for σ must exist below which conduction processes involving extended electronic states is not possible. By making certain plausible assumptions Mott determines σ_{min} to be $\sim 200\ \Omega^{-1}\text{cm}^{-1}$. For liquid conductivities above this value, but below $\sim 3000\ \Omega^{-1}\text{cm}^{-1}$, Mott argues that the density of states at the Fermi energy, $n(E_F)$, rather than Λ becomes the controlling factor. Setting $\Lambda \simeq d$ = constant, Mott obtains

$$\sigma = \frac{g^2e^2}{3hd} \tag{5}$$

and

$$S = \frac{\pi^2}{3}\frac{k_B{}^2T}{e}\,2\left[\frac{d\ \ln\ n(E)}{dE}\right]_{E=E_F} \tag{6}$$

where $g = \frac{n(E_F)}{n_o(E_F)}$ is in the range 1 to ~ 0.3, and $n_o(E_F)$ is the free-electron density of states.

It is not yet possible to go beyond the simple free-electron for the Hall coefficient with any degree of confidence. The central difficulty here is that the Hall coefficient involves the transport of *momentum* and this is not a well-defined quantity in situations in which $\Lambda \sim d$ (Edwards 1962). There is mounting theoretical evidence that $R/R_o = (g)^{-\alpha}$ Fukuyama *et al* 1969, Ziman 1967) although the appropriate value of α is disputed.

(c) $\Lambda \sim d$ with $g < 0.3$: For g-values less than ~ 0.3, Mott suggests that Anderson localization occurs. The conductivity arises from thermal excitation of carriers across the mobility gap, or by hopping processes. The semi-empirical relationship for the conductivity (Stuke 1969) is $\sigma = C\exp(-E/kT)$ where C is constant and E is a characteristic energy. According to Cutler and Mott (1969) the usual metallic equation can still be used to calculate the thermoelectric power.

Holstein (1961) and Holstein and Friedman (1968) are among the several investigators who have considered how the Hall coefficient should be calculated for systems in which the electronic wavefunctions are localized. Recent theoretical work by Friedman (1971), based on the Kubo formula, is of particular significance. Friedman derives formulae for the Hall coefficient in the short mean-free-path materials in both the degenerate and the non-degenerate limits. The Friedman formula for the liquid alloys under discussion (normally assumed to be degenerate) is

$$R/R_0 = kg^{-1}$$

where k is a constant of order unity.

B. Comparison of Theory with Experiment

If models proposed for the electronic and atomic structures are accepted, then the measured transport agrees well with the theory described in §5A. Let us again focus attention on the Mg-Bi system; the rapid increase in resistivity as bismuth is added to magnesium indicates a marked reduction in the mean free path. The liquid alloy, therefore, enters a regime in which R,S and σ are all dominated by $n(E_F)$. At the same time a peak (at resonance) in the density of states followed by a dip will appear in the manner illustrated in figure 6. The electrical properties of semiconducting *M-M* alloys (type II) will, as the stoichiometric composition is approached, reflect these progressive changes in n(E). Moreover, the observed CESR for *M-M* semiconducting liquids in which lithium is one component is entirely consistent with this model.

A recent experiment that further supports the theory has been carried out by Ioannides and Enderby (unpublished). The thermoelectric power of pure liquid lithium at the melting point is large and positive (+22 μVdeg^{-1}). According to the present picture, Li_3Bi with excess lithium (figure 6g) should have a large and *negative* S. Ioannides and Enderby found that for 5 at% bismuth in lithium, the sign change predicted in S had *already* occurred. Further experiments on this interesting system are at present underway.

Acknowledgments

I wish to acknowledge the financial support awarded for this work by the Science Research Council, and to thank Professor Sir Nevill Mott for his continued interest and encouragement.

REFERENCES

Allgaier, R.S. (1969); Phys. Rev. 185, 227.

Anderson, P.W. (1958); Phys. Rev. 109, 1492.

Animalu, A.O.E. and Heine, V. (1965); Phil. Mag. 12, 1249.

Asik, J.R., Ball, M.A., Slichter, C.P. (1969); Phys. Rev. 181, 645.

Cornell, E.K. and Slichter, C.P. (1969); Phys. Rev. 180, 358.

Cutler, M. (1971); Phil. Mag. 24, 381.

Cutler, M. and Field, M.B. (1968); Phys. Rev., 169, 632.

Cutler, M. and Mallon, C.E. (1966); Phys. Rev. 144, 642.

Cutler, M. and Mott, N.F. (1969); Phys. Rev. 181, 1369.

Dancy, E.A. (1965); Trans. Met. Soc. AIME, 233, 270.

Edwards, S.F. (1962); Proc. Roy. Soc. A267, 518.

Elliot, R.J. (1954); Phys. Rev. 96, 266.

Enderby, J.E. (1972); in "Amorphous and Liquid Semiconductors" (ed. J. Tauc) Plenum Press (London and New York) - in press.

Enderby, J.E. and Collings, E.W. (1970); Jnl. Non-Cryst. Solids 4, 161.

Enderby, J.E. and Hawker, I. (1972); Proc. Int. Conf. on Amorphous and Liquid Semiconductors, Journal of Non-Crystalline Solids 8/9, 687.

Enderby, J.E., North, D.M. and Egelstaff, P.A. (1966); Phil. Mag. 14, 961.

Enderby, J.E. and Nguyen, V.T. (1972); Proc. Int. Conf. on Amorphous and Liquid Semiconductors, Journal of Non-Crystalline Solids 8/9, 262.

Enderby, J.E. and Simmons, C.J. (1969); Phil. Mag. 20, 125.

Faber, T.E. and Ziman, J.M. (1965); Phil. Mag. 11, 153.

Freeman, J.F. and Robertson, W.D. (1961); J. Chem. Phys. 34, 769.

Friedman, L. (1971); to be published.

Fukuyama, H., Ebisawa, H. and Wada, Y. (1969); Proc. Theor. Phys. 42, 494.

Hanson, H.P. (1958); "Constitution of Binary Alloys", New York; McGraw-Hill.

Heine, V. and Weaire, D. (1970); Solid State Phys. 24, 249.

Holstein, T. (1961); Phys. Rev. 124, 1329.

Holstein, T. and Friedman, L. (1968); Phys. Rev. 165, 1019.

Hultgreen, R., Orr, R.L., Anderson, P.D. and Kelley, K.K. (1963); "Selected Values of Thermodynamic Properties of Metals and Alloys"; New York; Wiley.

Ilschner, B.R. and Wagner, C.N. (1953); Acta Met. 6, 712.

Joffe, A.F. and Regal, A.R. (1960); "Progress in Semiconductors", ed. A.F. Gibson; New York; Wiley.

Koster, G.F. and Slater, J.C. (1954); Phys. Rev. 95, 1167.

Mott, N.F. (1966); Phil. Mag. 13, 989.

Mott, N.F. (1967); Adv. Phys. 16, 49.

Mott, N.F. (1970); Phil. Mag. 22, 1.

Mott, N.F. (1971); Phil. Mag. 24, 1.

Nakamura, Y. and Shimoji, M. (1971); Trans. Farad. Soc. 67, 1270.

Sanderson, E.A. (1960); "Chemical Periodicity"; New York; Reinhold.

Stern, E.A. (1966); Phys. Rev. 144, 545.

Stuke, J. (1969); Festörperprobleme, 9, 46.

Velicky, B., Kirkpatrick, S. and Ehrenreich, H. (1968); Phys. Rev. 175, 747.

Wagner, C.N.J., Ocken, H. and Joshi, M.L. (1965); Naturf. 20A, 325.

Wignall, G.D., Enderby, J.E., Hahn, C.E.W. and Titman, J.M. (1965); Phil. Mag. 12, 433.

Ziman, J.M. (1961); Phil. Mag. 6, 1013.

Ziman, J.M. (1967); Proc. Phys. Soc. 91, 701.

X-RAY EMISSION AND ABSORPTION SPECTRA OF CRYSTALLINE AND AMORPHOUS SILICON

G. Wiech and E. Zöpf

Sektion Physik der Universität München, West Germany

1. INTRODUCTION

With the revival of interest in amorphous semiconductors during the last few years, attention has been focused on silicon; this is because solid silicon can be prepared as nearly perfect single-crystals, with pronounced long-range periodicity, as well as in the form of amorphous films with only short-range ordering. Thus it is possible to compare observations on the crystalline and amorphous forms of the same material, and to see which properties change on going from the ordered to the disordered state.

The x-ray emission spectra of crystalline silicon, the Kβ-emission band (Läuger 1968; also the present report) and the $L_{2,3}$-emission band (Wiech 1968, Wiech and Zöpf 1971a, 1971b) are well known. Moreover, in the case of crystalline silicon not only calculations of the band structure and of the density of states are available but also calculations of the intensity distributions of the Kβ and L emission bands (Klima 1970). These results are in good agreement with the experimental spectra. On the other hand, for amorphous silicon to-date only few experimental results are available.

In the present paper the K and L emission bands, and L-absorption measurements of both crystalline and amorphous silicon are reported.

2. EXPERIMENTAL

A. Spectrometers

The experiments were performed with the concave grating spectrometer (Wiech 1964, 1966) and the Johann-type spectrometer (Läuger 1968) already described, and under similar conditions to those reported by Wiech and Zöpf (this Volume, p.173).

The Si L-emission and the Si L-absorption were measured using a new continuous dynode multiplier, developed in our laboratories (Feser 1971, Wellmann 1971). In the wavelength region of the Si L-emission band this multiplier is more sensitive by a factor of 4 than the Bendix (model M306), which we have used so far.

B. Preparation of Samples

Films of amorphous silicon were prepared by evaporation, with the temperature of the substrate maintained roughly at room temperature. The samples were prepared and tested in two different ways:

In the first, silicon was deposited onto a copper substrata within a time duration of 2min at a pressure of $\sim 10^{-5}$ torr, using heavy tungsten boats. The thickness of the films, as measured by the oscillating quartz method, was 7000-8000 Å. The samples were examined by the Debye-Scherrer method. The diffraction photographs show two broad haloes around the 111, and the 220 and 311 reflections, and agree well with those reported by Brodsky *et al* (1970).

In the second case a layer of about 1 μm was deposited onto a plate of mono-crystalline silicon within $\sim$3min at a pressure of 10^{-6} torr. An electron gun was used for evaporation. The samples were examined by electron diffraction.

The samples in both cases were used to measure the $K\alpha_{1,2}$-doublet, the K and L emission bands, and the self-absorption spectra (Liefeld 1968, see next section).

For photon-absorption measurements thin foils were prepared by evaporation of silicon at 10^{-5} torr, onto glass plates covered with a thin layer of NaCl. The films were detached by floating in water and picked-up from the surface of the water, with thin foils of Pioloform serving as support. The thickness of the films used for absorption measurements varied from 1000 to 2200 Å. The thickness of the films was determined using multiple beam interferometry and a quartz-crystal thin film monitor.

The intensities I_0 (Pioloform only) and I (Pioloform and silicon) were measured using step-counting, and at each step at least 1000 pulses were recorded. In the vicinity of the Si L-absorption edge the radiations of the Si $L_{2,3}$-emission band and the Be K-emission band were used as light sources.

C. Self-absorption Spectra Measurements

X-rays excited by electrons (or photons) in bulk material are partially absorbed on their way to the surface of the sample (self-absorption). Liefeld (1968) who carefully studied the influence of self-absorption on the line-shape of emission bands, found that the point-by-point intensity-ratio of two curves measured at two different electron energies is similar to the photon absorption spectrum of the anode material. These intensity-ratio curves are called self-absorption spectra.

This method has been used here to study the absorption characteristics of crystalline and amorphous silicon. While it is easy to prepare samples of amorphous silicon and to study their photon-absorption spectrum, the preparation of thin films ($\sim$2000Å) of crystalline silicon is very difficult. It is possible that in the case of ordered alloys, which cannot be prepared by evaporation, the self-absorption technique is the only means of studying absorption characteristics. In our Institute at München some work on this subject is in progress.

The self-absorption measurements were performed at electron energies of 1.0 and 3.5 keV. We have investigated the penetration depth of low-energy electrons in aluminium, and for these energies found penetration depths of 740Å and 5650Å; for the average depth of production of characteristic x-rays we obtained 200Å and 2280Å respectively (Hoffmann *et al* 1969). For silicon similar values are to be expected. Thus it seems that the low voltage used (1.0 kV) is high enough to avoid considerable influence from target contamination or from oxides, and the higher voltage (3.5 kV) is low enough to ensure that electrons do not pass through the 1-μm film of amorphous silicon.

For energies higher than 3.5keV we observed that the L-emission band of amorphous silicon changed and became similar to the L-emission band of crystalline silicon. Since these changes are easily observed, the L-emission band is a sensitive detector for the quality and constancy of the amorphous films during the course of the measurements. At lower electron energies the intensity distribution of the L-emission band of amorphous silicon remained unchanged over many hours.

3. RESULTS AND DISCUSSION

A. Emission Bands

In figure 1 the Kβ and the $L_{2,3}$ emission bands of crystalline and amorphous silicon are shown. The corresponding K and L bands are correlated in energy by means of the $K\alpha_1$-lines (Si $K\alpha_1$ crystalline, 1739.91 eV; amorphous, 1739.92 eV).

In the $L_{2,3}$-emission bands of crystalline silicon we find two strongly marked maxima at 89.65eV and 92.15eV respectively. From the main peak towards higher energies the intensity

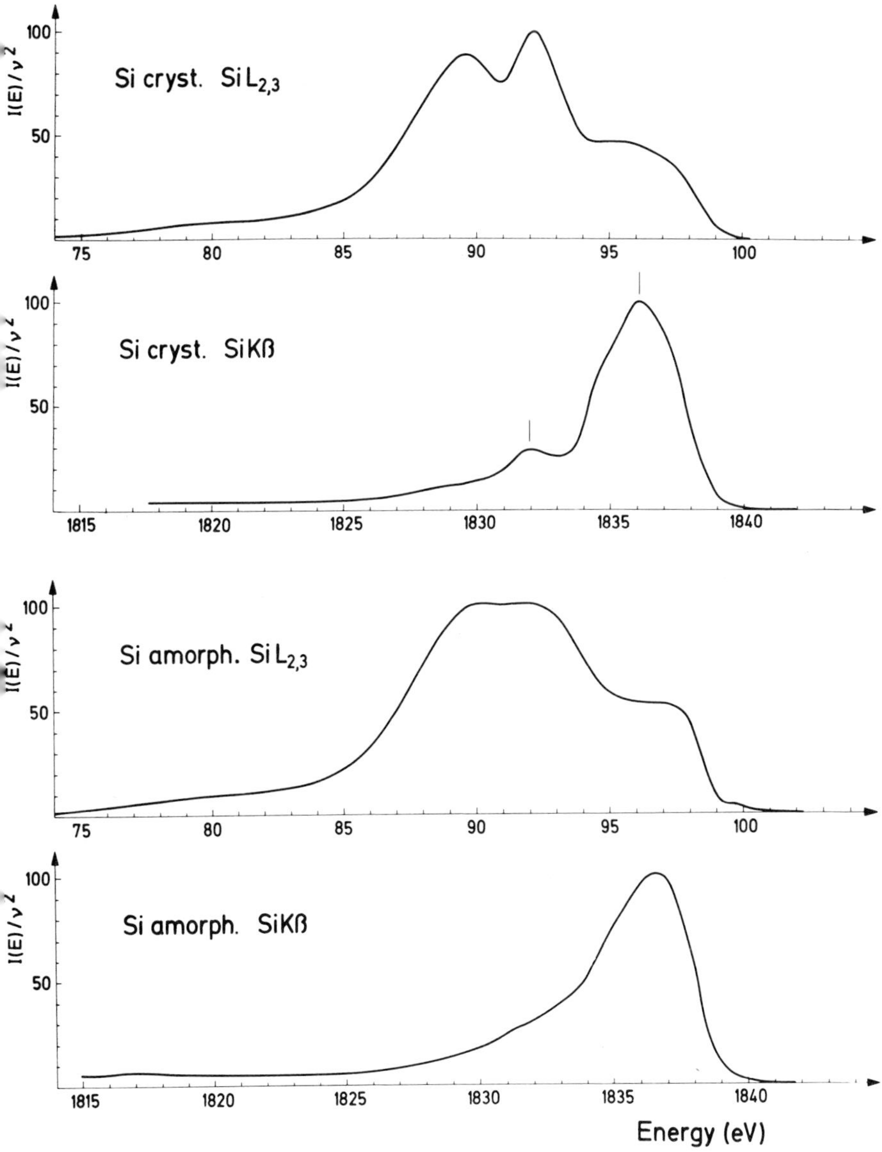

Figure 1 Kβ and $L_{2,3}$ emission bands of crystalline and amorphous silicon.

decreases gradually showing some weak fine-structure. The $L_{2,3}$-emission band of amorphous silicon is on the whole similar to that for the crystalline phase, but the curve is smoothed and differs in detail. The most obvious effect is the nearly complete disappearance of the two peaks; there remain only two flat maxima at 90.25 and 91.85 eV. The intensity of the L-emission band of amorphous silicon is at all wavelengths higher than that of crystalline silicon. There is a broad shoulder and a comparatively rapid fall-off in intensity at the high-energy edge. At the foot of the edge of the L-emission band of amorphous silicon a small peak (99.6 eV) is observed, while for the crystalline form we find a steady decrease in intensity to zero.

Turning now to the Kβ-emission bands, it can be seen that the structural features found for crystalline silicon have to a large extent disappeared for amorphous silicon. This is especially true for the peak at 1832.0 eV. The main peak has shifted from 1836.1 eV for crystalline silicon to 1836.65 eV for the amorphous phase. The high-energy K-emission edge of amorphous silicon, which is nearly parallel to that of crystalline silicon, has also shifted to higher energies by 0.2-0.3eV. A similar intensity distribution for the Kβ-emission band of evaporated silicon was previously reported by Krämer (1960).

At the low-energy end of the Kβ-emission band ($\sim$1817eV) a low-energy 'satellite' is observed. This indicates a small amount ($<$4%) of oxidized silicon in the sample, but does not however, effect the main band.

The L-emission bands of both crystalline and amorphous silicon show small rises in intensity in the low-energy tail, at $\sim$80eV. This may be due to the plasmon satellite which is superposed on the L-emission band.

B. Absorption Spectra

Figure 2 shows the absorption spectra of crystalline and

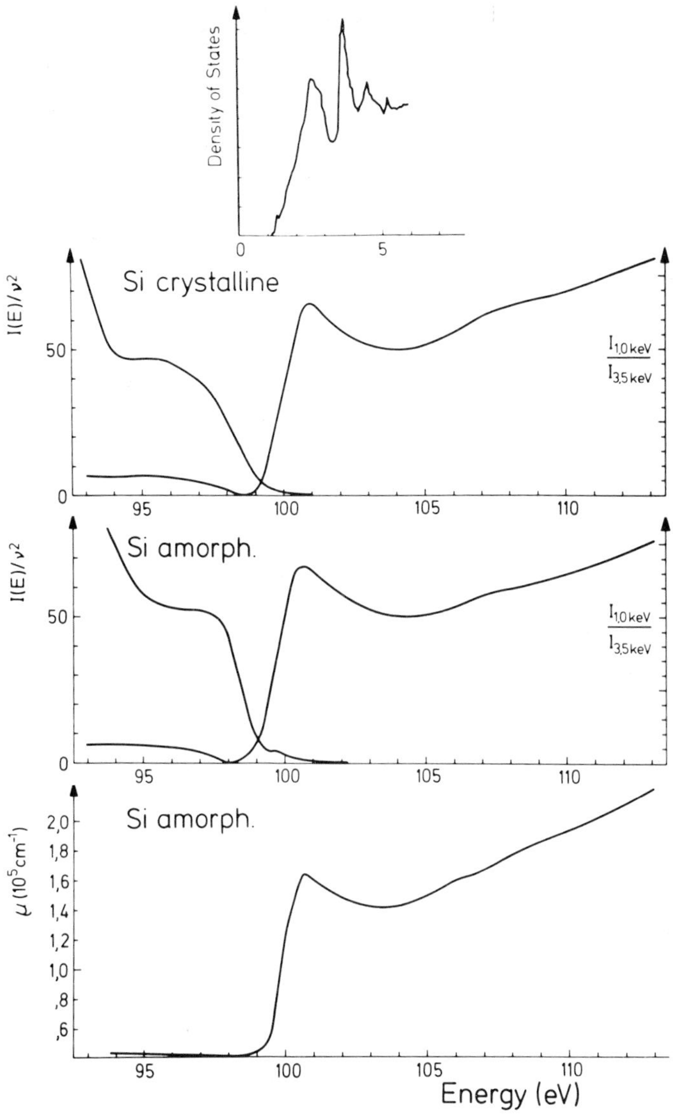

Figure 2 Valence and conduction band structure of crystalline and amorphous silicon:

a) Density of states of the conduction band of crystalline silicon (after Stukel *et al* 1971).

b) High-energy part of the $L_{2,3}$-emission band, and self-absorption spectrum, of crystalline silicon.

c) High-energy part of the $L_{2,3}$-emission band, and self-absorption spectrum, of amorphous silicon.

d) $L_{2,3}$-photon absorption spectrum of amorphous silicon.

amorphous silicon together with the high-energy part of the $L_{2,3}$-emission bands. For comparison the density of states of the conduction band of the crystalline phase has been included (2a). The absorption curves shown (2b and 2c) are self-absorption spectra obtained using 1.0 and 3.5 keV electrons. The curves were normalized at their minima, at ∿104eV. Both absorption curves show similar characteristics, but differ in detail. Crystalline silicon has a minimum of absorption at 98.6eV, and a maximum at 101.0eV; the energy at half-maximum intensity is 99.9eV. The corresponding energy values for amorphous silicon are all situated at rather lower energies (98.1, 100.7 and 99.65 eV). The absorption edge is shifted to lower energies by an average amount of 0.25eV.

The photon-absorption spectrum of amorphous silicon, obtain-by Metzger (1971), is shown in figure 2d. A maximum of absorption appears at 100.7eV in agreement with the self-absorption spectrum. The position of the absorption edges, together with the results obtained by other investigators, are listed in Table I. The data reported by Metzger are in good agreement with those of Ershov and Lukirskii (1967). Comparing the values of the linear absorption coefficient we find good agreement on the low-energy side of the edge, while on the high-energy side the present values are larger by a factor 1.5.

Table I

Position of the L_2 and L_3 absorption edges of amorphous silicon

Author	L_2-edge (eV)	L_3-edge (eV)	ΔE (eV)
Tomboulian and Bedo (1956)	101.4	100.7	0.7
Ershov and Lukirskii (1967)	100.2	99.6	0.6
Metzger (1971)	100.4±0.2	99.8±0.2	0.6

C. The Conduction Band of Crystalline Silicon

As pointed out above the Kβ and $L_{2,3}$ emission bands of crystalline silicon have previously been described and compared with theoretical results; good agreement between the theoretical and experimental results was found. These results, concerning the valence-band structure of crystalline silicon, will therefore only be mentioned here for completeness. For a detailed discussion we refer to Wiech (1968), and Wiech and Zöpf (1971a, 1971b).

The L-absorption spectra enable us to extend the discussion to the conduction band. Figure 2a shows the density of states of crystalline silicon (Stukel *et al* 1971). The density-of-states curve shows several marked peaks. However, the L-absorption curve shows only one peak just above the absorption edge, though the resolution of the spectrometer (0.37eV) was sufficient to resolve two peaks at a distance of about 1eV.

The absence of a second peak might be a consequence of the dipole selection rules; only conduction-band states with s-like or d-like symmetry contribute to the L-absorption. Klima (unpublished communication) has also calculated a density of states of the conduction band, and theoretical K and L absorption spectra, for crystalline silicon. The gross features of the density-of-states curve are similar to those found by Stukel *et al*. According to Klima's calculations the first peak in the density of states is predominantly s-like and is therefore observed in the L-absorption spectrum. The second peak is predominantly p-like but should nonetheless be present in the L-absorption, although with reduced intensity. Apparently its intensity is lower than predicted by theory.

D. The Band Structure of Amorphous Silicon

From the results presented in figures 1 and 2 we can draw some conclusions concerning the band structure of amorphous silicon. The most important result is that the gross features of the emission bands, and of the abosrption spectra, are similar for the crystalline and the amorphous states. This means that the band structure and the density of states for crystalline and amorphous silicon must show the same gross features. Break-down of the long-range ordering does not change the band structure completely. Long-range periodicity is therefore not responsible for the main features of the electronic structure of silicon. Since in amorphous silicon only short-range spacial ordering exists, and predominantly tetrahedrally arranged interatomic bonds are generally assumed, the electronic structure in the crystalline phase is essentially achieved also by special bonding of neighbouring atoms, and not so much by the periodic arrangement of the atoms. On the other hand, the similarity of the emission bands of amorphous and crystalline silicon strongly supports the assumption that the strong tetrahedral bonding orbitals of the crystalline phase are preserved in the amorphous state. The detailed structure in the spectra of amorphous silicon, compared with that of crystalline silicon, is smeared out. This is especially evident in the emission spectra, and not so much in the absorption spectra. Since this smearing-out is observed also for the marked structure in the middle and the lower part of the emission bands, we may conclude that the break-down of long-range order likewise affects high and low lying bands. This means that all van-Hove singularities, and the whole of the fine-structure in the density of states, is smeared out or disappears. In comparison with the crystalline phase the high-energy edges of the K and L emission bands of amorphous silicon are shifted to higher energies by some tenths of an electron volt; and the L-absorption edge - together with fine structure to the low and high energy sides - is shifted to lower energies by about the same amount.

Because only the p-like states of the valence band contribute to the K-emission, and only the s-like and d-like states contribute to the L-emission, the s-like and p-like states increase in the upper part and near the top of the valence band. On the other hand, s-like states of the conduction band extend to lower energies (no data are available to-date for p-like states). Whether this behaviour near the band gap is due to a steeper rise in the density of states or to an extension of states into the band gap is still an open question. Measurements performed with higher resolution may provide an answer. Following the model of Cohen (1969, 1971), and of Mott (1969), we should expect localized states throughout the band gap in the case of amorphous silicon, and consequently an extension of the density of states into the band gap. But it is questionable whether the density of such localized states is large enough to give rise to the observed effects.

Acknowledgments

The authors are greatly indebted to E. Döring (Forschungslabor, Siemens AG, München) for preparing samples of amorphous silicon, and to G. Melchart for his assistance during the course of the measurements.

REFERENCES

Brodsky, M.H., Title, R.S., Weiser, K. and Pettit, G.D. (1970); Phys. Rev. B1, 2632.

Cohen, M.H., Fritzsche, H. and Ovshinsky, S.R. (1969); Phys. Rev. Lett., 22, 1065.

Cohen, M.H. (1971); Physics Today, May 1971.

Ershov, O.A. and Lukirskii, A.P. (1967); Sov. Phys. Solid State 8, 1699.

Hoffmann, L., Wiech, G. and Zöpf, E. (1969); Z. Physik, 229, 131.

Klima, J. (1970); J. Phys. C - Solid State Physics, 3, 70.

Krämer, H. (1960); Thesis, Universität München.

Läuger, K. (1968); Thesis, Universität München.

Liefeld, R.J. (1968); in "Soft X-Ray Band Spectra", p.133; ed. Fabian, D.J.; Academic Press London.

Metzger, H. (1971); Diplomarbeit, Universität München.

Mott, N.F. (1969); in Festkörperprobleme, Vol. 9, p.22; Vieweg, Braunschweig.

Stukel, D.J., Collins, T.C. and Euwema, R.N. (1971); in Proc. 3rd IMR Symposium Electronic Density of States, Nat. Bur. Stand. (U.S.), Spec. Publ. 323, p.93, ed. Bennett, L.

Tomboulian, D.H. and Bedo, D.E. (1956); Phys. Rev. 104, 590.

Wellman, P. (1971); Diplomarbeit, Universität München.

Wiech, G. (1964); Thesis, Universität München.

Wiech, G. (1966); Z. Physik, 193, 490.

Wiech, G. (1968); in "Soft X-Ray Band Spectra", p.59; ed. Fabian, D.J.; Academic Press, London.

Wiech, G. and Zöpf, E. (1971a); in Proc. 3rd IMR Symposium Electronic Density of States; Nat. Bur. Stand. (U.S.), Spec. Publ. 323, p.335, ed. Bennett, L.

Wiech, G. and Zöpf, E. (1971b); Journal de Physique, 32, C4-200.

ON THE TEMPERATURE DEPENDENCE OF SOFT X-RAY SPECTRA AND KNIGHT SHIFT IN LIQUID METALS AND THEIR ALLOYS

B.L. Gyorffy

H.H. Wills Physics Laboratory, University of Bristol, Bristol

1. INTRODUCTION

As was argued elsewhere (Gyorffy and Stott, this volume, p385) calculation of the soft x-ray spectra, Knight shift and Mossbauer isomer shift of metallic systems, in the one-electron approximation, can be useful for understanding the gross features of these measurements even although the electron-electron interaction is neglected. Our purpose here is to present an approximation scheme that will allow such calculations for liquid metals and their alloys. The method is designed to work in the nearly-free-electron limit and predicts that the three quantities mentioned depend on temperature through the various partial structure factors that describe the two-particle correlations in the liquid.

The most general system we shall consider is a liquid binary alloy. Thus, in the one-electron picture, we must study the states of an electron in the alloy potential

$$V(\mathbf{r}) = \sum_i v_i\ (\mathbf{r} - \mathbf{R}_i) \qquad (1)$$

where $\mathbf{R}_i$ is the position of the ion i and v_i is the corresponding scattering potential. A convenient way of parameterizing this potential function is to write

$$v_i(\mathbf{r} - \mathbf{R}_i) = \xi_i\ v^A(\mathbf{r} - \mathbf{R}_i) + (1 - \xi_i)v^B(\mathbf{r} - \mathbf{R}_i) \quad (2)$$

where $\xi_i = 1$ when the scatterer is an A-type atom described by the potential $v^A(\mathbf{r} - \mathbf{R}_i)$, and $\xi_i = 0$ when it is B-type with a corresponding potential $v^B(\mathbf{r} - \mathbf{R}_i)$. Hence, to specify the one-electron Hamiltonian we must assign particular values to all the parameters $\xi_1, \xi_2, \xi_3 \ldots \xi_N$ and to the position coordinates $\mathbf{R}_1, \mathbf{R}_2, \ldots \mathbf{R}_N$.

However, experiments measure ensemble averages, and therefore we must treat these parameters as random variables. The accepted procedure is to calculate a given physical observable for a given configuration and average it over all configurations compatible with the measurement of that observable (Edwards 1958). In the present case of a liquid binary alloy the weight to be attributed to each configuration is given by the N-particle alloy distribution function $f^N(\xi_1\mathbf{R}_1, \xi_2\mathbf{R}_2, \ldots \xi_N\mathbf{R}_N)$ which describes all the static correlations of the liquid.

For instance, it was shown by Gyorffy and Stott (1971), that the soft x-ray emission intensity from an A atom at $\mathbf{R}$ in a binary liquid alloy may be written as

$$I^A(\varepsilon) \alpha \int dr^3 \int dr'^3 \phi_c^{A*}(\mathbf{r} - \mathbf{R}) \phi_c^A(\mathbf{r}' - \mathbf{R}) \nabla . \nabla' \left\langle\!\left\langle r | Im\ G^+(\varepsilon) | r' \right\rangle\!\right\rangle_{A,\mathbf{R}} \tag{3}$$

where $\phi_c^A(\mathbf{r})$ is the wavefunction of the empty core state, ∇ and ∇' occur because of the dipole approximation, $G^+(\varepsilon)$ is the Green-function operator for an electron moving in the alloy potential $V(\mathbf{r})$ and $\langle\ \rangle_{A,R}$ denotes averaging overall configurations for which there is an A atom at $\mathbf{R}$.

Similarly, the contact density P_F, which enters into the Knight-shift formula, is given by

$$P_F = -\frac{1}{\pi} \left\langle\!\left\langle \mathbf{r} | Im\ G^+(\varepsilon_F) | \mathbf{r} \right\rangle\!\right\rangle_{A,R} \Big|_{\mathbf{r} = \mathbf{R}} \tag{4}$$

Hence our task is to construct a theory for the partially averaged Green function $\left\langle\!\left\langle \mathbf{r} | Im\ G^+(\varepsilon) | \mathbf{r}' \right\rangle\!\right\rangle_{A,\mathbf{R}}$.

2. THE MODEL HAMILTONIAN

Experience with band structure calculations suggests that a fairly realistic crystal potential may be constructed in the form of non-overlapping muffin tins. We shall therefore assume, without much loss of generality, that this can be used for a liquid alloy and take $v^A(\mathbf{r})$ and $v^B(\mathbf{r})$ to be zero for $|\mathbf{r}|\rangle a$ where a is the radius of the muffin-tin well. This assumption simplifies matters enormously. Gyorffy and Stott (this volume, p) have shown for $\mathbf{r}$ and $\mathbf{r}'$ inside the muffin-tin well of the n -th atom at R_n,

$$\langle \mathbf{r}| Im\ G^+(\varepsilon)|\mathbf{r}'\rangle = \sum_{LL'} \Delta_L^A(\mathbf{r}_n)\Delta_{L'}^A(\mathbf{r}_n')\ Im\ \mathcal{T}_{LL'}^{nn}(\varepsilon) \tag{5}$$

In this equation

$$\Delta_L^A(\mathbf{r}_n) \equiv \Delta_L^A(\mathbf{r} - \mathbf{R}_n) = -\frac{\sqrt{\varepsilon}}{\sin\,\delta_L^A(\varepsilon)}\,R_L^A(|\mathbf{r} - \mathbf{R}_n|)Y_L(\widehat{\mathbf{r} - \mathbf{R}_n}) \tag{6}$$

where $\delta_L^A(\varepsilon)$ is the scattering phase-shift characteristic of the potential function $v^A(\mathbf{r}_n)$, $R_L^A(r_n)$ is the corresponding solution of the radial wave equation, with the boundary condition that for $|\mathbf{r}_n| > a$

$$R_L^A(|\mathbf{r}_n|) = \cos\delta_L^A(\varepsilon)\ j_L\ (\sqrt{\varepsilon}|\mathbf{r}_n|) - \sin\delta_L^A(\varepsilon)\ n_L\ (\sqrt{\varepsilon}|\mathbf{r}_n|); \tag{7}$$

and the $Y_L(\mathbf{r}\text{-}\mathbf{R}_n)$ are the real spherical harmonics for angular momentum L (ℓ,m). Furthermore, $\mathcal{T}_{LL'}^{n,n}(\varepsilon)$ is the generalization of the "on the energy shell matrix element" of a single scatterer t-matrix

$$t_L^\ell(\varepsilon) = -\frac{\sin\delta_L^\ell(\varepsilon)}{\sqrt{\varepsilon}}\,e^{\,i\delta_L^\ell(\varepsilon)} \tag{8}$$

to many non-overlapping muffin-tin potentials. It satisfies the multiple scattering equation (Gyorffy and Stott, this volume, p385)

$$\mathcal{T}_{LL'}^{\ell\ell'}(\varepsilon) = t_L^\ell(\varepsilon)\delta_{\ell\ell'}\delta_{LL'} + \sum_{\ell''\neq\ell}\sum_{L''} t_L^\ell(\varepsilon)G_{LL''}(\mathbf{R}_\ell-\mathbf{R}_{\ell''})\mathcal{T}_{L''L'}^{\ell''\ell'}(\varepsilon) \tag{9}$$

where, in the present case of a liquid binary alloy,

$$t_L^{\ell}(\varepsilon) = \xi_\ell \; t_L^A(\varepsilon) + (1 - \xi_\ell) \; t_L^B(\varepsilon) \tag{10}$$

and

$$G_{LL'}(\mathbf{R}_\ell - \mathbf{R}_{\ell'}) = \frac{2}{\pi} \int d^3k \; \frac{e^{i\mathbf{k}.(\mathbf{R}_\ell - \mathbf{R}_{\ell'})}}{\varepsilon - k^2 + i\eta} \; Y_L(\hat{k})Y_{L'}(\hat{k}) \tag{11}$$

Note that equation (5) does not give the Green function everywhere. It is valid only inside the muffin-tin well about $\mathbf{R}_n$. However, since the core state $\phi_c^A(\mathbf{r}-\mathbf{R}_n)$ is very well localized, it follows from equation (3) that the information provided in equation (5) is sufficient to calculate the soft x-ray spectra due to a core hole at $\mathbf{R}_n$ for a particular alloy configuration. Thus the emission intensity is determined by the parameters $\xi_1, \xi_2, \xi_3 \cdots \xi_N$, the position vectors $\mathbf{R}_1, \mathbf{R}_2, \ldots \mathbf{R}_N$, the solution of the Schrödinger equation inside the muffin-tin well surrounding the radiating atom and the phase shifts $\delta_L^A(\varepsilon), \delta_L^B(\varepsilon)$. It can be seen from equation (4) that similar remarks apply to the Knight-shift contact density P_F^A.

Thus, a realistic theory of both these effects would be obtained if, on iterating equation (9) to find $\mathcal{T}_{LL'}^{nn}(\varepsilon)$, we could carry out the averaging with respect to $f^N(\xi_1 R_1; \xi_2 R_2 \ldots \xi_N R_N)$. However, this is an impossible task since even in the most favourable cases, only the respective pair correlation functions are known. Nevertheless, equations (5) and (9) may be used as the starting point for a small-phase-shift approximation. This we shall examine in the next section.

3. THE NEARLY-FREE-ELECTRON LIMIT

As mentioned above, one way of calculating $\mathcal{T}_{LL'}^{nn}(\varepsilon)$ is to iterate equation (9). The result is a series for $\mathcal{T}_{LL'}^{nn}(\varepsilon)$, whose terms are of increasing order in $t_L^{\ell}(\varepsilon)$. When the phase-shifts are small, $\mathcal{T}_{LL'}^{nn}(\varepsilon)$ may be approximated by the first few terms in this expansion. Note that there will be an $\sqrt{\varepsilon}$ term coming from each $G_{LL'}(\mathbf{R}_\ell - \mathbf{R}_{\ell'})$ in the expansion and this will take care of the $\frac{1}{\sqrt{\varepsilon}}$ singularity in $t_L^\ell(\varepsilon)$. Taking the expansion to the cubic term only, the averaging discussed in the previous section can be trivially performed in terms of the various pair correlation functions (Ziman and Faber 1965). After some obvious rearrangement of terms one obtains

$$\langle\langle \mathbf{r}|\mathit{Im}\ G^+(\varepsilon)|r'\rangle\rangle_{\mathbf{AR}} = \sum_{LL'} Y_L(\hat{r})R_L^A(r)R_{L'}^A(r')Y_L(\hat{r}') \times$$

$$\Big[\sqrt{\varepsilon}\ \delta_{LL'} + \mathit{Im}\Big\{e^{i\delta_L^A(\varepsilon)} \sum_{L''} \int d^3R\ g^{AA}(\mathbf{R})G_{LL''}(\mathbf{R})t_{L''}^A(\varepsilon)G_{L''L'}(-\mathbf{R}) \times e^{i\delta_{L'}^A(\varepsilon)}\Big\} + \tag{12}$$

$$\mathit{Im}\Big\{e^{i\delta_L^A(\varepsilon)} \sum_{L''} \int d^3R\ g^{AB}(\mathbf{R})G_{LL''}(\mathbf{R})t_{L''}^B(\varepsilon)G_{L''L'}(-\mathbf{R})e^{i\delta_{L'}^A(\varepsilon)}\Big\}\Big]$$

where $g^{AA}(\mathbf{R})$ and $g^{AB}(\mathbf{R})$ are the appropriate partial radial distribution functions.

This is our central result, and we proceed now to discuss its consequences.

Equation (12) can be readily evaluated. Given the potentials v^A and v^B it is a straightforward numerical task to integrate the corresponding Schrödinger equation and obtain $R_L^A(r)$, $R_L^B(r)$, $\delta_L^A(\varepsilon)$ and $\delta_L^B(\varepsilon)$. The only additional information needed to calculate $\langle\langle \mathbf{r}|\mathit{Im}\ G^+(\varepsilon)|\mathbf{r}'\rangle\rangle_{RA}$ is the partial radial distribution functions $g^{AA}(R)$, $g^{AB}(R)$. These are sometimes available from x-ray diffraction or neutron scattering experiments. They can be generated also, to the accuracy that this kind of calculation is likely to require, from the solutions of the Percus-Yevick equations for a mixture of hard spheres (see Ashcroft and Langreth 1967). Clearly, due to the presence of these partial radial distribution functions, the Green function obtained from them will be both temperature and concentration dependent.

The validity of our expansion is controlled by the smallness of $\sqrt{\varepsilon}|t_L^{\ell}(\varepsilon)|$. At first sight this implies that the corresponding potentials v_{ℓ}, are weak; hence the reference to nearly-free electrons. However, it is clear from equation (8) that the magnitude of $\sqrt{\varepsilon}|t_L^{\ell}|$ is governed by $\sin\delta_L^{\ell}(\varepsilon)$. Therefore, all that is required of the potential v_l is to produce phase-shifts that are close to $n\pi$ where n is an integer. That is to say, the potential may be very strong; it may have several bound states and distort the incident plane wave beyond recognition in the vicinity of the core, but as long as the phase shift is small, apart from additive factors of $n\pi$, the above nearly-free-electron approximation will hold. The physical meaning of our approximation is that we treat the potential about the site of interest exactly while assuming that the scattering caused by the other atoms is weak. As the other atoms may influence only the wavefunction at the selected site by their scattering properties $t_L^{\ell}(\varepsilon)$, this effect can be treated only to the first order.

By now it should be evident that our approach is more or less equivalent to a pseudo-potential calculation. The major difference is that the usual pseudo-potential (Philips and Keinmon, 1959; Heine and Abarenkov, 1965; Animalu and Heine, 1965) unlike the muffin-tin potential, has a long-range tail. This means that it can describe an interstitial potential which fluctuates about its mean as the configuration changes. Since the conductivity depends on the fluctuations of the total t-matrix about its average value, a pseudo-potential is better suited to the calculation of conductivity than a muffin-tin potential, which gives a non-fluctuating interstitial potential (Evans, unpubld. communication). However, the muffin-tin model might work equally well for a quantity such as the partially averaged Green function which is related more closely to the average t-matrix than to its fluctuations. Furthermore, preliminary calculations for sodium (Jewsbury, unpublished communication) show that the phase-shifts for a self-consistent muffin-tin potential are a good deal smaller than the phase-shift associated with a pseudo-potential that produces good results for the conductivity. In fact, the smallness of phase-shifts involved in our work ensures that we have a well-controlled approximation scheme that is convergent. However, this is not the case with the pseudo-potentials that give large corrections when used in a higher-order approximation (Ashcroft and Schaich, 1970). Consequently, we may expect that for the physical properties considered here a muffin-tin model potential will give more reliable results.

Further comments are best made for the various physical quantities separately.

4. KNIGHT SHIFT

Combining equation (4) and equation (12) gives

$$P_F^A = \frac{1}{4\pi^2} \; R_o^A(0,\varepsilon_F)|^2\sqrt{\varepsilon}\Big[1 + 16\pi^2\, Im\{e^{i2\delta_o^A(\varepsilon)} \times \sum_L \left(t_L^A(\varepsilon_F)g_L^{AA}(\varepsilon_F) + t_L^B(\varepsilon_F)g_L^{AB}(\varepsilon_F)\right)\}\Big] \qquad (13)$$

where

$$g_\ell^{AA}(\varepsilon_F) = \int d^3R\; g^{AA}(R)\; h_\ell^{(+)}(\sqrt{\varepsilon_F}R), \qquad (14)$$

$h_\ell^{(+)}(\sqrt{\varepsilon_F}R)$ is the spherical Hankel function and $g_L^{AB}(\varepsilon_F)$ is given by an expression entirely analogous to (14).

Thus the calculation of the contact density P_F^A would consist of:

a) Constructing the muffin tin potentials $v^A(\mathbf{r})$ and $v^B(\mathbf{r})$. Unfortunately the present theory gives no guidance on the potential to be used. However, it is reasonable to follow the Mattheiss prescription usually employed in band structure calculations (Mattheiss, 1964).

b) Calculation of the phase-shifts $\delta_L^A(\varepsilon_F)$, $\delta_L^B(\varepsilon_F)$ and numerically integrating the Schrödinger equation inside the muffin-tin for $v^A(\mathbf{r})$ to give $R_0^A(0,\varepsilon_F)$.

c) Evaluation finally of the $g_L^{AA}(\varepsilon_F)$ and $g_L^{AB}(\varepsilon_F)$. Here one may use either the partial distribution function obtained experimentally (Orton *et al*, 1960; North *et al*, 1968; Greenfield Wellendorf, to be published), or follow Ashcroft and Langreth (1967) and use the solutions of the Percus Yevic equations for a binary mixture of hard spheres with suitably adjusted packing fractions.

For comparison with other work on Knight shift, consider a liquid of A atoms with a small amount of B impurities. Let us assume that the Fermi energy, ε_F, and the radial wavefunction, $R_0^A(0,\varepsilon_F)$, do not change on alloying. Then the fractional change on alloying may be calculated as follows

$$\frac{\Delta P_F^A}{P_F} = \frac{P_F^A(\text{alloy}) - P_F^A(\text{pure metal})}{P_F^A(\text{pure metal})} = \frac{Im\, e^{i2\delta_0^A} \sum_L g_L^{AB}(\varepsilon_F) t_L^B(\varepsilon_F)}{1 + Im\, e^{i2\delta_0^A} \sum_L g_L^{AA}(\varepsilon_F) t_L^A(\varepsilon_F)} \tag{15}$$

where it has been assumed the $g_L^{AA}(\varepsilon)$ also remains unchanged. To lowest orders in the phase shifts, this expression reduces to

$$\frac{\Delta P_F^A}{P_F} = \frac{1}{\sqrt{\varepsilon_F}} \sum_\ell (2\ell+1) \int dR^3 \left\{ \left[j_\ell^2(\sqrt{\varepsilon_F}R) - n_\ell^2(\sqrt{\varepsilon_F}R) \right] \sin^2\delta_\ell^B(\varepsilon_F) + j_\ell(\sqrt{\varepsilon_F}R)\, n_\ell(\sqrt{\varepsilon_F}R) \sin 2\delta_\ell^B(\varepsilon_F) \right\} g^{AB}(R) \tag{16}$$

where we have used the identity $h_1^+ = \frac{1}{2}(j_\ell + in_\ell)$, with j_ℓ and n_ℓ representing the spherical Bessel and Neuman functions respectively.

Equation (16) is the one used by Odle and Flynn (1966) to

discuss Knight-shifts in copper alloys. They obtained it by generalizing a theory for pure metals developed by Daniel and Blandin (1959). As this theory is based on the idea of local screening of impurity charges we see that the present theory properly includes an account of the Friedel oscillations. In fact it includes much more. When the Fermi energy is allowed to change the simple forms of equations (15) and (16) do not obtain even if the radial distribution function remains unaltered. The changes in the wavefunction $R_0^A(0,\varepsilon_F)$, due to changes in the Fermi energy on alloying, are presumably the 'rigid band' effects discussed by Watson, Bennett and Freeman (1968). Unfortunately, since they were working with OPW-wavefunctions, it is very difficult to make a detailed comparison.

Theories starting from a nearly-free-electron picture, and using pseudo-potentials, have been in existence for a long time. The first formula was derived by Edwards (1962) for weak potentials. Heine (1957) used pseudo-potentials explicitly. More recently Ashcroft and Scheich (1970) derived an expression that purports to be the best one can do within the context of the nearly-free-electron picture. Based on the pseudo-potential concept a number of calculations have been performed by Watanabe *et al* (1965), Halder (1970), Jena *et al* (1971) and Perdew and Wilkins (1970) for various liquid metals and alloys. In these calculations the pseudo-wavefunction is calculated to first or second order in the pseudo-potential (Setty and Mungruwade, 1971) and then othagonalized to the core states to yield the real wavefunctions. Frequently, the agreement with experiment is good. These theories are the same as the one represented by equations (13) and (14) in the sense that the scattering properties of far-away atoms are described by a weakly-scattering-model potential whose scattering matrix is evaluated in the first Born approximation. For small matrix elements this should give the same answer as the full scattering matrix. However, by integrating the Schrödinger equation inside the muffin tin we believe we are doing better than othogonalization of the pseudo-wave-function can be expected to do. However, proof of this assertion must come from numerical evaluation of equation (13). We hope to report such calculations in the near future.

5. MOSSBAUER: ISOMER SHIFT

A calculation most closely resembling the one advocated here has been presented by Ingelsfield (1970). He integrated the Schrödinger equation exactly inside the core of the atom whose nucleus was emitting the Mossbauer γ-ray photon and matched this wavefunction with the pseudo-wavefunction at the core radius. The pseudo-wavefunction was calculated to first order in the pseudo-potentials of the other atoms. By evaluating the wavefunction, obtained in this manner, at the nucleus, he was able to calculate the Mossbauer isomer shift which in our notation is proportional to

$$\int_0^{\varepsilon_F} d\varepsilon \left\langle\left\langle \mathbf{r} \middle| Im\ G^+ \middle| \mathbf{r}' \right\rangle\right\rangle_{A\mathbf{R}} \Big|_{r = R}$$

One would expect that equation (13) would yield very similar results, albeit with much less labour.

6. SOFT X RAY SPECTRA

Using equations (3) and (13) we can easily write down the relevant emission intensities. For K emission, for example we have

$$I_K^A(\varepsilon) \propto \frac{1}{3} \left|\int_0^\infty dr\ r^2\ R_0^{CA}(r) \frac{d}{dr} R_1^A(r,\varepsilon)\right|^2 \Big[\sqrt{\varepsilon} - 4\pi \sum_{\ell''} (2\ell''+1)\ Im\ \{e^{i2\delta_1^A(\varepsilon)}\ [g_{\ell''}^{AA}(\varepsilon) t_{\ell''}^A(\varepsilon) + g_{\ell''}^{AB}(\varepsilon) t_\ell^B(\varepsilon)]\}\Big] \tag{17}$$

The steps involved in evaluating equation (17) are the same as those discussed in connection with the Knight-shift formula equation (13). However, a new feature arises due to the fact that we must evaluate equation (17) for a whole range of energies not only the Fermi-energy ε_F. As mentioned already the soft x-ray spectra provide information about the states of the electron not only at the Fermi-energy but also at energies throughout the band. Therefore, comparison between the present theory and experimental soft x-ray spectra should provide a more stringent test of the nearly-free-electron model, as well as of the whole model potential concept afforded by calculations of P_F^A.

Unfortunately, there is little experimental information available at the moment on soft x-ray spectra of liquid metals and alloys. To our knowledge only the $L_{2,3}$-emission from liquid aluminium (Catterall and Trotter, 1963) has been been measured. It is hoped that the present work will stimulate experimental interest in this field, potentially so important for understanding electronic states in such disordered systems as liquid metals and alloys.

Theoretical efforts to understand the electronic structure in such systems are also scarce. Harrison (1968) has discussed the possibility of calculating soft x-ray spectra using a nearly-free-electron and pseudo-potential approach. Along similar lines, Ashcroft (1968) has carried out some interesting model calculations. However, to-date there seems to be no calculated emission intensities with realistic parameters for any system in the liquid state.

In conclusion we note that if we could study experimentally the fractional change of emission intensities on alloying; then in the limit, where very small phase-shifts and no change in ε_F occur, a formula similar to equation (16) would apply. In effect we would have

$$\frac{\delta I_K^A}{I_K^A} = \frac{3}{4\pi} \sum_{\ell} (2\ell+1)\int_0^{\infty} dR\ R^2 g^{AB}(R)\ \{\left[j_{\ell}^2(\sqrt{\varepsilon}R)-n_{\ell}^2(\sqrt{\varepsilon}R)\right]\sin^2 {}_{L}^{B}(\varepsilon)$$

$$+ j_{\ell}(\sqrt{\varepsilon}R)n_{\ell}(\sqrt{\varepsilon}R)\sin^2\delta_L^B(\varepsilon)\} \qquad (18)$$

In view of the success that Odle and Flynn had with equation (16) for the solvent Knight-shift, evaluating this formula for the same systems might be a reasonable first effort in calculating soft x-ray intensities for liquid alloys when experimental results become available.

REFERENCES

Animalu, A.O.E. and Heine, V. (1965); Phil. Mag., 12, 1249.

Ashcroft, N.W. and Langreth, D.C. (1967); Phys. Rev., 156,685.

Ashcroft, N.W. (1968); Soft x-ray Spectra, ed. D.J. Fabian, p.249.

Ashcroft, N.W. and Schaich, W. (1970); Phys. Rev. B1, 1370.

Blandin, A. and Daniel, E. (1959); Phys. Chem. Solids 10, 126.

Catterall, J.A. and Trotter, J. (1963); Phil. Mag. 8, 897.

Edwards, S. (1958); Phil. Mag. 3, 1021.

Edwards, S. (1962); Proc. Roy. Soc. A267, 518.

Evans, R. (1971); unpublished.

Greenfield, A.J. and Wellendorf J., J. Phys. C. (in press).

Gyorffy, B.L. and Stott, M.J. (1971); Sol. Stat. Comm. 9, 613-17.

Gyorffy, B.L. and Stott, M.J., present volume.

Harrison, W.A. (1968); Soft x-ray Spectra, ed. D.J. Fabian, p.227.

Halder, N.C. (1969); Pnys. Rev. 177, 471.

Halder, N.C. (1970); J. Chem. Phys. 52, 5450.

Heine, V. (1957); Proc. Roy. Soc. 240, 361.

Heine, V. and Abarenkov, I. (1965); Phil. Mag. 12, 529.

Ingelsfield, J.E. (1970); J. Phys. Chem. 31, 1435.

Jena, P., Das, T.P., Gaspari, G.D. and Halder, N.C. (1971); Phys. Rev. B3, 2158.

Jewsbury, P., to be published.

Mattheiss, L.F. (1964); Phys. Rev. 134, 192.

North, D.H., Enderby, J.E. and Egelstaff, P.A. (1968); J. Phys. C1, 784.

Odle, R.L. and Flynn, C.P. (1966); Phil. Mag. 13, 699.

Orton, B.R., Shaw, B.A. and Williams, G.I. (1960); Act Metolurgica 8, 177.

Perdew, J.P. and Wilkins, J.W. (1970); Sol. Stat. Comm. 8, 2041.

Philips, J.C. and Keinmon (1959); Phys. Rev. 116, 287.

Setty, D.L.R. and Mungruwade, B.D. (1971); Phys. Lett. 35A,11.

Watanabe, M., Tanaka, M., Endo, H. and Jones, B.K. (1965); Phil. Mag. 12, 347.

Watson, R.E., Bennett, L.H. and Freeman, A.J. (1968); Phys. Rev. Lett. 20, 653.

Ziman, J.M. and Faber, T.E. (1965); Phil. Mag. 11, 158.

THE APPLICATION OF MOLECULAR ORBITAL THEORY TO THE INTERPRETATION OF X-RAY EMISSION SPECTRA OF IRON COMPOUNDS

R.A. Slater and D.S. Urch

Department of Chemistry, Queen Mary College, London, England.

1. INTRODUCTION

Simple molecular orbital theory has been used successfully to interpret x-ray emission spectra from a variety of compounds of main group elements (Urch 1970). The purpose of the present paper is to report emission spectra for a selection of iron compounds - iron is chosen as a typical transition element - to see if the molecular orbital theory provides an equally suitable framework for the discussion of the main features of the x-ray emission spectra of a transition metal; in particular to see if changes associated with chemical bond formation can be rationalised. Fischer (1970) has already used this theory to interpret the x-ray emission spectra of titanium and vanadium oxides, borides, carbides etc.; the compounds discussed in this report include a range of complexes as well as simple oxides.

Iron is an element of great interest, not only for its technological applications but because of the wide variety of chemical compounds that it forms and the possibility of widely differing spin states for the ion in a complex. Iron readily forms two series of salts based on the ferrous Fe^{++}, d^6, and the ferric Fe^{+++}, d^5, cations. When surrounded by "weak" ligands whose orbitals only slightly perturb the central iron ion of a complex, then the spin state is that anticipated by a simple application of Hund's rule; i.e. the state of maximum multiplicity. Because there are five d-orbitals there are four unpaired spins in the ferrous cation, and five in the ferric case; also in the ferric cation each d-orbital will contain only one electron, a spherically symmetric allocation of electrons that enhances the stability of ferric high-spin complexes.

Conversely "strong" ligands will interact strongly with the central cation causing the d-orbitals to be divided into two groups of differing energy when the ligands are arranged, as they often are, at the vertices of the octahedron. The more tightly bound set of three orbitals belong to the t_{2g} irreducible representation of the octahedral point group, O_h, the less tightly bound have representation e_g (Orgel 1960). The energy separation between these sets of orbitals is such that in the ferrous case the t_{2g} orbitals are completely occupied and the e_g orbitals are empty. Thus, in strong-field ferrous complexes all the electrons are spin paired. In the corresponding ferric complexes all five d-electrons are in the t_{2g} orbitals, bearing one vacancy; i.e. in strong-field ferric complexes there is just one unpaired electron. Examples of all four possible types of complexes are considered in this report.

Hopefully, a study of iron atoms in relatively simple chemical environments will lead to general structural correlations with x-ray emission spectral features. Such correlations could then be used to increase understanding of bonding of iron atoms in alloys.

2. EXPERIMENTAL

The x-ray emission spectra of the following substances were studied: iron, ferrocene, potassium ferrocyanide, potassium ferricyanide, ferric oxide, ferrosoferric oxide, ferric sulphate, ferric alum, ferrous alum, potassium trisoxalatoferrate(III). Either commercially available samples of *Analar* purity were used or the compounds were prepared by standard techniques. Discs for irradiation were made with a ring press using terephthalic acid as a binder. The irradiations were carried out with a Philips PW-1410 x-ray fluorescence spectrometer, and using Harwell 2000-series counting equipment. A tungsten-anode x-ray tube (operated at 60KV and 45mA) was used for the irradiation. The K radiation was analysed by taking the third order diffraction from a germanium crystal, and detected with a scintillation counter via an auxiliary fine collimator.

3. RESULTS

The results are presented in Table I and typical Kβ spectra are shown in figures 1 and 2. In estimating the intensity of the $K\beta_5$ lines, a smooth 'background' tail has been extrapolated from the $K\beta_{1,3}$ peak to shorter wavelengths below each $K\beta_5$ line.

TABLE I

Sample	$K\beta_5/K\beta_{1,3}$ intensity ratio %	$K\beta'/K\beta_{1,3}$ intensity ratio %
Fe	1.98	<5 (a)
Fe_2O_3	1.44	30 (b)
Fe_3O_4	1.47	30 (b)
$Fe_2(SO_4)_3$	1.73	30 (c)
$K_3\{Fe^{III}(C_2O_4)_3\}$	1.15	25 (c)
Fe^{II} alum	1.35	22 (c)
Fe^{III} alum	1.32	28 (c)
$K_4\{Fe^{II}(CN)_6\}$	3.55	<5 (a)
$K_3\{Fe^{III}(CN)_6\}$	3.43	<5 (a)
$Fe(C_5H_5)_2$	2.09	<5 (a)

(a) no clearly defined Kβ' at all visible.
(b) 'shoulder' visible.
(c) distinct Kβ' peak resolved from $K\beta_{1,3}$.

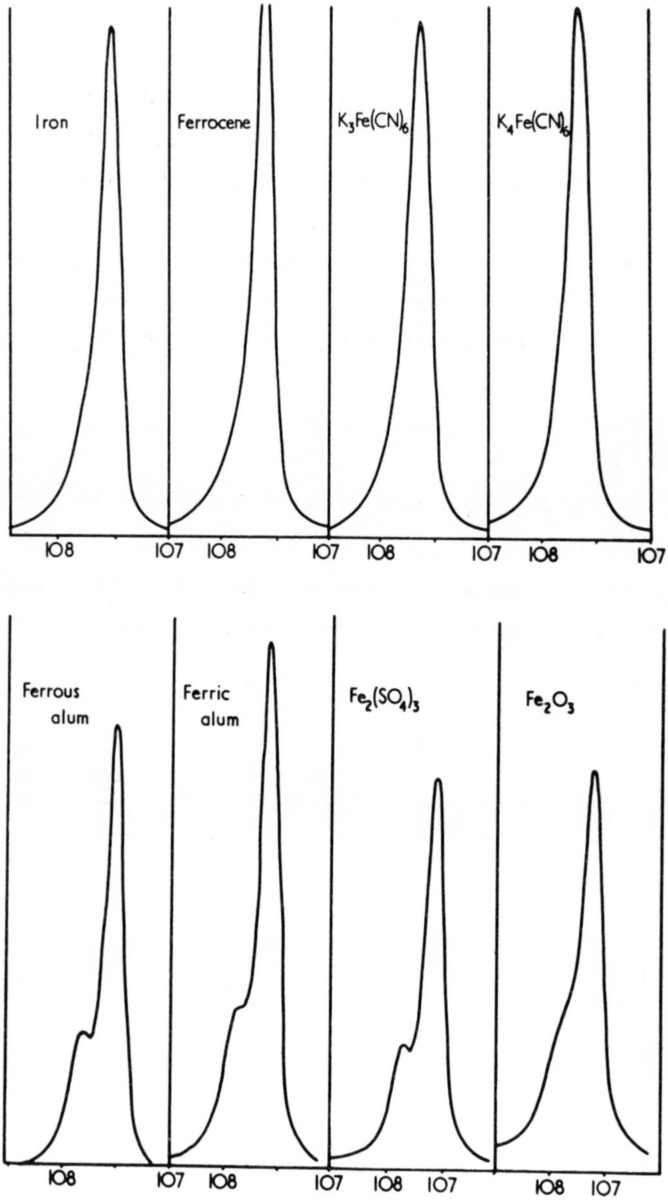

Figure 1 Kβ and Kβ' spectra of iron compounds. Horizontal scale: degrees 2θ (3rd order, germanium) (107°≡ 7062 eV, 108°≡ 7013 eV).

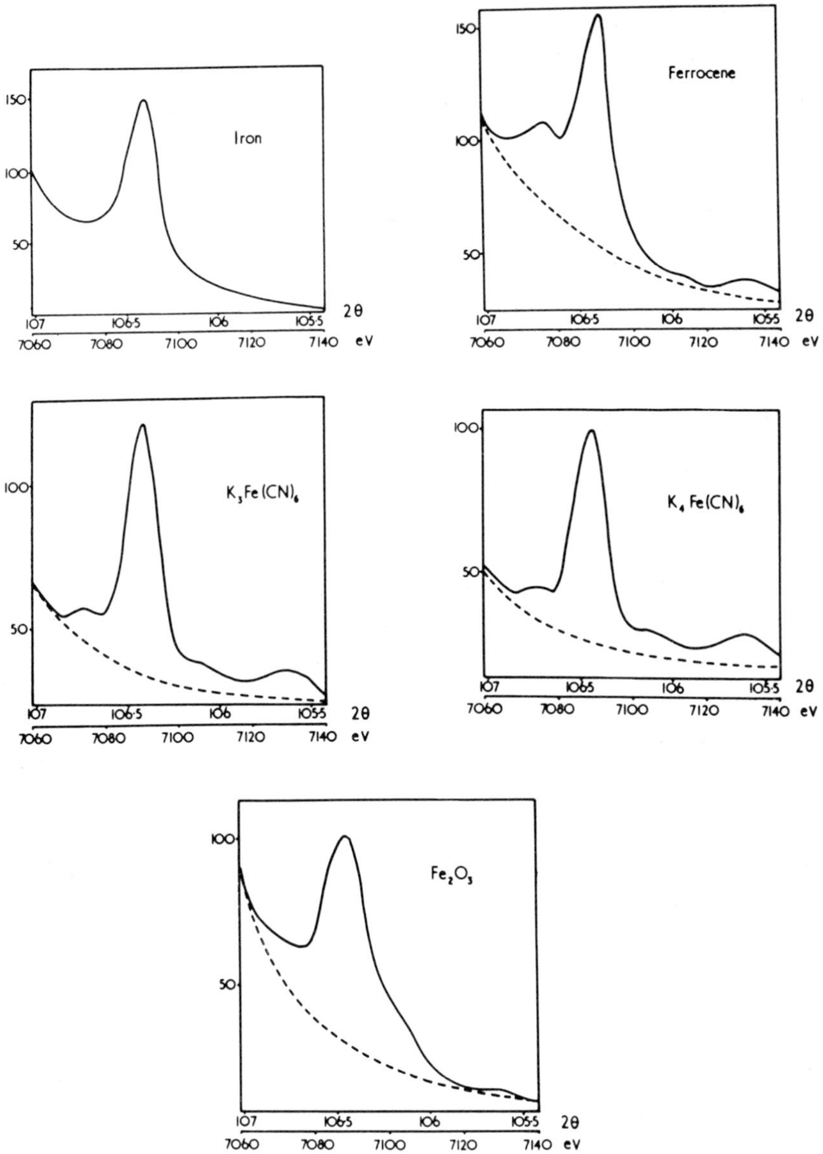

Figure 2 Typical spectra, in the $K\beta_5$ region. Horizontal scale: degrees 2θ (3rd order, germanium), and electron volts. Vertical scale: counts, $secs^{-1}$.

A striking feature is the nature of the $K\beta_{1,3}$ peak which for many complexes exhibits a clearly defined low-energy satellite $K\beta'$. Equally dramatic is the absence of this feature from other spectra. It is of interest to note that the lack of $K\beta'$ is associated with a relatively intense $K\beta_5$ peak.

4. M.O. MODEL AND DISCUSSION

A. Bonding

A qualitative molecular orbital energy level diagram for six ligand atoms octahedrally situated about a central atom is shown in figure 3.

A rough indication of the relative orbital participation in each m.o. is also given. The orbitals in group A are predominantly ligand in character and may be regarded as filled with ligand electrons. However, note the presence of some central atom character in all these orbitals. In particular those of t_{1u} symmetry will carry 4p character and so electrons in these orbitals will be able to participate in transitions to central atom 1s vacancies: the intensities of such transitions will reflect the amount of 4p character in these orbitals. It is proposed that such transitions are the origin of the $K\beta_5$ feature, rather than a forbidden $3d \rightarrow 1s$ transition. If this explanation is correct the relative intensity of the $K\beta_5$ peak (relative to a peak whose intensity might be expected to be unaffected by bond formation, e.g. $K\beta_{1,3}$) is a direct measure of the degree of covalent bonding between the ligand atoms and the central atom. This is indeed observed. The highest $K\beta_5:K\beta_{1,3}$ intensity ratios are found for the cyanide complexes; much lower ratios characterise ligand interactions with the central atom through oxygen. Cyanide groups are 'strong' ligands. The loosely bound orbital of CN^- interacts much more with the central atom than does the rather more tightly bound oxygen orbital: the oxygen bound ligands are all 'weak'.

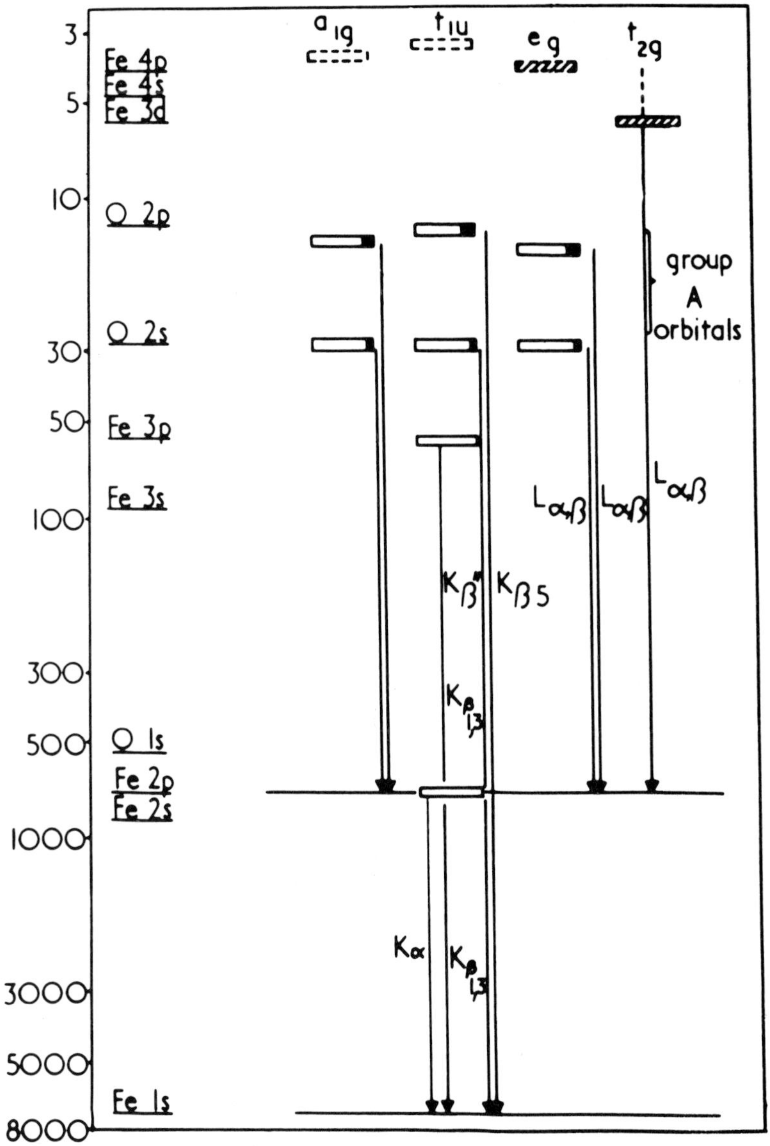

Figure 3 Approximate energy-level diagram for an octahedral iron complex. Only emission transitions for iron have been shown. Occupancy of the 3d orbitals depend upon valence and spin state, indicated by hatching. Vacant orbitals indicated by dotted lines. Iron atomic orbital participation in molecular orbitals shown in black.

It is tempting to ascribe some meaning to the variations in the $K\beta_5$:$K\beta_{1,3}$ intensity ratio for the various oxygen ligands. The ratio for the ligand H_2O (in the alums) is less than that for O^{--} (oxides) because of the inductive effect of the hydrogens in water; this causes the lone pairs of oxygen in water to be more tightly bound, and less available for covalent bond formation with iron. The oxalate-complex has the lowest intensity due to π-stabilisation within the ligand itself; the ratio is highest for a ferric sulphate because of the less efficient π-delocalization that obtains in the sulphate. Thus, electrons are respectively less available and more available for some degree of covalent bonding with the central atom.

B. Spin-State and X-ray Photoelectron Spectra

It is clear from the results presented in Table I that the structure of the $K\beta_{1,3}$ peak is directly related to the spin-state of the central atom and in no way connected with the formal valence state (ferrous or ferric). The spin-state of the atom is a function of the field generated by the ligands. Weak ligands, characterised by a relatively weak $K\beta_5$ peak, perturb the d orbitals only slightly so that a high-spin complex results. Similarly, strong ligands (relatively intense $K\beta_5$) give rise to low-spin complexes. The final state, after a $K\beta$ photon has been emitted, is an ion with a 3p hole. It is entirely reasonable that strong spin-orbit coupling should exist between a 3p vacancy and unpaired electron spins in 3d orbitals, as suggested by Tsutsumi and Obashi (1969) in their discussion of the $K\beta'$ line in chromium complexes. Thus, strong rather complex coupling, with the generation of spectroscopic states of quite different energies, is to be anticipated for high-spin complexes and little or no coupling for low-spin complexes. This of course is in complete agreement with observation; when the $K\beta_5$ is relatively strong, there is no $K\beta'$ and vice-versa.

The final state of the central atom, after Kβ emission, is the same as that for an XPS experiment after the direct removal of a 3p electron. The hypothesis advanced above can therefore be tested by a direct appeal to XPS data. Unfortunately such data exist for only two iron compounds, FeF_3 and $\{Fe(CN)_6\}^{4-}$ (Fadley and Shirley 1970). The former is a weak-field ferric complex and the XPS 3p-ionisation spectrum shows three peaks (two weak, one strong) over an energy range of ∿10-15 ev. While the energy separation is in accord with that observed for the Kβ-Kβ' separation the relative intensities are quite different; however, this is to be expected since peak intensities in XPS and x-ray fluorescence spectroscopy are governed by entirely different mechanisms. The XPS 3p-spectrum of the ferrocyanide complex shows only one line, as expected for a diamagnetic complex. These XPS results support the hypothesis that a broad Kβ'-$K\beta_{1,3}$ feature is caused by spin-orbit coupling in high-spin complexes.

C. Correlation with Mössbauer Results

It is reasonable to suppose that the degree of covalency measured by the relative intensity of the $K\beta_5$ peak should reflect the covalency to be found in the a_{1g} and e_g orbitals of group A. This proposal can be checked by an investigation of the lower-energy satellites in the L spectrum, which is being carried out at the present time (some preliminary spectra are shown in figure 4). If the idea is acceptable then the relative intensity of $K\beta_5$ can be related to the degree of 4s-participation in a_{1g} A orbitals. The presence of electronic charge in 4s orbitals will have a large effect upon the Mössbauer spectrum. Unfortunately this effect will be manifest in two ways: as a direct effect to increase the s-electron density at the iron nucleus, thereby causing a shift to low or negative δ values, and as an indirect effect that results from the reduction of local effective positive charge

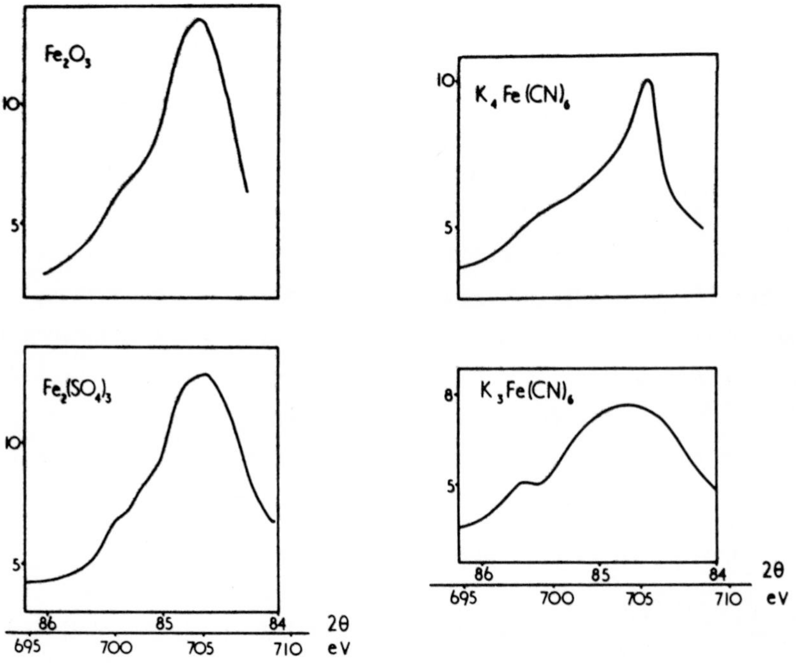

Figure 4 L_α spectra. Rubidium acid-phthalate used as analysing crystal. Vertical scale: counts, $secs^{-1}$.

at the iron atom, as a result of covalent bond formation. This decrease in charge about the iron atom will cause a slight expansion of the inner orbitals and so reduce the s-electron density causing a positive δ-shift.

The correlation, such as it is, established between the $K\beta_5$ intensity and the Mössbauer δ-shift values is shown in Table II (data taken from the compilation by Goldanskii 1964). The general qualitative conclusion is that the shift to low and negative δ values, exhibited by carbon ligands, is due to some degree of covalent bond formation involving 3d, 4p and also 4s orbitals; since low values of δ are associated with relatively intense $K\beta_5$ peaks and vice-versa.

TABLE II

Compound	Formal valence state of iron	Mössbauer shift from iron in stainless steel mm. sec^{-1}	Relative intensity of $K\beta_5$ (from Table I)
$K_4Fe(CN)_6$	II	+0.02	3.55
$K_3Fe(CN)_6$	III	-0.05	3.43
Ferrocene	II	+0.46	2.09
Fe_2O_3	III	+0.50	1.44
$Fe_2(SO_4)_3$	III	+0.53	1.73
Ferrous alum	II	+1.29	1.35
Ferric alum	III	+0.52	1.32

D. Ligand Identification

A study of $K\beta'$ satellite peaks in compounds of second-row main group elements has shown that the energy difference between the $K\beta'$ and $K\beta$ peaks can be related to the ligand atom (Urch 1971). The corresponding features in first-row transition metal complexes are $K\beta_5$ ($4p \rightarrow 1s$) and the low-energy satellite designated $K\beta''$. Extensive work in the USSR (Nemnonov *et al* 1969) has established a relationship that connects the $K\beta''$-$K\beta_5$ energy difference with the type of ligand. Unfortunately, such clear-cut evidence is not found in the results presented here. This is possibly due to the poor resolution of the various peaks involved from the high-energy 'tail' of the $K\beta_{1,3}$ peak. It is apparent that a hole in a molecular orbital with some 4p character will be affected by the spin state of the atom, but it is difficult to explain the energy separation of about 15eV observed between Kβ" and $K\beta_5$ for potassium ferrocyanide. In general compounds with carbon ligands have Kβ"-$K\beta_5$ separations of about 12-14eV and those with oxygen ligands about 20eV. These values are much larger than those observed by Nemnonov *et al* (1969) or by Urch (1970) for second-row compounds.

E. Other Special Features

The Kβ spectra of all the complexes or compounds, except Fe_3O_4, were characterised by a high-energy peak at 7130eV. This peak was not observed for pure iron, and for Fe_3O_4 it occurs at 7164eV. The intensity is about 1% of the $K\beta_{1,3}$ peak. The origin of the peak is obscure; it may be formed by transition in a double ionised molecule. It is in some way connected with bond formation, and yet is strangely indifferent to the nature of the ligands except in the special case of ferrosoferric oxide.

4. CONCLUSIONS

It is reasonable to conclude that simple molecular orbital theory provides a suitable framework for discussing the x-ray emission spectra of transition metal complexes and compounds. However, because many such complexes and compounds contain unpaired electrons the situation is more complicated than for main-group compounds, the vast majority of which are diamagnetic. Thus the basic molecular orbital energy-level diagram must be interpreted with caution; in many cases the exchange energy associated with spin-orbit coupling amounts to many electron volts. This causes changes in peak shape, and sometimes gives rise to 'new' peaks. Such effects can then be related to the spin state of the atom in the complex. In the investigation reported here 'low'spin complexes of iron exhibited a simple $K\beta_{1,3}$ peak whilst the high-spin complexes (four or five unpaired electrons) showed a pronounced low-energy Kβ' satellite.

The intensities of the $K\beta_5$ peaks are related to the degree of covalent bonding between the ligand and the central atom. This affects the spin state of the atom and is related to the chemical shifts observed in Mössbauer spectra.

Acknowledgments

The authors gratefully acknowledge financial support from The Royal Society, the Central Research Fund of the University of London, and Queen Mary College. One of us (R.A.S.) also wishes to thank the Science Research Council and Standard Telecommunications Laboratories for a CAPS research studentship.

REFERENCES

Fadley, C.S. and Shirley, D.A. (1970); Phys. Rev. A2, 1109.

Fischer, D.W. (1970); J. App. Phys. 41, 3561.

Goldanskii, V.I. (1964); "The Mössbauer Effect and its Application in Chemistry", Consultants Bureau, New York.

Nemnonov, S.A., Menshikov, A.Z., Kolobova, K.M., Kurmaev, E.Z. & Trapeznikov, V.A. (1969); Trans. Met. Soc. of AIME 245, 1191.

Orgel, L.E. (1960); "An Introduction to Transition-Metal Chemistry", Methuen, London.

Tsutsumi, K. and Obashi, M. (1969); Conference, "X-ray and Electronic Spectra", Ukr. Acad. Sci., Kiev, Vol.1 page 65.

Urch, D.S. (1970); J. Phys. C., (Solid State Phys), 3, 1275.

Urch, D.S. (1971); Advances in X-ray Analysis, 14, 250, Plenum Press, New York.

USE OF SOFT X-RAY BAND SPECTRA FOR DETERMINING VALENCE/ CONDUCTION BAND STRUCTURE OF TRANSITION-METAL COMPOUNDS

David W. Fischer

Air Force Materials Laboratory (LPA), Wright-Patterson Air Force Base, Ohio, USA.

1. INTRODUCTION

Soft x-ray valence band spectra have long been recognized for their potential use in determining the electronic structure of compounds (see, for example, O'Bryan and Skinner 1940). Over the years many correlations have been tried between the spectra from simple compounds and certain physical and chemical properties. Typically such a correlation must involve a direct relationship between a measured wavelength shift, or intensity variation in a certain spectral component, and a specific property such as bond character, bond distance, electrical conductivity, heat of formation, etc. While these properties are indeed manifestations of the electronic structure of the material, they provide a very incomplete picture of that structure.

Ideally, x-ray band spectra should be capable of yielding a more complete picture. This can be expected from the basic origin of the spectra, including both emission and absorption components. X-ray emission bands, according to classical description, are due to electron transitions from the occupied valence/conduction band to an inner-level vacancy. Absorption spectra are due to the ejection of an inner-level electron into one of the available vacant states in the outer regions of the atom. Since it is the structure of these outermost electronic levels that determines the properties of a material, the x-ray band spectra appear to be capable of providing considerable information on how a material behaves. Unfortunately, the number of cases in which the spectra have been successfully used in this manner is disappointingly few. There are several reasons for this, but in general they can be combined into two problem areas: (1) obtaining reliable band spectra, and (2) correct interpretation of the spectra. It is the intention in this paper to focus on these problems (primarily the second 'area') for some simple transition-metal compounds such as their oxides, nitrides, carbides and oxyanions.

Compounds of the first-row transition metals are especially interesting from an electronic structure standpoint, as is reflected in their remarkably varied physical and chemical properties (Adler 1968). In attempting to unravel the complexities of the electronic structure of some of the more fascinating compounds in this group, many different types of experiments have been performed. It is curious that x-ray band spectra, despite their potential value for determining significant features of electronic structure, have been virtually ignored in studying these materials. There are problems, to be sure, in obtaining and using the spectra but there are also distinct advantages. The present author has attempted to indicate this in some recent work on the soft x-ray spectra from some titanium, vanadium and chromium compounds (Fischer and Baun 1968, Fischer 1969, 1970a, 1970b, 1971a, 1971b).

One of the key points has been the use of a molecular orbital (MO) model to interpret the x-ray band spectra. Several other authors have also recognized the importance of MO theory for explaining features of x-ray band spectra that are otherwise difficult or impossible to rationalize (Dodd and Glen

1968, Best 1966, 1968, Seka and Hanson 1969, Manne 1970, Lawrence and Urch 1970, Urch 1970, Nagel 1970, Andermann and Whitehead 1971). On the other hand, there are some who accept that MO theory is useful for explaining the bonding in highly covalent materials, such as transition-metal complexes, but resist the present trend to apply it to predominantly ionic materials such as TiO_2 and Cr_2O_3, or to metal-like compounds such as TiN and VC. In fact, MO theory is quite flexible; it is capable of describing any degree of covalent-ionic bonding character, so that in principle it can be applied to simple inorganics such as oxides, nitrides and carbides. There is a high degree of interaction of metal-atom with nonmetal-atom orbitals in these compounds, therefore it seems reasonable to use a bonding model that takes these interactions into account. The model is also capable of explaining details in the spectra more accurately than any other model. One purpose of this paper is to demonstrate that the MO model is directly applicable to transition-metal compounds in which the metal ion has any allowable valence state and occupies either an octahedral or tetrahedral co-ordination site. X-ray band spectra will then be used to construct empirically a complete MO energy-level diagram, involving both occupied and vacant orbitals within ∿20eV of the Fermi-energy. No other experimental technique is capable of this.

To determine empirically the complete electronic structure of a transition-metal compound we require more information than is contained in any one emission or absorption band. The reason for this is rooted in the symmetry characters of the outer orbitals and in the dipole selection rules governing x-ray transitions. In forming one of the compounds, it is assumed that the 3d, 4s and 4p levels of the metal atom interact with the 2s and 2p levels of the nonmetal atom. According to MO theory, using a linear combination of atomic orbitals (LCAO), this interaction will result in a series of bonding and antibonding molecular orbitals, such as we show below in figures 5 and 6. The important point to note is that these outermost electron levels consist of admixed s,p, and d symmetries. Now a K-emission band results from transitions of the outermost electrons to a vacancy created in the 1s core level. According to the dipole selection rules only electrons in levels with p symmetry can make such a transition. The K-emission band therefore reflects only the distribution of p symmetry in the outer levels, and tells us nothing about the distribution of s and d symmetry. Conversely, the L and M bands reflect the distribution of s and d symmetry but not of p symmetry. Obviously, if we expect to obtain a complete picture of the outer electronic structure, it is necessary to combine the information present in both K and L band spectra. Most x-ray band structure investigations of compounds have not been made from this viewpoint, which makes the information obtained of limited scope. The importance of using the combined spectra has recently been demonstrated for titanium, vanadium and chromium oxides (Fischer 1970b, 1971a, 1971b). In the present report the following spectra will be examined for the transition-metal compounds: metal-ion (Me) L_3 (valence orbitals→Me2$p_{3/2}$), metal-ion L_2 (valence orbitals→Me2$p_{1/2}$), metal-ion $K\beta_{2,5}$ (valence orbitals→Me1s), non-metal ion (X) K (valence orbitals→X1s); and the corresponding absorption spectra.

The band spectra for the various transition-metal compounds are presented, with several objectives in mind. The primary objective is to determine empirically the complete outer electronic arrangement and energies of the bonding, antibonding and nonbonding orbitals. The resulting MO diagrams are compared with other types of experimental data and where possible with theoretical calculation. Variations in the spectra, on changing the metal ion, the non-metal ion or the co-ordination symmetry, follow logically from the MO interpretation. The x-ray results do not agree well with some of the theoretical calculations, and reasons for the disagreement are suggested. New interpretations, based on x-ray MO diagrams, are offered for certain optical absorption spectra. For some of the oxides two different types of 3d (t_{2g}) bonding are observed, and the measured orbital widths do not agree well with the narrow d-band model (Adler 1968, Adler and Brooks 1967).

2. EXPERIMENTAL

A. Instrumentation

The plane single-crystal vacuum spectrometer used to obtain the spectra has been described previously (Fischer and Baun 1968). Characteristic x-ray spectra are produced by direct electron bombardment of the target materials. The interchangeable anode assembly (brass, copper or aluminium) is constructed so that the x-ray takeoff angle is variable from 0° to 90°. A flow-proportional detector is used (with Formvar window and argon-methane flow at ∿120 torr). The spectrometer vacuum under normal operation is $\sim 10^{-6}$ torr. Measured wavelengths of the spectral features recorded in this investigation have a probable error of ±0.02Å (±0.3eV), but wavelength differences could be measured to ±0.005Å (±0.1eV). The data points in the spectra have a statistical deviation of 2-3% at the peak, and <1% in the tails.

B. Spectrometer Resolution and Dispersing Crystals

The Ti $L_{2,3}$, V $L_{2,3}$, Cr $L_{2,3}$, and oxygen K spectra were all obtained using either a rubidium acid-phthalate crystal (RAP, 2d=26. 118Å) or a clinochlore crystal (2d=28.393Å). The spectrometer resolution calculated for each spectrum is given in Table I.

Because the window width of the spectrometer varies with the wavelength, and the exact shape of the window is not known, the

Table I
Spectrometer 'window' width for various crystals and spectra.

Spectrum	Crystal	'Window' (eV)
$TiL_{2,3}$ (∼27.5Å)	Clinochlore	0.48
$VL_{2,3}$ (∼24.5Å)	RAP	0.51
$CrL_{2,3}$ (∼21.5Å)	RAP	0.81
Oxygen K (23,6Å)	RAP	0.53
Oxygen K	Clinochlore	0.94

band spectra in this report have not been corrected for instrumental broadening. This is not considered to have any significant effect on the interpretations offered.

The metal-ion K spectra were not obtained by the author, but are taken from the literature (with references given in the discussion of specific compound), and these spectra are also uncorrected for instrumental effects.

C. Sample Preparation and Measurement of Emission Spectra

Target specimens of the compounds were prepared by mixing a fine powder into a slurry with a solvent, such as ethanol, and painting it as a thin film onto the anode surface. Non-conductors were sprayed with a fine film of graphite. For each spectrum shown here, at least 20 complete runs were made under a wide variety of excitation conditions, varying parameters such as the electron-beam voltage and current, the sample chamber vacuum, and duration of sample bombardment by primary electron beam. Spectra were considered characteristic of the sample material only when completely reproducible. All metal-ion L and oxygen K spectra shown here satisfied this requirement, which is important because a typical bombarding electron beam voltage (3-4 kV) will probe only the first ∿100 atomic layers. If chemical change occurs at or near the surface during excitation, the spectra will not be characteristic of the bulk material.

For the metal-ion L and oxygen K wavelength region, it is extremely difficult to make absorption specimens that are both uniform and thin enough to transmit the continuum radiation, and a differential self-absorption method (described elsewhere, Fischer and Baun 1968, Fischer 1969) was used to obtain absorption spectra. Although one cannot obtain absolute absorption coefficients from such self-absorption spectra, they have the advantage of being obtained from the same specimens and at the same time as the emission bands. This considerably simplifies the matching of emission and absorption energy scales, and ensures that both forms of spectra represent the same chemical state of the target material.

D. Unfolding the Spectra

Most of the metal L_3 and oxygen K band spectra given here have been unfolded into constituent components using the Dupont (Model 310) Curve Resolver. Complicated spectra can be resolved into almost as many components as the operator has the patience to attempt, therefore two points were adhered to: (1) Both pure Gaussian and pure Lorentzian deconvolutions were tried for each spectrum, and (2) the simplest solution was sought; i.e. the spectra were resolved into as few components as possible.

It was found that the L_3 spectra (both in emission and absorption) were best approximated by Gaussian components, and the oxygen K spectra by Lorentzian components. This was the case for every compound, including many that are not shown here. It is not clear why the spectra should unfold into different symmetry components; inner levels are believed generally to be of Lorentzian shape, while outer levels (molecular-orbitals) are commonly assumed to be Gaussian (Jorgensen 1962). Notwithstanding, the unfolding procedure used results in precisely the correct number of components predicted by the MO model, for both octahedral and tetrahedral compounds. With certain exceptions, the unfolded components can be used not only to position accurately the individual electronic orbitals, but also to give some indication of their width.

3. RESULTS AND DISCUSSION

A. General Comments

In interpretation of x-ray band spectra of transition-metal compounds, especially those of titanium, vanadium and chromium, one is aided by the fact that many of them are isostructural or isoelectronic. Some also show variable valence states, and can occupy both octahedral and tetrahedral symmetry sites with respect to the same anion. This permits variation of specific structural parameters such as the identity of the metal ion, the anion, the co-ordination symmetry or the valence state; and the influence of each on the x-ray spectrum can be assessed. A few examples are listed in Table II.

However, before attempting to analyze the spectra on this basis, it is important to have reliable and reproducible band shapes. This is not trivial because the transition-metal $L_{2,3}$-emission bands are subject to serious distortion from satellite and self-absorption effects (Liefeld 1968, Fischer 1968, 1969, 1971b). In the vicinity of the absorption edges strong emission components, which are actually multiple-ionization satellites, can be misinterpreted as part of the parent emission band. These fall to the high-energy side of the emission edge, and it is important to know the position of the edge. Here the self-absorption effect becomes an advantage: from the self-absorption spectrum the position of the absorption edge can be

Table II

Series of transition metal compounds in which a particular structure parameter is varied.

Series type	STRUCTURE PARAMETERS constant	varied	Examples
Series 1.	metal ion	valence state	TiO, Ti_2O_3, TiO_2
	anion	crystal structure	
	coordination number		
Series 2.	anion	metal ion	Ti_2O_3, V_2O_3, Cr_2O_3
	coordination number		
	crystal structure		
Series 3.	metal ion	anion	TiO, TiN, TiC
	coordination number		
	crystal structure		
Series 4.	metal ion	coordination number	$Cr_2O_3, CrO_3, CrO_4^=$
	anion	valence state	
		crystal structure	

accurately located (Fischer 1968, 1969, 1971b). For all of the compounds discussed in this report it is assumed that the emission and absorption edges coincide. This holds for electrical conductors but may not be true for insulators. The only way to observe the true emission edge is to measure the L_3 band at threshold voltage (Liefeld 1968), which for compounds it is too difficult, especially with a flat crystal spectrometer. The assumption of emission-edge absorption-edge coincidence is reasonable and appears to work for the compounds reported here. All the L_3-emission bands shown were obtained under conditions of negligible self-absorption. For interpreting the spectra, all of the emission intensity to the high-energy side of the absorption edge is assumed to be satellite emission and has been subtracted-off.

All of the metal L_3 and oxygen K absorption spectra from the compounds are self-absorption replicas. It should also be noted that in the oxygen K emission band, obtained with a RAP or KAP crystal, there is a strong peak at 23.3Å which is due to the crystal and is not part of the true emission band (Liefeld *et al* 1969, Fischer 1969). This anomalous peak does not occur with the clinochlore crystal which was used for the oxygen K spectra shown here.

B. Chemical Effects in Band Spectra

It is well known that by changing the chemical environment of an atom, very large differences can often be observed in its x-ray band spectrum. However, in most cases the exact causes of these changes are not so well known. Interpretations made for the spectra of one type of atom or compound often break down when extended to a different atom or compound. What is needed is a uniform model for interpretation of all x-ray band spectra of simple compounds in terms of the chemical interactions between the constituent elements. Transition metal compounds provide a particularly fertile group of materials for studying chemical effects in spectra. Large changes are observed in the $L_{2,3}$ spectra of titanium, vanadium and chromium compounds on varying for example the anion, the cation, the valence state, or the co-ordination symmetry (Fischer and Baun 1968, Fischer 1969, 1971a, 1971b). In order to evaluate these chemical effects, it is of interest to examine the spectra from systematically chosen compounds, such as those listed in Table II. Figures 1-4 correspond to series 1-4 of this table.

Series 1 comprises binary compounds of the same two elements, but involving different valence states of the metal ion. The oxides of titanium, vanadium and chromium form a good example. Figure 1 shows the Ti L_3-emission and L_3-absorption spectra for the three major oxidation states of titanium. In these spectra, as for all spectra presented in this report, each of the

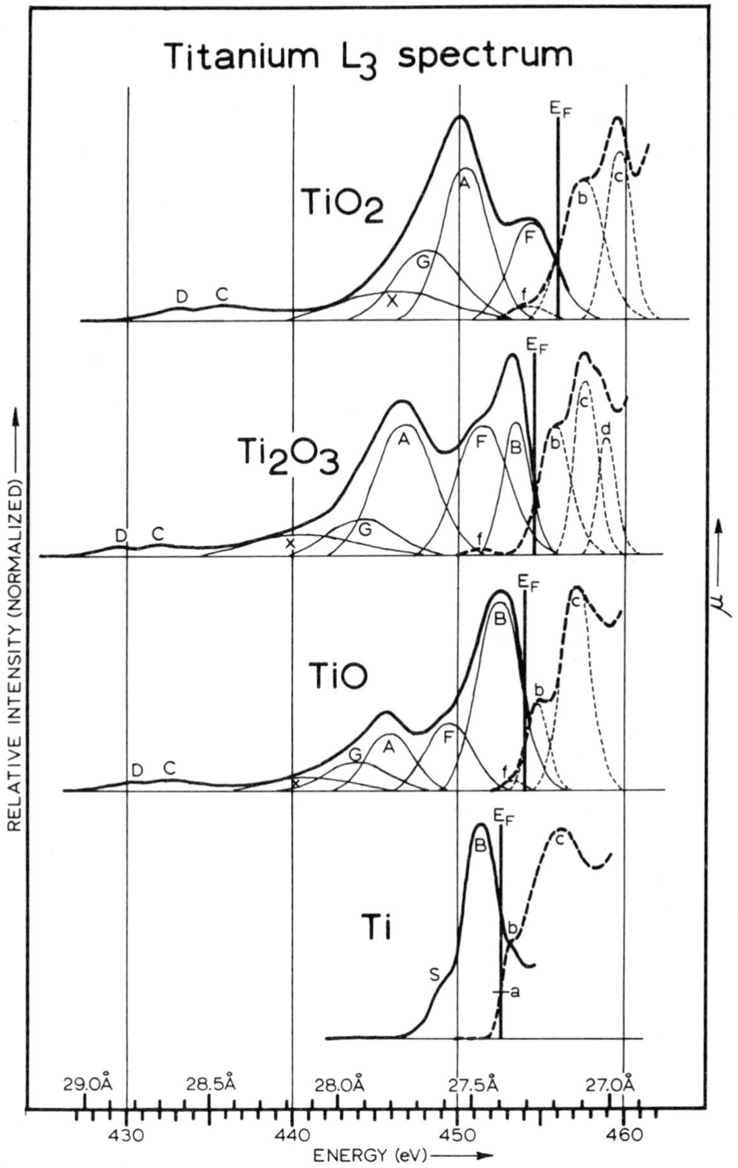

Figure 1 Titanium L_3 emission and absorption spectra for the pure metal and its oxides. Emission spectra (full lines) were obtained under conditions of negligible self-absorption and are not corrected for instrumental window. Dotted lines are the absorption spectra. Peak intensities are normalized for comparison.

emission maxima is denoted by a capital letter and each of the absorption maxima by a lowercase letter.

Several peaks appear in the oxide spectra not present at all in the pure metal spectrum. The more prominent of these are labelled F,A,X,C and D. Notice that as the oxidation state is increased, peak A grows markedly in intensity and shifts to higher energy. The change is not so much a growth of peak A, but rather a decrease in intensity of peak B (it might be more correct to plot the spectra with peak A normalized to the same height in each oxide), and as emission peak B fades absorption peak b increases. The same effects are observed for vanadium and chromium oxides.

The obvious interpretation is that peaks F,A,X,C and D arise from the presence of the anion (oxygen), and peaks B and b primarily from the metal ion (titanium). The decrease in intensity of peak B reflects the loss of 3d electrons with increased oxidation state. This loss of 3d electrons causes an increase in density of vacant 3d-states and an increase in the absorption peak b. The same effect could be approximated by decreasing the atomic number of the metal ion instead of changing the oxidation state; which is the case in series 2 of Table II.

This case is exemplified by the oxide series Ti_2O_3, V_2O_3 and Cr_2O_3; and the metal L_3 spectra from these oxides are shown in figure 2. If peak A were normalized for each spectrum it would be readily observed that the major changes involve emission peak B and absorption peak b. Their intensities are proportional to the number of 3d electrons as described for figure 1. Peaks A,X,C and D remain virtually unchanged indicating likewise that they arise from the oxygen atoms, and if this is correct they should show significant dependence on the anion.

Change of anion corresponds to series 3 of Table II, and

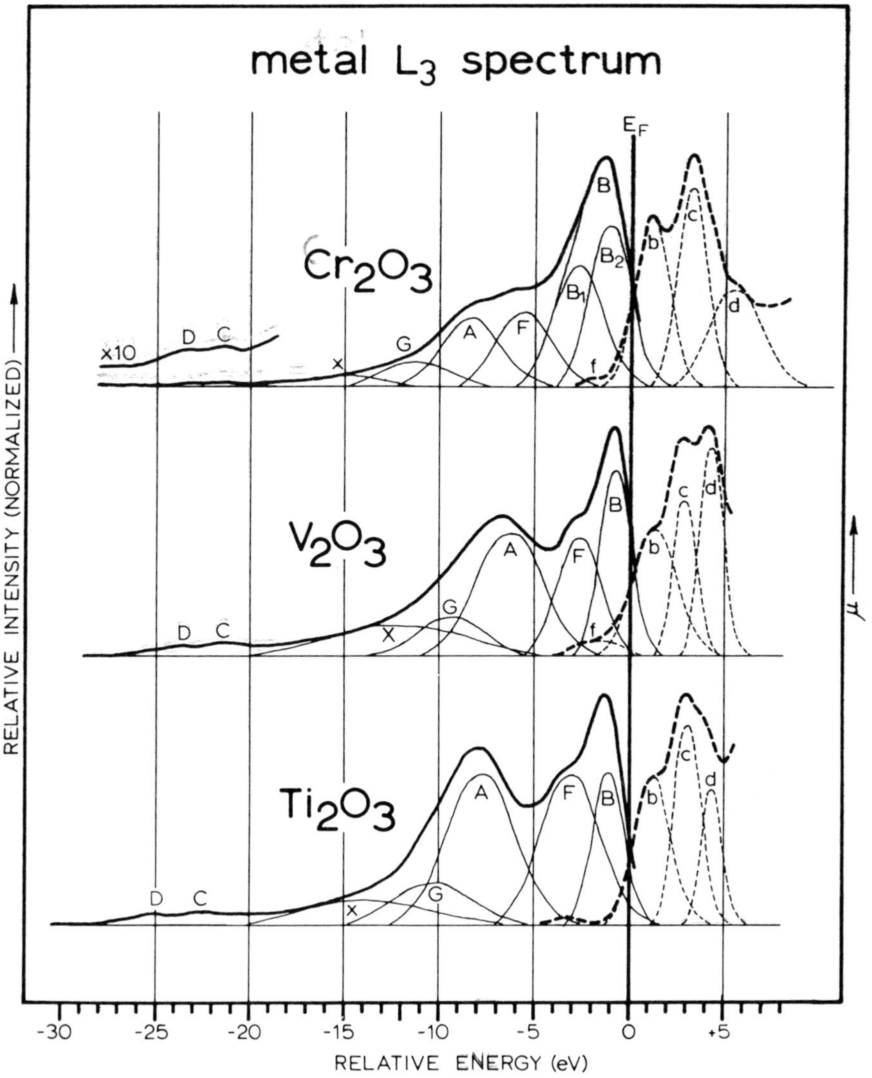

Figure 2 Metal L_3 emission and absorption spectra for Ti_2O_3, V_2O_3 and Cr_2O_3. Peak intensities normalized.

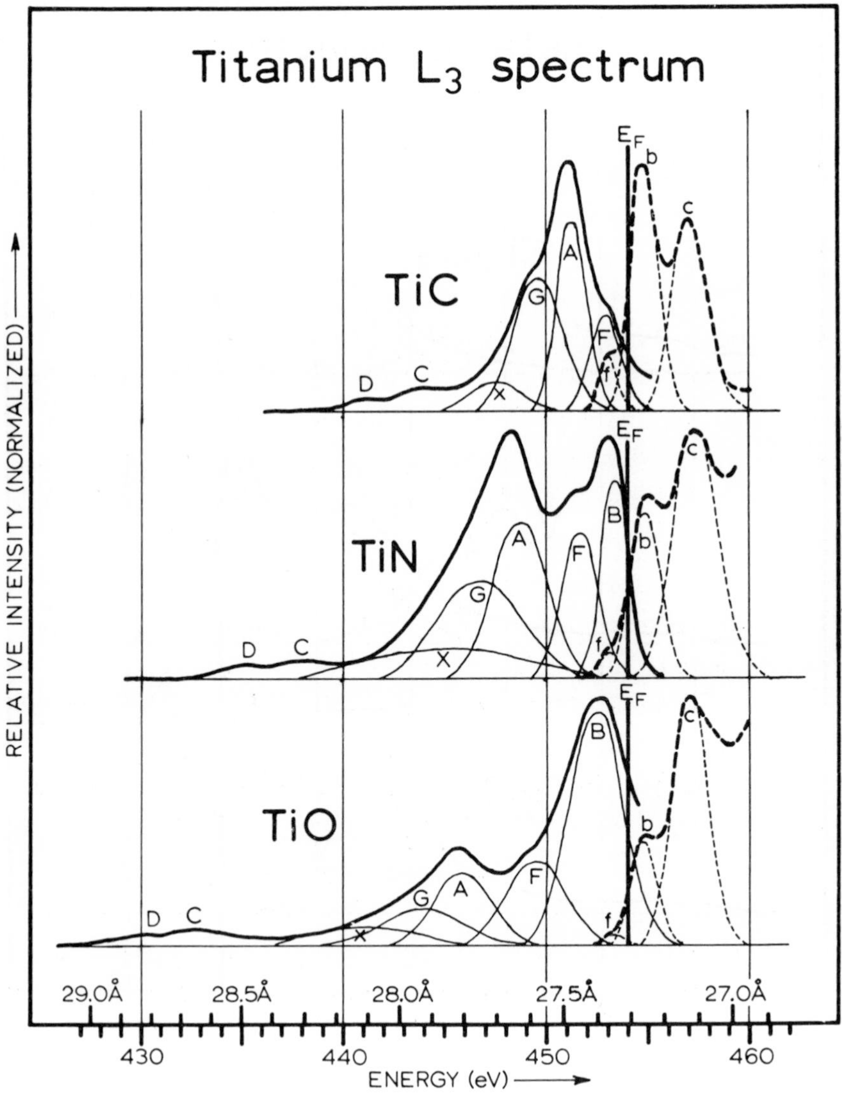

Figure 3 Titanium L_3 emission and absorption spectra for TiO, TiN and TiC. Peak intensities normalized.

here we choose compounds with the same metal ion and same crystal structure but differing anions. The series TiO, TiN and TiC forms a good example and their Ti L_3 spectra are shown in figure 3. The overall change in the spectra is obvious. All of the emission peaks vary both in energy and relative intensity. Peaks B and b again reflect the variation in number of 3d electrons, and the low-energy peaks clearly reflect the change in energy of the 2p and 2s levels of the anion. On going from oxygen through nitrogen to carbon the 2p-2s energy separation decreases from approximately 16eV through 13eV to 9eV. This suggests that peaks C and D arise from the anion 2s level, and peaks A and G from the anion 2p level.

A fourth major change that can be made in the structure parameters of simple transition-metal compounds involves the co-ordination symmetry of the metal ion. In most of their binary compounds the transition metal is octahedrally co-ordinated to the anion. However, in for example the vanadates and chromates the metal ion is tetrahedrally co-ordinated. Series 4 of Table II comprises compounds that have the same metal ion and same anion, but different co-ordination symmetry. An example is the series Cr_2O_3, $CrO_4^=$ and CrO_3. Cr_2O_3 has octahedral co-ordination, $CrO_4^=$ has regular tetrahedral co-ordination and CrO_3 has a much distorted tetrahedral structure. The chromium L_3 spectra of these compounds are shown in figure 4. There is a considerable difference in the spectra on going from octahedral to tetrahedral symmetry, although part of the change is due to the valence state. For octahedral symmetry, chromium has a +3 valence, while for tetrahedral symmetry it has a +6 valence.

It is apparent from figures 1-4 that the immediate chemical environment of an atom plays an important rôle in determining the x-ray band spectrum; parts of the spectra arise from the anion and others from the metal ion. To formulate a model to

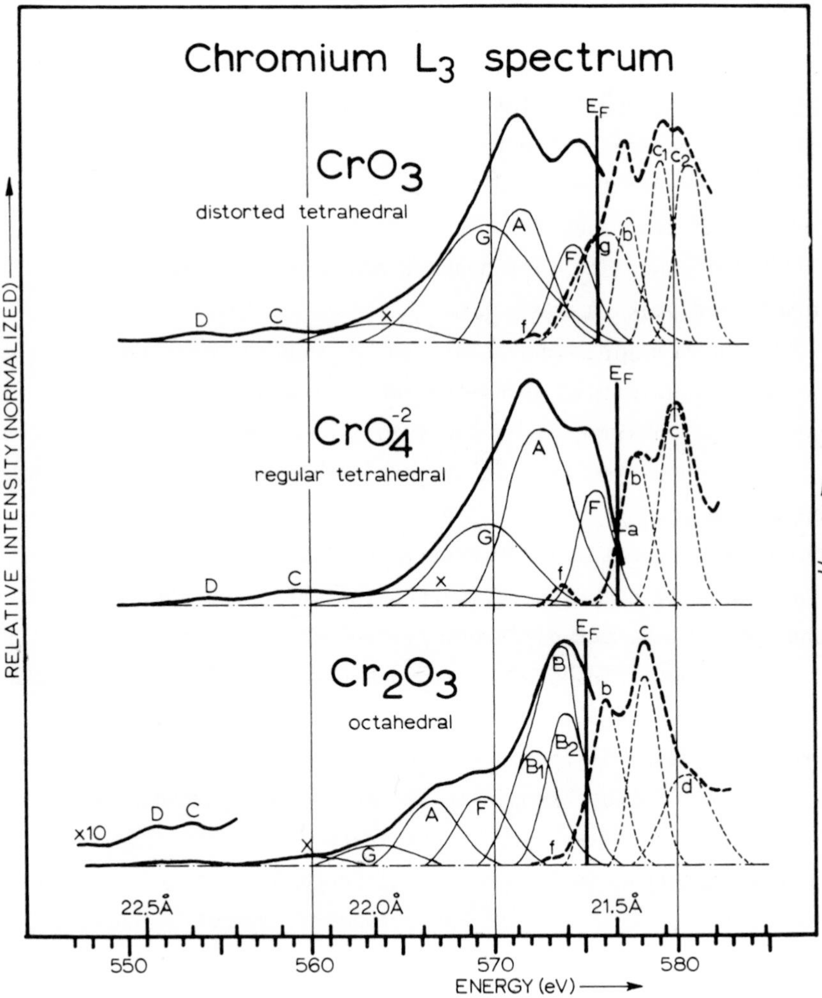

Figure 4 Chromium L_3 emission and absorption spectra for Cr_2O_3, $CrO_4^{=}$ and CrO_3. Peak intensities normalized.

account for the observed effects, several approaches were tried. At first the cross-transition model appeared to show promise (Fischer and Baun 1968). Peak A (figures 1-3) was assumed to arise from a cross-transition originating in the anion 2p level and terminating in the metal-ion L_3 level. Peaks C and D were regarded as analogous transitions from the anion 2s states. However, as data on more and more compounds were accumulated it became apparent that the simple cross-transition model would not satisfactorily account for many of the observed spectral changes. It is now believed that a molecular orbital model offers a better interpretation.

C. The Molecular Orbital Model

To apply the molecular orbital (MO) model to transition metal compounds we assume that the 3d, 4s and 4p atomic orbitals of the metal interact with the 2s and 2p atomic orbitals of the anion, and use the linear combination of atomic orbitals (LCAO) approximation. The derivation of the molecular orbitals (e.g. Ballhausen and Gray 1964, Figgis 1966) is not detailed here. However, it is of interest to know the type of MO diagram to be expected for a typical transition metal compound: the relative energies and wave symmetries of the molecular orbitals depends primarily on the nature of the metal ion and anion and on their co-ordination symmetry.

Schematic AO-MO-AO diagrams for a transition metal ion octahedrally and tetrahedrally co-ordinated to oxygen are shown in figures 5 and 6 (see Ballhausen and Gray 1964). The diagrams are only qualitatively correct but they provide a starting point for interpreting the individual x-ray spectra. The method is now to assign each spectral peak to an electron transition between a specific molecular orbital and core level. This is considerably easier to do for x-ray spectra than for optical spectra because in the x-ray case the inner level is essentially atomic in character and can be considered as having a con-

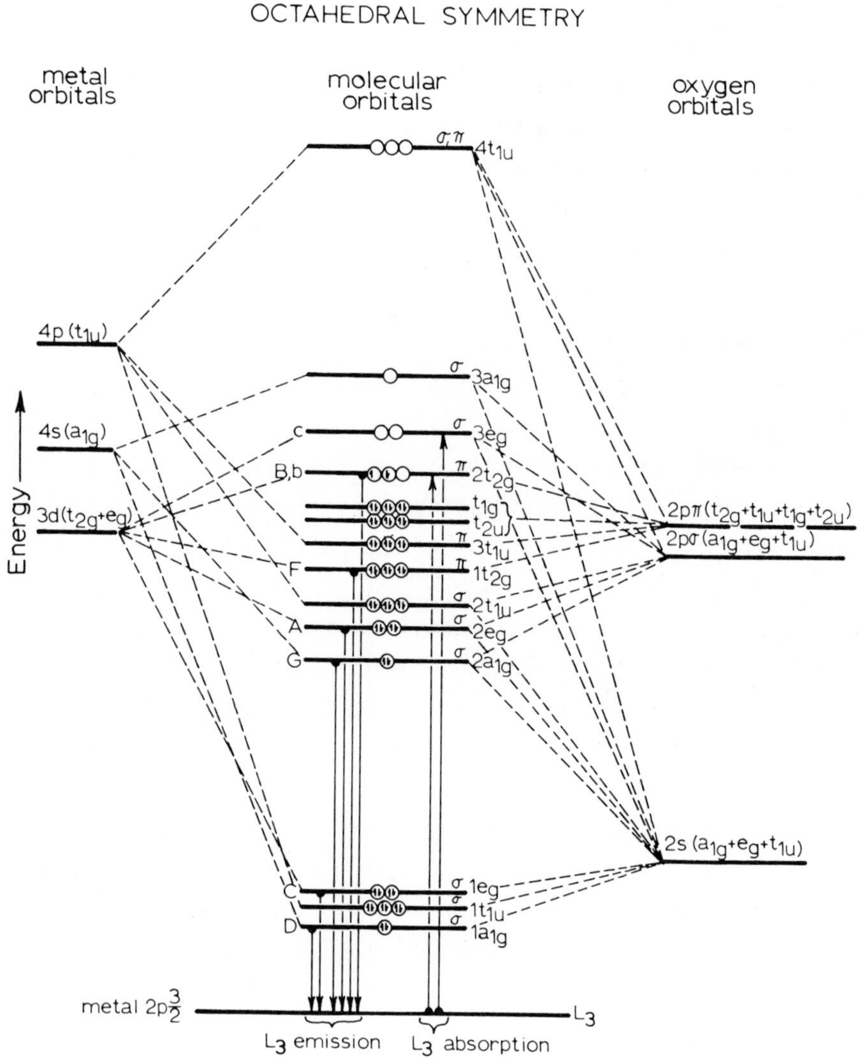

Figure 5 AO-MO-AO energy-level diagram for transition metal octahedrally co-ordinated to oxygen. (Energies not to scale).

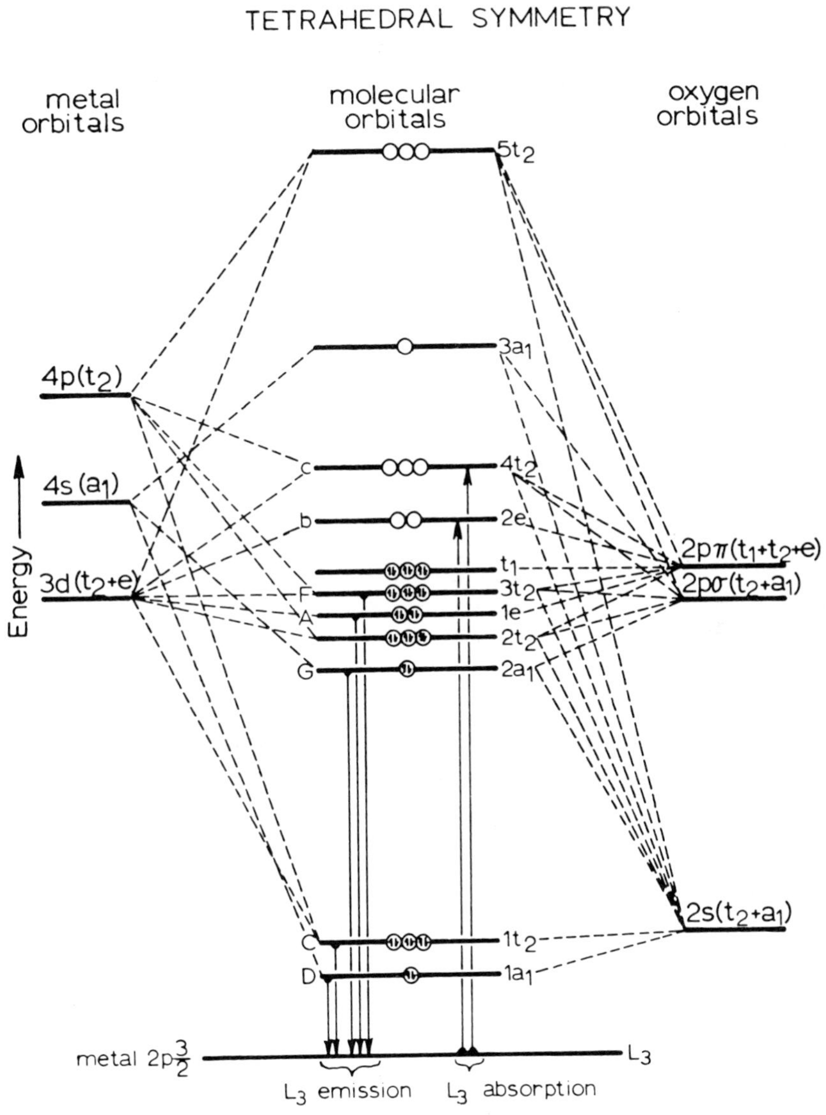

Figure 6 AO-MO-AO energy-level diagram for transition metal tetrahedrally co-ordinated to oxygen. (Energies not to scale).

stant energy for a given compound. If such MO assignments can be made with reasonable confidence for both emission and absorption then the spectra can form a foundation for deducing a complete molecular-orbital structure for each compound. No other experimental technique has been shown capable of doing this.

The MO assignments assume the usual dipole selection rules. In figures 5 and 6 vertical arrows indicate the MO's most likely to contribute to the metal ion L_3 spectrum. These contain at least some 3d or 4s admixture. Clearly not all orbitals will contribute to the L_3 band because some are predominantly of p symmetry. Furthermore, there are two kinds of p symmetry; metal-ion 4p, and oxygen 2p. Thus no one spectrum can give a complete picture of the MO-structure, and to obtain the complete picture it is necessary to combine the information from several spectra. For transition metal compounds, the emission and absorption spectra required are the metal-ion L_3, the metal-ion K, and the anion K.

In the following sections the spectra of TiO, Cr_2O_3 and $CrO_4^{=}$, are examined and interpreted on the basis of the MO model. These three compounds were chosen for several reasons: the complete metal-ion K spectra for all three are available; MO calculations available for TiO and $CrO_4^{=}$ permit comparison of experiment with theory; Cr_2O_3 and $CrO_4^{=}$ represent different co-ordination symmetries, for the same two elements, and should provide a good test of the MO interpretation. Furthermore, although TiO and Cr_2O_3 are both octahedrally co-ordinated, they have different crystal structures and different properties which should mean a different bonding mechanism involving the 3d orbitals.

D. Titanium Oxide

TiO crystallizes in the rocksalt structure and is a good electrical conductor. The titanium atoms are at the centre of regular octahedra of oxygen atoms. The expected MO-structure is indicated in figure 5, and its relation to the TiO band spectra is illustrated in figure 7, in which the Ti L_3 and oxygen K spectra (Fischer and Baun 1968) and the TiK spectrum (Chirkov *et al* 1967) are shown. The zero of energy is arbitrarily placed at the Fermi energy E_F, which is assumed to lie at the Ti L_3 absorption edge. The spectra are positioned, in relative energy, by lining up absorption peaks b and c (for the reason outlined below).

The Ti L_3 spectrum is resolved into Gaussian components, and the oxygen K spectrum into Lorentzian components. Before unfolding the L_3-emission band, background emission to the high-energy side of the L_3 absorption edge is subtracted; we assume that these emission components are the L_3 multiple-ionization satellites and the L_2-emission band. Absorption features occurring to the high-energy side of maximum c are also ignored, since they form the L_2-absorption and involve transitions from the same outer orbitals as the L_3. The Ti K spectrum has not been unfolded; it has been taken from Chirkov *et al* (1967) and not enough is known about the experimental conditions.

Referring to figures 5 and 7, we examine the Ti L_3-emission band. This spectrum should reflect primarily the distribution of 3d states and we assume that the main components are due to transitions from the occupied orbitals which contain significant 3d-character. Peaks B, F and A are therefore assigned respectively to the $2t_{2g}$, $1t_{2g}$ and $2e_g$ molecular orbitals. The $2t_{2g}$ orbital is only partially occupied, so it should also be involved in the absorption spectrum. In fact, it is assumed that the first two absorption maxima represent the two lowest vacant MO's, i.e. $2t_{2g}$ and $3e_g$. These consist mostly of 3d character but are also expected to have some p-character due to the titanium-oxygen orbital overlap. Both the Ti K and oxygen K absorption spectra contain the peaks b and c, and it is the alignment of these for the three spectra that dictates their relative positions on the energy scale. In the Ti K-emission band the two strongest peaks are assumed to originate from orbitals consisting of some 4p-symmetry; in this case, the $2t_{1u}$ and $3t_{1u}$ orbitals. It is further assumed that the main oxygen K-emission components arise from the t_{2u} and t_{1g} non-bonding 2p 'lone-pairs'. Peaks C and D in the L_3 band, and the Kβ" satellite in the K band, are then due to the $1e_g$, $1a_{1g}$, and $1t_{1u}$ levels which are associated primarily with the oxygen 2s states. Component X was introduced into the unfold for the sole purpose of exactly matching the shape in the tail of the band (broadened by Auger effects, plasmon satellites, etc.). It does not correspond to any normal single-electron transition. The MO assignments for all the peaks are summarized in Tables III and IV, including both octahedral and tetrahedral symmetries. These tables will also apply to Cr_2O_3 and $CrO_4^=$, discussed later.

According to the MO interpretation illustrated in figure 7, all the individual molecular orbitals of TiO can be placed empirically on a relative energy scale. The spectra indicate that the oxygen 2s orbitals are definitely involved in bonding. There has been discussion on whether or not the oxygen 2s orbitals are too tightly bound to interact with the metal 3d, 4s and 4p levels; if these oxygen 2s orbitals did not in fact participate in the bonding, then peaks C and D in the L_3 emission and peak Kβ" in the K emission could not easily be accounted for. This same argument is applied later to Cr_2O_3 and $CrO_4^=$.

We must ask how meaningful is the empirically deduced MO diagram of figure 7. Such empirical deductions can sometimes be erroneous and misleading. The question, in the case of TiO, is largely one of whether theoretical calculations can support the empirical results. No strict MO calculations have been made for TiO, but band structure calculations have been made by

Table III

Proposed electronic transitions to correspond to the peaks observed in the L_3 emission and absorption spectra of transition-metal compounds.

EMISSION

Peak	Electron Transition Octahedral Site	Electron Transition Tetrahedral Site
B	$2t_{2g} \rightarrow 2p3/2$	not observed
F	$1t_{2g} \rightarrow 2p3/2$	$3t_2 \rightarrow 2p3/2$
A	$2e_g \rightarrow 2p3/2$	$1e \rightarrow 2p3/2$
G	$2a_{1g} \rightarrow 2p3/2$	$2a_1 \rightarrow 2p3/2$
C	$1e_g \rightarrow 2p3/2$	$1t_2 \rightarrow 2p3/2$
D	$1a_{1g} \rightarrow 2p3/2$	$1a_1 \rightarrow 2p3/2$
X	'extended tail' broadening effects	

ABSORPTION

Peak	Electron Transition Octahedral Site	Electron Transition Tetrahedral Site
a	L_3 edge	L_3 edge
b	$2p3/2 \rightarrow 2t_{2g}$	$2p3/2 \rightarrow 2e$
c	$2p3/2 \rightarrow 3e_g$	$2p3/2 \rightarrow 4t_2$
g	---	$2p3/2 \rightarrow en$
f	exciton (?)	exciton (?)

Ern and Switendick (1965) using the augmented-plane-wave (APW) method; these results, shown in figure 7, qualitatively agree with the empirical MO diagram. In each case regions consisting primarily of 2s, 2p, 3d or 4s symmetry fall approximately within the same energy interval. However, the individual MO's are not sharp levels as indicated, and broadening and overlapping of the orbitals results in bands similar to the Ern and Switendick calculated structure (Fischer 1970a, 1970b, 1971b).

A further comparison of theory with experiment, for TiO, is

Table IV

Proposed electronic transitions corresponding to the peaks observed in the metal K and oxygen K spectra of transition-metal oxides.

Spectrum	Peak	Electron Transition Octahedral Site	Electron Transition Tetrahedral Site
Metal K Emission	$K\beta_{2,5}$ (or $K\beta_5$)	$2t_{1u}\rightarrow 1s$	$2t_2\rightarrow 1s$
	$K\beta_5'$	$3t_{1u}\rightarrow 1s$	not observed
	$K\beta''$	$1t_{1u}\rightarrow 1s$	$1t_2\rightarrow 1s$
Metal K Absorption	b	$1s\rightarrow 2t_{2g}$	not observed
	c	$1s\rightarrow 3e_g$	$1s\rightarrow 4t_2$
	d	$1s\rightarrow 3a_{1g}$	$1s\rightarrow 3a_1$
	e	$1s\rightarrow 4t_{1u}$	$1s\rightarrow 5t_2$
O K Emission	A	several possibilities	?
	B	$t_{1g}, t_{2u}^{\cdot}\rightarrow 1s$	$t_1\rightarrow 1s$
	C	several possibilities	?
	D	$2t_{2g}\rightarrow 1s$	---
O K Absorption	a	K edge	K edge
	b	$1s\rightarrow 2t_{2g}$	$1s\rightarrow 2e$
	c	$1s\rightarrow 3e_g$	$1s\rightarrow 4t_2$
	f	exciton (?)	exciton (?)

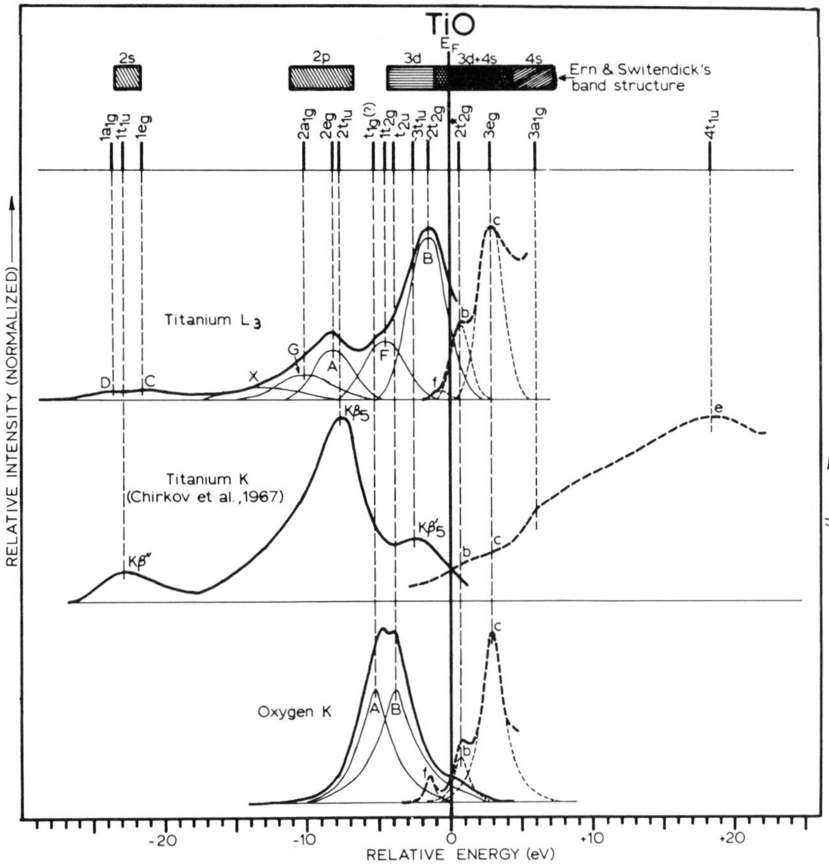

Figure 7 Empirical derivation of MO-structure of TiO, obtained by combining the Ti L_3, Ti K and oxygen K band spectra. The calculated band structure (Ern and Switendick 1965) is shown above.

seen in the emission band shapes and density-of-states histograms. It is known that the emission band is not a direct picture of the density-of-states, due to the influence of transition probabilities; nonetheless it is common to assume that features in the x-ray spectra correspond to characteristics in the density of states. Since the band structure of TiO consists of admixed s, p and d states, the Ti L_3 and Ti K bands should contain complimentary information. This, to some extent, is observed in figure 7. It has already been shown (Fischer

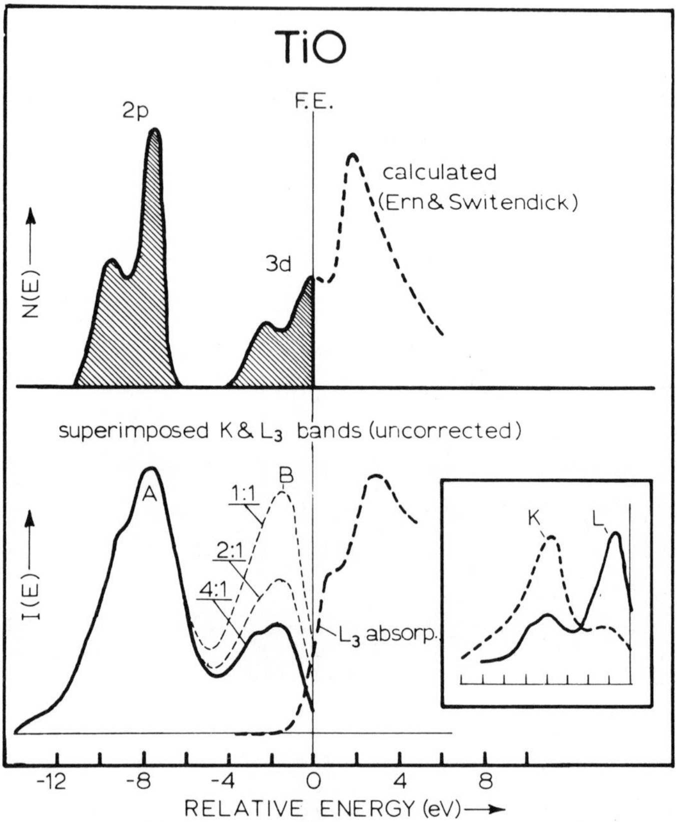

Figure 8 Comparison of superimposed Ti K and L3 bands with the calculated density of states for TiO (Ern and Switendick 1965). Inset are the individual emission bands. Peak heights normalized to same intensity. (Reproduced, with permission, from Fischer 1970b).

1970b) that by combining the L_3 and K bands, in an arbitrary ratio, the density-of-states curves could be closely approximated for TiO, TiN and TiC. The results for TiO are shown in figure 8. The 'experimental' curve composed of 4 parts K to 1 part L is remarkably close to the Ern and Switendick (1965) calculated density of states.

From the interpretation in figure 7 we can understand the chemical effects observed in figures 1-4. For example the

Ti L_3 spectra of TiO, TiN and TiC (figure 3) resolved into their individual components and the electron-transition assignments for these components are listed in Table III. Emission peak B corresponds to the occupied portion of the $2t_{2g}$ orbital, which contains two electrons in TiO, one in TiN and none in TiC, and correspondingly peak B becomes weaker on going from TiO to TiN and disappears in TiC. On the other hand, components G, A and F remain virtually unchanged because the corresponding orbitals contain the same number of electrons for each compound. The spectra in figures 1-4 can all be explained on a similar basis (Fischer 1970a, 1971a, 1971b).

With reservations, the half-widths and relative intensities of each of the resolved components in the metal-ion L_3 band (figures 3 and 7) can be related to the electron-orbital widths and the relative amount of admixed 3d-character. This particular analysis is easier to perform for Cr_2O_3 than for TiO and is therefore explained more fully in the next section.

E. Chromic Oxide

The co-ordination symmetry of Cr_2O_3 like TiO is octahedral, but in this case the octahedra are slightly distorted (the Corundum crystal structure), with a different relationship between neighbouring octahedra and between nearest-neighbour metal atoms (Goodenough 1960). This affects the x-ray spectra, as we shall see below. Also, Cr_2O_3 is an insulator both above and below its Neel temperature ($\sim 45^{\circ}C$), whereas TiO is a conductor at all temperatures

The band spectra of Cr_2O_3, and their MO relationships, are illustrated in figure 9 where we show the Cr L_3 and oxygen K spectra (Fischer 1971b), and the CrK spectrum (Menshikov and Nemnonov 1963). These MO assignments are listed in Tables III and IV.

In general, the spectra of TiO and Cr_2O_3 are similar but

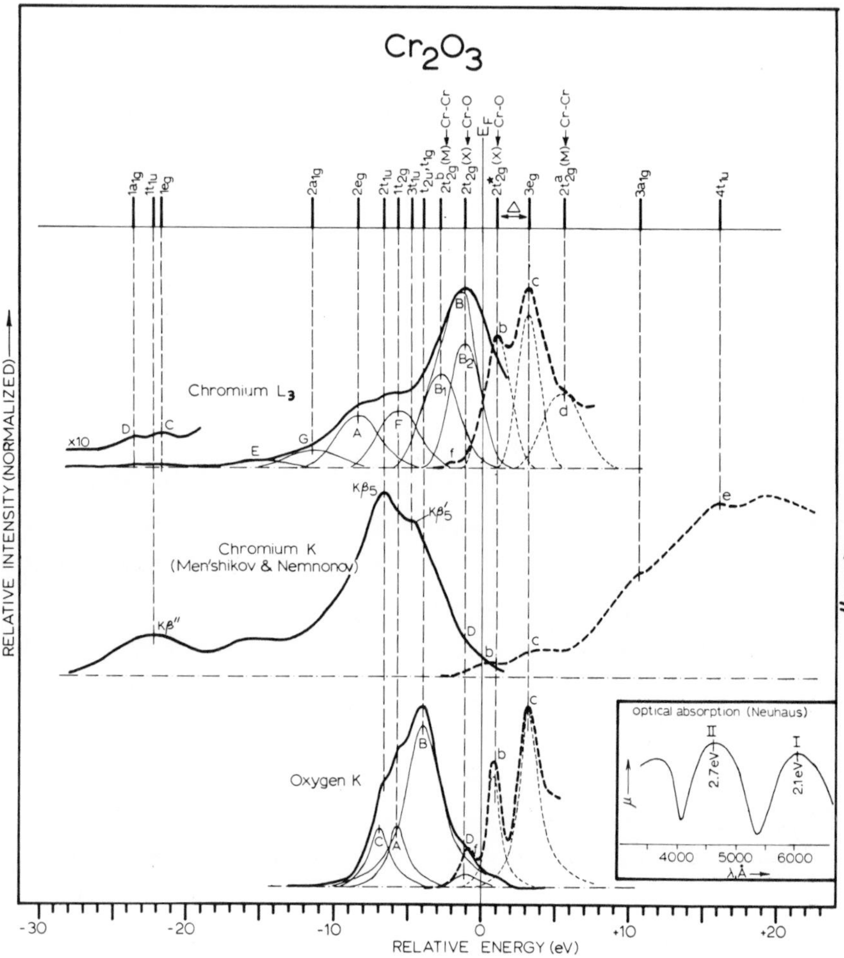

Figure 9 Empirical derivation of MO-structure of Cr_2O_3 by combining the chromium L, chromium K and oxygen K band spectra. (Reproduced, with permission, from Fischer 1971b).

there are two differences to note. For TiO the oxygen K-emission band unfolds into two components of equal intensity. For Cr_2O_3 the unfolding results in three components, one of which (peak B) is clearly predominant. The Cr_2O_3 band is typical of the transition-metal oxides investigated to date; the main component at ∿-4eV is assigned to the non-bonding 2p orbitals t_{2u} and t_{1g}, and the weaker peaks (A and C in figure 9) to bonding orbitals, which also contribute to the metal-ion spectra. These bonding orbitals result from interaction between the O 2p and metal 3d, 4s and 4p atomic orbitals, and are believed to be localized more on the oxygen ions than on the metal ions. However, we cannot explain why the oxygen K band of TiO is different from that of most other oxides.

More interesting, in the Cr_2O_3 and TiO spectra, is the difference observed in peak B. This peak is assigned to the occupied portion of the $2t_{2g}$ orbital, which is the highest occupied orbital and consists primarily of 3d-symmetry. Similar differences are found in L_3-absorption components b and d, associated with the unoccupied portion of the $2t_{2g}$ orbital. For Cr_2O_3, peak B consists of two components, B_1 and B_2 (figure 9). We suggest that all the four components B_1, B_2, b and d involve $2t_{2g}$ orbitals of two different types. This is a consequence of the co-existence of two different kinds of bond involving the 3d (t_{2g}) electrons. First, in the Corundum structure c-axis cation-cation pairs are formed and strong t_{2g}-t_{2g} covalent bonding can occur between the (chromium) cations (Adler *et al* 1967, Goodenough 1960, Morin 1961); and second, we have the t_{2g}-$p\pi$ (chromium-oxygen) bond. For these two types of bond to be observed in the L_3 spectrum, the $2t_{2g}$ orbitals would have to be non-degenerate and split. Some splitting will be caused by the trigonal field in the Corundum structure. Also, the chromium atoms forming the c-axis pairs have the closest cation-cation distances, and the t_{2g} orbitals associated with the covalent bond for this pair could exhibit a large bonding-antibonding splitting (Adler *et al* 1967, Goodenough 1960). Antiferromagnetic ordering could further contribute to this splitting. Therefore we suggest that the Cr L_3 components, B_1 and d, could represent the bonding and antibonding set of the $2t_{2g}$ orbital associated with chromium-chromium covalent bond. These components are labelled $2t_{2g}{}^b(M)$ and $2t_{2g}{}^a(M)$ respectively in figure 9, and are assumed to be one-electron states. The $2t_{2g}{}^{b}(M)$ orbital is occupied and the $2t_{2g}{}^{a}(M)$ orbital is vacant. Components B_2 and b would then represent the occupied and vacant two-electron states associated with the 3d-2pπ bond. They are labelled $2t_{2g}(X)$ and $2t_{2g}{}^{*}(X)$ in figure 9.

The three outermost electrons in Cr_2O_3 are therefore involved in two distinct bonding mechanisms. One of the electrons is localized in a metal-metal covalent bond and the other two are associated with the metal-oxygen π bond. Whether this orbital has a true energy gap, between the occupied and vacant states is not clear from the x-ray spectra, but the Fermi energy is assumed to be in this region.

This interpretation is supported by the fact that since components B_1 and d represent the metal-metal covalent bond, they should consist of almost pure 3d-character and should therefore not contribute to either the Cr K or oxygen K spectra. We can see from figure 9 that there are indeed no components in the K spectra corresponding to B_1 and d. On the other hand, since components B_2 and b in the L_3 spectrum are interpreted as due to the 3d-2pπ bond, they should also contribute to the oxygen K spectrum and perhaps also to the Cr K spectrum. We see in figure 9 that these contributions do occur, evidenced by peaks D and b in both the O K and Cr K spectra. These relationships can be seen in Table V.

The above interpretation of peaks B, b and d in the L_3 spectrum is not based solely on the results obtained for Cr_2O_3. Much consideration has been given also to the Ti L_3 spectrum of Ti_2O_3 and the V L_3 spectrum of V_2O_3, (figure 2). We noted that the intensities of emission component B and absorption component b for these oxides is directly proportional to the number of 3d electrons, while the intensities of components F, A and G remain relatively unchanged. Therefore components B and b are associated with the partially occupied $2t_{2g}$

Table V

Line widths and relative intensities of the unfolded components of the band spectra of Cr_2O_3. The values have not been corrected for spectrometer window or inner-level width.

Spectrum	Component	Assigned MO	Half-width (eV)	Relative Integrated Intensity (Emission)
Cr L_3	B_1	$2t_{2g}^{b}$ (M) cation-cation	3.0	89
(Gaussian)	B_2	$2t_{2g}$ (X) cation-anion	2.4	100
3d-character	F	$1t_{2g}$	3.3	62
	A	$2e_g$	3.4	60
	G	$2a_{1g}$	3.8	21
	b	$2t_{2g}^{*}$ (X) cation-anion	2.4	--
	c	$3e_g$	2.1	--
	d	$2t_{2g}^{a}$ (M) cation-cation	3.0	--
oxygen K	B	t_{2u}	2.8	100
(Lorentian)	A	$1t_{2g}$	1.6	21
2p-character	C	$2t_{1u}$	1.7	23
	D	$2t_{2g}$ (X)	1.8	5
	b	$2t_{2g}^{*}$ (X)	1.1	--
	c	$3e_g$	1.5	--

orbital and components F, A and G are associated with filled orbitals.

The Cr K band shown in figure 9 has been interpreted by Menshikov and Nemnonov (1963) in terms of two different types of d electrons. They concluded that the $K\beta'_5$ peak was associated with the collectivized (conduction) d electrons, the $K\beta_5$ paeak with the localized d electrons, and the $K\beta''$ peak with crossover transitions of oxygen valence electrons to the chromium K level. However, according to the MO model (figure 9) all three of the $K\beta$ peaks are associated with localized orbitals partly consisting of chromium 4p-character. Peak $K\beta'_5$ arises from the $4p\text{-}2p\pi$ bond, peak $K\beta_5$ from the $4p\text{-}2p\sigma$ bond, and peak $K\beta''$ from the 4p-2s bond. This interpretation is more in accord with the dipole selection rules than that offered by Menshikov and Nemnonov. It provides a good example of the advantage to be gained from using combined K and L spectra for structure determinations.

Adler and Brooks (1967) postulated that the d bands in transition-metal oxides, such as Ti_2O_3 and V_2O_3, are extremely narrow, of width a few tenths of an eV. We would then expect this 'narrow band' model to apply to Cr_2O_3. In fact the d bands in Cr_2O_3 are likely to be even narrower than for the corresponding titanium and vanadium oxides because on moving across the 3d series the d orbitals are contracted by the increased nuclear charge, so that nearest-neighbour overlap would be less (Morin 1961). However, the spectra do not appear to support the 'narrow band' model. In the case of Cr_2O_3, the Cr L_3-emission band has been resolved into five components and the L_3 absorption into three (figure 9). The measured half-widths and relative integrated intensities of these components are listed in Table V. Before relating the measured component widths to the actual electron-orbital widths, the measured widths must first be corrected for instrumental and core-level broadening. The spectrometer window was ∿0.8eV. The chromium L_3 level has a width of ∿0.4eV (Parratt 1959). Thus the total full-width at half-maximum is ∿0.9eV. Molecular vibrations and spin-orbit coupling may also cause orbital broadening, but amounting probably to no more than a few tenths of an eV. We can see from Table V that the uncorrected component widths are considerably larger than the total broadening. The B_1 and d components, associated with the d orbitals of the cation-cation bond, have a measured half-width of 3.0eV. Assuming a conservatively large correction of 1.3eV, a width of at least a 1.7eV remains. Similarly, the d orbitals of the cation-anion bond (components B_2 and b) would have a corrected half-width of about 1eV. These values may contain uncertainties but it is nonetheless apparent that the d orbitals in Cr_2O_3 are considerably broader than is proposed in the Adler-Brooks model. This is also the case for the spectra of the titanium and vanadium oxides (figure 2). Recent photoemission studies of TiO_2 and VO_2 (Derbenwick 1970) also fail to support the 'narrow d-band' model.

The components of the oxygen K spectrum tend to be narrower than those of the Cr L_3 spectrum as indicated in Table V. This is mainly due to the fact that both the spectrometer window (0.5eV) and inner-level width (0.2eV) are smaller for the oxygen spectrum. All of the individual molecular orbitals in Cr_2O_3 appear to have half-widths on the order of 1 to 2 eV. Some 'solid state' broadening is expected due to electron interaction between atoms from neighbouring octahedra. Also, the distorted octahedral symmetry may cause unresolved splitting of degenerate orbitals, causing apparent broadening.

If the method of unfolding the spectra, especially the Cr L_3 band, is accepted as reasonably correct, then the relative intensities of the components can be used to provide a general indication of the amount of 3d-character in the e_g and

Table VI

Relative percentages of 3d-character in e_g and t_{2g} valence orbitals of Cr_2O_3, determined from the unfolded Cr L_3 band spectrum.

Component	Number of electron states (n)	Relative integrated intensity	I/n = % d character	
B_1 ($2t_{2g}$ Cr-Cr)	1	100	100	occupied states (total 3.5 d electrons)
B_2 ($2t_{2g}$ Cr-O)	2	112	56	
F ($1t_{2g}$)	6	70	12	
A ($2e_g$)	4	67	17	
b ($2t_{2g}$ Cr-O)	2	176	88	vacant states
c ($3e_g$)	4	205	51	
d ($2t_{2g}$ Cr-Cr)	1	100	100	

t_{2g} valence orbitals. The results are given in Table VI. In determining the values in the last column of the table, it was first assumed that the one-electron $2t_{2g}$ orbitals associated with the chromium-chromium covalent bond (components B_1 and d) represent 100% d-character. These components were assigned an arbitrary intensity 100, and the others scaled accordingly. Relative intensities were then divided by the number of electron states appropriate to the orbital. It is difficult to assess the accuracy of the results because they depend primarily on the unfolding procedure. First it must be assumed that the transition probability remains constant through the band, and second we have uncertainties in relating the intensities of absorption to those of emission. Nonetheless the results (Table VI) are interesting and confirm, qualitatively, the expected cation and anion contributions to the t_{2g} and e_g orbitals. The bonding orbitals ($2e_g$ and $1t_{2g}$) appear to be strongly polarized toward the oxygen ions, whereas the antibonding orbitals ($3e_g$ and $2t_{2g}$) are polarized toward the chromium ions.

The intensities in Table VI indicate 3.5 d-electrons in the occupied orbitals of Cr_2O_3.

Further evidence for the empirical MO-structure proposed in figure 9 is found in the optical absorption spectrum of Cr_2O_3 (Neuhaus 1960). A prominent peak at 2.1eV is believed to be a measure of the ligand field splitting parameter, Δ. For Cr_2O_3, Δ would correspond to the energy separation between the $2t_{2g}^*$ and $3e_g$ orbitals. However, from figure 9, we see that both these orbitals (absorption peaks b and c) are vacant and no electrons are available for optical transition between them. We see also from figure 9 that the energy separation between the highest occupied orbital (B_2) and the lowest vacant orbital (b), both of which have $2t_{2g}$ symmetry, is 2.2eV. This is in good agreement with the optical value, and apparently it is this separation and not Δ that the optical absorption spectrum measures.

F. The $CrO_4^=$ ion

In $CrO_4^=$ the chromium atom is surrounded by a regular tetrahedral arrangement of oxygen atoms. The tetrahedral field results in a molecular orbital structure different from that of octahedral Cr_2O_3 (figure 6) and the differences are observed in the Cr L_3 and Cr K spectra (Fischer 1971b) and the oxygen K spectrum (Best 1966); see figures 4 and 10. Relative energies for the spectra are obtained by aligning the absorption maxima c. As for TiO, we assume that dipole selection rules dictate the peak assignments, using the tetrahedral orbital arrangement and the term symbols shown in figure 6. The results are listed in tables III and IV and are illustrated in figure 10 (Fischer 1971b). All of the $CrO_4^=$ molecular orbitals can be accounted for.

$CrO_4^=$ uhlike Cr_2O_3 has no partially occupied orbital. The Cr has a +6 valence state so that all the bonding orbitals are exactly filled and all the antibonding orbitals completely

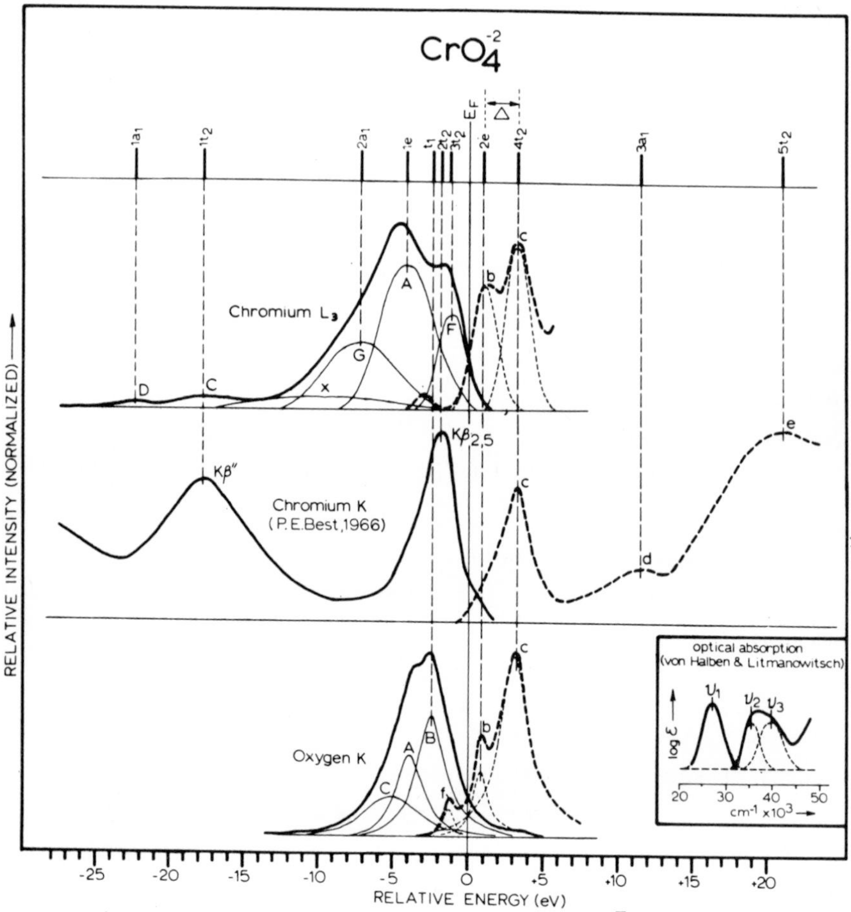

Figure 10 Empirical MO-structure of $CrO_4^=$ obtained by combining the chromium L, chromium K and oxygen K band spectra of Na_2CrO_4. (Reproduced, with permission, from Fischer 1971b).

empty. This is reflected in the Cr L_3 emission bands (figure 4); the peak labelled B in the Cr_2O_3 spectrum does not appear in the spectra of the tetrahedral compounds. The peak B always signifies that the lowest antibonding orbital is partially occupied.

The optical absorption spectrum of $CrO_4^=$ (Carrington and Symons 1960) has two primary peaks at 3.3 and 4.5eV. At least

Table VII

Electron transition assignments for two principle maxima in optical spectrum of $CrO_4^{=}$

Peak Position (eV)	Wolfsberg and Helmholz (1952)	Ballhausen and Liehr (1958)	Viste and Gray (1964)	Oleari *et al* (1965)	Present work x-ray
$3.32(\nu_1)$	$t_1 \rightarrow 4t_2$	$t_1 \rightarrow 2e$	$t_1 \rightarrow 2e$	$\begin{bmatrix} t_1 \rightarrow 4t_2 \\ t_1 \rightarrow 2e \end{bmatrix}$	$t_1 \rightarrow 2e$(3.3eV)
$4.54(\nu_2)$	$3t_2 \rightarrow 4t_2$	$t_1 \rightarrow 4t_2$	$3t_2 \rightarrow 2e$	$\begin{bmatrix} 3t_2 \rightarrow 4t_2 \\ 3t_2 \rightarrow 2e \\ 3t_2 \rightarrow 3a_1 \end{bmatrix}$	$3t_2 \rightarrow 4t_2$(4.5eV)
Highest occupied orbital	t_1	t_1	t_1	t_1	$3t_2$
Lowest vacant orbital	$4t_2$	2e	2e	$3a_1$	2e
Δ(eV)	1.6	1.2	3.1	0.5	2.3

four different interpretations of these peaks have been proposed; see Table VII. Wolfsberg and Helmholz (1952) calculated an MO-structure for $CrO_4^=$ in which the highest filled orbital was t_1 and the lowest empty orbital was $4t_2$. On this basis, they interpreted the optical spectrum as indicated in column 2. Ballhausen and Liehr (1958) suggested that the lowest empty orbital was 2e instead of $4t_2$. Their interpretation is shown in column 3. ESR measurements (Carrington and Schonland 1960) indicate also that the lowest empty orbital is 2e. A further MO calculation was made by Viste and Gray (1964) using simplifying assumptions that caused them to call the structure 'Pseudo' $CrO_4^=$. Their interpretation of the optical spectrum is indicated in column 4. Lastly Oleari *et al* (1965) performed a self-consistent MO calculation and interpreted the optical spectrum in terms of multiple transitions for each peak (column 5). They justified these multiple assignments by pointing out that the absorption bands are quite broad, and that the second band in fact has a shoulder. Surprisingly, the x-ray spectra do not agree completely with any of these interpretations. All assume that the highest filled orbital is the nonbonding t_1. The spectra (figure 10) indicate that the highest filled orbital is actually $3t_2$. The deduced structure is compared with the calculated structures by Viste-Gray and Oleari *et al* in figure 11. The zero energy point is arbitrarily placed at the t_1 non-bonding level. The Wolfsberg-Helmholz calculation is not included; it is generally regarded as incorrect, and it does entirely disagree with the spectra. The calculation by Oleari *et al* appears also to disagree, despite the fact that it is supposedly self-consistent. Oleari *et al* made calculations for chromium charge numbers of 0 and +1; the diagram in figure 11 is for the charge +1, but the zero-charge diagram shows even worse agreement. A strange feature is the large energy separation ($\sim$9eV) between the highest filled and

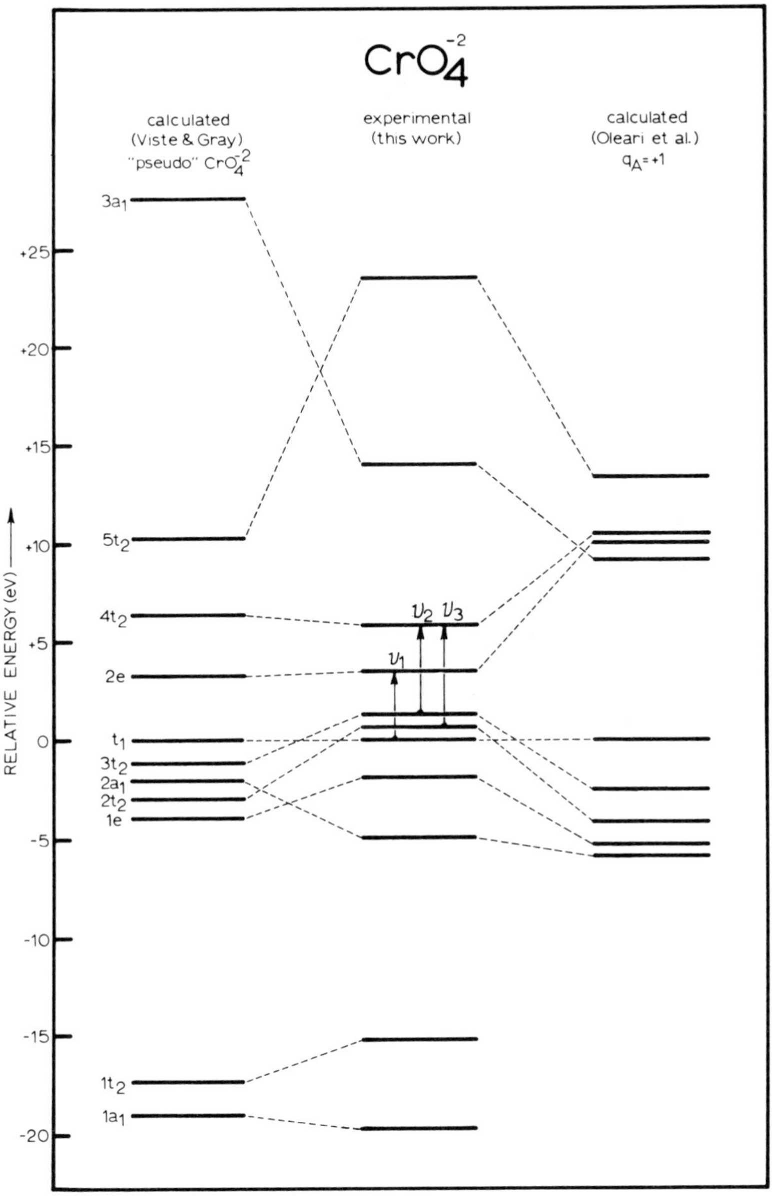

Figure 11 Comparison of relative MO energy positions of $CrO_4^{=}$ determined in this work and calculated by Viste and Gray (1964) and Oleari *et al* (1965). Zero of energy arbitrarily placed at t_1 non-bonding level to facilitate comparison (Reproduced, with permission, from Fischer, 1971b).

lowest empty orbitals.

Better agreement is found, surprisingly, between the x-ray spectra and the Viste-Gray pseudo-$CrO_4^=$ calculation. Perhaps we should take note of the Fensky and Sweeny (1964) conclusion that such calculations depend strongly on assumptions made in determining the symmetry character of hybridized orbitals.

The deduced MO-structure for $CrO_4^=$ is supported by a different interpretation of the optical absorption spectrum. On the basis of the orbital positions determined figure 10, the optical peaks at 3.3(ν_1) and 4.5eV(ν_2) can be assigned respectively to the transitions $t_1 \rightarrow 2e$ and $3t_2 \rightarrow 4t_2$. Furthermore the second peak can be resolved into two components, ν_2 and ν_3, using a DuPont curve resolver, as illustrated inset in figure 10. The component ν_3 is at 5.0eV and is assigned to the transition $2t_2 \rightarrow 4t_2$ (figure 11). We emphasized that the orbital energy differences agree *exactly* with the optical peak positions. According to the Ballahusen and Liehr interpretation the 2e and $4t_2$ orbitals would be separated by only 1.2eV. This is certainly too small, as indicated by the Cr L_3 absorption spectrum and the oxygen K-absorption spectrum, and by the optical spectra of similar compounds (Carrington and Jorgensen 1961). The energy separation is known as the ligand field splitting parameter Δ. The x-ray spectra shown here indicate a value of 2.3eV for Δ. The results indicate also that each of the optical peaks has a unique transition assignment, and that multiple assignments such as those suggested by Oleari *et al* are incorrect. Table VII summarizes the situation for $CrO_4^=$.

The component A in the oxygen K emission band (figure 10) has not been assigned to any particular orbital. This is because this component appears not to have anything to do with the $CrO_4^=$ ion. In the chromates, the oxygens are involved in bonding to two different metal ions. In, for example, Na_2CrO_4

we have the bonds Cr-O and Na-O; the latter is probably highly ionic. On observing the oxygen K band for several different chromates it appears that spectral component A is associated with the oxygen bond to the metal cation; though it is more intense than would be expected, and the reason is not clear.

The Cr L_3 band (figure 10) was resolved into Gaussian components. Since there should be no significant interaction between neighbouring tetrahedral units, the individual orbitals for $CrO_4^{=}$ should not be broadened to the extent observed for Cr_2O_3. With the exception of component G, the $CrO_4^{=}$ orbitals do appear to be narrower. The width of this component is probably doubtful for most of the unfolded L_3 bands because of the low-energy tailing. Component X was in fact introduced for the purpose of matching the shape of the tail.

A similar MO-model can be applied to the distorted tetrahedral structures of CrO_3 and $Cr_2O_7^{-2}$ (Fischer 1971b).

4. SUMMARY AND CONCLUSIONS

Several points support our MO-interpretation of the band spectra of TiO, Cr_2O_3, and $CrO_4^{=}$:

(1) Individual spectra. The main peaks follow the expected selection rules; that is, peaks in the L band arise only from orbitals with considerable d or s character; those in the K band arise only from orbitals having considerable p character.

(2) Combined spectra. Orbitals that consist of admixed metal 3d and oxygen 2p symmetries (1e, 2e, $2t_{2g}$ . . .) contribute to both the metal L and oxygen K bands. Orbitals that are virtually atomic in character (t_{2u}, cation-pair t_{2g}) contribute only to the appropriate spectrum of the element involved.

(3) The intensities of L_3 spectral components associated with the bonding and antibonding t_{2g} and e_g orbitals show expected relative contributions of 3d and 2p symmetries.

(4) The deduced MO-structures for Cr_2O_3 and $CrO_4^{=}$ can accurate-

ly account for the optical absorption spectra, and new interpretations are suggested for these.

(5) There are no 'left-over' components in the unfolded spectra. Each component can be logically assigned to a specific molecular orbital for both octahedral and tetrahedral co-ordination symmetries.

An important result of correct interpretation of band spectra, regardless of the model used, is the information that can be gained from the spectra. Some examples based on the MO-model have been indicated in this paper, including:

(1) Accurate relative energies of electronic orbitals.

(2) Widths of individual orbitals and the amount of admixed d-character, obtained by unfolding spectra.

(3) Observation of two distinct types of $3d(t_{2g})$ bonding in the Corundum structure - a metal-metal covalent bond and a metal-oxygen π bond.

(4) Direct determination of the ligand field-splitting parameter Δ.

(5) Direct evidence that the anion 2s-orbitals are involved in bonding.

The power of the MO-interpretation, illustrated in figures 7, 9, and 10, is the way it effectively ties together the various cation and anion emission and absorption spectra with chemical interaction in a compound; and with various solid state phenomena such as co-ordination symmetry, bond distance, valence state, bond character, etc. More complete work is needed, particularly in careful MO calculations for direct comparison with experimental data. However, it is apparent that soft x-ray band spectroscopy is a powerful tool for probing the electronic structure of compounds.

REFERENCES

Adler, D. (1968); Rev. Mod. Phys. 40, 714.

Adler, D. and Brooks, H. (1967); Phys. Rev. 155, 826.

Adler, D., Feinleib, J., Brooks, H. and Paul, W. (1967); Phys. Rev. 155, 851.

Andermann, G. and Whitehead, H.C. (1971); Advan. X-Ray Anal. 14, 453.

Ballhausen, C.J. and Gray, H.B. (1964); Molecular Orbital Theory: Benjamin, New York.

Ballhausen, C.J. and Liehr, A.D. (1958); J. Mol. Spectr. 2, 342.

Best, P.E. (1966); J. Chem. Phys. 44, 3248.

Best, P.E. (1968); J. Chem. Phys. 49, 2797.

Carrington, A. and Jorgensen, C.K. (1961); Mol. Phys. 4, 395.

Carrington, A. and Schonland, D.S. (1960); Mol. Phys. 3, 331.

Carrington, A. and Symons, M.C.R. (1960); J. Chem. Soc. (London), 889.

Chirkov, V.I., Blokhin, S.M. and Vainshtein, E.E. (1967); Fiz. Tverd. Tela 9, 1116.

Derbenwick, G.F. (1970); Stanford Electronics Laboratories, Tech. Rept. 5220-2.

Dodd, C.G. and Glenn, G.L. (1968); J. Appl. Phys. 39, 5377.

Ern, V. and Switendick, A.C. (1965); Phys. Rev. 137, A1927.

Fensky, R.F. and Sweeny, C.C. (1964); Inorg. Chem. 3, 1105.

Figgis, B.N. (1966); Introduction to Ligand Fields: Wiley, New York.

Fischer, D.W. (1969); J. Appl. Phys. 40, 4151.

Fischer, D.W. (1970a); J. Appl. Phys. 41, 3561.

Fischer, D.W. (1970b); J. Appl. Phys. 41, 3922.

Fischer, D.W. (1971a); Appl. Spectr. 25, 263.

Fischer, D.W. (1971b); J. Phys. Chem. Solids 32, 2455.

Fischer, D.W. and Baun, W.L. (1968); J. Appl. Phys. 39, 4757.

Goodenough, J.B. (1960); Phys. Rev. 117, 1442.

Jorgensen, C.K. (1962); Absorption Spectra and Chemical Bonding in Complexes: Pergamon, New York.

Lawrence, D.F. and Urch, D.S. (1970); Spectrochim. Acta. B25, 305.

Liefeld, R.J. (1968); in Soft X-Ray Band Spectra, ed. D.J. Fabian, pp 133-149: Academic, London.

Liefeld, R.J., Hanzely, S., Kirby, T.B. and Mott, D. (1969); Adv. X-Ray Anal. 13, 373.

Manne, R. (1970); J. Chem. Phys. 52, 5733.

Menshikov, A.Z. and Nemnonov, S.A. (1963); Bull. Acad. Sci. U.S.S.R. (Phys. Ser.) 27, 402.

Morin, F.J. (1961); J. Appl. Phys. 32, 2195.

Nagel, D.J. (1970); Advan. X-Ray Anal. 13, 182.

Neuhaus, A. (1960); Z. Krist. 113, 195.

O'Bryan, H.M. and Skinner, H.W.B. (1940); Proc. Roy. Soc. (London) A176, 229.

Oleari, L. DeMichelis, G. and DiSipio, L. (1965); Mol. Phys. 10, 111.

Parratt, L.G. (1959); Rev. Mod. Phys. 31, 616.

Seka, W. and Hanson, H.P. (1969); J. Chem. Phys. 50, 344.

Urch, D.S. (1970); J. Phys. Soc. C 3, 1275.

Viste, A. and Gray, H.B. (1964); INorg. Chem. 3, 1113.

VonHalben, H. and Litmanowitsch, M. (1941); Helv. Chim. Acta. 24, 44.

Wolfsberg, M. and Helmholz, L. (1952); J. Chem. Phys. 20, 837.

EFFECT OF SURFACE OXIDE ON SOFT X-RAY EMISSION BANDS OF METALS AND ALLOYS

J.E. Holliday*

Edgar C. Bain Laboratory, U.S. Steel Corporation Research Center, Monroeville, Pennsylvania, USA.

1. INTRODUCTION

In the past several years there has been a large amount of work done on changes in soft x-ray emission bands with alloying. Since the number of publications in this area is extensive, the reader may refer to the reviews by Curry (1968), Holliday (1970) and Fabian (1971). There has been considerable effort to determine which spectral changes are due to alloying and which are due to electronic structure changes (Liefeld 1967, Fischer and Baun 1968). However, there has been little effort to determine how much the surface oxide of the metal and alloy contributes to the observed changes in the emission bands. The experimental results reported here were performed in order to study systematically the effects of surface oxide on the emission bands of alloys.

2. EXPERIMENTAL RESULTS

The emission bands were measured using a soft x-ray, grazing incidence, grating spectrometer with a grating of one meter radius of curvature, 3600 grooves/mm, 1° blaze angle and a platinum surface. The spectrometer has been described in de-

*Now with: McDonnell Douglas Research Laboratories, McDonnell Douglas Corporation, St. Louis, Missouri.

tail elsewhere (Holliday 1968). The x-ray chamber was pumped to a vacuum of 1 x 10^{-7} torr. In order to reduce carbon contamination, the x-ray chamber was pumped with a titanium sublimation pump, and a cold finger was placed around the target.

Table I Fe L_2/L_3 Intensity ratios of Fe-Cr alloys at 1.5kV (from Holliday and Frankenthal 1972)

%Cr in Alloy	Fe L_2/L_3
0	0.35
5	0.34
10	0.35
14	0.32
19	0.32
24	0.32

One emission band parameter that is strongly affected by alloying is the L_2/L_3 intensity ratio for the first-series transition metals. In Table I, the Fe L_2/L_3 intensity ratio is tabulated as a function of increasing chromium content for a target potential of 1.5kV. It will be observed that there is a reduction in the Fe L_2/L_3 ratio with increasing chromium content. Fischer (1965) observed an increase in the Fe L_2/L_3 ratio with decreasing target voltage and attributed this to self absorption (figure 1). By measuring the Fe L_2/L_3-emission band with and without a 500Å iron film between the grating and detector, Holliday has shown that the Fe L_2/L_3 intensity ratio is not affected by self absorption. Skinner *et al* (1954) have shown that the L_2/L_3 intensity ratio of the first series transition metals is greater for the oxides than for the corresponding pure metal. For the spectrometer used for the present measurements, the Fe L_2/L_3 intensity ratio was 0.26 for pure iron and 0.45 for Fe_2O_3. The increase observed by Fischer in the L_2/L_3 intensity ratio with decreasing target potential, could be due to surface oxide because a greater percentage of

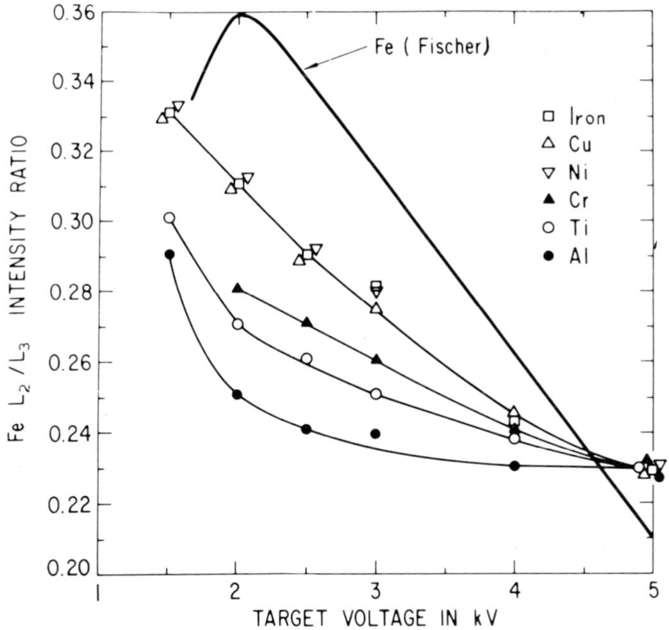

Figure 1 Comparison of the Fe L_2/L_3 intensity ratio for evaporated films on iron with that for uncoated iron as a function of target voltage. Fischer (1965). Results for iron show higher Fe L_2/L_3 ratio.

the surface will be sampled as the potential is reduced. If the reduction in the Fe L_2/L_3 intensity ratio with increasing chromium content is due to surface oxide, then there must be a decrease in the amount of iron oxide with increasing chromium content.

The basic reason for the reluctance of x-ray spectroscopists to accept the oxide explanation is the belief that the metal surface can be made sufficiently oxide-free, in vacuum of 10^{-7} to 10^{-8} torr, to produce no contribution of surface oxide to the x-ray spectra of the bulk material at low accelerating voltages. In measuring the oxygen K-emission band, the present investigator has found it impossible to eliminate the oxygen K-emission intensity for transition metals. The oxygen K-emission band intensity for iron targets in a vacuum of 10^{-7} torr indicates ∿30Å of oxide on the surface. From free-energy data we can see that aluminium will replace iron in iron oxide-also Al_2O_3 has one of the highest negative free energies of formation for an oxide. Thus an aluminium film evaporated onto iron would reduce the iron.

A. Multilayer Films

In figure 1 the Fe L_2-emission to L_3-emission intensity ratio is shown as a function of accelerating voltage for a 50Å aluminium film evaporated onto one half of an iron target. This curve is compared with the Fe L_2/L_3 ratio measured from the uncoated half of the iron target (□ points on the curve) and to the Fe L_2/L_3 measurement of Fischer and Baun (1968). It may be seen that at low accelerating voltages the Fe L_2/L_3 ratio is well below that of the uncoated iron. This result indicates that the aluminium has reduced the iron oxide. It is believed the surface of the iron is heated sufficiently during the evaporation of aluminium to cause this reaction to occur. Further support for this oxide explanation was obtained by evaporating in turn approximately 50Å of titanium, chromium, nickel and copper, which have progressively lower negative free energies of formation of their oxides, onto iron surfaces. It may be seen from figure 1 that for copper, nickel, chromium, titanium and aluminium films there is a progressively greater reduction in the Fe L_2/L_3 ratio corresponding to the progressively larger negative free energies of formation of their oxides. Actually, copper and nickel oxide have a lower negative free energy of formation than iron oxide and we would expect there to be no change in the Fe L_2/L_3 ratio. From figure 1 it can be seen that, within the experimental error of measuring the Fe L_2/L_3 ratio (±0.5), there is no change in the Fe L_2/L_3 ratio for nickel or copper evaporated onto iron.

If self-absorption is the explanation for the increase in Fe L_2/L_3 ratio with decreasing accelerating voltage, then we would expect the Fe L_2/L_3 ratio to increase relative to that for pure iron with increasing atomic number of the evaporated film. This is because an increase in the density of the evaporated surface film would have the effect of decreasing the accelerating voltage. However, it can be seen from figure 1 that this does not occur. The evaporated films with highest densities show no change in Fe L_2/L_3 ratio, and the Fe L_2/L_3 ratio becomes progressively smaller, relative to that for the uncoated iron, with decreasing density of the film. Further evidence that the reported changes in Fe L_2/L_3 ratio are due to surface oxide is provided by the results that at 5kV the uncoated iron and the coated iron specimens all have the same Fe L_2/L_3 ratio. This is because at high accelerating voltages the surface contribution becomes negligible.

The curve of Fe L_2/L_3 ratio *vs* accelerating voltage for evaporated aluminium on iron is nearly flat above 2.5kV. Below 2.5kV there is a distinct rise in the curve indicating that the aluminium has not replaced all of the iron in the iron oxide. However, for target voltages above 2.5kV the flatness of the curve indicates that the contribution of the iron oxide to the Fe L_2/L_3 ratio is negligible.

The Fe L_2/L_3 intensity ratio reported by Fischer (1965), shown in figure 1, is greater than that for the uncoated iron at accelerating voltages below 4.5kV indicating that the iron target used by Fischer had a larger amount of surface oxide than that of the present uncoated iron specimen. The lower Fe L_2/L_3 value at 5kV could be due to a difference in spectrometer analyzers. Fischer used a crystal analyzer while the present results were obtained with a grating spectrometer. The reduction in the Fe L_2/L_3 ratio he obtained below 2kV was not observed in the present measurements.

Further support for the above conclusions is obtained from the oxygen K-emission spectra of the composite films in figure

1. Figure 2 shows the oxygen K-emission band for aluminium on iron, for copper on iron, and for Fe_2O_3 and Cu_2O. It may be seen that for copper on iron the oxygen K emission has the same intensity distribution as that from Fe_2O_3 and not as that from Cu_2O, indicating that the oxygen is combined with the iron and has not been reduced by the copper. For aluminium on iron, the oxygen K emission intensity distribution is not the same as from Fe_2O_3, but is that from Al_2O_3 showing that the aluminium has replaced the iron in Fe_2O_3.

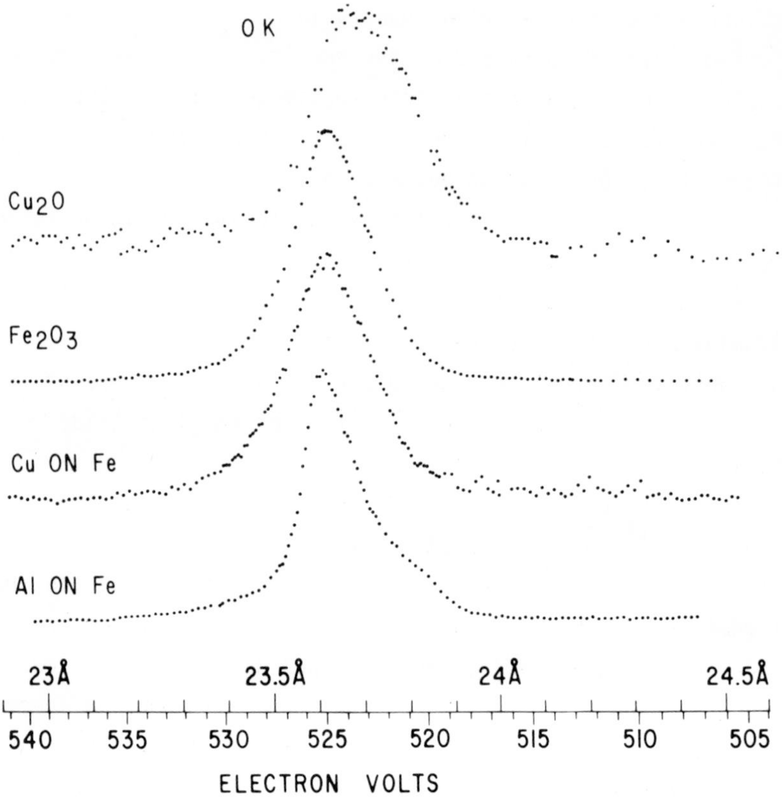

Figure 2 The O K-emission bands for Cu_2O and Fe_2O_3, and for multilayer films of copper and aluminium evaporated on iron.

B. Oxides on Alloys

Clearly oxides of thicknesses found typically on metals

and alloys in vacua such as that commonly used in x-ray spectrometers, do contribute significantly to the shape of emission bands below 4kV target potential. The results also indicate the possibility that with alloys, the alloying element with the highest electronegativity difference will be oxidized to a greater amount than the other alloying elements.

It has already been observed from the data in Table I that there is a reduction in the Fe L_2/L_3 intensity ratio which could be due to a reduction in the amount of iron oxide with increasing chromium content. Since chromium should be oxidized to a greater extent than the iron, because of its higher electronegativity, a better understanding of the contribution of surface oxides to alloy spectra should be obtained from the Cr $L_{2,3}$-emission band of Fe-Cr alloys. Figure 3 shows the Cr $L_{2,3}$-band for iron with respectively 10, 14 and 24 at% chromium measured at 3kV target potential. We see a significant change in the Cr $L_{2,3}$-emission band with decreasing chromium content. In order to determine that these changes are due to surface oxides rather than electronic structure changes, the Cr $L_{2,3}$ band from Fe-24at%Cr was measured at 2kV target potential. The Cr $L_{2,3}$ band at 2kV is compared with that measured at 3kV in figure 4. There is a pronounced change in the intensity distribution of the Cr $L_{2,3}$ band from Fe-24at%Cr, on going from 3 to 2 kV, and is in the same direction when the chromium content is reduced with the target potential kept constant (figure 3). These results suggest strongly that the changes in Cr $L_{2,3}$ emission are due to chromium oxide near the surface of the Fe-Cr alloy.

The chromium L_2 peak from Cr_2O_3 is shifted approximately 1.8eV towards higher energies relative to the Cr L_3 peak from metallic chromium. Since the 1-meter spectrometer did not give sufficient resolution to separate these peaks, the spectrometer was modified to a two-meter instrument, using a 2400 grooves/mm

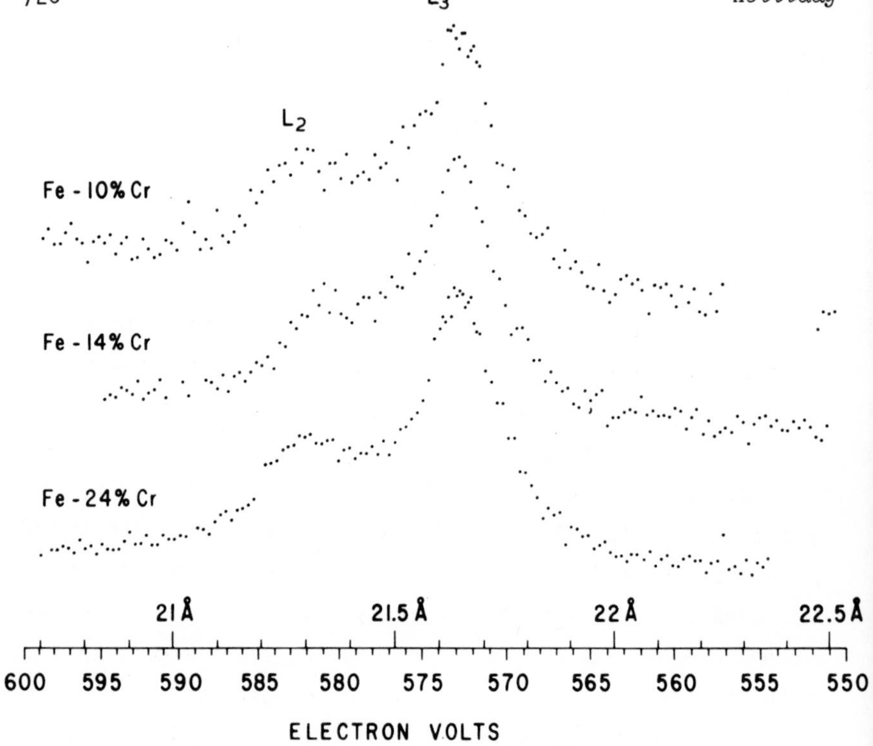

Figure 3 The Cr $L_{2,3}$ emission bands for iron with respectively 10, 14 and 24 at% chromium. The target voltage was constant at 3KV. (Reproduced with permission from Holliday and Frankenthal [1972])

grating. In figure 5 we show the Cr $L_{2,3}$ band from Fe-5at%Cr measured using the 2-meter spectrometer. There are two peaks at the top of the band and they are separated by 1.8eV indicating that the high-energy peak is from Cr_2O_3. The above results provide convincing evidence that the increase in the Cr L_2/L_3 ratio, and the broadening of the Cr L_3 peak with decreasing chromium content (figure 3), is due to the Cr $L_{2,3}$ band from the Cr_2O_3 on the surface.

Knowing the relative positions of the Cr $L_{2,3}$ bands from chromium and Cr_2O_3, and their intensity distributions, it was possible to resolve the emission bands of figure 3 into two separate Cr L_2 and L_3 bands using a Dupont curve resolver.

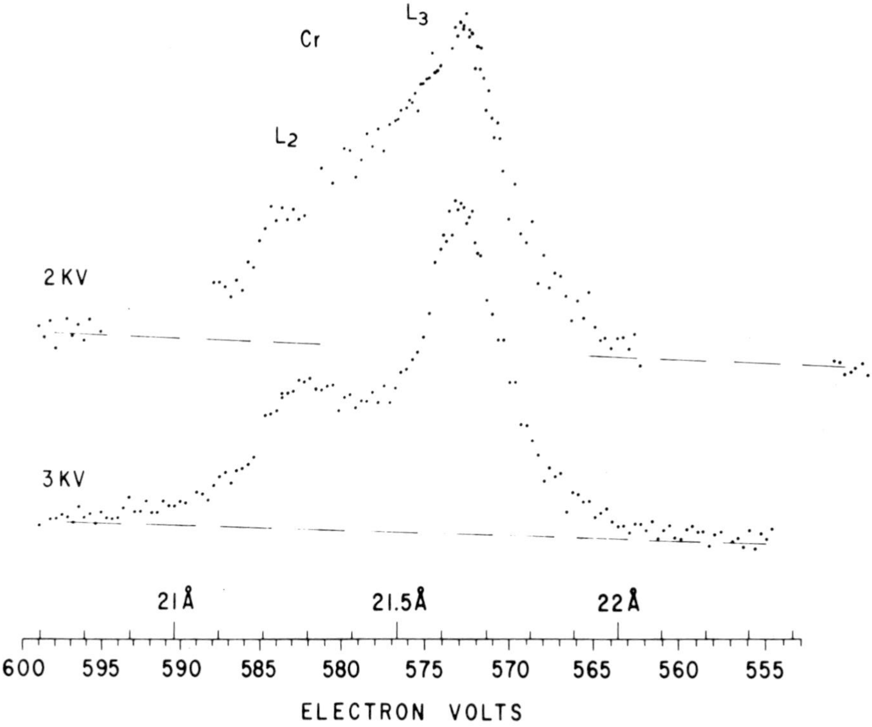

Figure 4 The Cr $L_{2,3}$-emission band from Fe-24at%Cr alloy for a target voltage of 2 and 3 kV.

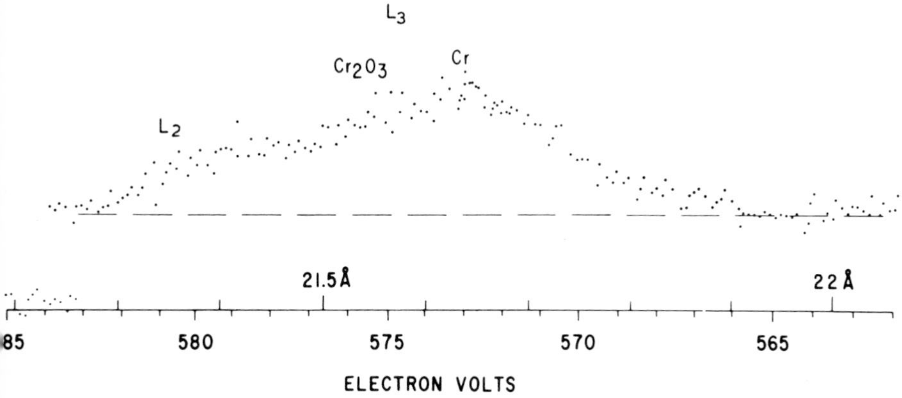

Figure 5 The Cr $L_{2,3}$-emission band for Fe-5at%Cr alloy using a 2-meter grating spectrometer.

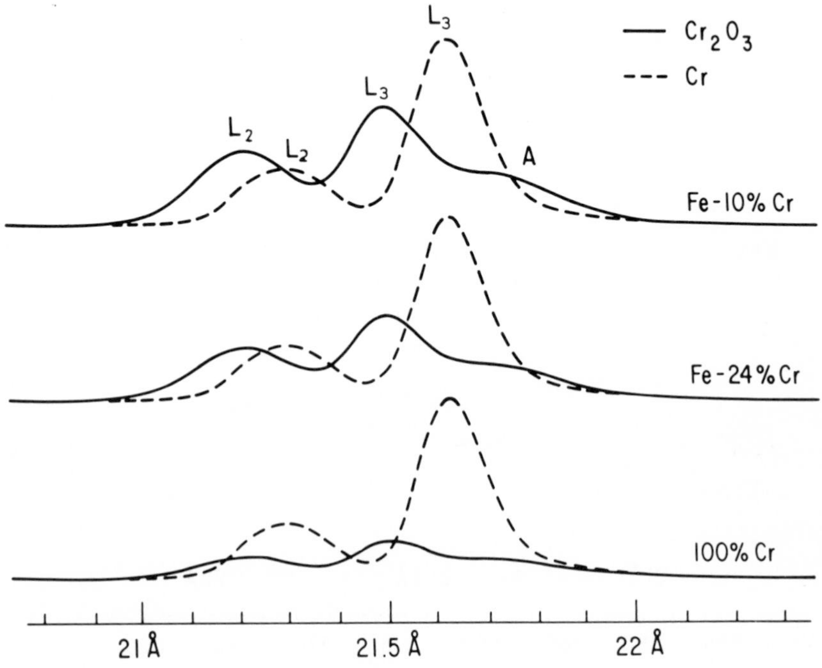

Figure 6 The Cr $L_{2,3}$-emission band from chromium and Fe-10at%Cr and 24at%Cr. The spectra have been resolved into the Cr $L_{2,3}$ band from Cr_2O_3 and that from chromium using a Dupont curve analyzer.

The results are shown in figure 6 for iron with respectively 10, 24 and 100 at% chromium.The Fe-14at%Cr was not plotted since it was very similar to Fe-24at%Cr. On going from 100% chromium to 10% chromium there is an increase in the Cr^{3+} L_3/Cr^0 L_3 intensity ratio. This increase is due to an increase in the number of chromium atoms combined with oxygen, relative to the uncombined chromium atoms for the volume of the alloy from which the x-rays are emitted. This means that there is either an increase in the thickness of the Cr_2O_3 with decreasing chromium content or the oxide thickness remains constant and the Cr^{3+} concentration increases due to the chromium atoms (Cr^0) diffusing into the surface layer where they replace the iron in Fe_2O_3. The depletion of the chromium atoms in the bulk increases with decreasing chromium content due to the greater number of iron atoms that must be replaced in the surface oxide. These conclusions appear to be supported by the results in Table I. The increase in the Fe L_2/L_3 ratio with decreasing chromium content shows an increase in iron oxide indicating that reduction is becoming less efficient at low chromium contents due to there being insufficient numbers of uncombined chromium near the surface to replace the iron.

3. CONCLUSION

For target potentials below 4kV the present experiments show that contribution from the surface oxide must be considered when interpreting the changes in emission bands from alloys. Since the emission bands measured in this investigation were first-series transition metal L-emission bands, in the 15 to 30 Å region, the M-emission bands of the first series transition metals (150-250 Å) will be even more affected by the surface oxide. The results also show that in order properly to correct alloy emission bands for surface-oxide effects, a knowledge of the nature of the oxidation process in alloys is necessary.

The importance of soft x-ray emission as a tool in surface studies can also be seen from the above measurements. The ability to measure light elements, both qualitatively and quantitatively, and to determine the nature of the chemical bond from various band features, makes soft x-ray spectroscopy ideally suited for investigating complex multilayer films on metal surfaces. At present there is no other satisfactory method of investigating these.

Acknowledgments

The author wishes to thank L. Helwig and L. Lewis of the Applied Research Laboratory of U.S. Steel Corporation for preparing the evaporated films, and W.H. Hester of the Fundamental Research Laboratory, U.S. Steel Corporation for assisting in the measurements.

REFERENCES

Curry, C. (1968); in "Soft X-Ray Band Spectra and Electronic Structure of Metals and Materials", ed. D.J. Fabian; Academic Press, London, p.173.

Fabian, D.J. (1971); CRC Reviews in Solid State Sciences, 1, 255.

Fischer, D.W. (1965); J. Appl. Physics, 36, 2048.

Fischer, D.W. and Baun, W.L. (1968); J. Appl. Physics, 39, 4757.

Holliday, J.E. (1968); in "Handbook of X-rays", (ed. by E.F. Kaelbe), p.38-1: McGraw-Hill, New York.

Holliday, J.E. (1970); in "Techniques of Metals Research" Vol. III part I, (ed. by R.F. Bunshah), p.325: John Wiley, New York.

Holliday, J.E. and Frankenthal, R.P. (1972); J. Electrochem Soc., to be published.

Liefeld, R.J. (1968); in "Soft X-Ray Band Spectra and Electronic Structure of Metals and Materials", ed. D.J. Fabian; Academic Press, London, p.133.

Skinner, H.W.B., Bullen, T. and Johnston, J.E. (1954); Phil. Mag., 45, 1070.

ON THE INTENSITIES OF LONG WAVELENGTH SATELLITES OF Kβ SERIES

Masao Sawada*

Osaka Electro-Communication University, Osaka, Japan.

ABSTRACT†

The Kβη-band has been interpreted by the author (Sawada 1932) as a two-electron transition x-ray satellite. This designation was mainly based on its wavelength position and on the governing selection rules. With regard to its intensity the interpretation was only qualitative and in the present calculation the prediction is examined quantitatively.

In the author's earlier paper a calculation was developed, following Bloch (1935) but starting with a helium-like wavefunction; exchange effects are now included. The terms are separated into singlet and triplet states, and the perturbations are electrostatic interactions. Many terms appear in the expression for the perturbed wavefunction, but the only ones retained are those relating to large energy difference in the denominator. The final expression is interpreted in terms of two electron jumps, or the collision process of two holes, since the formula is the product of the dipole moment and the electrostatic interactions.

A plausible model for this two-hole collision is presented:

*Professor Emeritus of Osaka University.

†This paper was presented at the Conference on which this Volume is based and has been published in full elsewhere (Sawada 1972).

an electron hole in the 1s state rises to a 2p-state and collides with a 2p-hole already existing there. These positively charged holes repel one another and one of them sinks to a 2s-level, creating a 2s-hole, and the other rises to form a 3s-hole. The final state is thus 2s3s.

References

Bloch, T. (1935); Phys. Rev. 48, 187.

Sawada, M. (1932); Memoirs Coll. Sci. Kyoto Univ. A.15, 43.

Sawada, M. (1972); Rep. Osaka Electro-Communication Univ. (Eng. Nat. Sci.) No. 8, 82.

Author Index

Numbers in italics are the pages on which the references are listed

A

B

G

L

M

N

T

Subject Index

D

E

F

G

Z